PROBLEMS IN ADVANCED ORGANIC CHEMISTRY

Problems in Advanced Organic Chemistry

JERRY MARCH

Professor of Chemistry
Adelphi University

MARCEL DEKKER, INC.

95 Madison Avenue, New York, New York

MARCEL DEKKER, INC.
95 Madison Avenue, New York, New York 10016

Library of Congress Catalog Card Number 70-176119

ISBN: 0-8247-1449-0

PRINTED IN THE UNITED STATES OF AMERICA

To Herbert Meislich and Samuel Wilen

Preface

The problems in this book have been designed to accompany my textbook, "Advanced Organic Chemistry," published by the McGraw-Hill Book Company. However, they will also be useful to students in courses which do not use my textbook as well as to those who are studying for qualifying examinations. The majority of the problems are devoted to synthesis and mechanisms, but there are also problems on other topics dealt with in the textbook.

By and large, the problem sets correspond to the chapters of the textbook. There are 21 chapters (including two appendix chapters) in the textbook. Thirteen of the problem sets correspond to single chapters, and two to groups of chapters (Set 1 corresponds to Chapters 1, 2, and 3 and Set 3 to Chapters 7, 8, and 9). I have not included any problems for Chapter 5 ("Carbonium Ions, Carbanions, Free Radicals, and Carbenes") or Chapter 6 ("Instrumental Methods for the Determination of Structure"). In the former case, these intermediates occur frequently in the mechanism problems in Sets 4-14 and additional problems are unnecessary. In the latter case, a number of excellent problem books devoted entirely to spectral interpretation are available, and the inclusion of problems in this area would have increased the size of the book to little useful purpose.

The problems in synthesis and mechanisms (Sets 4-14) are not hypothetical or invented. All of them are actual examples taken from the literature. In general, the problems in any of these Sets can be solved by using reactions and principles covered in the corresponding chapter of the textbook or in any previous chapter.* Because of the correspondence of the problem sets to specific chapters, the student knows that each solution has something to do with the material covered in that chapter. Therefore it was deemed desirable to include an extra set (Set 14) of problems in synthesis and mechanism which are not tied to any specific chapter of the textbook.

Answers are given for about one-third of the problems. For all other problems, literature references are provided. The student may easily determine whether or not the answers to a given problem will be found at the back of the book. In all cases, if the answer is in the book, there will be no literature reference at the end of the problem. If the answer is not in the book, there will be a literature reference at the end of the problem. Answers are not provided for all problems for reasons of

*However, there are a few synthesis problems where it may be necessary to use one or more reactions from future chapters. These are all very common reduction reactions, such as the reduction of a nitro to an amino group, or the reduction of a ketone to an alcohol, with which students should be familiar from their first year of Organic Chemistry.

expedience (they would have unduly increased the size and price of the book) and for reasons of policy. It is a good idea for the student to look up the answers in the literature himself. This increases his familiarity with the literature and helps him to develop the art of extracting a specific piece of useful information from a paper which usually contains many other things as well.

The answers given in this book are, for the most part, the same as those in the original papers though in some cases explanations have been supplied where they are lacking in the original. It should be emphasized that neither the answers in this book nor in the original papers are infallible. It is quite possible that the solution devised by the student will be better than the one supplied by the authors of the paper or of this book. Of course, in the case of synthesis problems, we know that the reactions used by the original investigators have actually succeeded in producing the desired product (this cannot always be certain with hypothetical synthetic schemes), but even here it is quite possible that the student will devise a better way, though he would have to go into the laboratory to prove it is better. At this point the basic ground rule for synthesis problems should be mentioned: You must use the given starting material, but in addition you may use any readily available compound, organic or inorganic.

The student will note as he skims through this book that several different conventions have been used in drawing structural formulas. This policy is deliberate and was adopted because these differing conventions are found in the literature, and it was deemed best to draw most of the formulas in the same way as in the papers they come from. The student should recognize that uniformity in drawing formulas, particularly with respect to stereochemistry, has not been achieved in the literature and that he must learn to deal with all of the types in use.

I would like to thank the students and faculty of the Organic Seminar at Adelphi University who allowed many of these problems to be tried out on them. In addition, I would like to acknowledge helpful conversations with Dr. Sung Moon, Dr. Joseph Landesberg, and Dr. Lawrence Katz.

Jerry March

Garden City, N. Y.

Contents

Preface v
Problem Sets
1. Bonding
Problems 1
Answers 271
2. Stereochemistry
Problems 13
Answers 280
3. Acidity, Mechanisms, and Reactivity
Problems 29
Answers 293
4. Aliphatic Nucleophilic Substitution
Problems 37
Answers 299
5. Aromatic Electrophilic Substitution
Problems 62
Answers 308
6. Aliphatic Electrophilic Substitution
Problems 80
Answers 316
7. Aromatic Nucleophilic Substitution
Problems 92
Answers 322
8. Free-radical Substitution
Problems 102
Answers 326
9. Addition to Carbon-carbon Multiple Bonds
Problems 112
Answers 331
10. Addition to Carbon-hetero Multiple Bonds
Problems 135
Answers 342
11. Eliminations
Problems 163
Answers 356

12. Rearrangements
Problems 183
Answers 363
13. Oxidations and Reductions
Problems 215
Answers 376
14. Additional Problems in Synthesis and Mechanisms
Problems 233
Answers 384
15. Nomenclature
Problems 259
Answers 395
16. Literature
Problems 264
Answers 398
Author Index 401

PROBLEM SET 1

BONDING

Corresponds to Chapters 1, 2, and 3 of Advanced Organic Chemistry.

1. Predict molecular shapes and bond angles.

   a. CO_2 b. $AlCl_3$ c. $BeCl_2$ d. $CH_2{=}C{=}CH_2$ e. PCl_3

   f. $\overset{+}{N}(CH_3)_4$ g. $(CH_3)_2O$ h. $(CH_3)_3O^+$

2. Draw Lewis structures. If resonance exists draw all significant canonical forms.

   a. HONO b. CH_2N_2 (CH_2 N N) c. $(CH_3N_2)^+$ (CH_3 N N)

   d. HNO_3 $\left(\begin{matrix}H\ O\ N\ O\\ \quad\ \ O\end{matrix}\right)$ e. PO_4^{-3} $\left(\begin{matrix}O\\ O\ P\ O\\ O\end{matrix}\right)$ f. N_2O (N N O)

   g. CO h. CO_2 (O C O) i. NO_2 (O N O) j. PhCOOH

   k. $(NC{-}CH{-}CN)^-$ l. ClO_2^- (O Cl O) m. $CH_2{=}CHCl$

   n. $(CH_3)_3S^+$ o. H_3BNH_3 p. PCl_5 q. PhO^-

   r. *p*-$O_2NC_6H_4NH_2$ s. ethyl free radical (consider hyperconjugation)

   t. pentalene dianion u. phenylsydnone

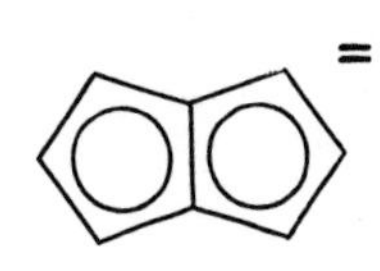

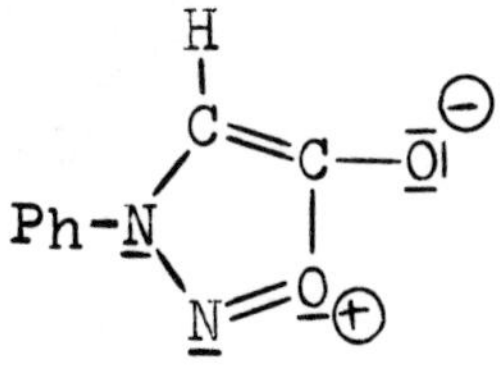

3. Which of the following have dipole moments, and in which direction?

   a. cyclopentane b. $CHCl_3$ c. d.

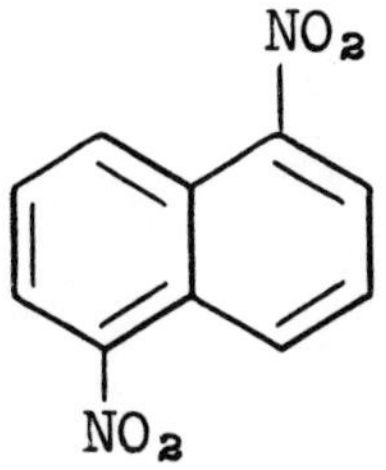

3. Cont.

e. 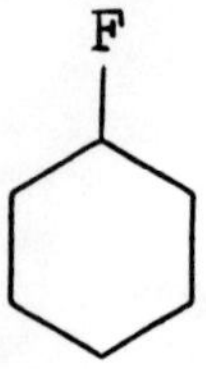 F

f. NC–CN

g. 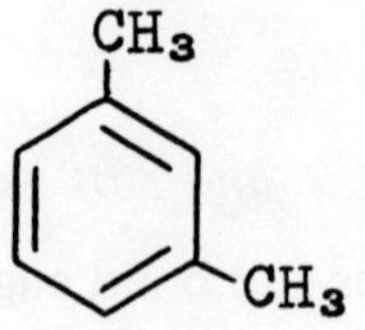 CH_3 CH_3

h. 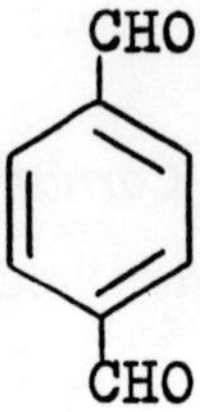 CHO CHO

4. Predict which compound has the larger dipole moment.

a.  Cl CN (i) or Cl CN (ii)

b.  NH_2 NO_2 (i) or Cl NO_2 (ii)

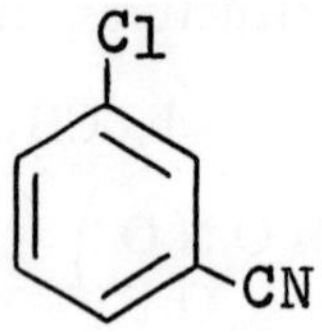

c. 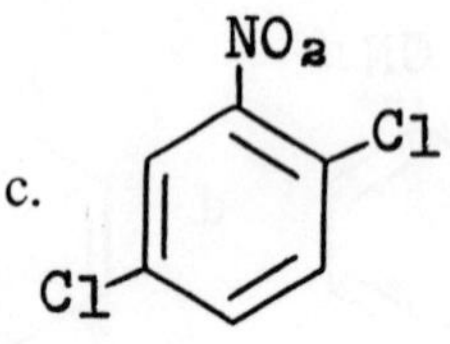 NO_2 Cl Cl (i) or NO_2 Cl Cl (ii)

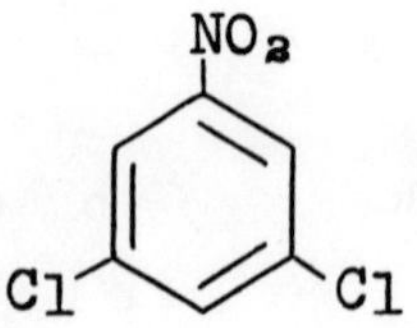

4. Cont.

d. i or ii

e. i or ii

f. i or ii

5. Given the following dipole moments, in which direction is the negative end in each case?

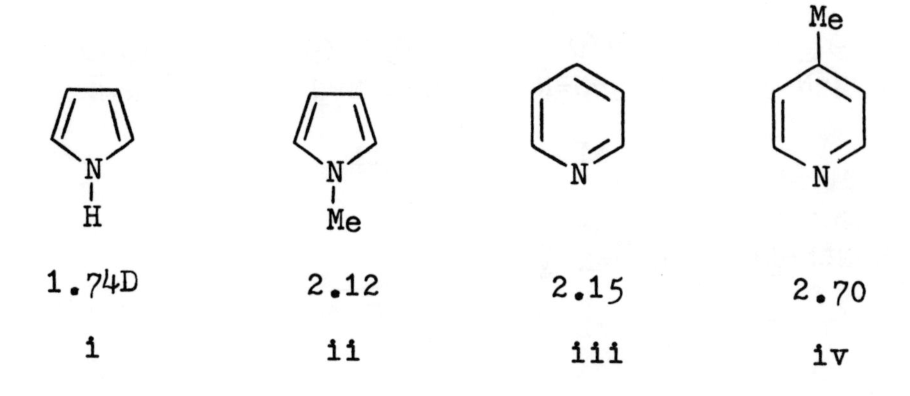

6. If the heat of combustion (to liquid H_2O and gaseous CO_2) at 25° for 1-butene gas is 649.5 kcal/mole, calculate the bond energy of the C=C bond in this compound, given the following values (all at 25°):

Heat of combustion of graphite	94.1 kcal/mole
Heat of combustion of hydrogen (to liquid H_2O)	68.3
Heat of sublimation of graphite	−171.7
Heat of atomization of hydrogen	−104.2
Bond energy for a C–C bond	84
Bond energy for a C–H bond	98

7. In each case, arrange the canonical forms in order of their contributions to the resonance hybrid.

a.

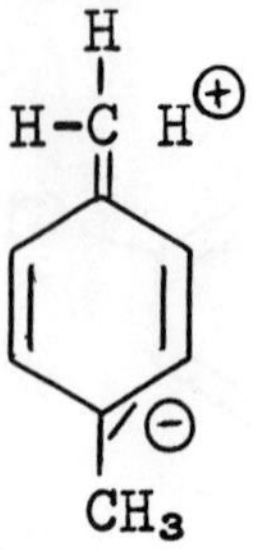

H–C–H⊕ ; ⊖ ; CH_3 CH_3 ; CH_3 CH_3 ; CH_3 CH_3 ; CH_3

i    ii 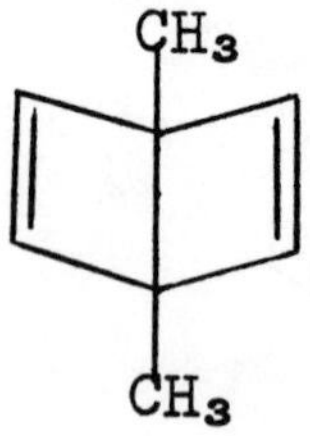   iii    iv

b.

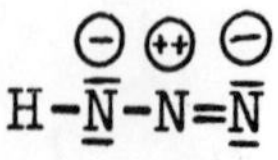

$H\text{–}\overset{\ominus}{N}\text{–}\overset{\oplus\oplus}{N}\text{=}\overset{\ominus}{N}$ (i) $H\text{–}\overset{\ominus}{N}\text{–}\overset{\oplus}{N}\text{≡}N$ (ii) $H\text{–}\overset{\ominus}{N}\text{–}N\text{=}\overset{\ominus}{N}$ (iii) $H\text{–}N\text{=}\overset{\ominus}{N}\text{=}\overset{\oplus}{N}$ (iv)

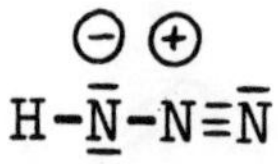

$H\text{–}\overset{\oplus}{N}\text{≡}\overset{\oplus}{N}\text{–}\overset{=}{N}$ (v) $H\text{–}N\text{=}\overset{\oplus}{N}\text{=}\overset{\ominus}{N}$ (vi)

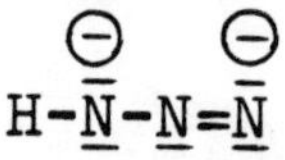

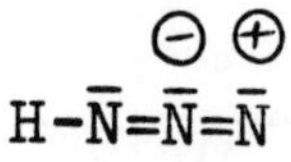

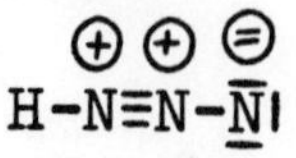

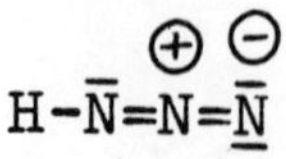

7. Cont.

c.

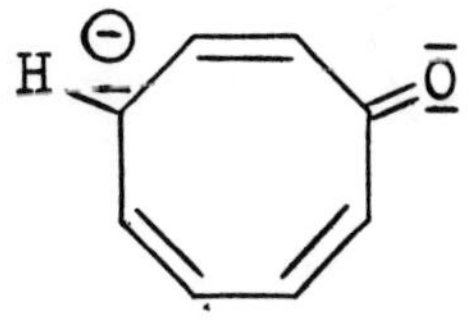

H O
i

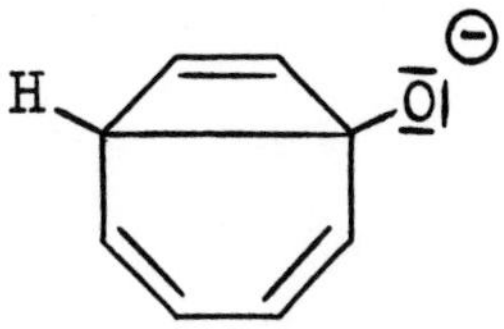

H O
ii

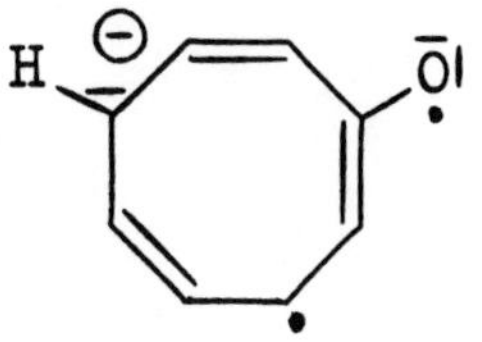

H O
iii

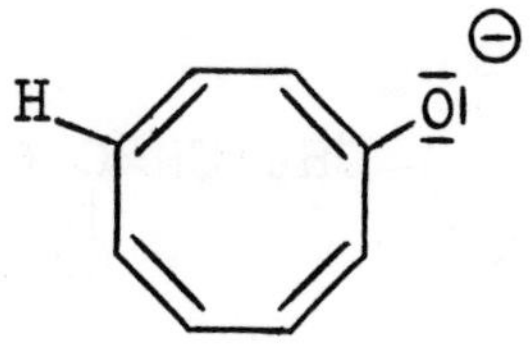

H O
iv

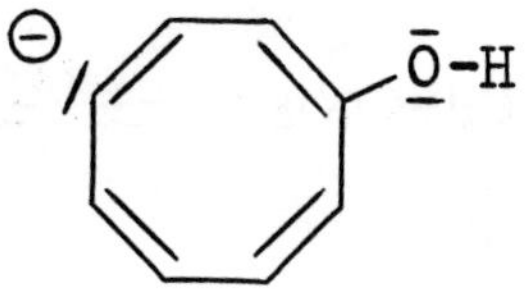

O-H
v

8. Predict the order of acidity.

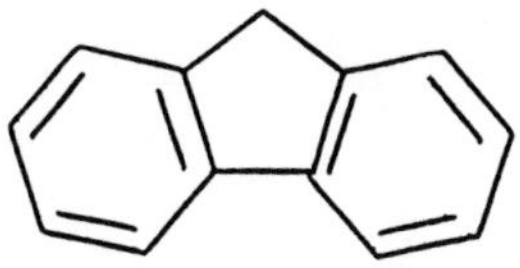

i

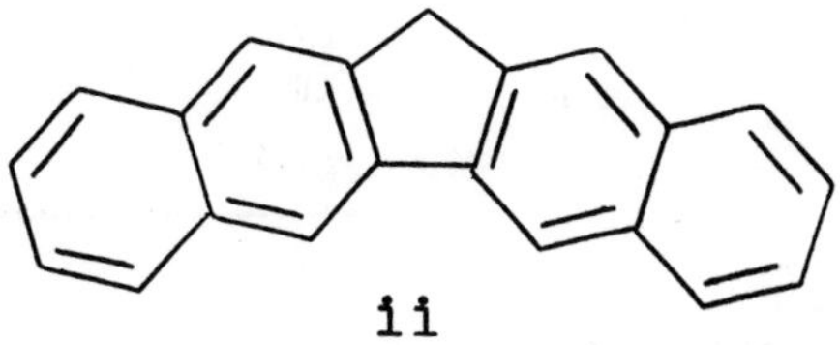

ii

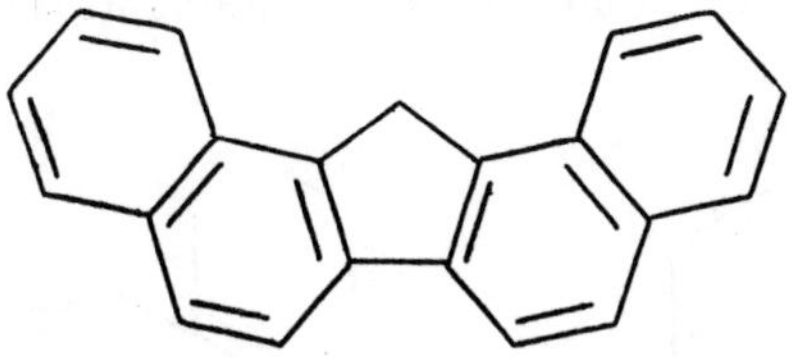

iii

9.

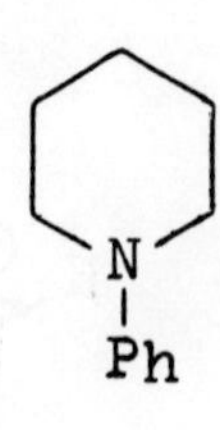

i ii

Explain why ii is a stronger base than i.
Wepster, *Rec. Trav. Chim.* **71**, 1171 (1952).

10.

R–C_6H_4–CH=CH–CH(CH_3)–COOEt $\xrightleftharpoons{\text{base}}$ R–C_6H_4–CH_2–CH=C(CH_3)–COOEt

i ii

The two isomers shown were equilibrated by base. The percentage of i at equilibrium was found to depend on R as follows:

R	% i
Me	45
Et	68
i-Pr	78
t-Bu	86

Explain.
Dolby and Riddle, *J. Org. Chem.* **32**, 3481 (1967).

11. [10]-Annulene (i) has been prepared at –190°, but on warming it isomerizes to 9,10-dihydronaphthalene.* However, compound ii is stable at room temperature. Explain.

i ii

*van Tamelen and Burkoth, *J. Amer. Chem. Soc.* **89**, 151 (1967). The *cis-trans-cis-trans-cis* isomer of i is shown, though actually van Tamelen and Burkoth did not know which isomer was present.

12. Which of the following would you expect to show aromatic properties?

a.

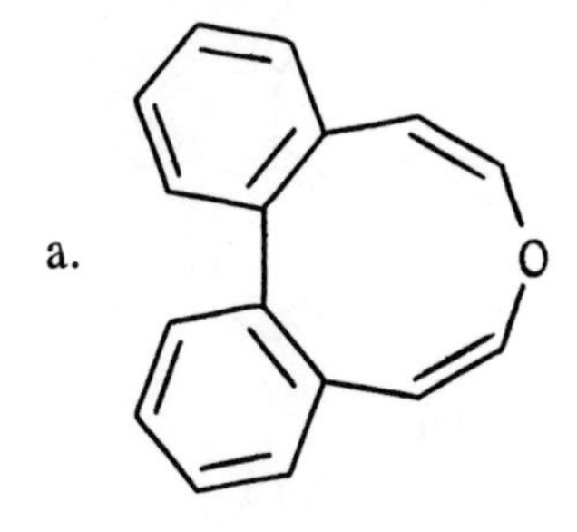

b. Ph Ph S S Ph Ph

c. NMe2 C NMe2

d. ++

e. O H H

f. calicene

g.

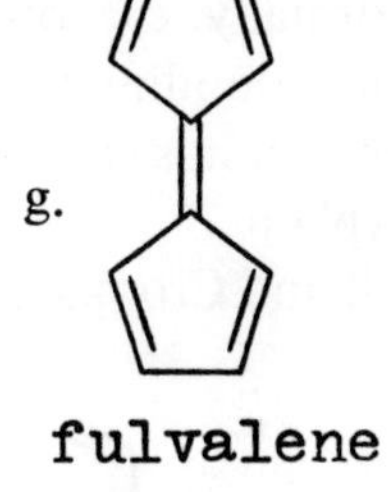

fulvalene

h. s-indacene

i. Ph C N-Ph ⊕ CH3-N C C-O ⊖ OMe

13.

O O i $\xrightarrow{H^+}$ OH OH

13 Cont.

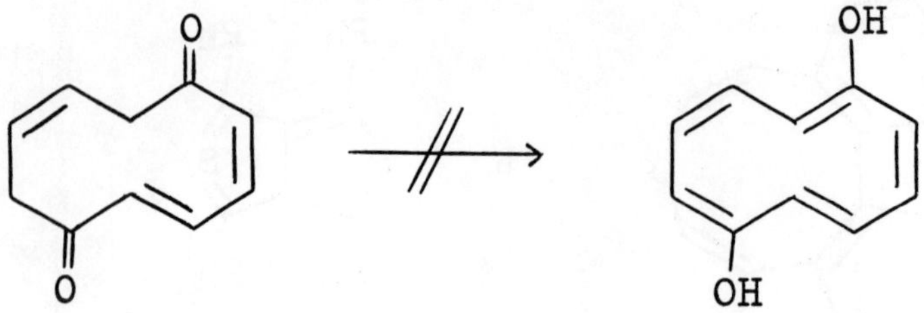

O O → (not) OH OH

ii iii

Compound i readily isomerized to hydroquinone when treated with dilute acid, but ii could not be isomerized to iii. Explain.

Mulligan and Sondheimer, *J. Amer. Chem. Soc.* **89**, 7118 (1967).

14.

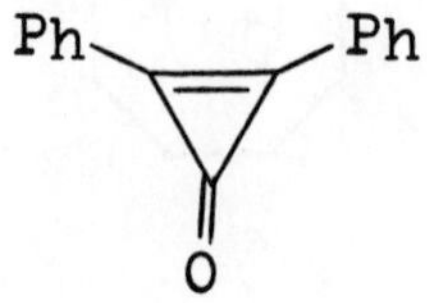

Ph Ph O Ph Ph SO_2

i 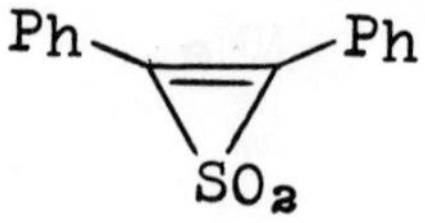 ii

Normally, carbonyl compounds are attacked by base more readily than the corresponding SO_2 compounds. But in the case of the three-membered ring compounds shown above, ii was attacked by NaOMe 5000 times faster than i. Explain.

Bordwell and Crooks, *J. Amer. Chem. Soc.* **91**, 2084 (1969).

15.

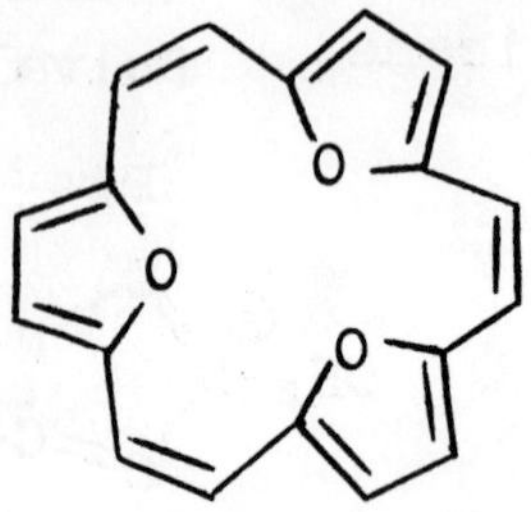

O O O S S S

i 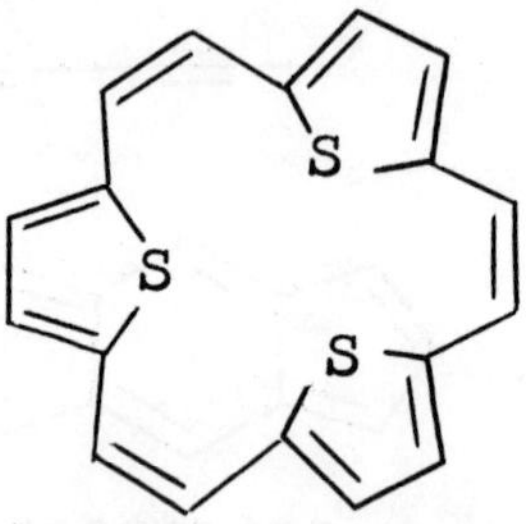 ii

Compound i shows properties of an aromatic compound. Compound ii does not. Account for this difference in properties.

Badger, Elix, and Lewis, *Aust. J. Chem.* **19**, 1221 (1966).

16.

Fulvene (i) displays aromatic properties. Would these properties be enhanced or diminished in the case of ii?

17.

Ph Ph Ph Ph

*H COO-t-Bu *H COO-t-Bu

i ii

Compound i, upon treatment with *t*-BuOK in *t*-BuOD for 31 hr gave 5% D exchange at the starred hydrogen. Compound ii, similarly treated, gave 22% exchange in 30 min. Explain.

Breslow and Battiste, *Chem. Ind.* (London) **1958**, 1143.

18. Which of the following are alternant and which are nonalternant hydrocarbons? Of the alternant hydrocarbons, which will have a nonbonding energy level?

a. 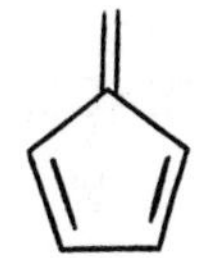

b. $H_2C{=}CH{-}\dot{C}H_2$

c. $\overset{\oplus}{C}H_2$

d. 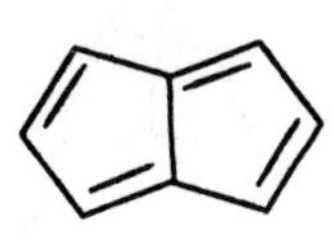

e.

f. 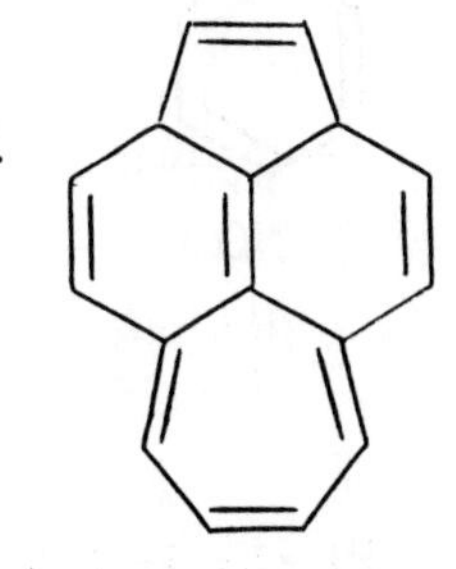

18. Cont.

g.

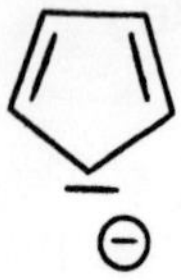

h.

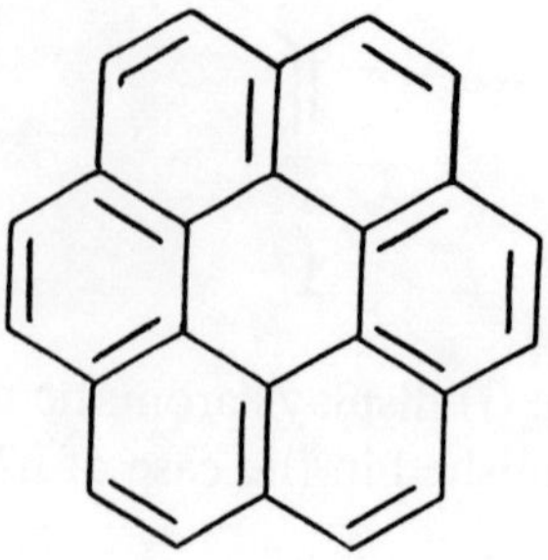

19. In each case, does the compound shown exhibit significant tautomerism? If so, draw the other tautomer.

a. 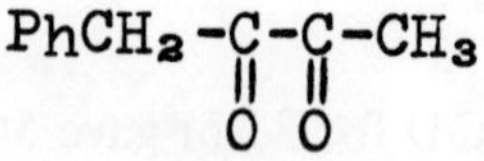 $PhCH_2-C-C-CH_3$ (O O)

b. $(CH_3)_3C-C-CH-C-CH_3$ (Ph, C=O; O O)

c. HO, OH, OH, HO, O, CH_2OH

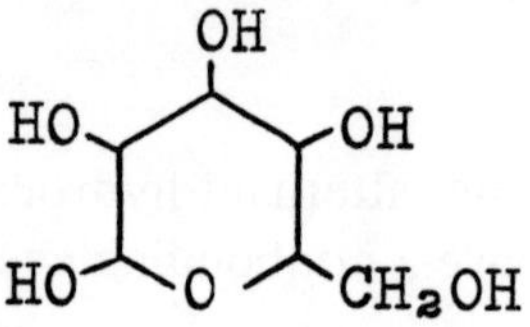

d. 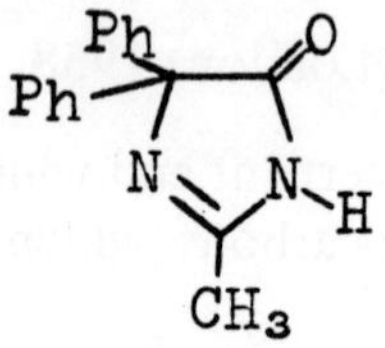 Ph, Ph, O, N, N-H, CH_3

e. CH_2COOH, O, O

f. $CH_3-C=N-N-C-NH_2$ (CH_3, Ph; O)

g. $Ph-C-NHCH_3$ (NH)

h. $PhCCH=CCH_2Ph$ (O, F)

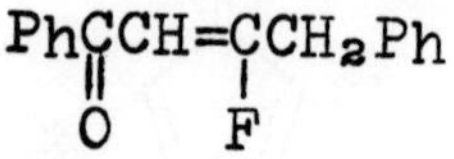

i. 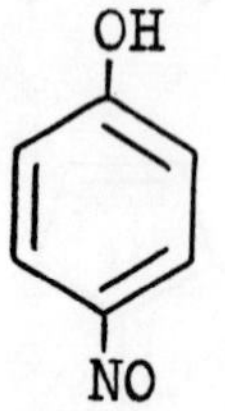 OH, NO

j. 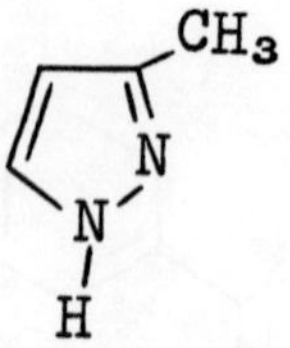 CH_3, N, N-H

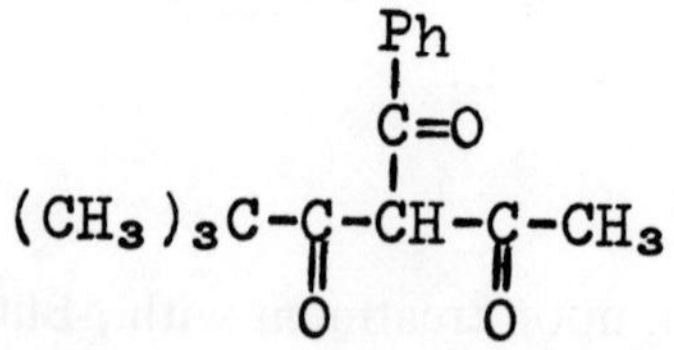

k. 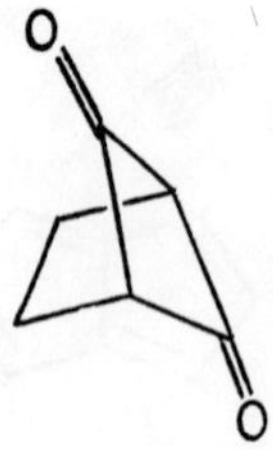 O, O

l. 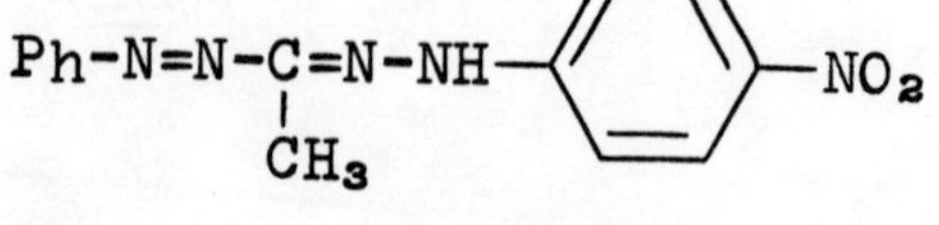 $Ph-N=N-C=N-NH-C_6H_4-NO_2$ (CH_3)

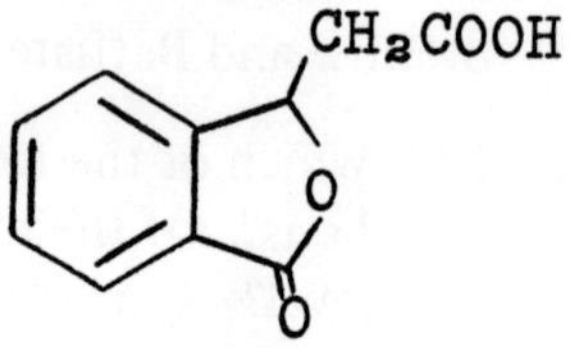

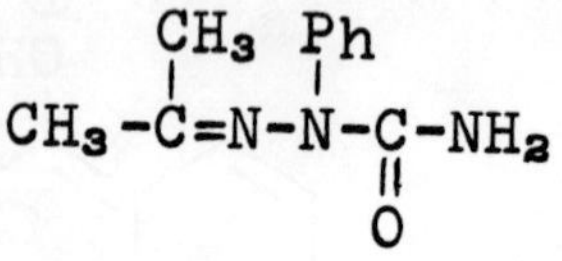

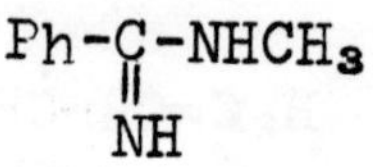

20. Which of the following will form intramolecular hydrogen bonds?

a. 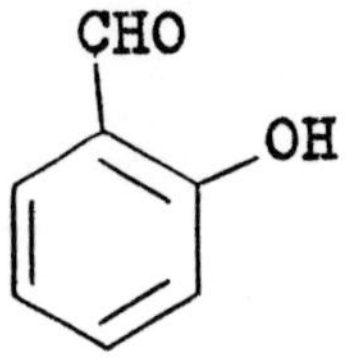 (benzene ring with CHO and OH on adjacent positions)

b. 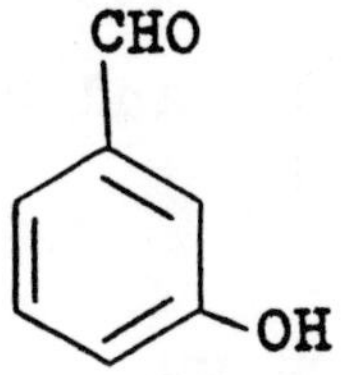 (benzene ring with CHO and OH in meta positions)

c. 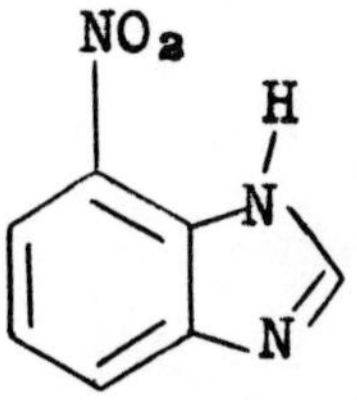 (benzimidazole ring with NO_2, N–H, N)

d. $BrCH_2CH_2CH_2OH$

e. 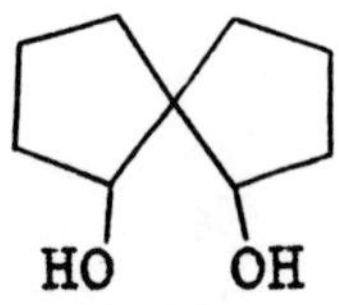 (spiro compound of two cyclopentane rings, each with HO / OH)

f. $HOCH_2C{\equiv}CH$

g. 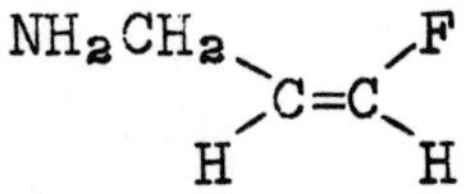 $NH_2CH_2(H)C{=}C(F)H$

21. Will intermolecular hydrogen bonds form? If so, show which atoms are most likely to bond.

a. HCN and HCN

b. HCHO and HCHO

c. HCHO and H_2O

d. Cl^- and H_2O

e. 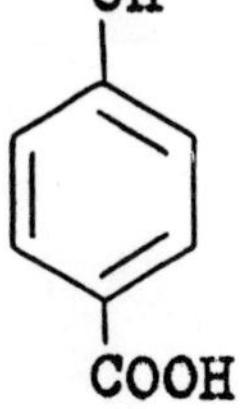 (benzene ring with OH and COOH in para positions) and acetone

f. 1,4-dioxane and $HCFCl_2$

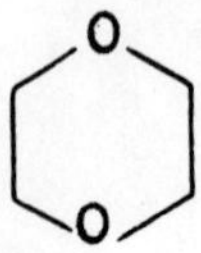

g. $HOCH_2CH_2NH_2$ and $HOCH_2CH_2NH_2$

h. CH_3Br and H_2O

22. Of the three possible internal hydrogen-bonded structures for *a*-hydroxyisobutyric acid, only i and ii are actually found. Explain.

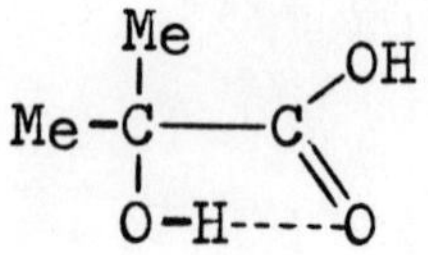

i: $Me_2C(O{-}H\cdots O{=})C{-}OH$

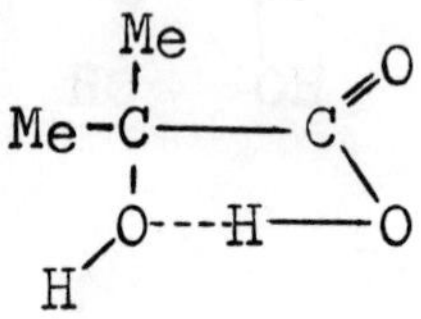

ii: $Me_2C(OH\cdots H{-}O)C{=}O$

iii: $Me_2C(O{-}H\cdots O{-}H)C{=}O$

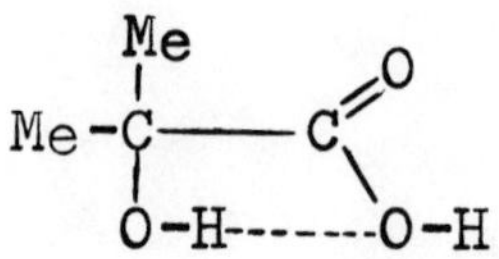

PROBLEM SET 2
STEREOCHEMISTRY

Corresponds to Chapter 4 of Advanced Organic Chemistry.

1. In each of the following cases, is the compound shown resolvable?

a.

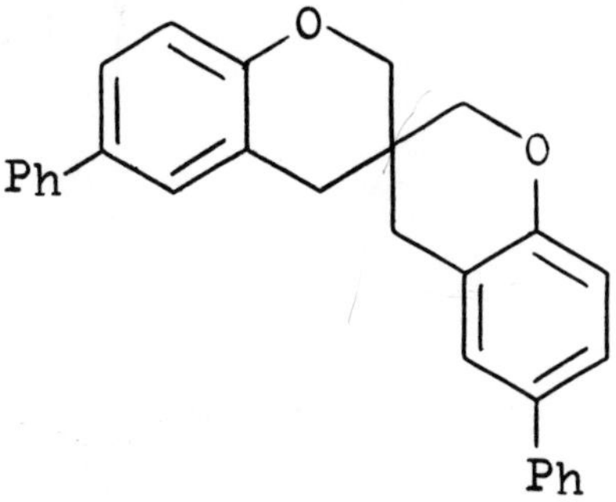

b.

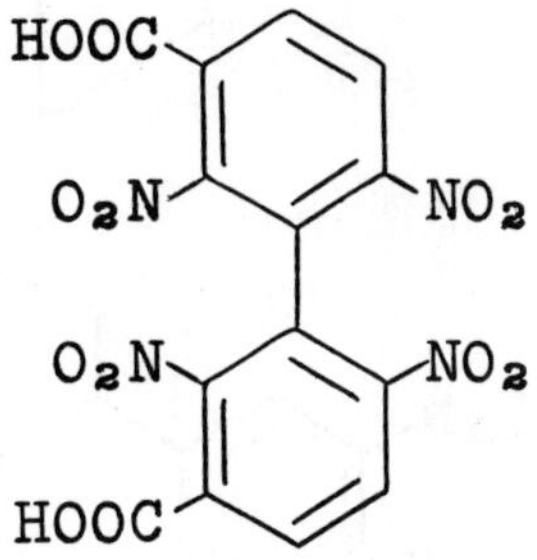

c.

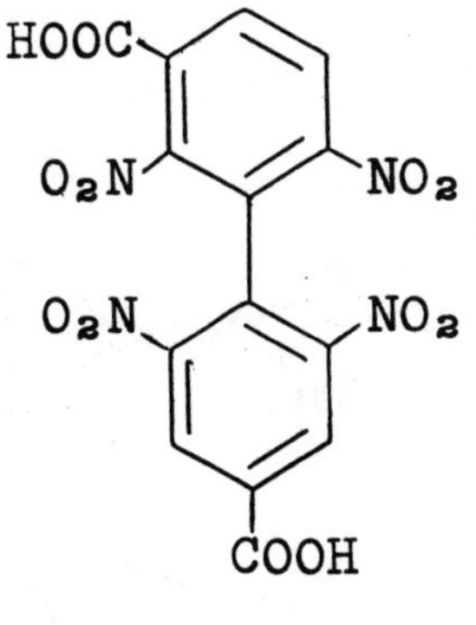

d.

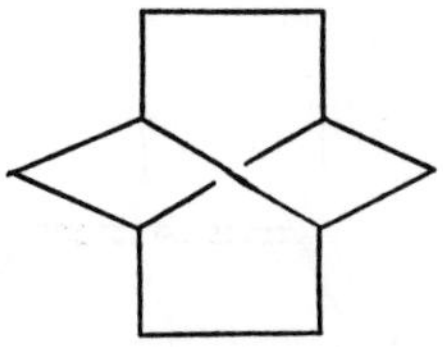

twistane

e.

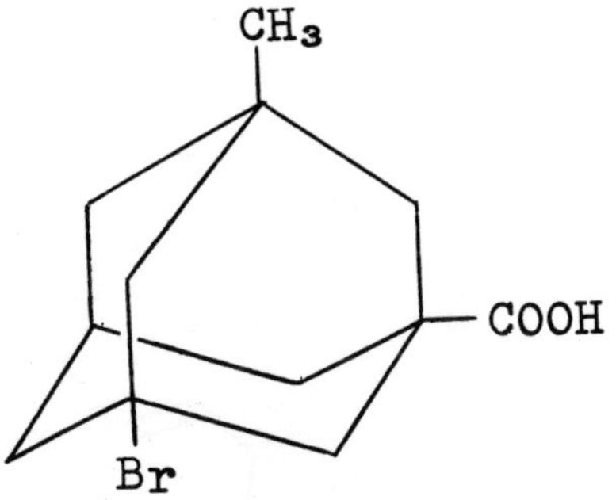

f.

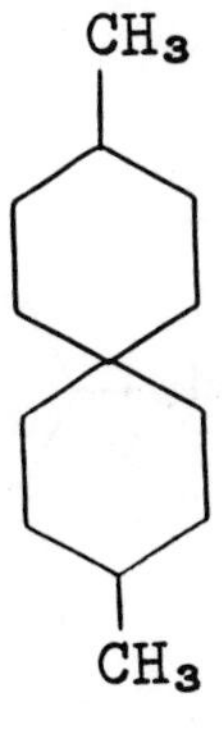

1. Cont.

g.

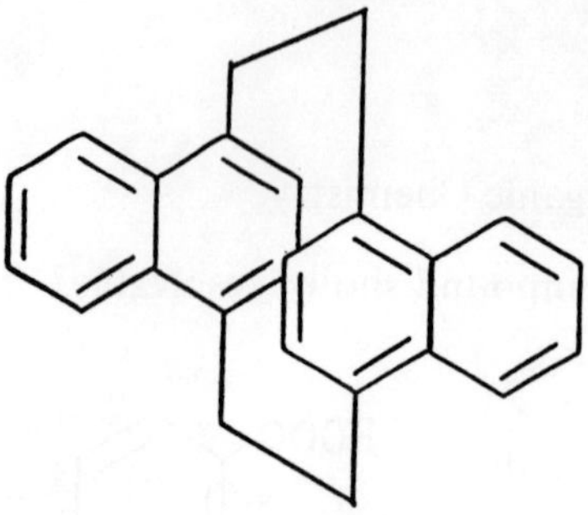

h.

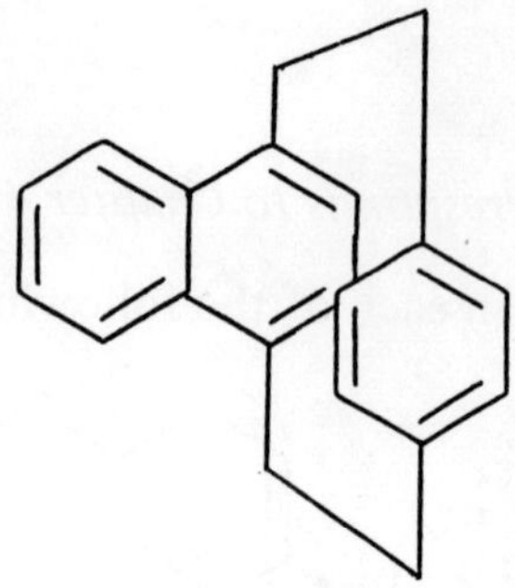

i.

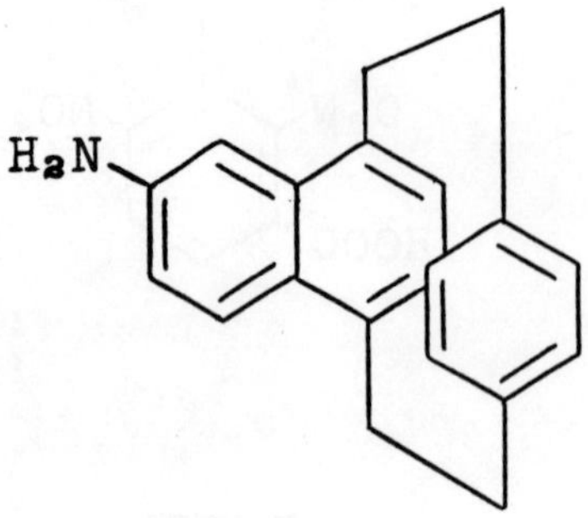

H_2N

j.

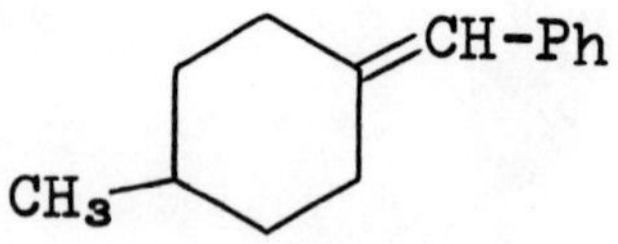

CH-Ph

CH_3

k.

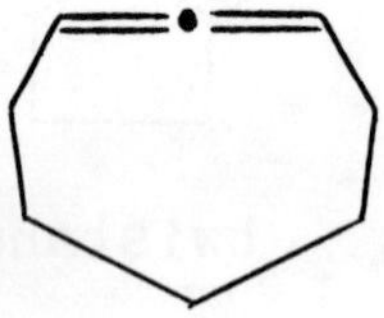

l.

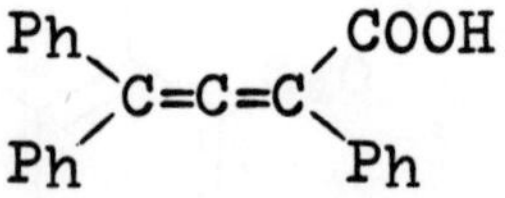

Ph, COOH

C=C=C

Ph, Ph

m.

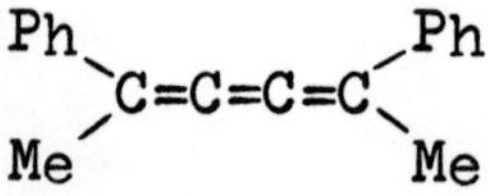

Ph, Ph

C=C=C=C

Me, Me

n.

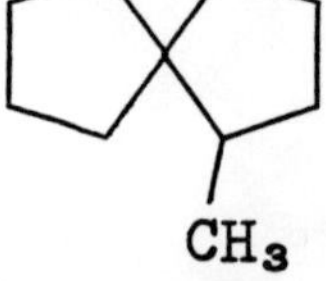

CH_3

o.

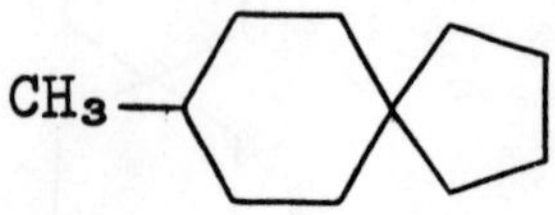

CH_3

p.

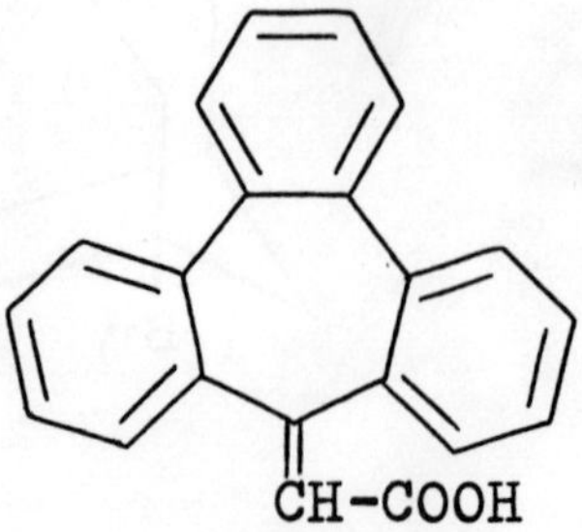

CH-COOH

1. Cont.

q.

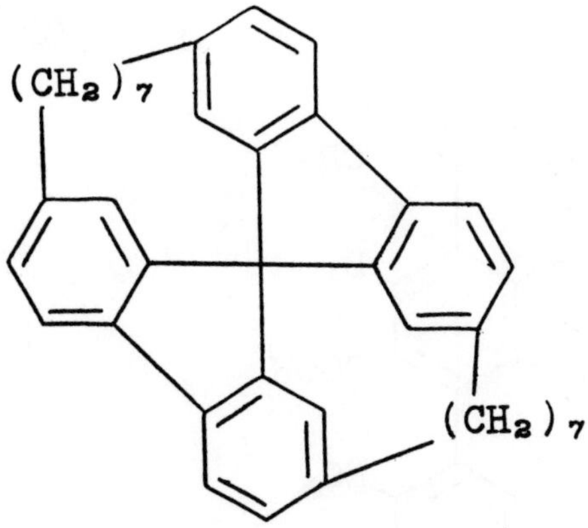

$(CH_2)_7$

$(CH_2)_7$

2. Draw all possible stereoisomers of each of the following. Label *dl* pairs, cis, trans, meso, etc.

a.

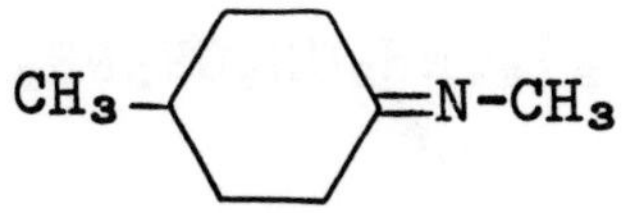

CH_3—⬡=N—CH_3

b.

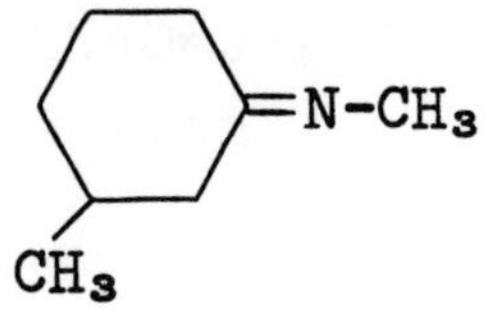

⬡=N—CH_3

CH_3

c.

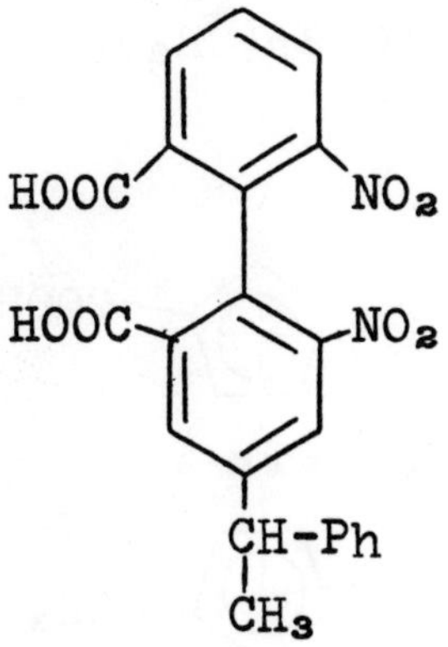

HOOC NO$_2$

HOOC NO$_2$

CH—Ph

CH_3

d.

CH_2OH

CH—CH_3

O_2N I

O_2N I

CH—CH_3

CH_2OH

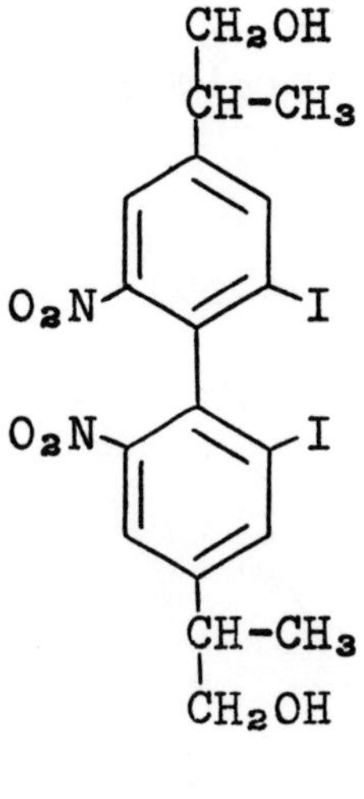

2. Cont.

e.

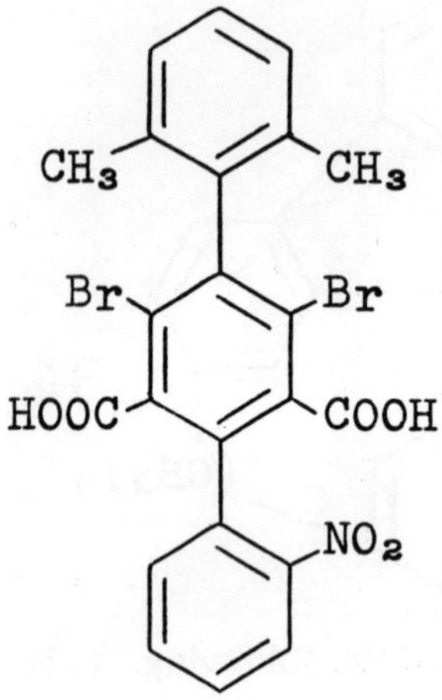
CH3
CH3
Br
Br
HOOC
COOH
NO2

f. $CH_3-CH(OH)-CH=CH-CH_3$

g. $CH_3-C(CH_3)=C=CH-CH_3$

h. $Co(NH_3)_2(NH_2CH_2CH_2NH_2)_2^{+3}$

i. $Co(NH_3)_4Cl_2^{+}$

j. $Co(NH_2CH_2CH_2NH_2)_3^{+3}$

k. $PtBr_2(NH_3)_2$

l. $SiClBr(CH_3)_2$

m. EtGeMeClBr

n.

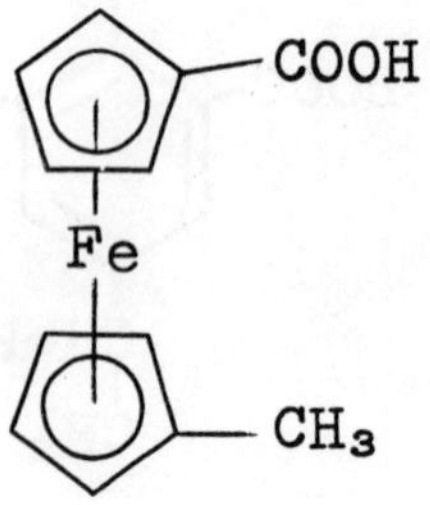
COOH
Fe
CH3

2. Cont.

o.

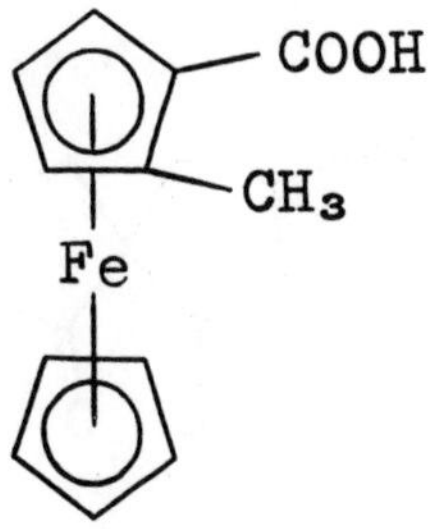

COOH
CH_3
Fe

p.

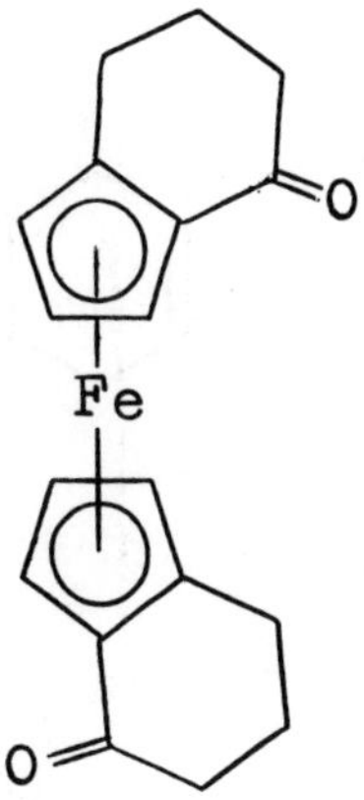

O
Fe
O

q. 1,4,7,10,13-Cyclopentadecapentaene

r.

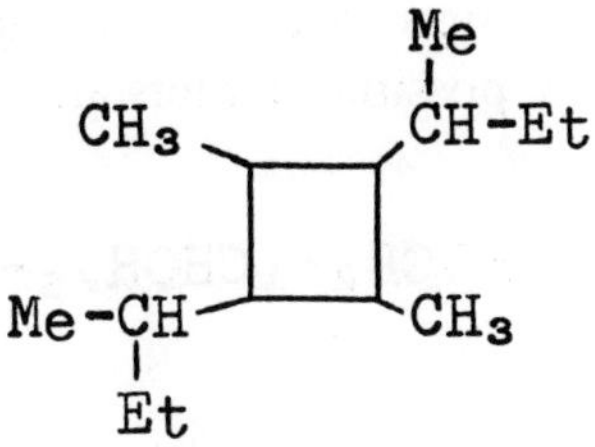

Me
CH_3 CH-Et
Me-CH CH_3
Et

s.

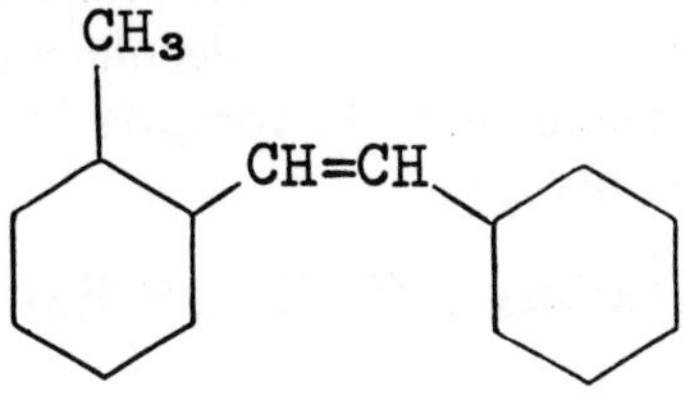

CH_3
CH=CH

t.

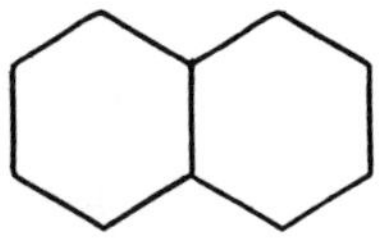

u.

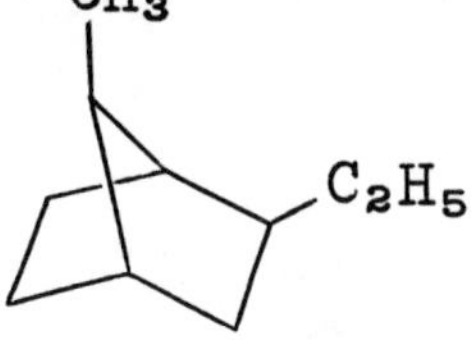

CH_3
C_2H_5

v.

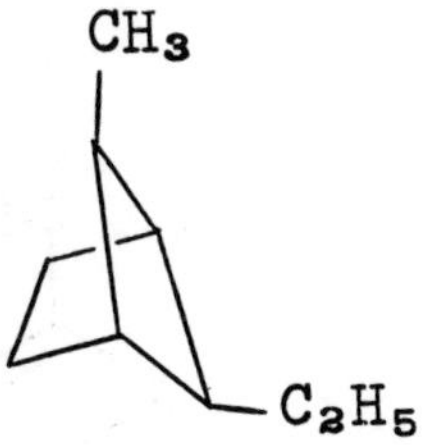

CH_3
C_2H_5

w.

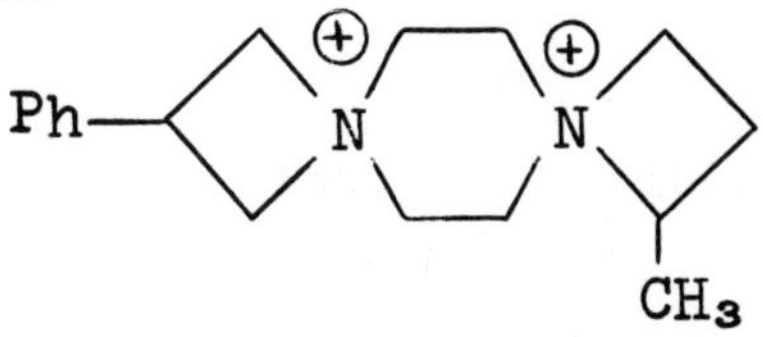

Ph N N
CH_3

2. Cont.

x. CN, CN

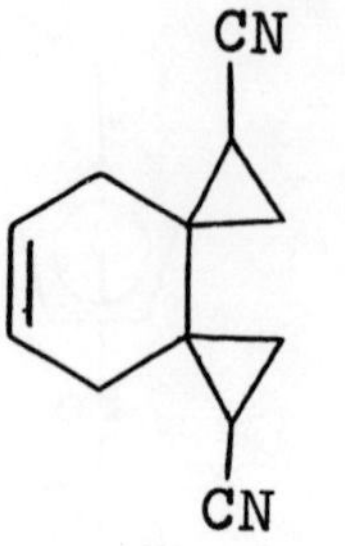

y. CH_3, CH_3, O, O

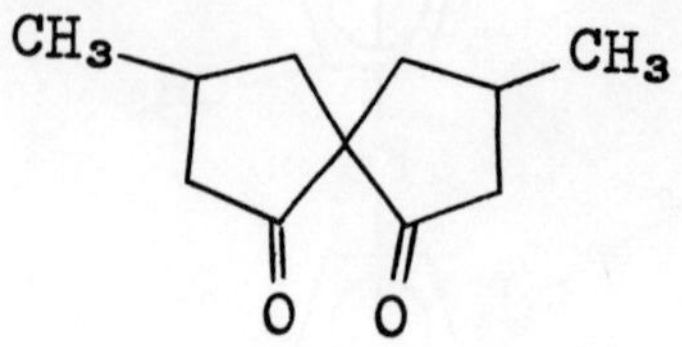

z. CH_3, CH_3, CH_3

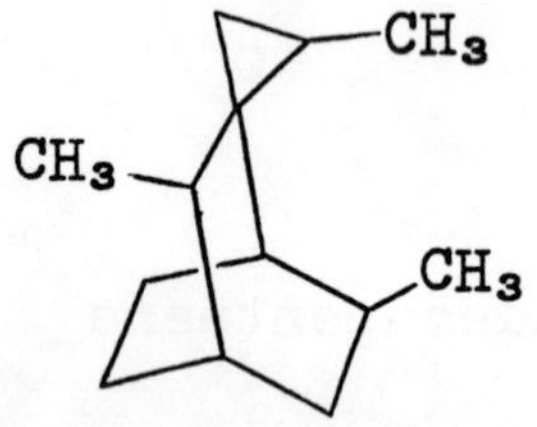

3. In each case answer these three questions: 1. How many asymmetric carbon atoms does the molecule contain; 2. Therefore, how many isomers are predicted from the formula 2^n; and 3. How many possible isomers are there actually?

a. $HOCH_2-(CHOH)_4-CH_2OH$

b. $HOCH_2-(CHOH)_5-CH_2OH$

c. $(Ph-CHCH_3)_3-CH$

d. $(Ph-CHCH_3)_4-C$

e. Cl, Br, I

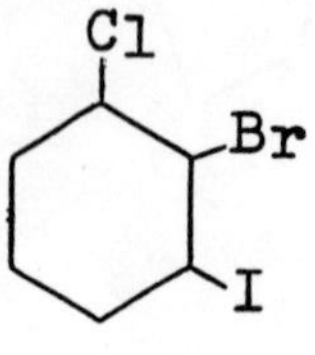

f. Cl, Br, Br

g. Cl, Br, Cl

h. OH, HO, OH, OH

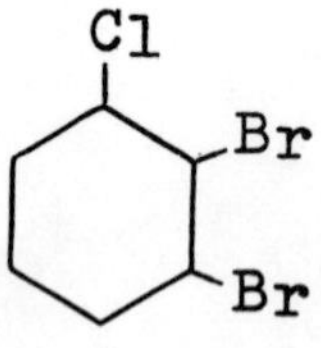

i. CH_3

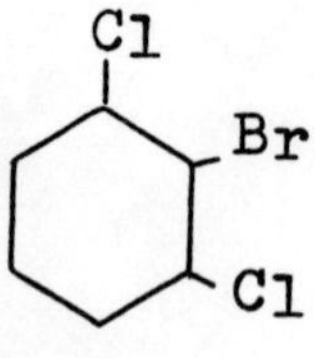

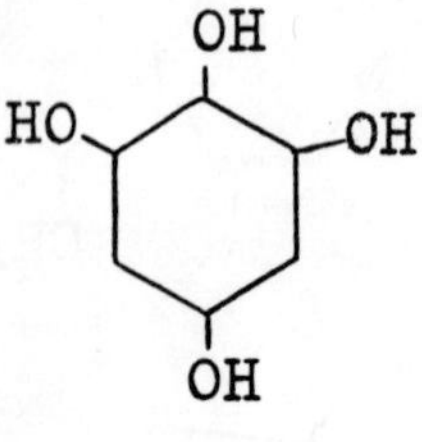

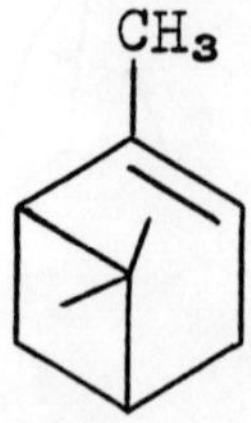

α-pinene

3. Cont.

j.

abietic acid

k.

cholesterol

l.

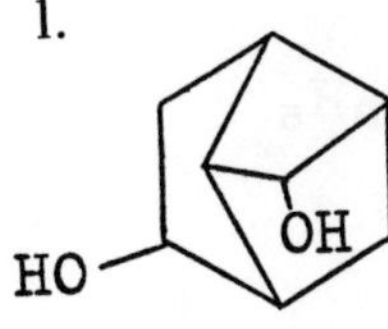

m.

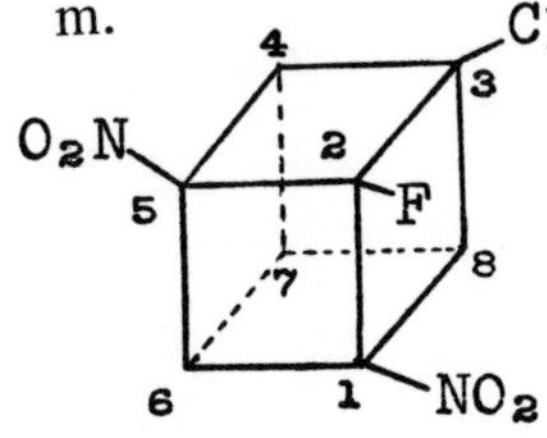

n.

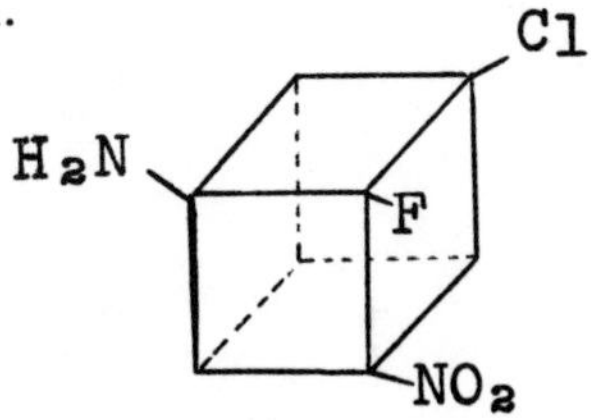

o.

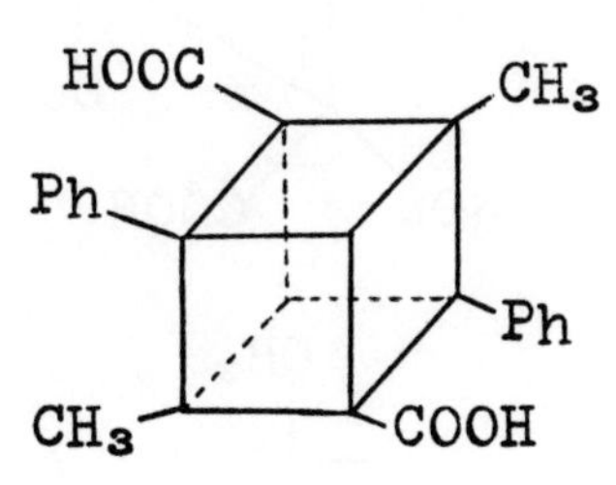

p.

4. Designate each asymmetric carbon as R or S:

a.

b.

c.

d.

e.

f.

g.

4. Cont.

h. C_2H_5 / CH_3—C—$CH(CH_3)_2$ / H

i. CH_2OH / CH_3—C—$COCH_3$ / OH

j. CH_2SH / H_2N—C—CH_2SCH_3 / CH_3

k. $COCH_3$ / $CH_2{=}CH$—C—$CH{=}CH{-}CH_3$ / C_2H_5

l. Br, COOH, H, ii, H, i, CH_3, OH

m. HO, H, H, ii, i, COOH, OH, COOH

n. H, OH, H, ii, i, COOH, OH, COOH

o. CH_3, H, O

p. H, H, i, ii, Br, Br

q. H, H, ii, i, H_3C, iii, H

5. Beginning in 1968, *Chemical Abstracts* has used a new system to designate cis-trans isomers. The two substituents at each carbon of the double bond are ranked according to the Cahn-Ingold-Prelog system. Then, that isomer which has the 2 higher-ranking substituents on the same side of the double bond is called *Z* (for zusammen); the other is *E* (for entgegen). For example, *cis*-2-pentene is now called *Z*-2-pentene, because CH_3 and C_2H_5 are on the same

5. Cont.

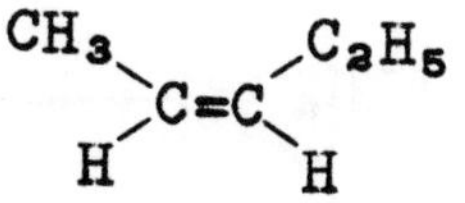

$CH_3(H)C=C(C_2H_5)(H)$

side, and each ranks higher than its partner. Designate as *Z* or *E* (for each double bond):

a. 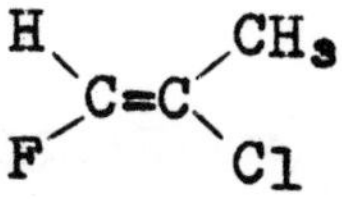 $(H)(F)C=C(CH_3)(Cl)$ b. $(Br)(Cl)C=C(F)(Cl)$ c. $(Ph)(Me)C=C(COOH)(Ph)$

d. $(H)(CH_3)C^2=C^3(CH_3)-C^4(H)=C^5(CH_3)(H)$

e. $(ClCH_2)(HOOC)C=C(CH_2B(OH)_2)(C_2H_5)$

f. $(CH_3)(C_2H_5)C=C(C_2H_5)(CH_2CH_2CH_3)$

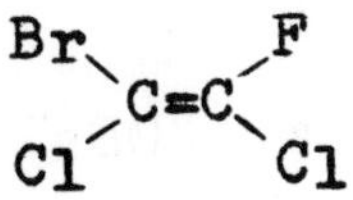

g. $(Ph)(p\text{-}H_2NC_6H_4)C=C(SO_2Cl)(SO_3H)$

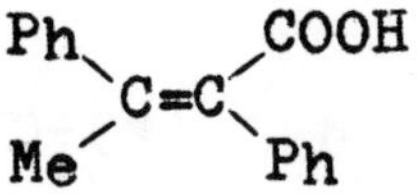

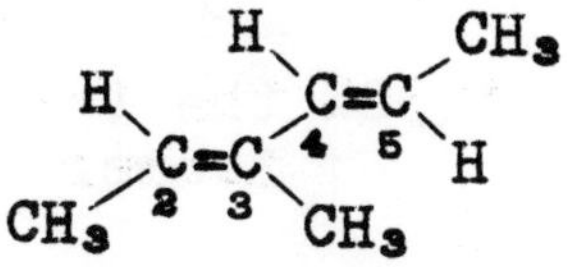

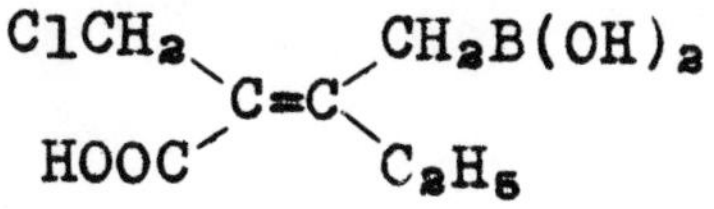

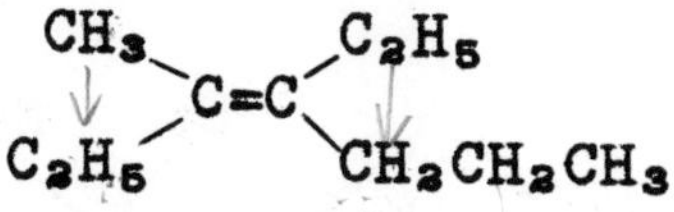

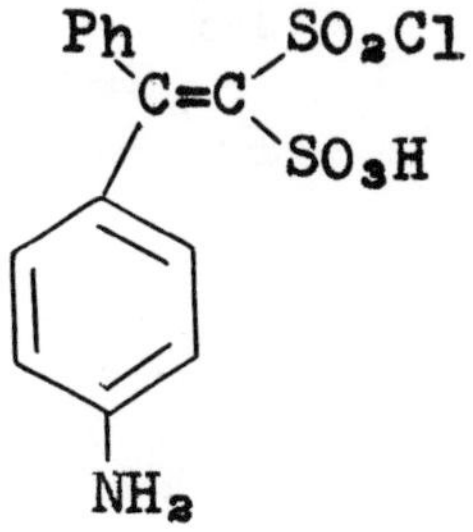

h. $(HCO)(CH_3CO)C^3=C^4(t\text{-}Bu)-C^5(H)=C^6(N\text{-morpholino})(NMe_2)$

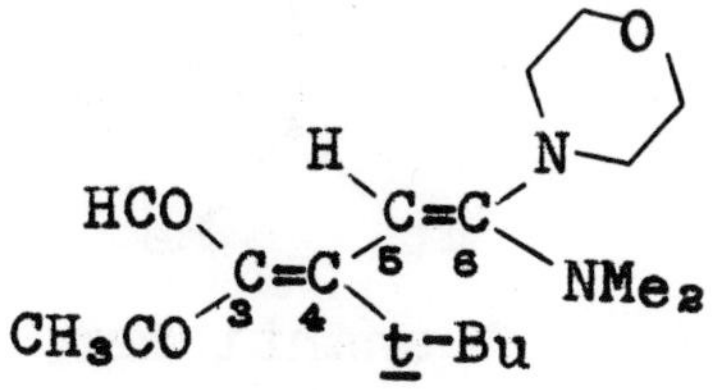

6. Classify the following hypothetical reactions as i. stereospecific; ii. stereoselective but not stereospecific; or iii. neither stereospecific or stereoselective.

a. 1-methylcyclopentene + Br_2 ⟶ 100% 1,2-dibromo-1-methylcyclopentane (Br, CH_3 on C-1; H, Br on C-2)

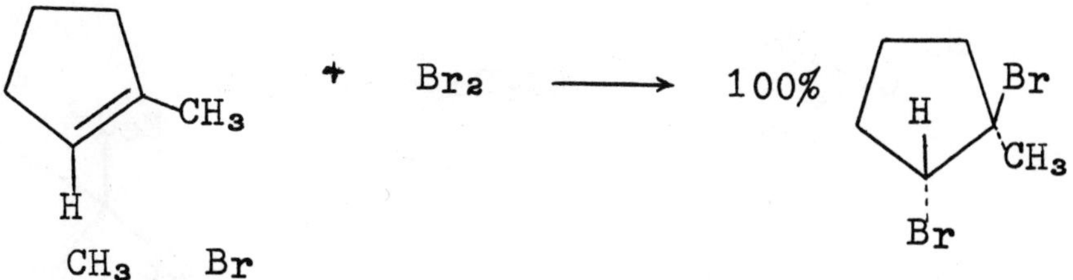

b. $CH_3CHBr-CHBrCH_3$ + Zn ⟶ 80% $(CH_3)(H)C=C(CH_3)(H)$ + 20% the trans isomer

$CH_3CHBr-CHBrCH_3$ + Zn ⟶ 80% $(CH_3)(H)C=C(H)(CH_3)$ + 20% the cis isomer

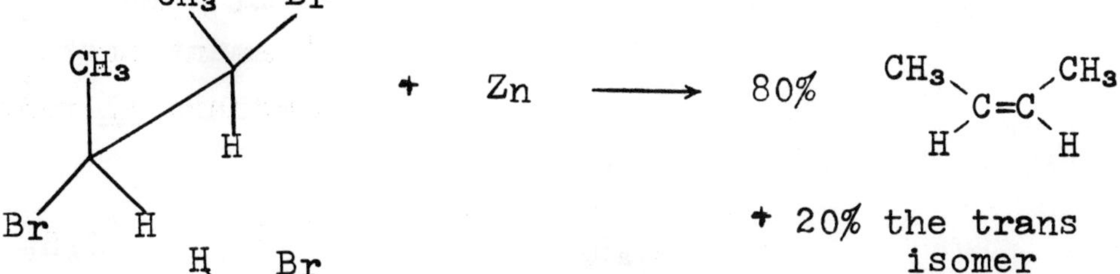

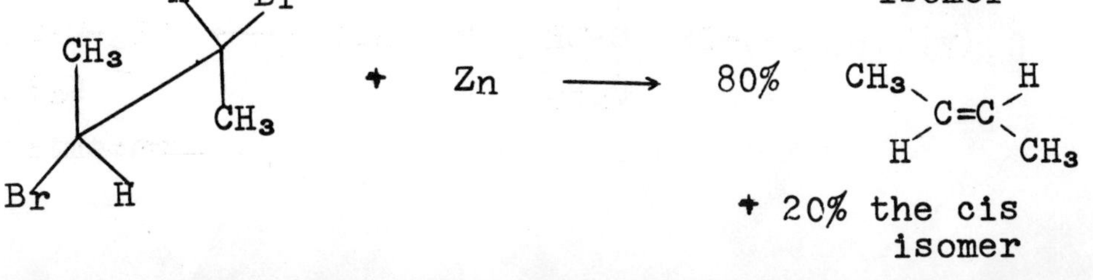

6. Cont.

c. $Ph-\underset{\underset{O}{\|}}{C}-CH_3$ + $LiAlH_4$ $\longrightarrow$ racemic $Ph-\underset{\underset{CH_3}{|}}{CH}-OH$

d. $(+)-CH_3-\underset{\underset{O}{\|}}{C}-\overset{\overset{Ph}{|}}{\underset{\underset{Cl}{|}}{C}}-Et$ + OEt^- $\longrightarrow$ 70% $(+)-EtOOC-\overset{\overset{Ph}{|}}{\underset{\underset{Me}{|}}{C}}-Et$

+ 30% the (−) isomer

e. $(Me)(Et)C{=}C(Me)(Et)$ (Me and Me on the same side) + CH_3SH $\longrightarrow$ 100% [sawhorse structure: front carbon bearing MeS, Et, Me; rear carbon bearing Et, Me, H]

+ enantiomer

threo <u>dl</u> pair

$(Et)(Me)C{=}C(Me)(Et)$ (Et and Me on the same side) + CH_3SH $\longrightarrow$ 100% [sawhorse structure: front carbon bearing MeS, Et, Me; rear carbon bearing Me, Et, H]

+ enantiomer

erythro <u>dl</u> pair

f. (+) or $(-)-CH_3-\overset{\overset{C_2H_5}{|}}{\underset{\underset{C_3H_7}{|}}{C}}-OH$ + HCl $\longrightarrow$ $CH_3-\overset{\overset{C_2H_5}{|}}{\underset{\underset{C_3H_7}{|}}{C}}-Cl$

racemic

6. Cont.

g.

$$(+)\text{-}CH_3\text{-}\overset{\displaystyle C_2H_5}{\underset{\displaystyle C_3H_7}{\overset{|}{\underset{|}{C}}}}\text{-}OH \quad + \quad PCl_5 \longrightarrow 60\%\ (-)\text{-}CH_3\text{-}\overset{\displaystyle C_2H_5}{\underset{\displaystyle C_3H_7}{\overset{|}{\underset{|}{C}}}}\text{-}Cl$$

+ 40% the (+) isomer

$$(-)\text{-}CH_3\text{-}\overset{\displaystyle C_2H_5}{\underset{\displaystyle C_3H_7}{\overset{|}{\underset{|}{C}}}}\text{-}OH \quad + \quad PCl_5 \longrightarrow 60\%\ (+)\text{-}CH_3\text{-}\overset{\displaystyle C_2H_5}{\underset{\displaystyle C_3H_7}{\overset{|}{\underset{|}{C}}}}\text{-}Cl$$

+ 40% the (−) isomer

h.

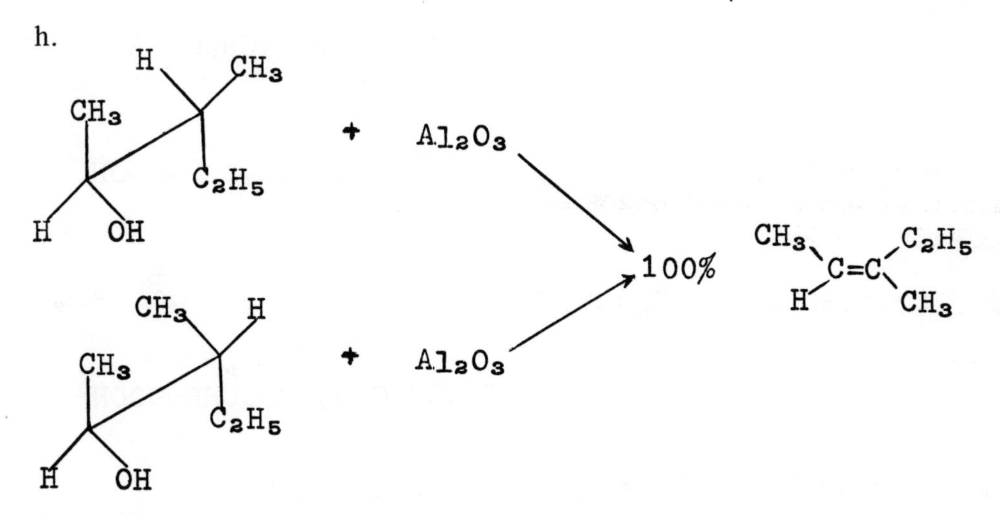

7. Write in the Fischer projection:

   a. R-*sec*-butyl alcohol
   b. S-α-chlorophenylacetic acid
   c. R-$C_6H_5CHDCH_3$
   d. *erythro*-PhCHClCHBrPh (either enantiomer)
   e. *threo*-2,3-dichloro-2,3-dibromopentane (either enantiomer)

8. Write Newman projection formulas and sawhorse formulas for:

   a. meso-2,3-diphenylbutane
   b. either (+) or (−) 2,3-dibromo-2,3-diphenylbutane
   c. *erythro*-2,3-diaminopentane (either enantiomer)
   d. 2(S),3(S)-2,3-dichlorosuccinic acid

9. Draw the formula of 3(S)-methyl-1(R),2(S)-cyclohexanediol.

10. Only from the data given, and without any knowledge of the mechanism in each case, for each reaction given, pick one answer from part A and one from part B. For reaction d, answer part B for *each* asymmetric carbon atom.

a. $\text{R-(+)-}CH_3CH_2CH(CH_3)\text{-}MgBr + O_2 \xrightarrow{H_2O}$

$CH_3CH_2CH(CH_3)\text{-O-O-H}$

b. $\text{R-(-)-}CH_3CH_2CH(CH_3)CH_2COOH \xrightarrow[H^+]{EtOH}$

$CH_3CH_2CH(CH_3)CH_2COOEt$

c. $CH_3CH_2COOH + Br_2 \xrightarrow{PCl_5} CH_3CHBrCOOH$

d. $\text{S-(+)-}CH_3CH(C_2H_5)CH_2CH_2COOH + Br_2 \xrightarrow{PCl_5}$

$CH_3\overset{4}{C}H(C_2H_5)\overset{3}{C}H_2\overset{2}{C}HBrCOOH$

Part A:

i. The product would rotate the plane of polarized light (+).

ii. The product would rotate the plane of polarized light (−).

iii. The product would rotate the plane of polarized light, but without doing the experiment it is not possible to predict in which direction.

iv. The product would not rotate the plane of polarized light.

v. It is not possible to decide, from the data given, whether or not the product would rotate the plane of polarized light.

Part B:

i. The product would have the R configuration.

ii. The product would have the S configuration.

iii. Exactly half (within the capacity for experimental measurement) would have the R configuration and the other half the S.

iv. Some of the product would have the R configuration, and the rest S, but it would not, except fortuitously, be equal amounts.

Part B Cont.

v. It is not possible, from the information given, to predict anything about the configuration.

11. In each case, predict the most stable conformation:

a. $HOCH_2CH_2F$

b. *meso*-2,3-dibromobutane

c. isobutyl chloride

d. *threo*-2-(N,N-dimethylamino)-1,2-diphenylethanol

e. 4-methyl-*t*-butylcyclohexane

f. 4-methylcyclohexanol

g. 3,4-dimethyl-1-methoxycyclohexane

h. cycloheptane

i.

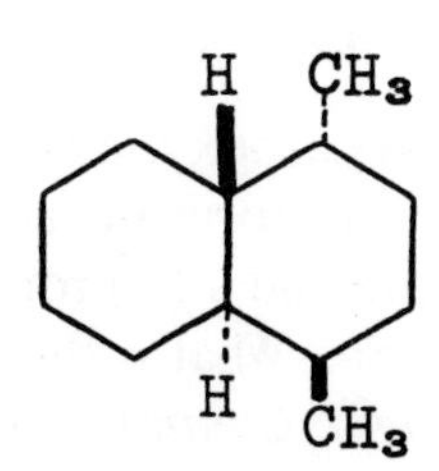

12. There are 6 all-staggered nonsuperimposable conformations of *n*-pentane. Draw them. Which is the most stable?

13. Draw and label all the stereoisomers of perhydrophenanthrene, and rank them in order of stability.

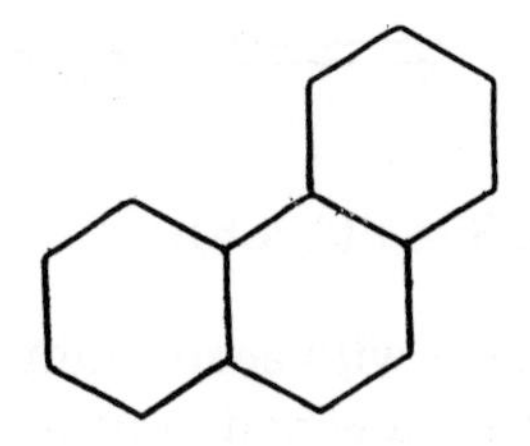

perhydrophenanthrene

14. What kinds of strain are found in the boat form of cyclohexane?

15. Two isomers, one m. 257-65° and the other m. 241-58°, were isolated by treatment of the corresponding diacyl dihalide with dimethylamine. Both of these isomers, by all spectral analyses, had this formula:

15. Cont.

CONMe$_2$; I ; I ; H_2N ; CONMe$_2$; I

Explain.
Ackerman, Laidlaw, and Snyder, *Tetrahedron Lett.* **1969**, 3879.

16.

i + $CH_3NHCHCH_3$ (Ph) ⟶ iii

i ii iii

A sample of (-)-γ-butyrolactone-γ-carboxylic acid (i, configuration unknown) was treated with excess racemic ii to form the amide iii. The ii which was recovered from the reaction mixture had a (+) rotation. The (+) isomer of ii is known to have the R configuration. What is the configuration of (−)–i?
Červinka and Hub, *Collect. Czech. Chem. Commun.* **33**, 2927 (1968).

17. In cases where an electron-withdrawing group (X) is located adjacent to a hetero atom (Y) in a 6-membered ring, the group X generally prefers the axial position (this is known as the anomeric effect). Explain.

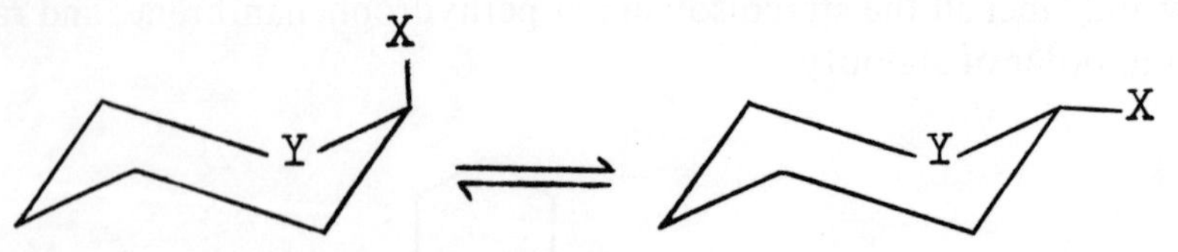

Eliel and Giza, *J. Org. Chem.* **33**, 3754 (1968).

18. When methyl cholate is treated with acetic anhydride and pyridine each of the three β-hydroxy groups can be acylated in turn.

HO ; COOMe ; 12 ; 3 ; 7 ; HO ; OH

methyl cholate

18. Cont.

Predict the order of acylation of the groups and explain your prediction.
Blickenstaff and Orwig, *J. Org. Chem.* **34,** 1377 (1969).

19.

Ph ⊕ CH_3 / N / BF_4^- / CH_3 / Ph ⇌ CH_3 ⊕ Ph / N / BF_4^- / CH_3 / Ph

i　　ii

CH_3 ⊕ Ph / N / C / ClO_4^- / CH_3 / $CHMe_2$

iii

The compound i is much more easily converted to its isomer ii (half-life at 27° is 322 minutes) than open chain analogs, e.g. iii. Explain.
Krebs and Breckwoldt, *Tetrahedron Lett.* **1969**, 3797.

20. In most cases where six-membered rings are fused, trans ring junctions are more stable than cis, and base treatment of methyl 11-oxooleanolate (ii), a derivative

E, D, H, O, H, COOMe, HO, H — i

E, D, H, O, H, COOMe, HO, H — ii

of β-amyrin, results in a change in the D/E ring junction from cis to trans, as expected. However, similar treatment of methyl 11-oxoursolate (i), a derivative of α-amyrin, gives no change; that is in this case, the cis ring junction is more stable. Explain.

21.

CH_3 / Ph--C--CH_2-C-CH_3 (C=O) / H + $LiAlH_4$ ⟶ PhCH(CH_3)CH_2CH(OH)CH_3

What is the configuration of the product?
Brienne, Ouannès, and Jacques, *Bull. Soc. Chim. Fr.* **1967**, 613; **1968**, 1036.

22. (−)-Menthone (which has the configuration shown in i) shows a weak positive Cotton effect, while (+)-isomenthone (ii) shows a strong positive Cotton effect. Draw conformational structures for i and ii.

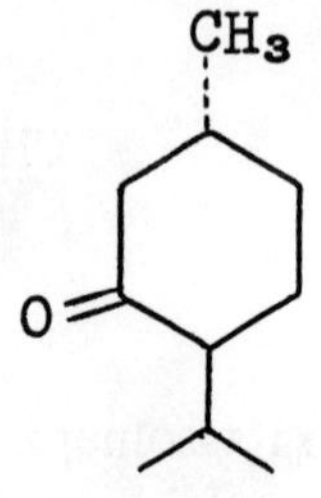

CH_3 CH_3

O O

i 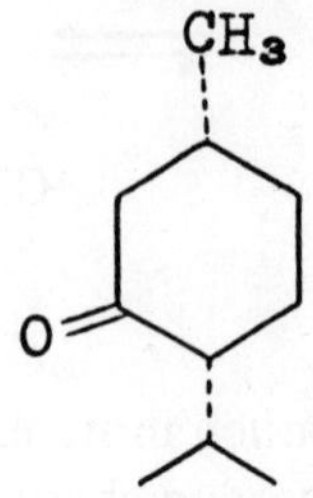 ii

23. The 10-methyldecalone shown has a positive Cotton effect. What is the configuration of the methyl group?

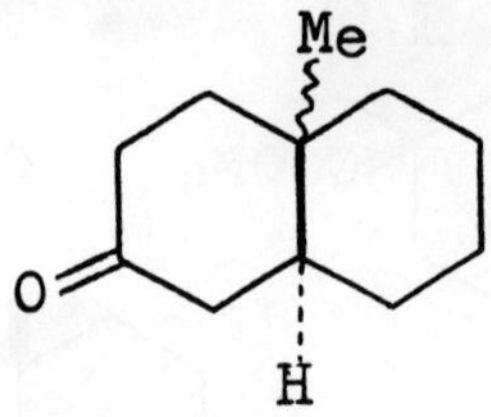

Me

O

H

PROBLEM SET 3

ACIDITY, MECHANISMS, AND REACTIVITY

Corresponds to Chapters 7, 8, and 9 of Advanced Organic Chemistry.

1. Two mechanisms have been suggested for the alkaline decomposition of 3-nitroso-1-oxa-3-azaspiro[4.5] one (i); one (path a) involving a vinyl cation and the other (path b) a vinyl carbene. Devise an experiment to distinguish between these mechanisms.

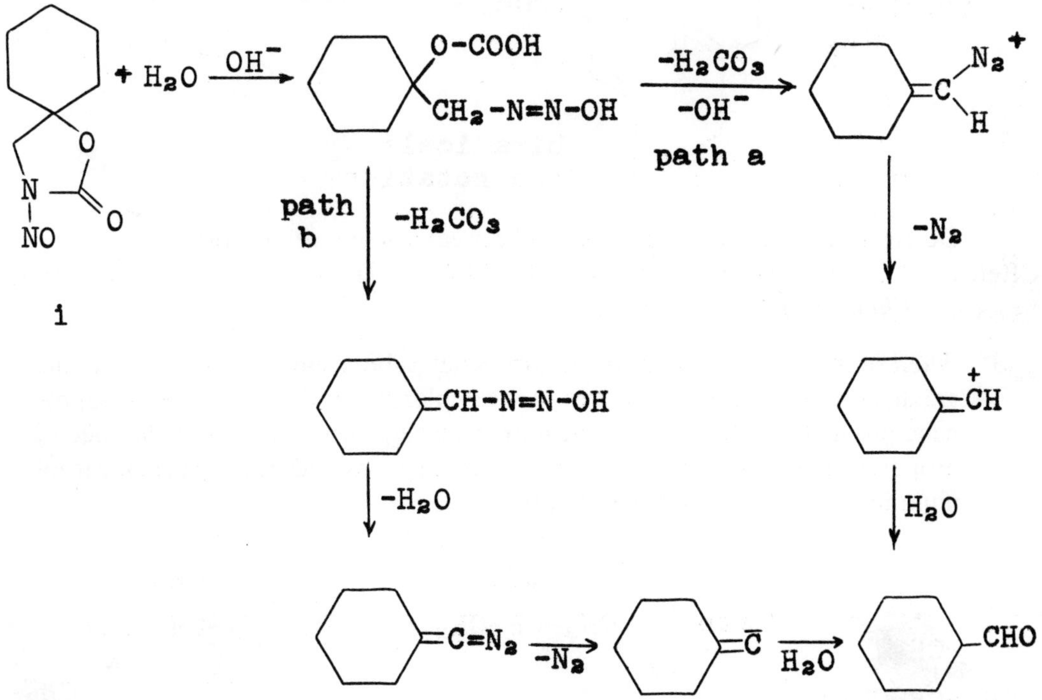

+ H_2O $\xrightarrow{OH^-}$ O-COOH / CH_2-N=N-OH $\xrightarrow[-OH^-]{-H_2CO_3}$ path a =C(N_2^+)H

N, NO, O, O

i

path b $-H_2CO_3$ $-N_2$

=CH-N=N-OH $\qquad$ =$\overset{+}{C}$H

$-H_2O$ $\qquad$ H_2O

=C=N_2 $\xrightarrow{-N_2}$ =$\bar{C}$ $\xrightarrow{H_2O}$ CHO

Newman and Okorodudu, *J. Amer. Chem. Soc.* **90**, 4189 (1968).

2. The (–) enantiomer of 4-carbomethoxy[2.2] paracyclophane (i) was racemized

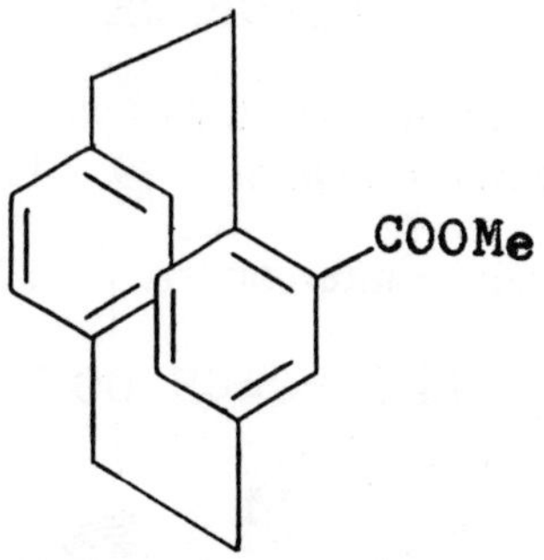

COOMe

i

when heated to 200°. Two mechanisms can be envisioned:

2. Cont.

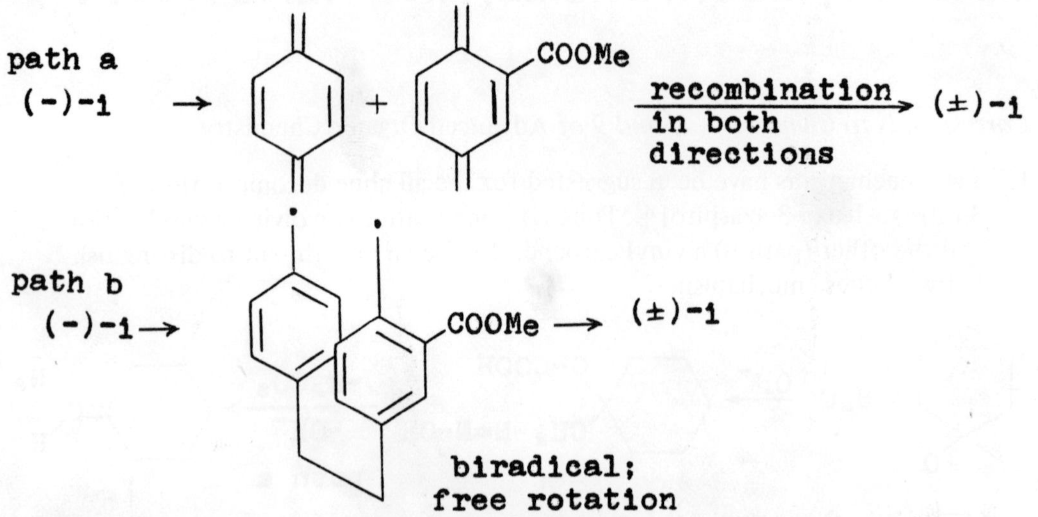

path a

(−)-i → [biradical] + [biradical, COOMe] $\xrightarrow{\text{recombination in both directions}}$ (±)-i

path b

(−)-i → [biradical, COOMe] → (±)-i

biradical;
free rotation

Devise an experiment to distinguish between these mechanisms.
Reich and Cram, *J. Amer. Chem. Soc.* **91**, 3517 (1969); See also Delton and Cram, *J. Amer. Chem. Soc.* **92**, 7623 (1970).

3. When simple primary aliphatic amines, such as butylamine or isobutylamine, are diazotized in aprotic solvents such as chloroform there is more cyclopropane product isolated than in the more usual aqueous solvents. It has been proposed that these products arise via carbenes, instead of the normal carbonium ions, i.e. path b instead of path a:

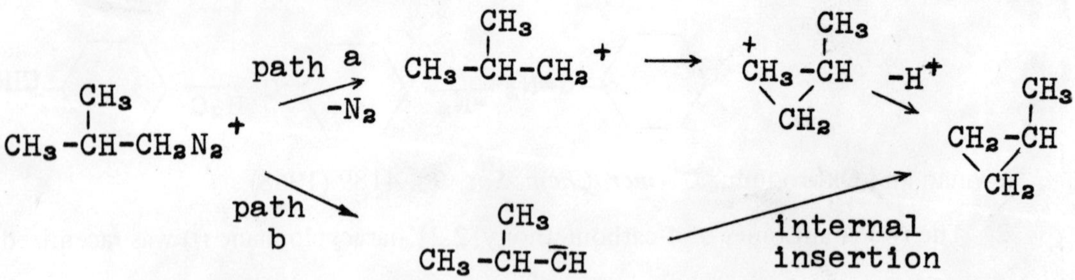

path a: $CH_3\text{-}CH(CH_3)\text{-}CH_2N_2^+ \xrightarrow{-N_2} CH_3\text{-}CH(CH_3)\text{-}CH_2^+ \longrightarrow$ [protonated cyclopropane: $\overset{+}{C}H_3\text{-}CH(CH_3)$, bridged by CH_2] $\xrightarrow{-H^+}$ [$CH_2\text{-}CH(CH_3)$, bridged by CH_2: methylcyclopropane]

path b: $CH_3\text{-}CH(CH_3)\text{-}CH_2N_2^+ \longrightarrow CH_3\text{-}CH(CH_3)\text{-}\bar{C}H \xrightarrow{\text{internal insertion}}$ [methylcyclopropane]

Devise an experiment to decide which pathway is operating.
Jurewicz and Friedman, *J. Amer. Chem. Soc.* **89**, 149 (1967).

4. Compounds i, ii, and iii can be interconverted by base treatment:

$$\underset{\text{i}}{MeCH_2C{\equiv}CCOO^-} \underset{k_{-1}}{\overset{k_1}{\rightleftharpoons}} \underset{\text{ii}}{MeCH{=}C{=}CHCOO^-} \underset{k_{-2}}{\overset{k_2}{\rightleftharpoons}} \underset{\text{iii}}{MeC{\equiv}CCH_2COO^-}$$

At 65° in 6.25 *M* aqueous NaOH, the equilibrium concentrations are (i) 1.28%; (ii) 16.5%; (iii) 82.2%. The solvent isotope effect $k_1(D_2O)/k_1(H_2O)$ for the conversion of i to ii was found to be 1.4; while $k_2(D_2O)/k_2(H_2O)$ was 1.6. Which of the following two mechanisms is better supported by the isotope effect evidence:

path a
carbanion mechanism

$$\text{i} \underset{}{\overset{\text{base}}{\rightleftharpoons}} \left[\text{Me}\bar{\text{C}}\text{HC}\equiv\text{CCOO}^- \longleftrightarrow \text{MeCH=C=}\bar{\text{C}}\text{COO}^-\right]$$

$$\rightleftharpoons$$

$$\text{iii} \rightleftharpoons \left[\text{MeC}\equiv\text{C}\bar{\text{C}}\text{HCOO}^- \longleftrightarrow \text{Me}\bar{\text{C}}\text{=C=CHCOO}^-\right] \rightleftharpoons \text{ii}$$

path b
simultaneous mechanism

$$\text{Me-CH(H)-C}\equiv\text{C-COO}^- \;(\text{with } \bar{\text{O}}\text{H}^- \text{ and H-OH}) \rightleftharpoons \text{Me-C=C=CH-COO}^- \;(\text{with } \bar{\text{O}}\text{H}^- \text{ and H-OH}) \rightleftharpoons \text{iii}$$

Bushby and Whitham, *J. Chem. Soc., B* **1969**, 67.

5.

$$\text{PhCH}_2\text{-C(=O)-N}_3 \xrightarrow{\Delta} \text{PhCH}_2\text{-N=C=O}$$

In the above Curtius rearrangement, the $PhCH_2$ group migrates from the carbon to nitrogen. Devise an experiment to test whether the $PhCH_2$ completely breaks away from the C to form two fragments which then come together, or whether it always remains attached.

6.

$$\text{C}_6\text{H}_5\text{-}\overset{\oplus}{\text{N}}\text{H}_2\text{-CH}_3 + \text{HCl} \longrightarrow p\text{-CH}_3\text{C}_6\text{H}_4\text{NH}_2$$

Devise an experiment to test whether this rearrangement is intra or intermolecular. That is, does the CH_3 attack the same ring or a different one.

7.

$$(\text{Ph})_2\text{C=C(H)Br} \xrightarrow[\text{HOR}]{\text{OR}^-} \text{Ph-C}\equiv\text{C-Ph}$$

In this reaction, the question arises: which phenyl group migrates? That is, does the mechanism require that the phenyl trans to the bromine migrate, or the cis, or both? Devise an experiment to settle this question.

8. Devise an experiment to distinguish between these two possible mechanisms for the elimination reaction given:

path a: 2 steps

$$CH_3-CH_2-\overset{\oplus}{N}Me_3 \xrightarrow[\text{base, slow}]{-H^+} \overset{\ominus}{C}H_2-CH_2-\overset{\oplus}{N}Me_3 \longrightarrow CH_2=CH_2$$

path b: 1 step

$$H-CH_2-CH_2-\overset{\oplus}{N}Me_3 \xrightarrow[\text{base}]{} CH_2=CH_2$$

9. For the reaction $PhCH_2Cl + OH^- \rightarrow PhCH_2OH$ the following two mechanisms can be proposed:

S_N2: 1 step

$$HO^- + H_2C(Ph)-Cl \longrightarrow PhCH_2OH$$

S_N1: 2 steps

$$PhCH_2Cl \xrightarrow{\text{slow}} Ph\overset{\oplus}{C}H_2 \xrightarrow{OH^-} PhCH_2OH$$

Devise at least three different experiments to help determine which mechanism is actually operating.

Note: it is not acceptable to replace one or both of the CH_2 hydrogens by an alkyl or aryl group, since that may well change the mechanism.

10. The following mechanism has been proposed for the oxidation of NO to NO_2:

$$1. \quad 2NO \underset{k_2}{\overset{k_1}{\rightleftharpoons}} N_2O_2$$

$$2. \quad N_2O_2 + O_2 \xrightarrow{k_3} 2NO_2$$

a. Applying the steady-state assumption, and assuming that N_2O_2 goes back to starting material much more rapidly than it goes on to product, show that the rate law is:

10. Cont.

$$\frac{-d\,[NO]}{dt} = \frac{k_3 k_1}{k_2}\,[NO]^2\,[O_2]$$

b. What is the rate law if the contrary assumption is true: that N_2O_2 goes to product much more rapidly than to starting material?

11. The mechanism of the base-catalyzed bromination of ketones may be written as follows:

$$1.\quad R\text{-}\underset{\underset{O}{\|}}{C}\text{-}CH_3 + B \underset{k_2}{\overset{k_1}{\rightleftharpoons}} R\text{-}\underset{\underset{O}{\|}}{C}\text{-}\overset{\ominus}{\bar{C}}H_2 + BH^+$$

$$2.\quad R\text{-}\underset{\underset{O}{\|}}{C}\text{-}\overset{\ominus}{\bar{C}}H_2 + Br_2 \xrightarrow{k_3} R\text{-}\underset{\underset{O}{\|}}{C}\text{-}CH_2Br + Br^-$$

a. Derive the rate law $\dfrac{-d[\text{ketone}]}{dt}$, applying the steady state assumption.

b. Show that, if step 1 (in the forward direction) is very slow compared to step 2, the reaction is general base catalyzed.

c. Show that if step 1 is a rapid attainment of equilibrium followed by a much slower step 2, then the reaction (in water) is specifically OH^- catalyzed.

12. Arrange in order of decreasing acidity.

a. 1. H_2O 2. CH_4 3. HF 4. NH_3

b. 1. HBr 2. HI 3. HCl 4. HF

c. 1. C_2H_2 2. C_2H_4 3. C_2H_6 4. CH_4

d. 1. CH_3COOH 2. $H_3\overset{+}{N}CH_2COOH$ 3. $^-O_2CCH_2COOH$ 4. $HOCH_2COOH$

e. 1. HCOOH 2. CH_3COOH 3. $CH_3C{\equiv}CH$

4. $NCCH_2CN$ 5. H_2O 6. $PhCH_3$ 7. H_2CO_3

8. H_3O^+ 9. CH_3OH 10. PhOH 11. CH_4

12. $EtOOCCH_2COOEt$ 13. HF 14. NH_3

15. p-$O_2NC_6H_4OH$ 16. H_2SO_4 17. CH_3COCH_2COOEt

18. CH_3SH 19.

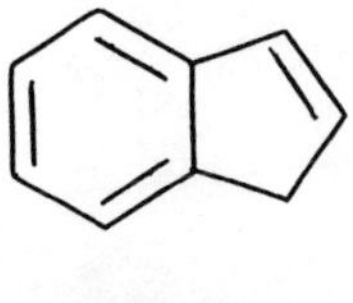

13. Arrange in order of decreasing basicity (reference acid: H^+).

a. 1. NH_3 2. $PhNH_2$ 3. CH_3NH_2 4. $(CH_3)_2NH$ 5. Ph_2NH

b. 1. $PhNH_2$ 2. 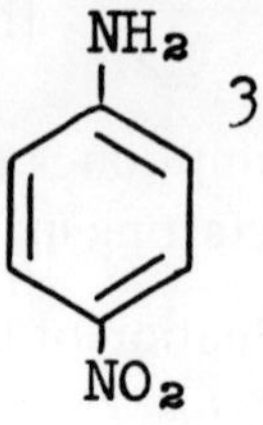 p-nitroaniline (NH_2, NO_2) 3. 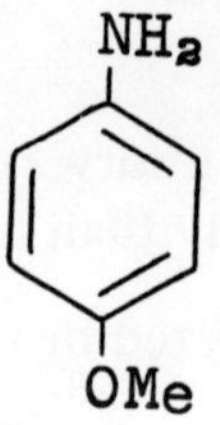 p-methoxyaniline (NH_2, OMe) 4. 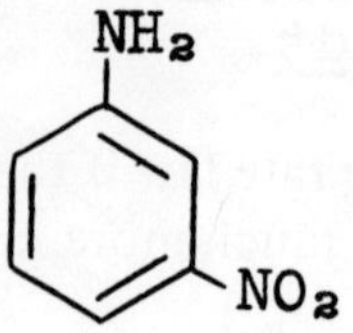 m-nitroaniline (NH_2, NO_2) 5. 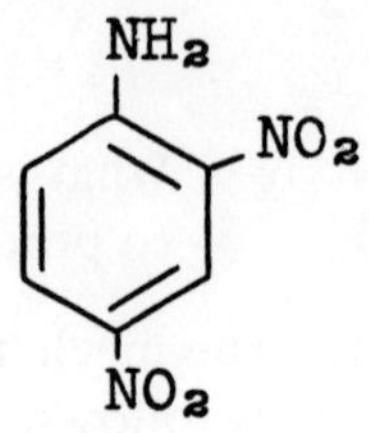 2,4-dinitroaniline (NH_2, NO_2, NO_2)

c. 1. PhCOOH 2. $PhCONH_2$ 3. $PhCOCH_3$ 4. $PhCOOCH_3$

14. In each case one acid is stronger than the other. Decide which is stronger and why.

a. 1. 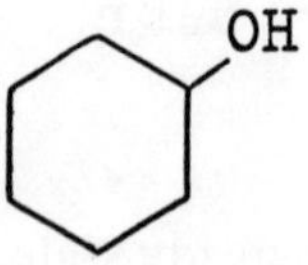 cyclohexanol (OH) 2. PhOH

b. 1. 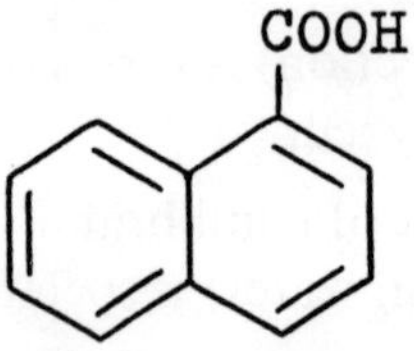 1-naphthoic acid (COOH) 2. 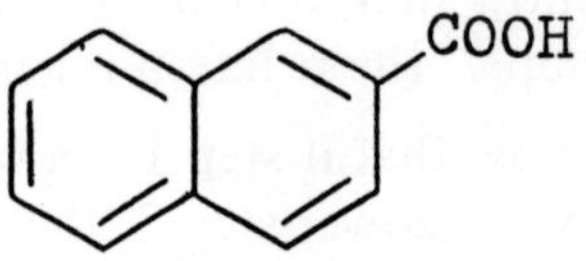 2-naphthoic acid (COOH)

c. 1. 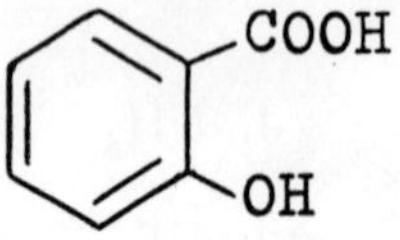 o-hydroxybenzoic acid (COOH, OH) 2. 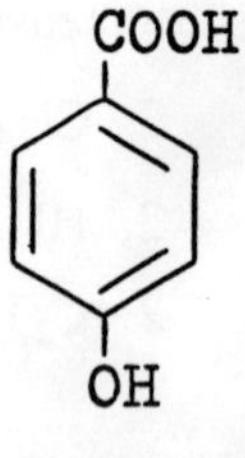 p-hydroxybenzoic acid (COOH, OH)

d. 1. 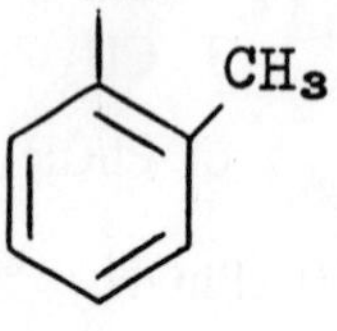 o-cresol (OH, CH_3) 2. 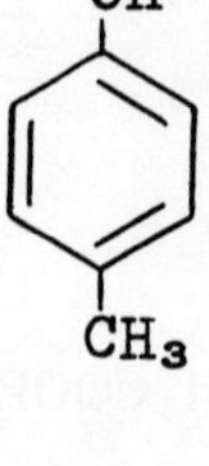 p-cresol (OH, CH_3)

15. In each case one base is stronger (reference acid: H^+) than the other. Decide which is stronger and why.

15. Cont.

a. 1. $PhNEt_2$ 2. $PhNMe_2$

b. 1. $PhNH_2$ 2. CH_3, NH_2, CH_3 (substituted benzene ring)

c. 1.

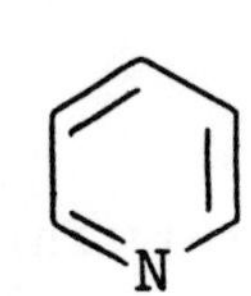

2.

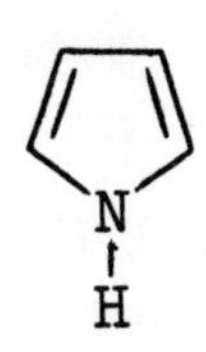

d. 1. $NH_2-\overset{\displaystyle O}{\overset{\|}{C}}-NH_2$ 2. $NH_2-\overset{\displaystyle NH}{\overset{\|}{C}}-NH_2$

16. Arrange in order of Lewis acid strength:

a. CH_4 b. $AlCl_3$ c. HCHO   d. CH_3^+ e. CMe_3^+

f. $ZnCl_2$

17. How could you convert i to ii?

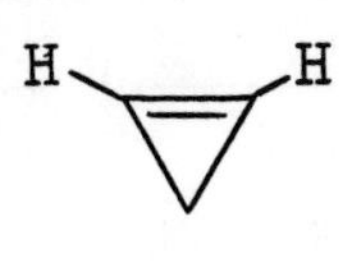

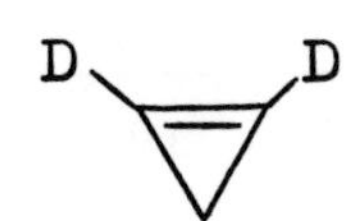

i   ii

Dorko and Mitchell, *Tetrahedron Lett.* **1968**, 341.

18. Which of these two equilibria would you expect to lie further to the right and why?

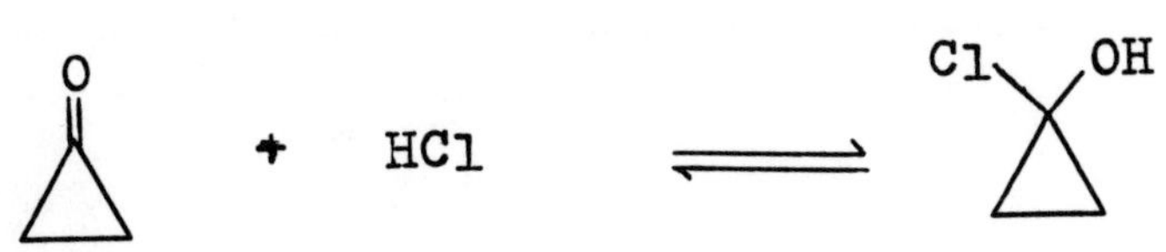

O + HCl ⇌ Cl, OH (cyclohexane)

18. Cont.

Turro and Hammond, *J. Amer. Chem. Soc.* **89**, 1028 (1967); *Tetrahedron* **24**, 6029 (1968).

19. For the reaction:

$$\text{ArN=CH-NHAr} \xrightarrow[\text{aq. 0.415N HCl}]{\text{20\% dioxane -}} \text{ArNH}_2 + \text{ArNHCHO}$$

at 39.7°, ρ is 3.78. If k for the unsubstituted case (Ar = Ph) is 1.60×10^{-4}, calculate k for the case where Ar = p-BrC_6H_4. σ_p for Br is 0.232. Note: The *measured* value for k in this case was 1.27×10^{-3}

20. $$XC_6H_5 + HOBr \xrightarrow[\text{50\% aq. dioxane}]{HClO_4} XC_6H_4Br$$ For the reaction at 50°,

rate constants for meta and para substitution are given in the table for several substituents.* σ and σ^+ values are also given.‡ Calculate the value of ρ corresponding to the σ^+ values. Show that no clear ρ value can be obtained when σ values are used.

	k	σ	σ^+
p-CH_3	356	−0.170	−0.311
m-CH_3	15.0	−0.069	−0.066
p-CMe_3	232	−0.197	−0.256
m-CMe_3	15.7	−0.10	−0.059
p-Ph	94	−0.01	−0.179
m-Ph	1.69	+0.06	+0.109
H	5.98	0	0

*The rate constants are calculated from data given by de la Mare and Harvey, *J. Chem. Soc.* **1956**, 36; **1957**, 131; and de la Mare and Hassan, *J. Chem. Soc.* **1957**, 3004.

‡These values, taken from Brown and Okamoto, *J. Amer. Chem. Soc.* **80**, 4979 (1958), are somewhat different from the best values accepted today.

PROBLEM SET 4

ALIPHATIC NUCLEOPHILIC SUBSTITUTION

Corresponds to Chapter 10 of Advanced Organic Chemistry.

The first 44 problems are synthetic. In each one you are asked to convert one compound into another. In every case the conversion has actually been carried out, and the reference is given.

1. $PhSO_2(CH_2)_5Cl \longrightarrow PhSO_2$–(cyclopentyl)

Truce, Hollister, Lindy, and Parr, *J. Org. Chem.* **33**, 43 (1968).

2.

H_3C O
CH_2–N–C–Ph
CH_3–C–CHOH–Ph
CH_3

$\longrightarrow$

CH_3
$\oplus$ N–CH_3
CH_3 Ph
CH_3

Anderson and Wills, *J. Org. Chem.* **33**, 536 (1968).

3. CH_3–C(=O)–NH_2 $\longrightarrow$ Ph–C(=O)–CH_2–C(=O)–NH–C(=O)–Ph

Wolfe and Trimitsis, *J. Org. Chem.* **33**, 894 (1968).

4.

EtOOC CH_2CH_2CN
C
EtOOC CH_2CH_2CN

$\longrightarrow$

O
CH_3 COOH

Odom and Pinder, *Chem. Commun.* **1969**, 26.

5.

MeO
MeO
Br

$\longrightarrow$

HO
HO
CHO

Berney and Deslongchamps, *Can. J. Chem.* **47**, 515 (1969).

6.

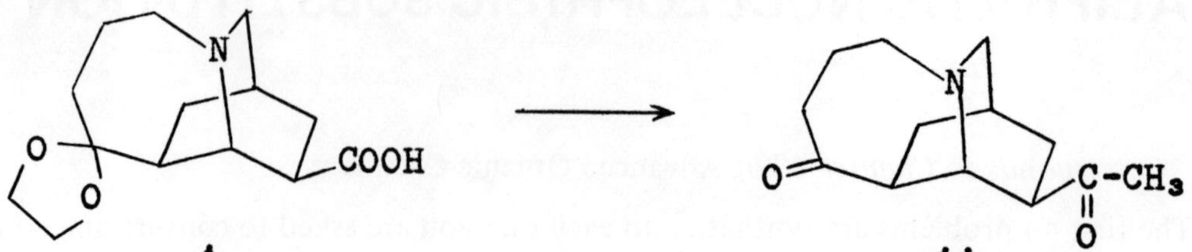

i → ii

COOH; $C(=O)-CH_3$

7.

$TsOCH_2-C(CH_2OTs)_3$ →

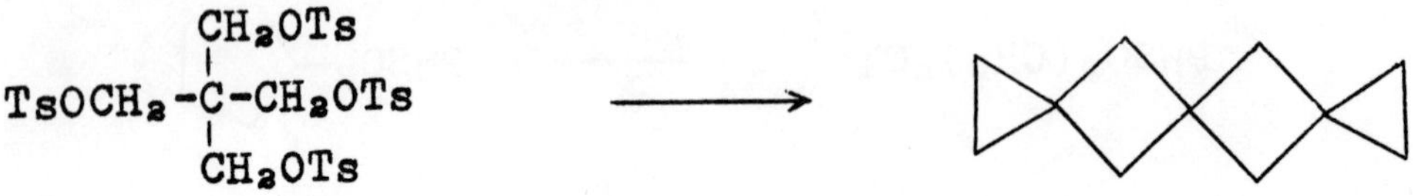

Buchta and Merk, *Justus Liebigs Ann. Chem.* **694**, 1 (1966); **716**, 106 (1968).

8.

$NH_2-C(=^{35}S)-NH_2$ ⟶ $CH_3CH_2CH_2CH_2{}^{35}SH$

Walling, *J. Amer. Chem. Soc.* **70**, 2561 (1948).

9.

NH_2OH ⟶ Et-NH-OMe

Kreutzkamp and Messinger, *Chem. Ber.* **100**, 3463 (1967).

10.

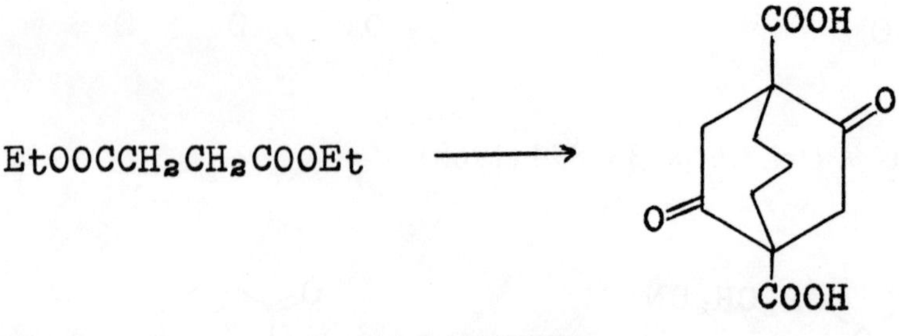

$EtOOCCH_2CH_2COOEt$ ⟶

COOH

O

O

COOH

Wood and Woo, *Can. J. Chem.* **46**, 3713 (1968).

11.

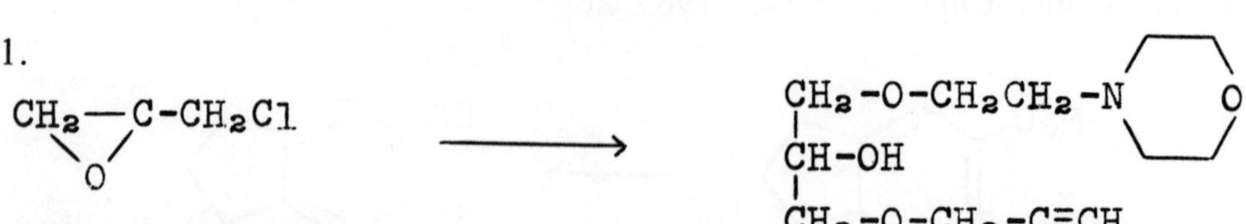

CH_2-C-CH_2Cl (epoxide O)

epichlorohydrin

i

⟶

$CH_2-O-CH_2CH_2-N$ (morpholine O)

CH-OH

$CH_2-O-CH_2-C\equiv CH$

ii

12.

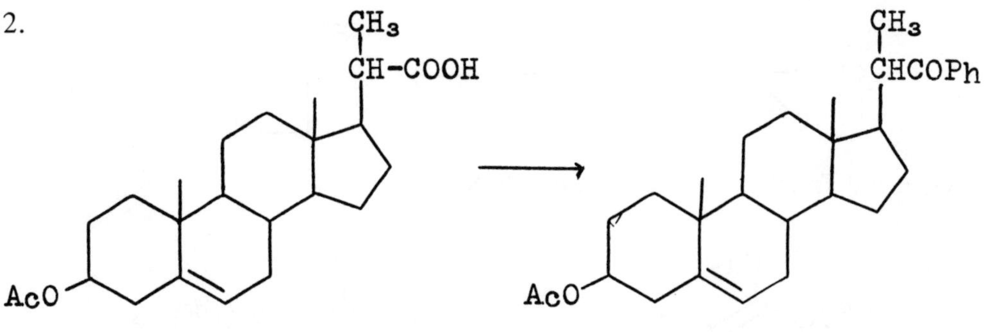

CH_3, CH–COOH, AcO → CH_3, CHCOPh, AcO

Cole and Julian, *J. Amer. Chem. Soc.* **67**, 1369 (1945).

13.

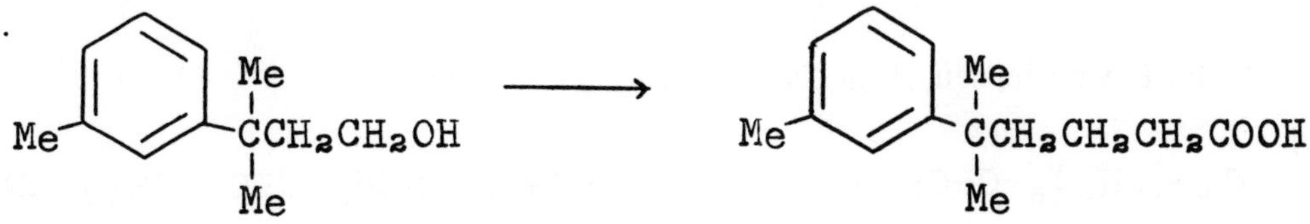

Me, Me, CCH_2CH_2OH, Me → Me, Me, $CCH_2CH_2CH_2COOH$, Me

Pandey and Dev, *Tetrahedron* **24**, 3829 (1969).

14.

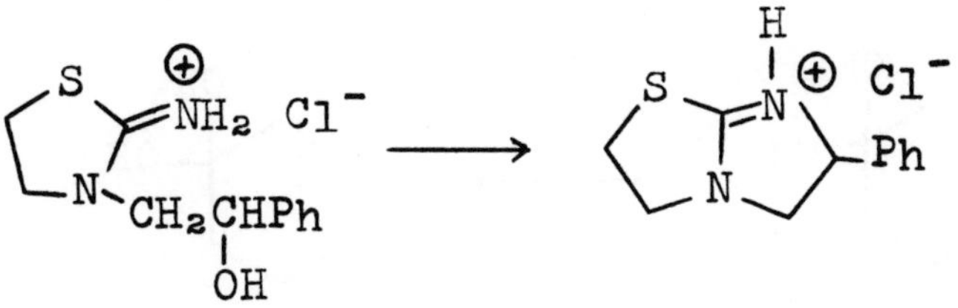

S, $\oplus$, NH_2 Cl^-, N, CH_2CHPh, OH → H, S, $N^{\oplus}$ Cl^-, N, Ph

For the reference, see Problem 15.

15.

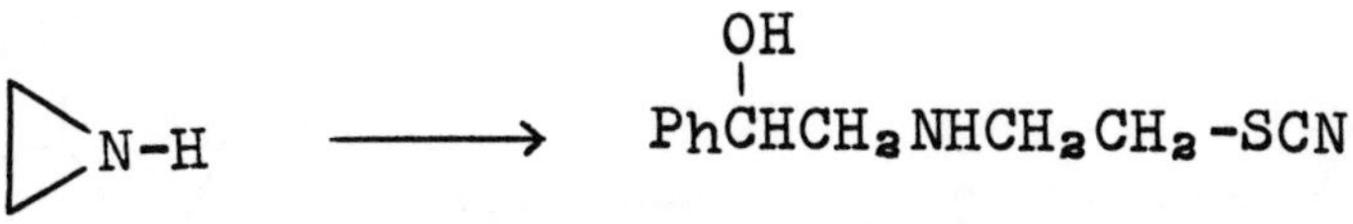

N–H → OH, $PhCHCH_2NHCH_2CH_2$-SCN

Spicer, Bullock, Garber, Groth, Hand, Long, Sawyer, and Wayne, *J. Org. Chem.* **33**, 1350 (1968).

16.

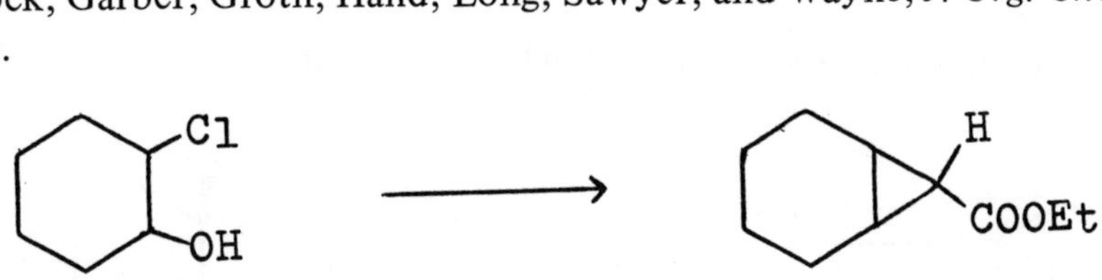

Cl, OH → H, COOEt

i ii

17. $PhCHCH_2OH$ (OH) → $PhCH_2CH_2OH$

Baltzly and Buck, *J. Amer. Chem. Soc.* **65**, 1984 (1943).

18.

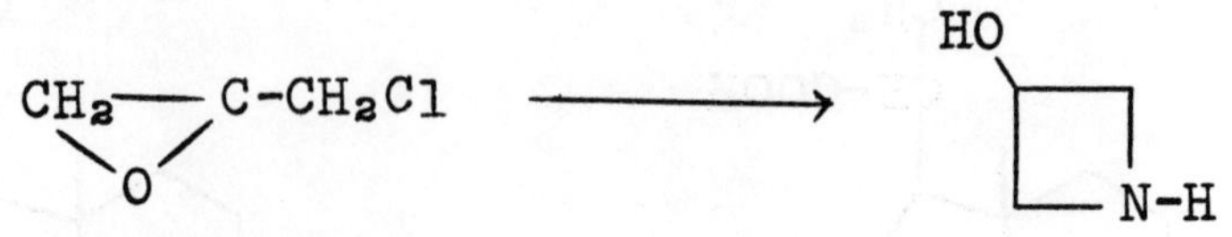

CH_2—C-CH_2Cl (epoxide, O) ⟶ 3-hydroxyazetidine (HO, N-H)

Chatterjee and Triggle, *Chem. Commun.* **1968**, 93.

19.

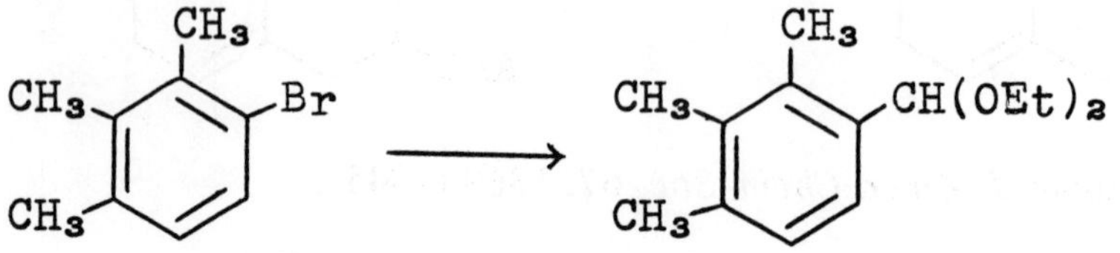

CH_3, CH_3, CH_3, Br (aryl bromide) ⟶ CH_3, CH_3, CH_3, $CH(OEt)_2$

Aasen and Liaaen Jensen, *Acta Chem. Scand.* **21**, 970 (1967).

20.

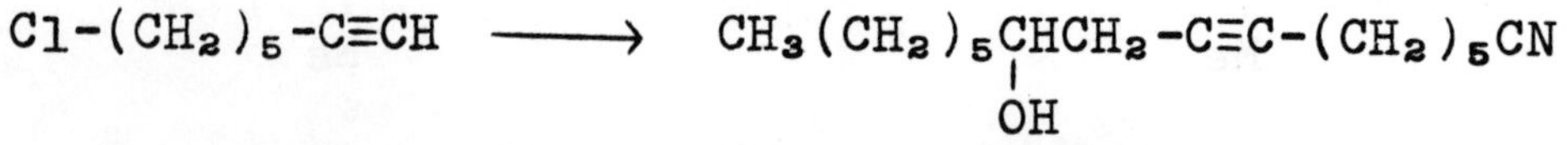

$Cl-(CH_2)_5-C{\equiv}CH \longrightarrow CH_3(CH_2)_5CH(OH)CH_2-C{\equiv}C-(CH_2)_5CN$

Eiter, Truscheit, and Boness, *Justus Liebigs Ann. Chem.* **709**, 29 (1967).

21.

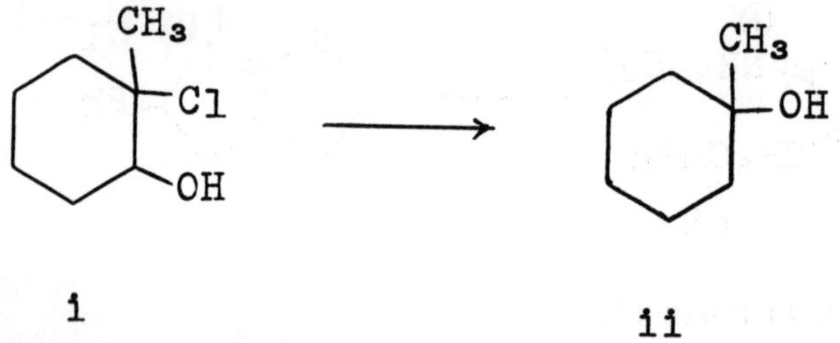

CH_3, Cl, OH (i) ⟶ CH_3, OH (ii)

i ii

22.

MeO—C₆H₄—CH—CH-COOEt (epoxide, O) ⟶ MeO—C₆H₄—(lactone, O, O) COOEt, COOEt

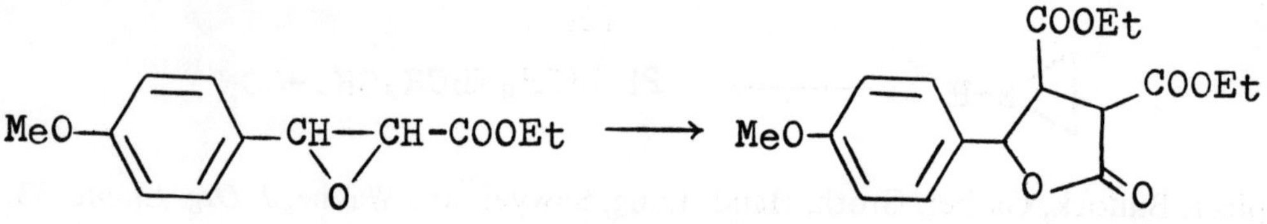

Takeda and Torii, *Bull. Chem. Soc. Jap.* **41**, 1468 (1968).

23.

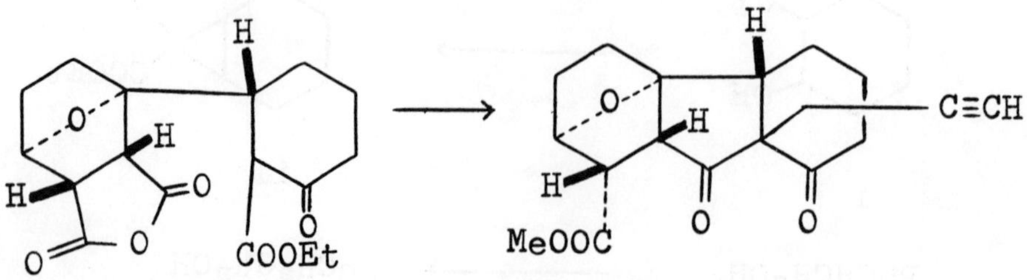

H, O, H, H, O, O, O, COOEt ⟶ H, O, H, H, MeOOC, O, O, C≡CH

Kitahara, Kato, Funamizu, Ototani, Inoue, and Izumi, *Chem. Commun.* **1968**, 1632.

24. $Cl_2CHCOOH \longrightarrow (EtO)_2CHCOOEt$

Moffett, *Org. Syn.* **IV**, 427.

25.

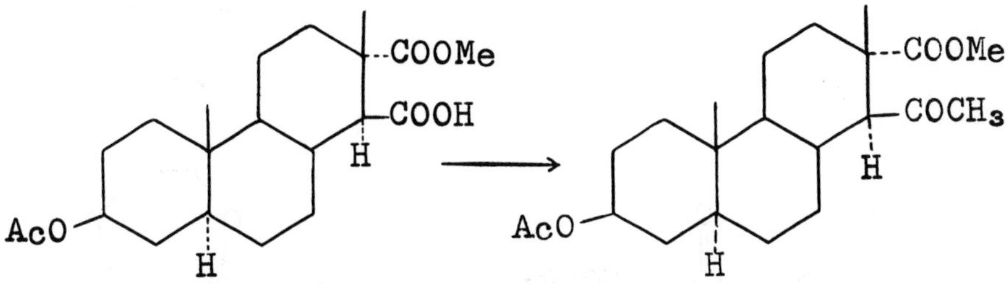

COOMe
COOH
H
AcO
H
⟶
COOMe
$COCH_3$
H
AcO
H

Banerjee and Gut, *Tetrahedron Lett.* **1969**, 51.

26.

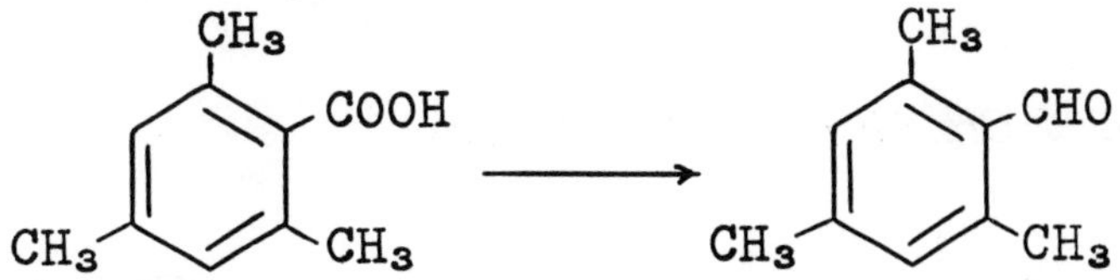

CH_3
COOH
CH_3
CH_3
⟶
CH_3
CHO
CH_3
CH_3

Barnes, *Org. Syn.* **III**, 551, 555.

27. $Ph-C(=O)-NH-C(=O)-CH_3 \longrightarrow Ph-C(=O)-NH-C(=O)-CH_2-C(=O)-Ph$

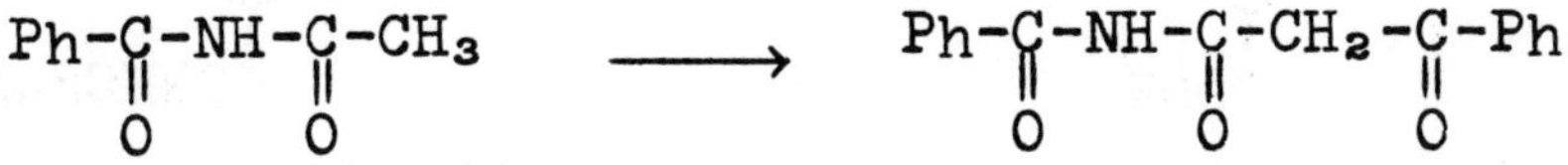

i ii

Wolfe, Trimitsis, and Hauser, *Can. J. Chem.* **46**, 2561 (1968); Work, Bryant, and Hauser, *J. Amer. Chem. Soc.* **86**, 872 (1964).

28.

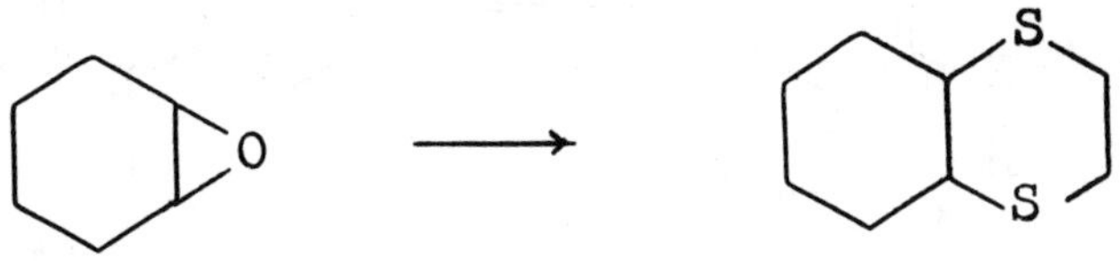

O
⟶
S
S

Culvenor, Davies, Hawthorne, Macdonald, and Robertson, *Aust. J. Chem.* **20**, 2207 (1967).

29.

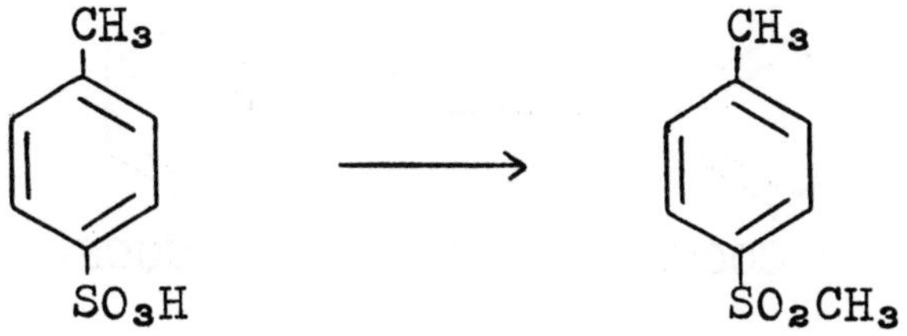

CH_3
SO_3H
⟶
CH_3
SO_2CH_3

i ii

30.

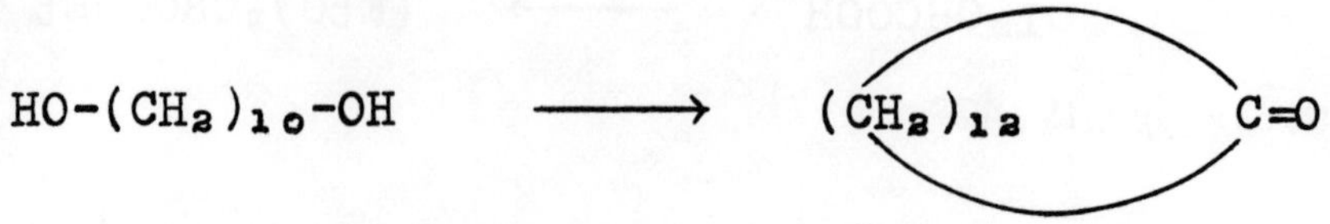

$$HO-(CH_2)_{10}-OH \longrightarrow (CH_2)_{12}\ C{=}O$$

i ii

31.

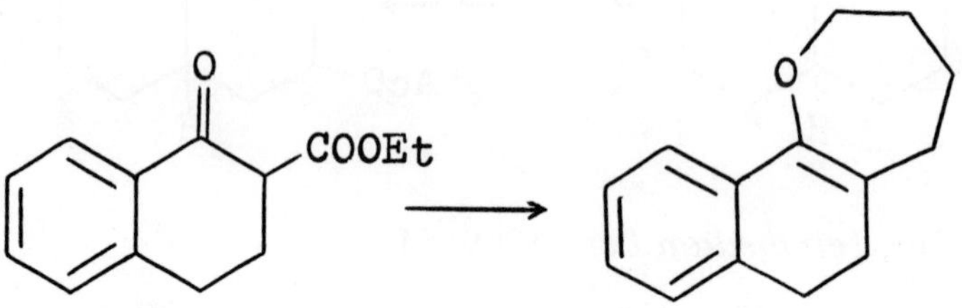

COOEt

Immer and Bagdi, *J. Org. Chem.* **33**, 2457 (1968).

32.

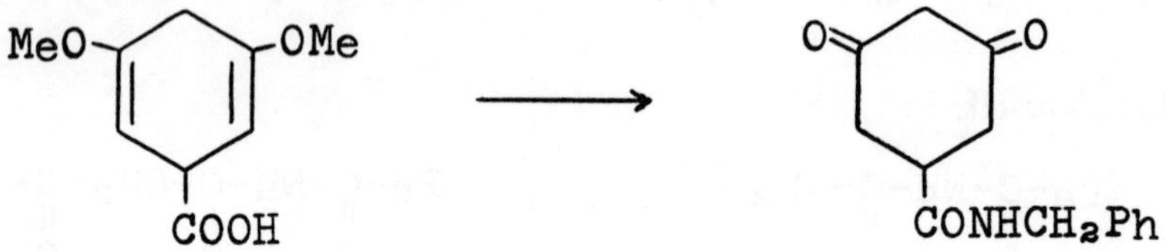

MeO OMe

COOH

O O

CONHCH$_2$Ph

Gensler, Gatsonis, and Ahmed, *J. Org. Chem.* **33**, 2968 (1968).

33.

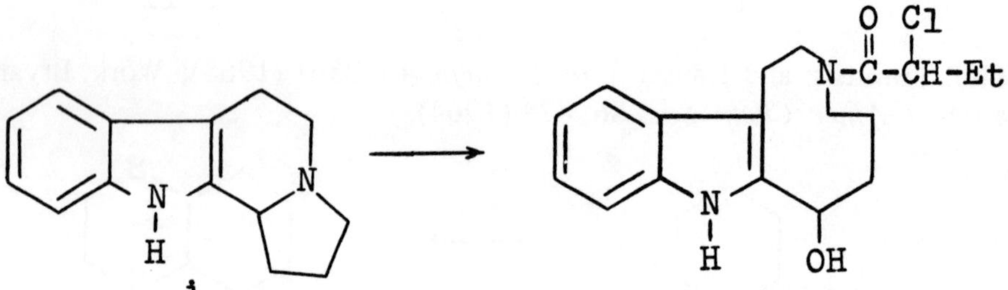

N H N

i

O Cl

N-C-CH-Et

N H OH

ii

34.

COOH COOH

$COCH_3$ $COCH_3$

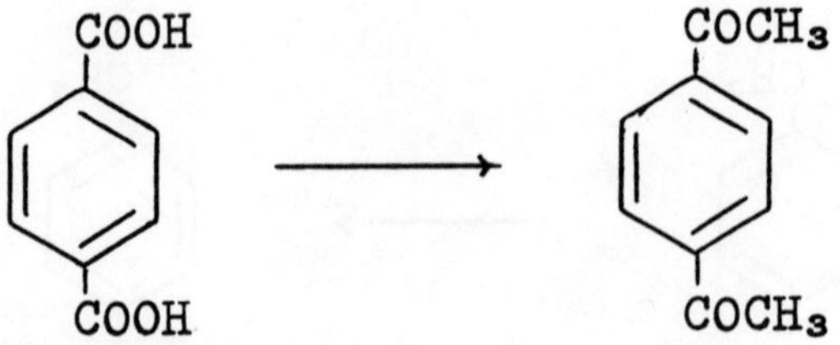

Pojer, Ritchie, and Taylor, *Aust. J. Chem.* **21**, 1375 (1968).

35.

Mahalanabis, Mukhopadhyay, and Dutta, *Chem. Commun.* **1968**, 1641.

36.

Bordwell, Frame, and Strong, *J. Org. Chem.* **33**, 3385 (1968).

37. $CH_3SCH_2CH_2C(OMe)_3 \longrightarrow CH_3SCH_2CH_2CHO$

Claus and Morgenthau, *J. Amer. Chem. Soc.* **73**, 5005 (1951).

38.

39.

Troxler, Stoll, and Niklaus, *Helv. Chim. Acta* **51**, 1870 (1968).

40.

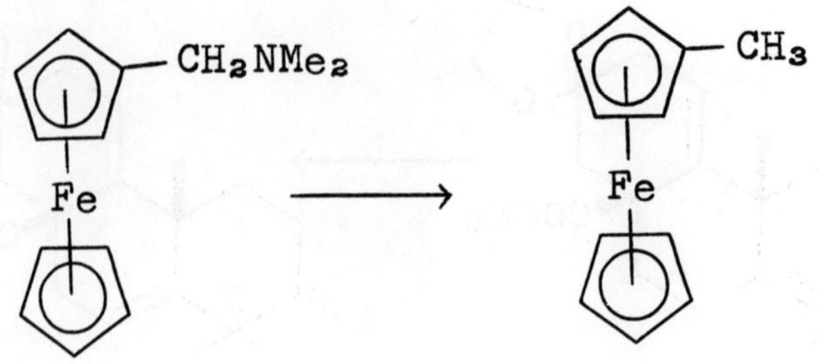

CH_2NMe_2 Fe $\longrightarrow$ CH_3 Fe

i ii

41.

O $\longrightarrow$ O COOEt

CH_3 CH_3

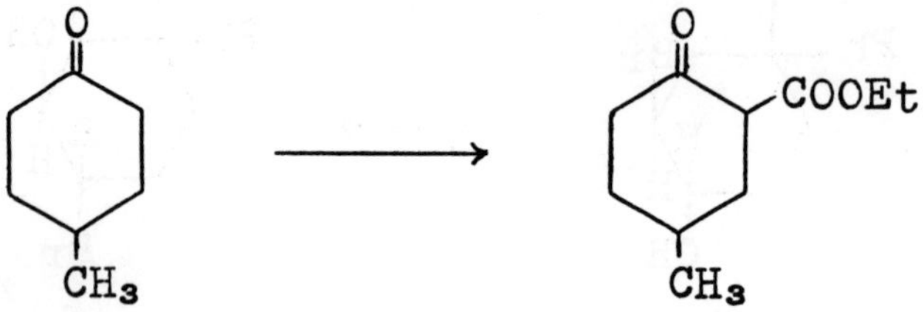

Saraf, *Aust. J. Chem.* 22, 2025 (1969).

42.

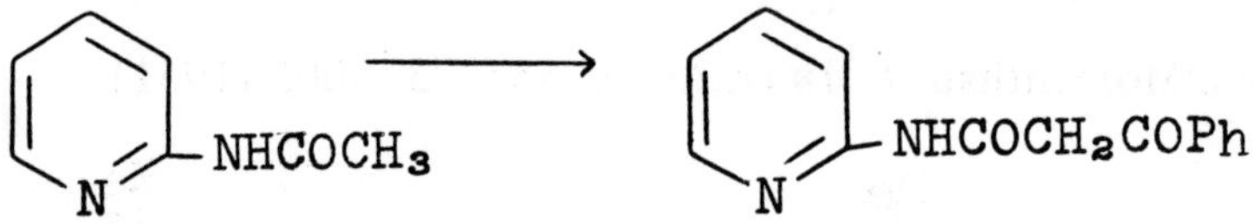

NHCOCH$_3$ (N) $\longrightarrow$ NHCOCH$_2$COPh (N)

Barnish, Hauser, and Wolfe, *J. Org. Chem.* 33, 2116 (1968).

43.

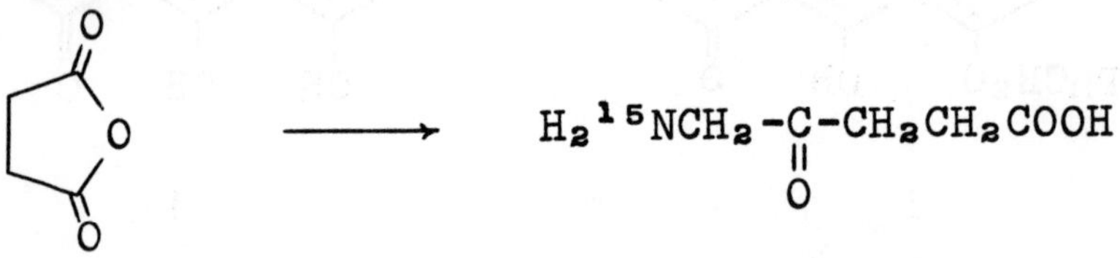

O O O $\longrightarrow$ $H_2{}^{15}NCH_2-\underset{\parallel}{C}(=O)-CH_2CH_2COOH$

i ii

44.

PhCOOH $\longrightarrow$ $Ph-\underset{\underset{O}{\parallel}}{C}-O-\underset{\underset{O}{\parallel}}{C}-OEt$

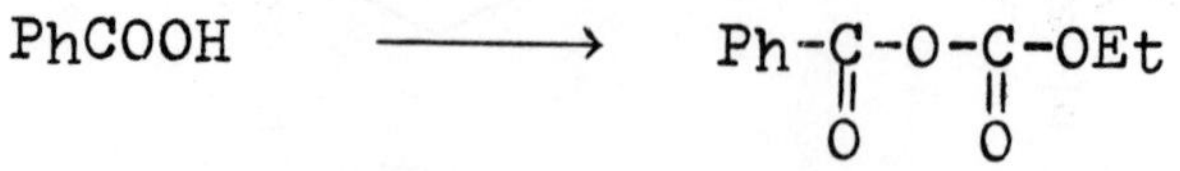

Price and Tarbell, *Org. Syn.* **IV**, 285.

The following conversions (problems 45 to 73) have been reported in the literature. In each case suggest a reasonable mechanism.

45.

$\xrightarrow{H^+}$

de Groot, Boerma, and Wynberg, *Tetrahedron Lett.* **1968**, 2365.

46.

$\xrightarrow{Et_2NH}$

i ii

47

$\xrightarrow{LiAlH_4}$

Kalsi, Wadia, Maheshwari, and Bhatia, *Chem. Ind.* (London) **1968**, 221.

48.

$\longrightarrow$

Bravo, Gaudiano, Ticozzi, and Umani-Ronchi, *Tetrahedron Lett.* **1968**, 4481.

49.

$$Ph\text{-}S\text{-}\underset{\underset{O}{\|}}{C}\text{-}OMe \xrightarrow[\text{MeCN, } 35^\circ \text{ 64 hr}]{Et_4N^+ F^-} 97\%\ Ph\text{-}S\text{-}Me + CO_2$$

Note: Other catalysts (e.g. Et_3N, Et_3P) can replace $Et_4N^+F^-$ with equally good results.

Jones, *J. Org. Chem.* 33, 4290 (1968).

50.

$$\underset{\textbf{i}}{Ph_2C(Cl){-}C({=}O){-}NHPh} + NaNH_2 \xrightarrow{\text{liq. } NH_3} \underset{\textbf{ii}}{Ph_2CH{-}N(Ph){-}C({=}O){-}NH_2} + \underset{\textbf{iii}}{Ph_2C(NHPh){-}C({=}O){-}NH_2} + \underset{\textbf{iv}}{Ph_2C(NH_2){-}C({=}O){-}NHPh}$$

ii, iii, and iv are formed in the ratio: 5:1:1.

51.

$$Cl_3C{-}CH(OH){-}CH{=}CCl_2 + SOCl_2 \xrightarrow[\text{5 hr}]{\text{reflux}} 95\%\ (Cl_3C)HC{=}CH(CCl_3)$$

Pilgram and Ohse, *J. Org. Chem.* **34**, 1586 (1969).

52.

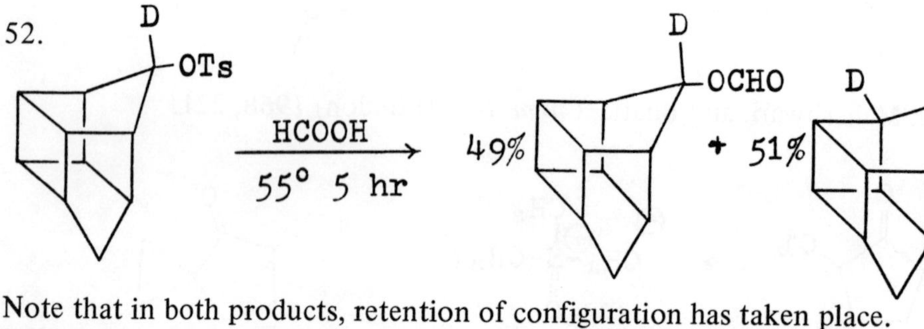

Note that in both products, retention of configuration has taken place.
Dilling, Plepys, and Kroening, *J. Amer. Chem. Soc.* **91**, 3404 (1969).

53.

$$\underset{\textbf{i}}{F_2C{=}CCl{-}CCl_2F} + \underset{\textbf{ii}}{NEt_4^+\ Cl^-} \xrightarrow[25^\circ]{CHCl_3} 100\%\ ClFC{=}CClCClF_2$$

Note: The reaction is first order each in i and ii.
Goldwhite and Valdez, *Chem. Commun.* **1969**, 7.

54. HO, H, H, CH_2Br (i) $\xrightarrow[\text{reflux overnight}]{\underline{t}\text{-BuOK}, \ \underline{t}\text{-BuOH}}$ O, H, CH_3 (ii) + HO, H, CH_2 (iii)

3:1 ratio

iii is the normal elimination product. Explain the formation of ii.
Gwynn and Skillern, *Chem. Commun.* **1968**, 490.

55.

$\underline{p}$-$MeC_6H_4SO_2CH_2CH_2SO_2Cl$ + NaOEt (molar ratio 1:5) $\xrightarrow[20^\circ \ 2 \text{ hr}]{\text{THF}}$

42% $\underline{p}$-$MeC_6H_4SO_2H$ + 36% $\underline{p}$-$MeC_6H_4SO_2CH_2CH_2OMe$

+ 18% $MeOCH_2CH_2SO_2Cl$

Miller and Stirling, *J. Chem. Soc., C* **1968**, 2895.

56.

$ClCH_2CH_2$-N (O, O) + $\underline{n}$-$C_8H_{17}NH_2$ → 86% $HOCH_2CH_2$-N N-C_8H_{17} (O)

Arnold, Pahls, and Potsch, *Tetrahedron Lett.* **1969**, 137.

57.

O, O, N, CH_3 (i) + H_2N-NH-C(=NH)-NH_2 (ii) → 39% NH_2, N, N, N, N, CH_3 (iii)

58.

COO^-, Ph-C(=O) → (MeO-S(=O)-Cl) 35% O, O, MeO, Ph

Newman and Courduvelis, *J. Amer. Chem. Soc.* 88, 781 (1966).

59. 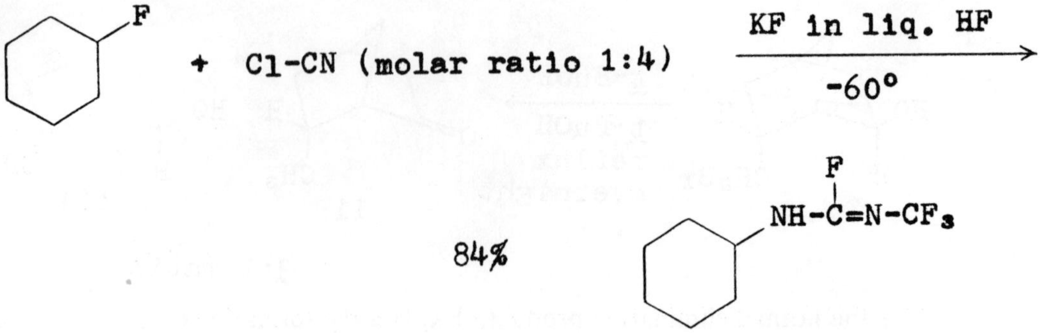

Alt and Weis, *Helv. Chim. Acta* **52**, 812 (1969).

60. 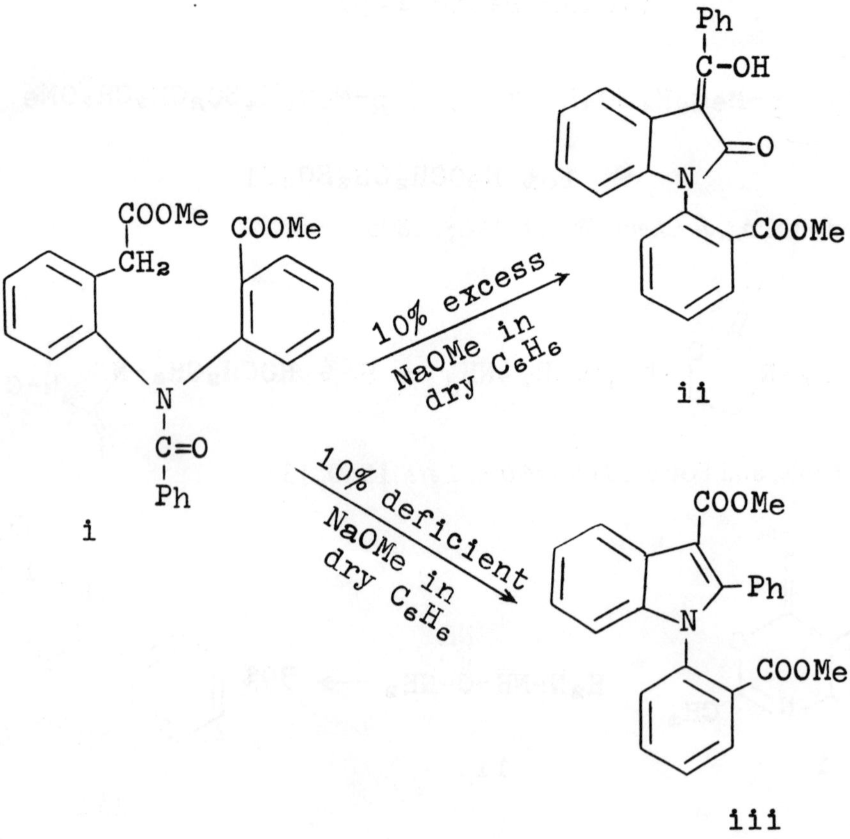

Notes: 1. ii and iii could not be interconverted by treatment with either excess or deficient NaOMe. 2. iv gave v with excess NaOMe, but with deficient NaOMe it gave vi, a stable crystalline enol.

60. Cont.

COOMe
CH2
COOMe
N
C=O
Cl
iv

excess OMe⁻

Cl
C-OH
=O
N
COOMe
v

deficient OMe⁻

COOMe
C=C
OH
Cl
NH
MeOOC
vi

Schulenberg, *J. Amer. Chem. Soc.* **90**, 7008 (1968).

61.

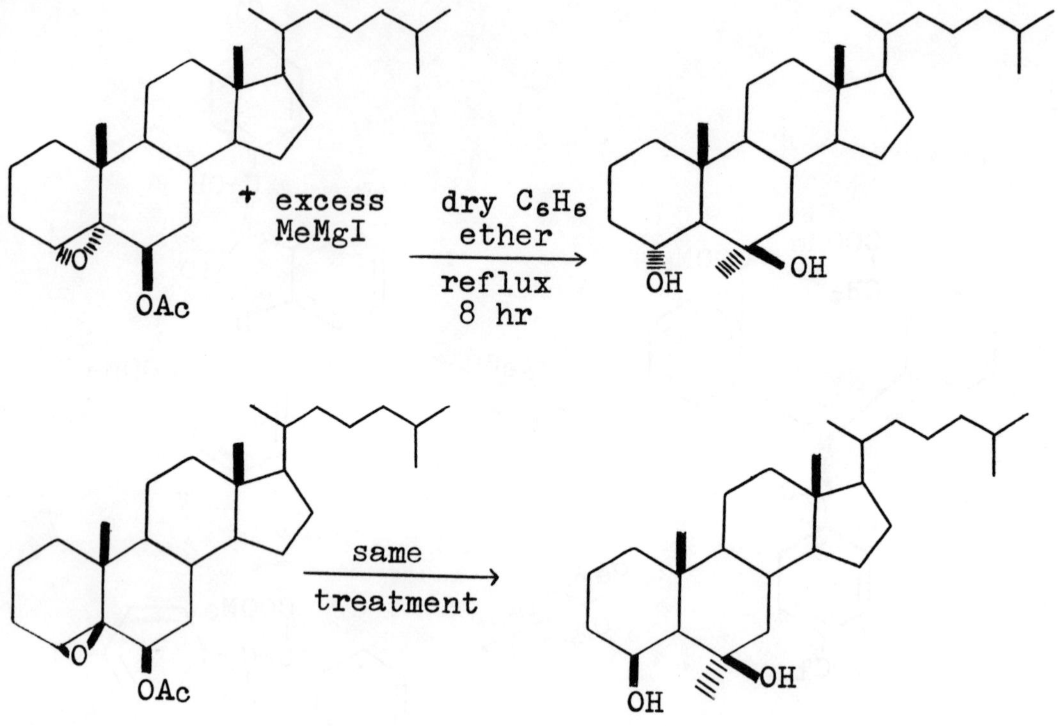

+ excess MeMgI

dry C_6H_6 ether, reflux 8 hr

OAc

OH OH

same treatment

OAc

OH OH

Bull, *J. Chem. Soc., C* **1969**, 1128.

62.

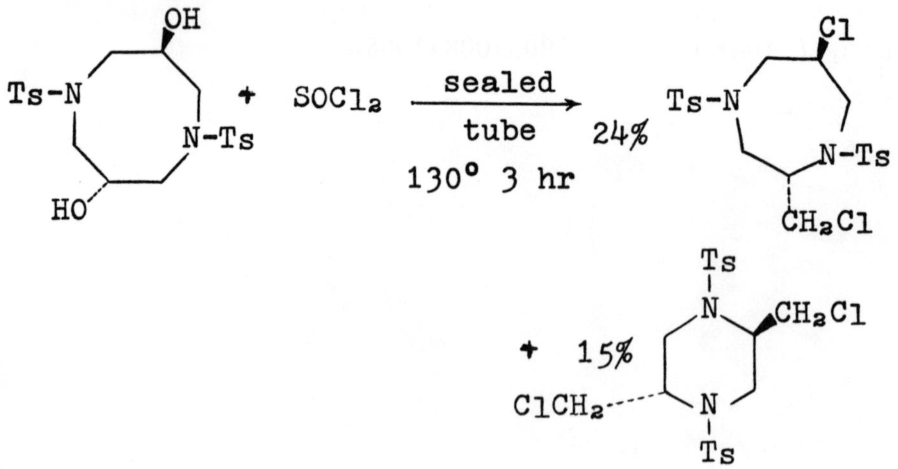

Ts-N, N-Ts, OH, HO

+ $SOCl_2$

sealed tube 130° 3 hr

24%

Cl, Ts-N, N-Ts, CH_2Cl

+ 15%

Ts, N, CH_2Cl, $ClCH_2$, N, Ts

Paudler, Zeiler, and Gapski, *J. Org. Chem.* **34**, 1001 (1969).

63.

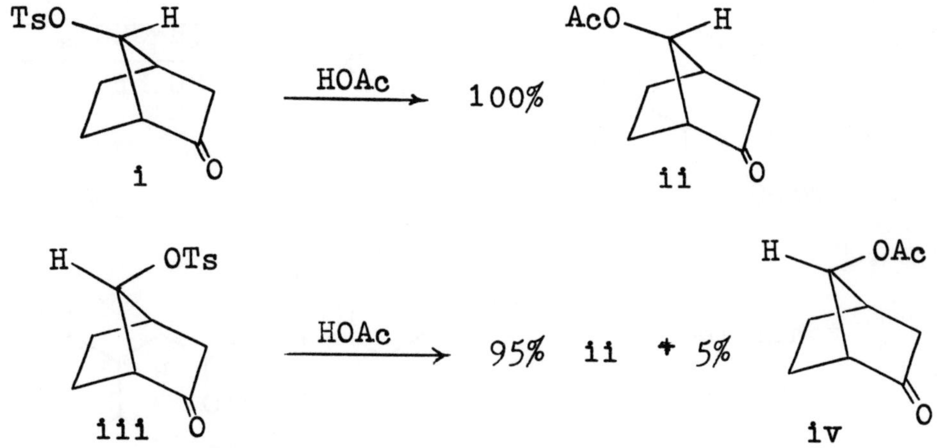

i $\xrightarrow{\text{HOAc}}$ 100% ii

iii $\xrightarrow{\text{HOAc}}$ 95% ii + 5% iv

Note: Acetolysis of i was much faster than that of iii.
Gassman and Hornback, *Tetrahedron Lett.* **1969**, 1325.

64.

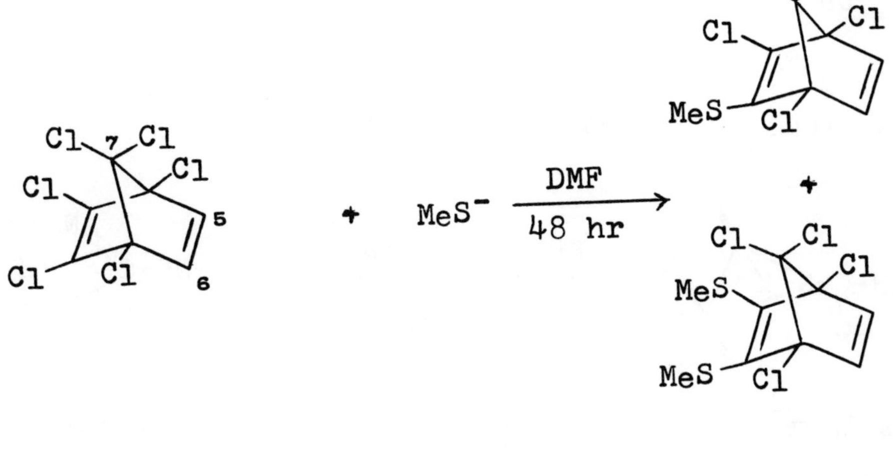

+ MeS^- $\xrightarrow[\text{48 hr}]{\text{DMF}}$ +

Note: When the 5-6 double bond was hydrogenated, or the 7-chlorines removed, the reaction rate was greatly reduced.
Davies and Rowley, *J. Chem. Soc., C* **1969**, 288.

65.

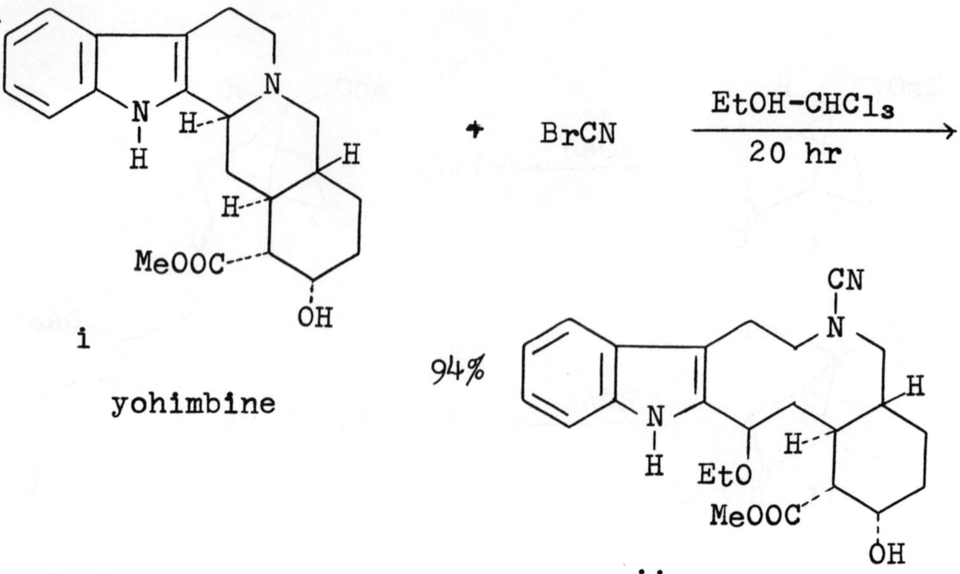

i

yohimbine

\+ BrCN $\xrightarrow[\text{20 hr}]{\text{EtOH-CHCl}_3}$ 94% ii

66.

$\xrightarrow[\text{25° 75 min}]{\text{SOCl}_2\text{-NaH-THF}}$ 33%

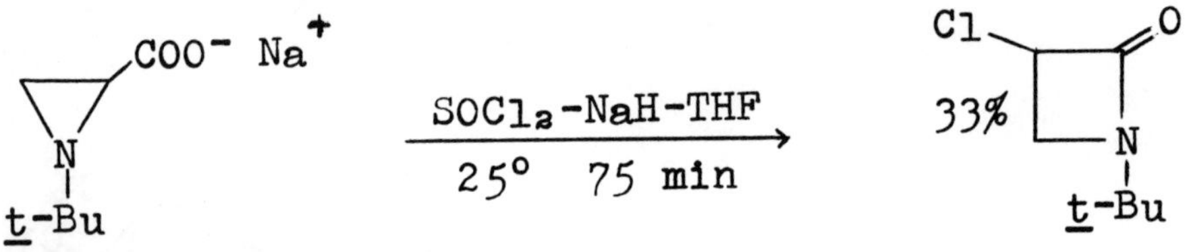

Deyrup and Clough, *J. Amer. Chem. Soc.* **91**, 4590 (1969).

67.

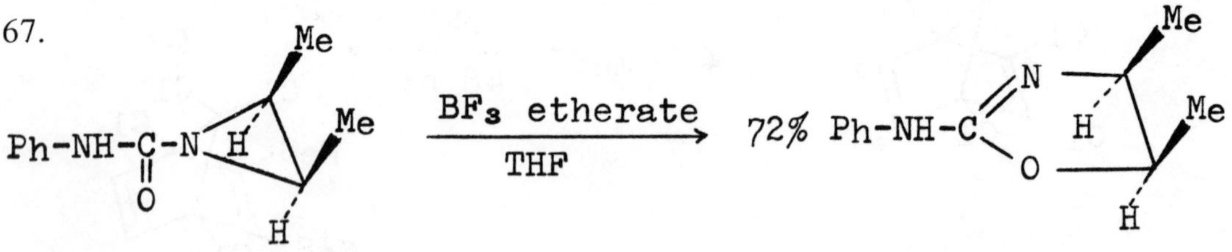

$\xrightarrow[\text{THF}]{\text{BF}_3 \text{ etherate}}$ 72%

cis isomer

cis isomer

Note: Five other solvents were used in place of THF. In all cases only the cis isomer was obtained.

Nishiguchi, Tochio, Nabeya, and Iwakura, *J. Amer. Chem. Soc.* **91**, 5835, 5841 (1969).

68.

$(NC)_2C{=}C(CN)Cl$ (i) + PhNHMe (ii) ⟶ $(NC)_2C{=}C(CN)-C_6H_4-NHMe$ (iii)

$(NC)_2C{=}C(CN)_2$ (iv) + PhNHMe ⟶ $(NC)_2C{=}C(CN)-N(Me)Ph$ (v)

69.

2,2-Dimethyl-1-(SO_2Ph)-1-(COOMe)cyclopropane $\xrightarrow[150^\circ\ 1.5\ hr]{MeOH}$ $Me-C(Me)(OMe)-CH_2-CH(SO_2Ph)-COOMe$

Cram and Ratajczak, *J. Amer. Chem. Soc.* **90**, 2198 (1968). See also Yankee and Cram, *J. Amer. Chem. Soc.* **92**, 6328, 6329, 6331 (1970).

70.

Br-substituted lactone bearing CH(OH)Me $\xrightarrow[150\text{-}160^\circ,\ 1\ hr]{H_2O}$ HO-substituted tetrahydrofuran bearing Me and COOH

Matsumoto, Ichihara, and Itō, *Tetrahedron Lett.* **1968**, 1989.

71.

$p\text{-}O_2N-C_6H_4-CH{=}CHCl$ (Cl trans to aryl) $\xrightarrow{MeO^-}$ $p\text{-}O_2N-C_6H_4-CH_2CH(OMe)_2$

$p\text{-}O_2N-C_6H_4-CH{=}CHCl$ (Cl cis to aryl) $\xrightarrow{MeO^-}$ $p\text{-}O_2N-C_6H_4-C{\equiv}CH$

Marchese, Naso, and Modena, *J. Chem. Soc., B* **1968**, 958.

72.

$$Ph{-}N{=}C(Cl){-}Cl + H_2NCH_2CH_2CH_2NH_2 \xrightarrow[70^\circ\ 1\ hr]{EtOAc-Et_3N}$$

Ph, N, N, N, Ph-N

Burkhardt and Hamann, *Chem. Ber.* **101**, 3428 (1968).

73.

$$CH_3(CH_2)_9{-}CH(CH_2Cl){-}O{-}C(=O){-}N\triangleleft \xrightarrow[in\ vacuo\ \ \Delta]{distn.}$$

i

CH_3, $(CH_2)_9$, O, N, CH_2CH_2Cl, O

ii

In the following problems (74 to 82) give the products and suggest reasonable mechanisms.

74.

$$Ph{-}N(Me){-}Et \xrightarrow[\underline{n}\text{-BuI}]{\underline{n}\text{-BuLi}}$$

Lepley and Khan, *J. Org. Chem.* **33**, 4362 (1968).

75.

$$NH_2CN + CH_3{-}C(=S){-}SEt \xrightarrow{MeOK} i \xrightarrow{CH_3I} ii\ (C_4H_6N_2S)$$

76.

$$\xrightarrow{Me_2SO_4} i \xrightarrow{NaOMe} ii\ (C_8H_{17}NO_3)$$

N, O, O, Me

Feinauer, *Angew Chem. Intern. Ed. Engl.* **7**, 731 (1968) [*Angew. Chem.* **80**, 703 (1968)].

77.

$$Ph_2C(OH){-}C{\equiv}C{-}Br + HBr \longrightarrow ii$$

i

78.

$$Ph{-}C{\equiv}C{-}CH_2CH_2CH_2CH_2CH_2Br \xrightarrow[THF\ -10^\circ]{LiNEt_2}$$

Crandall and Keyton, *Chem. Commun.* **1968**, 1069.

79.

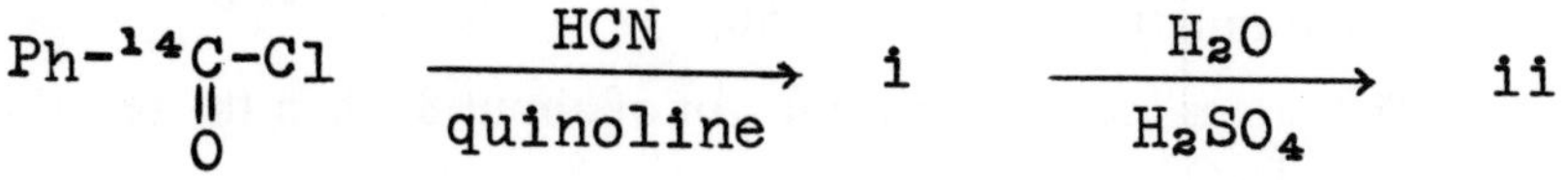

$$Ph-{}^{14}C(=O)-Cl \xrightarrow[\text{quinoline}]{HCN} \text{i} \xrightarrow[H_2SO_4]{H_2O} \text{ii}$$

Where will the label be in i and ii?

80.

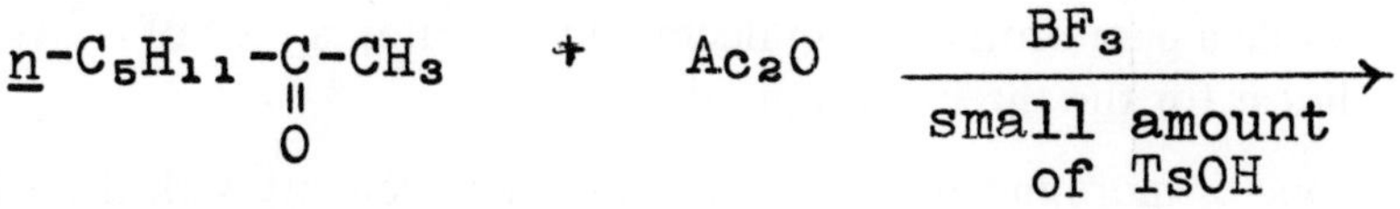

$$\underline{n}-C_5H_{11}-C(=O)-CH_3 + Ac_2O \xrightarrow[\text{small amount of TsOH}]{BF_3}$$

Hauser, Swamer, and Adams, *Org. React.* **8**, 59 (1954); p. 133.

81.

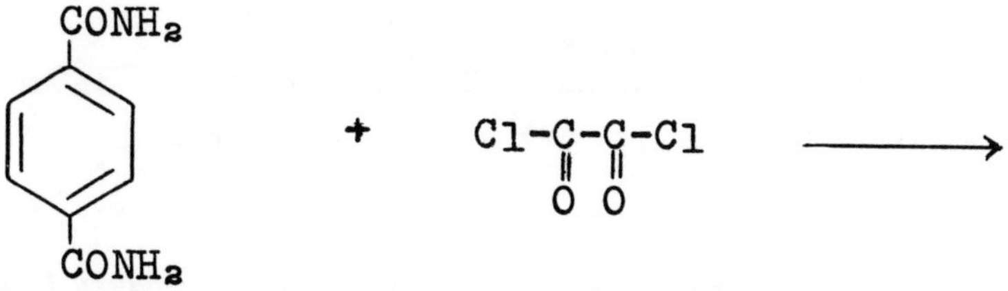

$$p-C_6H_4(CONH_2)_2 + Cl-C(=O)-C(=O)-Cl \longrightarrow$$

Tsuge, Itoh, and Tashiro, *Tetrahedron* **24**, 2583 (1968).

82.

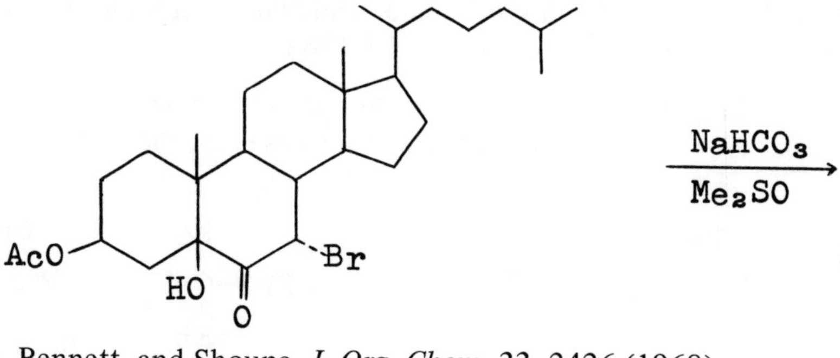

AcO, HO, Br, O

$$\xrightarrow[Me_2SO]{NaHCO_3}$$

Rowland, Bennett, and Shoupe, *J. Org. Chem.* **33**, 2426 (1968).

83.

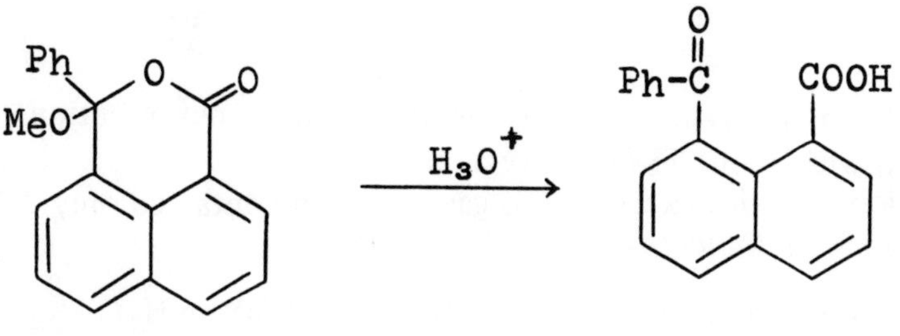

Ph, MeO, O, O

$$\xrightarrow{H_3O^+}$$

Ph-C(=O), COOH

i ii

Methyl pseudo-8-aroyl-1-naphthoates (i) react in aqueous acid to form 8-aroyl-1-naphthoic acids (ii). In a study of the mechanism, the following results were obtained:

1. The rate is pseudo first order.

83. Cont.

2. The Bunnett w value is -0.50.
3. The rate is increased by a factor of about 3 when the reaction is conducted in D_2O.
4. The entropy of activation is about zero.
5. A methyl group in the para position of the benzene ring increased the rate, while a para bromo group decreased the rate; a plot of σ^+ vs log k/k_0 was linear for the three compounds.

Write one or more mechanisms which are in agreement with these facts.

84. For the reaction:

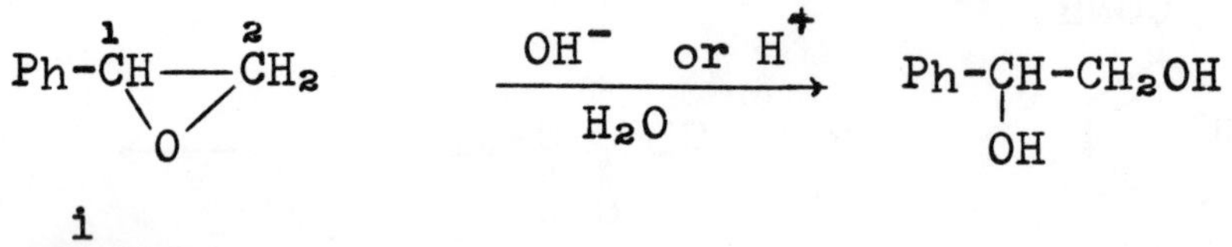

$$\underset{\textbf{i}}{\text{Ph}-\overset{1}{\text{CH}}\!-\!\overset{2}{\text{CH}_2}\ (\text{epoxide, }-\text{O}-)} \xrightarrow[\text{H}_2\text{O}]{\text{OH}^- \text{ or H}^+} \text{Ph}-\underset{\text{OH}}{\text{CH}}-\text{CH}_2\text{OH}$$

a. Devise an experiment to determine which bond of i breaks: C_1-O or C_2-O.

b. Predict which bond will break under acid and under basic conditions.

Audier, Dupin, and Jullien, *Bull. Soc. Chim. Fr.* **1968**, 3844, 3850.

85. The reaction between phenol and potassium *sec*-butyl diazotate (i) gave 9% ii and 10% iii. However, when optically active i was used, the ii obtained was

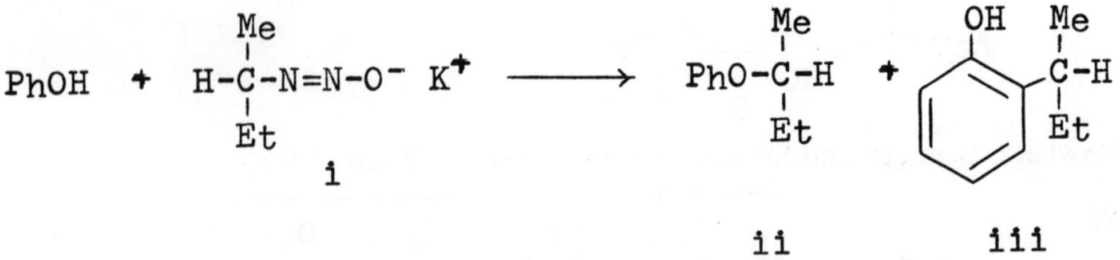

$$\text{PhOH} + \underset{\textbf{i}}{\text{H}-\overset{\text{Me}}{\underset{\text{Et}}{\text{C}}}-\text{N}=\text{N}-\text{O}^-\ \text{K}^+} \longrightarrow \underset{\textbf{ii}}{\text{PhO}-\overset{\text{Me}}{\underset{\text{Et}}{\text{C}}}-\text{H}} + \underset{\textbf{iii}}{o\text{-HO-C}_6\text{H}_4-\overset{\text{Me}}{\underset{\text{Et}}{\text{C}}}-\text{H}}$$

almost completely racemized (a small amount of net retention), while iii was formed with about 75-80% inversion. Explain.

Note: Reaction of i with excess PhOD gave ii which contained only 7.2% D.
Moss and Temme, *Tetrahedron Lett.* **1968**, 3219.

86. When 3-bromopropanoic acid was solvolyzed at 50° in H_2O-Me_2SO mixtures, the rate remained about the same from 0 to 80% Me_2SO (actually, a slight decrease); however, when the corresponding anion (the 3-bromopropionate ion) was solvolyzed, the rate in 80% Me_2SO was 150 times the rate in H_2O. Suggest an explanation.

Kingsbury, *J. Amer. Chem. Soc.* **87**, 5409 (1965).

87. Acetolysis at 25° of 4-methoxy-1-pentyl brosylate or of 5-methoxy-2-pentyl brosylate leads in either case to the same product mixture: 60% 5-methoxy-2-pentyl acetate and 40% 4-methoxy-1-pentyl acetate. Explain.

88.

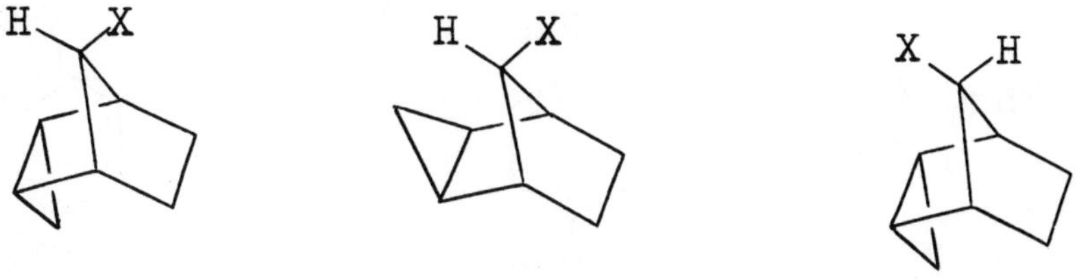

i ii iii

i solvolyzes about 10^{14} times faster than ii and about 10^{13} times faster than iii. Explain.

Tanida, Tsuji, and Irie, *J. Amer. Chem. Soc.* **89**, 1953 (1967); Battiste, Deyrup, Pincock, and Haywood-Farmer, *J. Amer. Chem. Soc.* **89**, 1954 (1967).

89. Dicyclohexylcarbodiimide (DCC) is a common reagent for combining an N-protected and a C-protected amino acid to obtain a protected dipeptide. When N-*p*-tolylsulfonyl-N′-cyclohexylcarbodiimide (i) was used instead of DCC, however, the product obtained was not the expected protected dipeptide (ii), but the heterocycle iii, and the same product was obtained in the absence of iv. Explain.

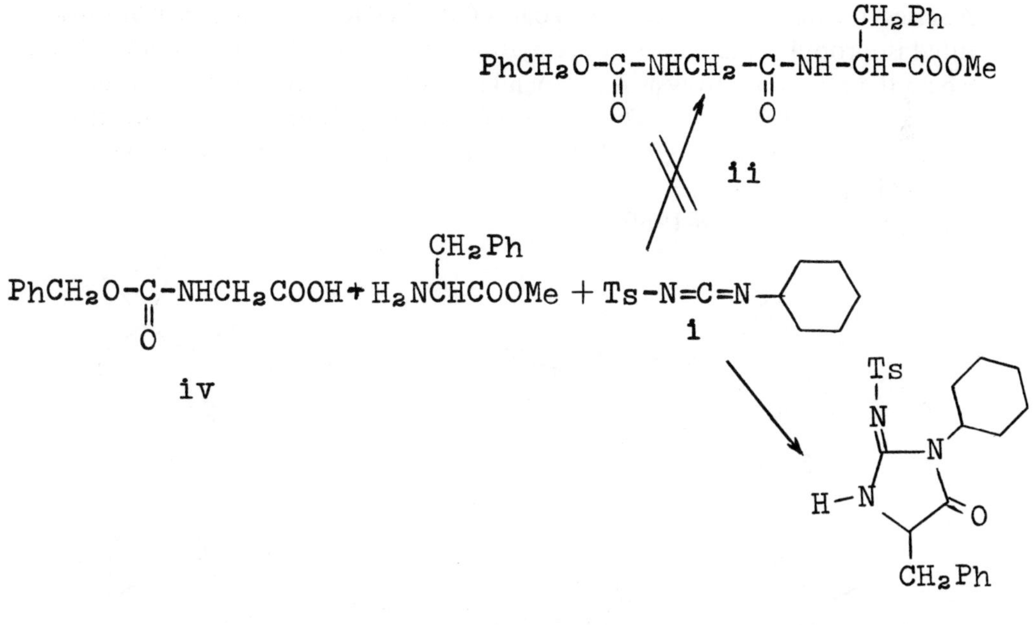

iii

Gupta and Stammer, *J. Org. Chem.* **33**, 4368 (1968).

90. In the reaction:

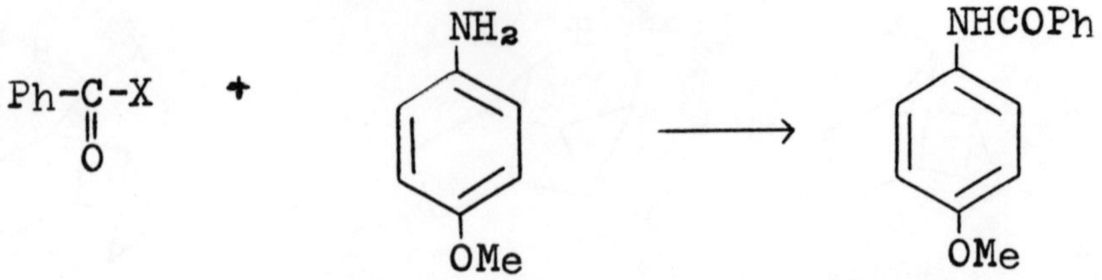

Ph-C(=O)-X + p-MeO-C_6H_4-NH_2 ⟶ p-MeO-C_6H_4-NHCOPh

the rate constants for different X groups were: X = F 0.000225; X = Cl 1.21; and X = Br 57.9. What does this imply about the rate determining step of the mechanism?

91. In the solvolysis of optically active 1-phenylethyl chloride in 60 or 70% aqueous dioxane, the rate of the reaction followed by a polarimeter was greater than the rate followed by titration of the HCl produced with standard base. Explain.

de la Mare, Hall, and Mauger, *Rec. Trav. Chim.* **87**, 1394 (1968).

92. The reactions of methyl iodide with simple aliphatic amines (e.g. $EtNH_2$, Et_2NH, Et_3N, $BuNH_2$, etc.) are slower in methanol than in benzene, although this is an S_N2 type II reaction which should be aided by polar solvents. Explain.

93. A series of substituted methyl benzoates (ArCOOMe, i) and another of substituted isopropyl benzoates (ii) were hydrolyzed in 98% H_2SO_4 to dilute oleum where the rate of hydrolysis is independent of both of the possible nucleophiles, H_2O and HSO_4^-. The ρ values for hydrolysis of i and ii, respectively, were −0.825 and +1.991. What conclusions can be drawn about the mechanism?

Hopkinson, *J. Chem. Soc., B* **1969**, 203.

94.

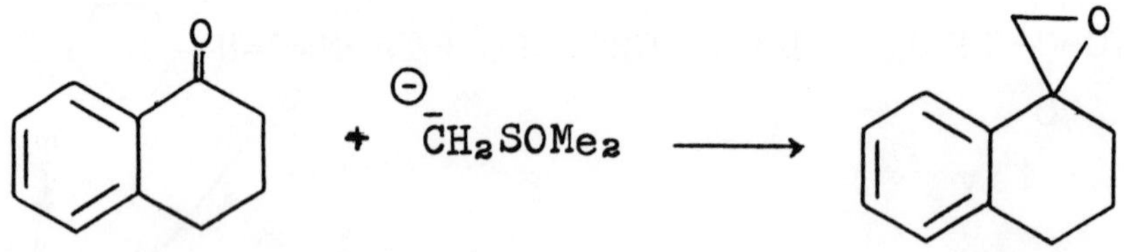

i + $\bar{C}H_2SOMe_2$ ⟶ ii

iii + $\bar{C}H_2SOMe_2$ ⟶ iv

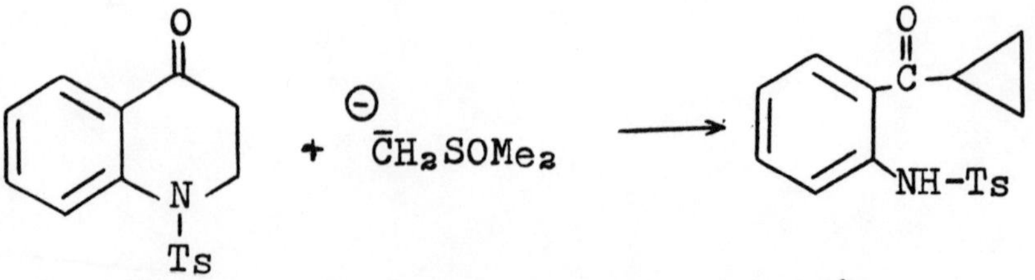

94. Cont.

The reaction of i with dimethyloxosulfonium methylide yields the normal product, ii, by addition of CH_2 to the C=O bond. However, when the same reagent is added to iii, the reaction takes a different course, and iv is produced instead. Explain.

Speckamp, Neeter, Rademaker, and Huisman, *Tetrahedron Lett.* **1968**, 3795.

95. Solvolysis of i and ii gave the products shown. Explain.

Note: none of the five products were interconvertible under the reaction conditions.

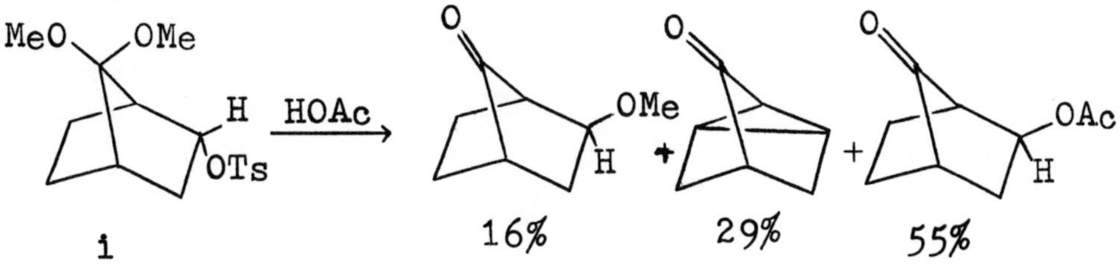

i 16% 29% 55%

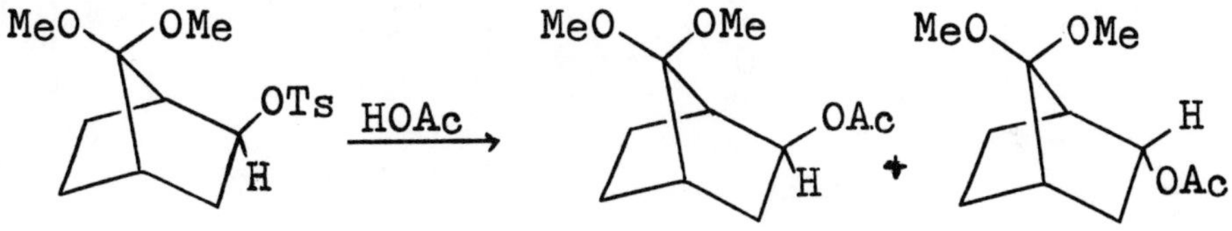

ii 95.5% 4.5%

Gassman and Marshall, *Tetrahedron Lett.* **1968**, 2429.

96

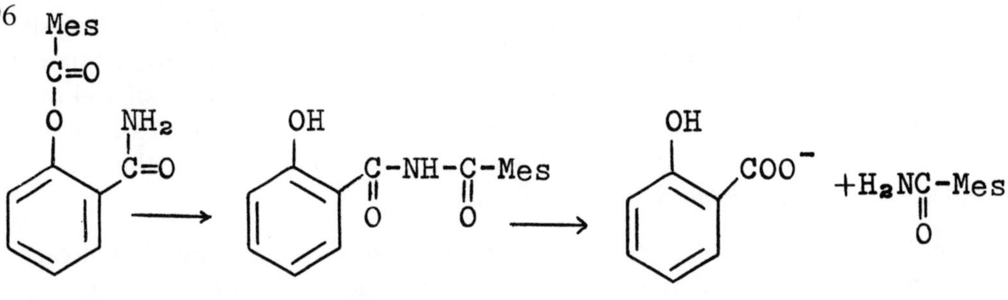

i ii iii iv

96. Cont.

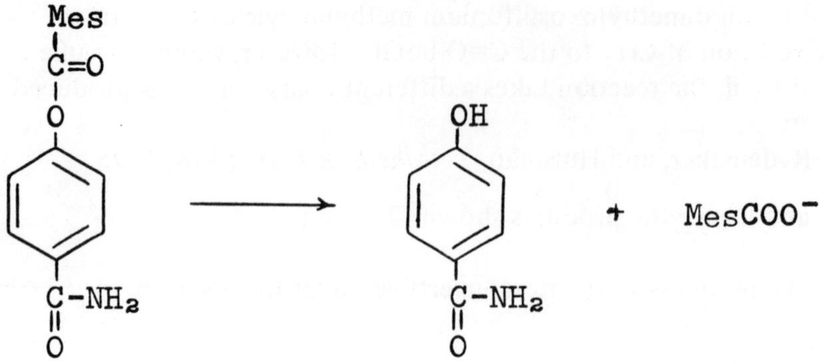

Mes
C=O
O
C-NH$_2$
O

OH
C-NH$_2$
O

+ MesCOO$^-$

v

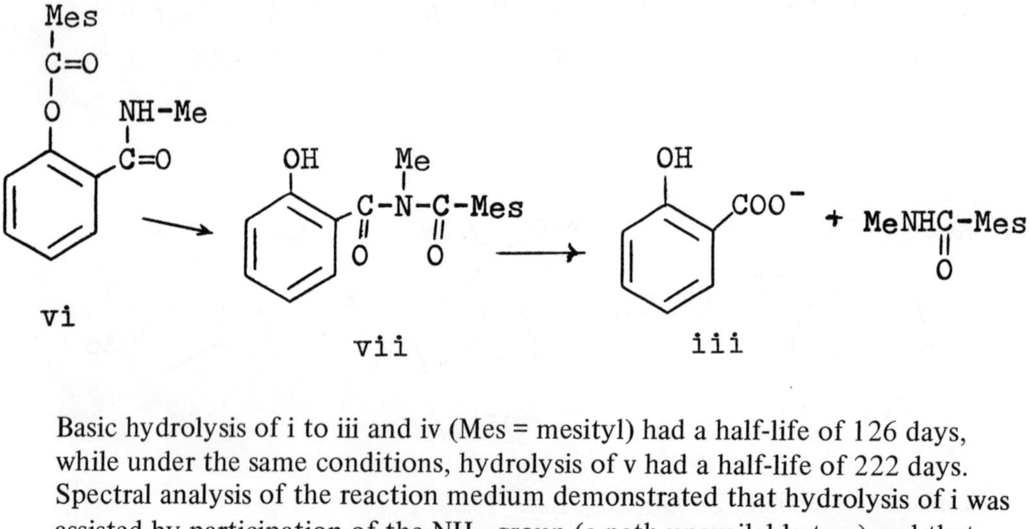

Mes
C=O
O NH-Me
C=O

vi

OH Me
C-N-C-Mes
O O

vii

OH
COO$^-$

iii

+ MeNHC-Mes
O

Basic hydrolysis of i to iii and iv (Mes = mesityl) had a half-life of 126 days, while under the same conditions, hydrolysis of v had a half-life of 222 days. Spectral analysis of the reaction medium demonstrated that hydrolysis of i was assisted by participation of the NH_2 group (a path unavailable to v) and that, in fact, conversion of i to the imide intermediate ii was very fast (half-life 23 minutes) and that the overall reaction was slowed by the fact that hydrolysis of ii was slow. However, vi, which also went through an imide intermediate (vii) was hydrolyzed much faster. The conversion to vii had a half-life of 144 minutes (slower than the corresponding reaction of i) but for hydrolysis of vii the half-life was only 77 seconds. Explain why vii hydrolyzed so much faster than ii.
Topping and Tutt, *J. Chem. Soc., B* **1969**, 104.

97. In S_N1 ethanolysis of 1-arylethyl chlorides at 34.8°, the introduction of each ortho or para methyl group raises the rate by a factor of about 15-40, *except* when the methyl group is a second ortho group, in which case the increase is much less. Explain.

97. Cont.

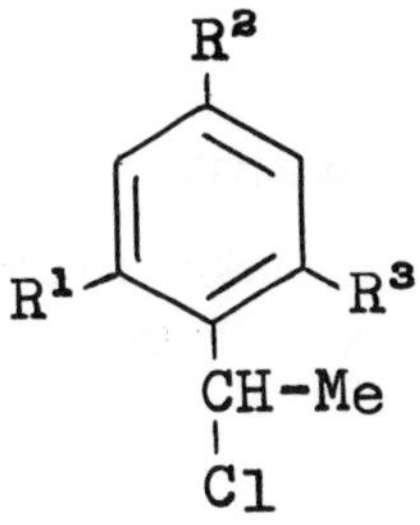

R^1	R^2	R^3	Relative rate
H	H	H	1
H	Me	H	40
Me	H	H	16
Me	Me	H	560
Me	H	Me	30
Me	Me	Me	1440

98. Dialkoxyalkyl halides (i) are usually not stable. They decompose.

a. Show the mechanism and products of decomposition.

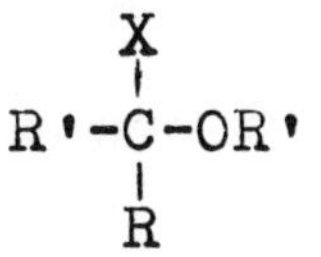

i

b. Which groups R′ and X would be most likely to give a stable dialkoxyalkyl halide?

99. The ratio for the rate of alkaline hydrolysis (attack by OH^-) of diethyl succinate ($EtOOCCH_2CH_2COOEt$) in $EtOH$-H_2O (60-40) at 35° (k_1/k_2) is 5.21. That is, the first ester group is hydrolyzed 5.21 times as fast as the second. But in Me_2SO-H_2O (60-40) at 35°, k_1/k_2 is 9.96. Explain the increase.
Venkatasubramanian and Rao, *Tetrahedron Lett.* **1967**, 5275.

100. For the esters NC-CMe_2-COOR and EtOOC-CMe_2-COOR (where R = *o* or *p*-nitrophenyl) hydrolysis by OH^- in H_2O takes place by the ordinary tetrahedral mechanism, and follows the rate law:

$$k_{observed} = k_{OH}[HO^-] + k_{H_2O}[H_2O] \quad (1)$$

That is, log k_{obs} is constant from pH 1 to 6, and then linearly rises from pH 6 to 14 as the concentration of OH^- increases. The esters NC-CH_2-COOR and EtOOC-CH_2-COOR (R as before), however, do not show this kinetic behavior. For these esters log k_{obs} increases only to about pH 10 and then levels off. That is, at high pH values there is zero order dependence on $[OH^-]$. Both of these esters hydrolyze faster than would be predicted from Eq. (1). Suggest a mechanism which would account for this.
Bruice and Holmquist, *J. Amer. Chem. Soc.* **90**, 7136 (1968).

PROBLEM SET 5

AROMATIC ELECTROPHILIC SUBSTITUTION

Corresponds to Chapter 11 of Advanced Organic Chemistry.

The first 25 problems are synthetic. In each case you are asked to convert one compound into another. In every case the conversion has actually been carried out and the reference is given.

1.

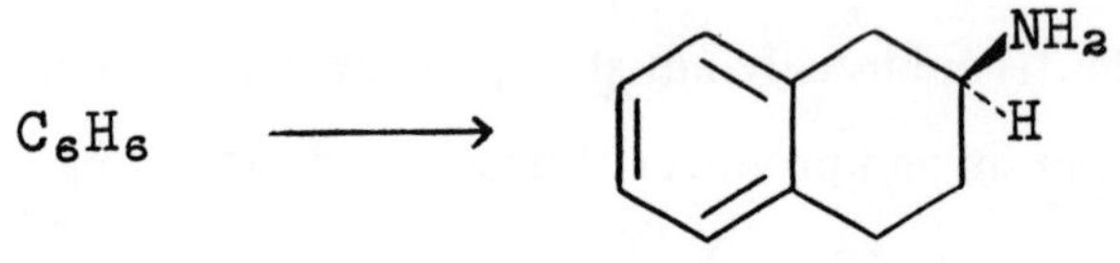

C_6H_6 ⟶ [structure: 2-amino-1,2,3,4-tetrahydronaphthalene, NH_2, H]

S enantiomer

Zymalkowski and Dornhege, *Tetrahedron Lett.* **1968**, 5743.

2.

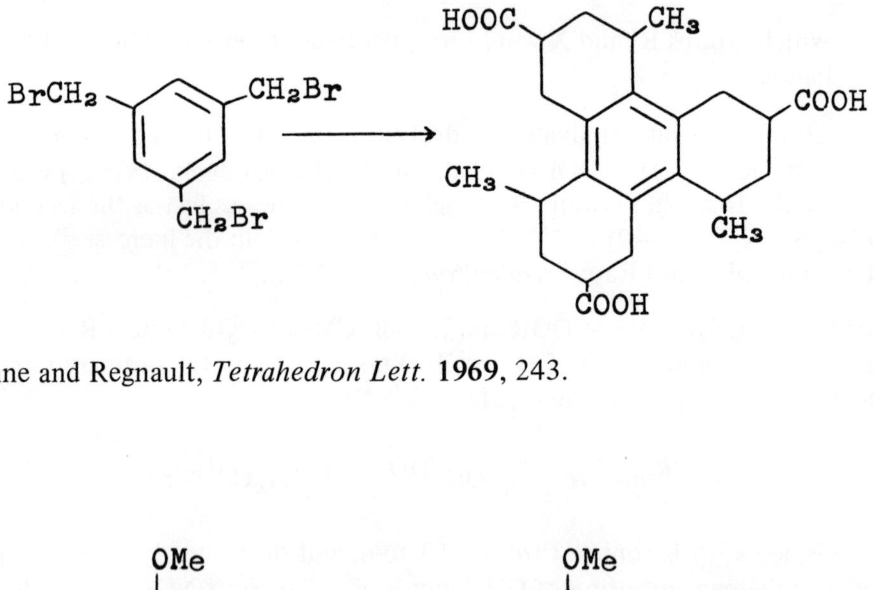

[structure: 1,3,5-tris(bromomethyl)benzene, $BrCH_2$, CH_2Br, CH_2Br] ⟶ [structure: HOOC, CH_3, COOH, CH_3, CH_3, COOH]

Canonne and Regnault, *Tetrahedron Lett.* **1969**, 243.

3.

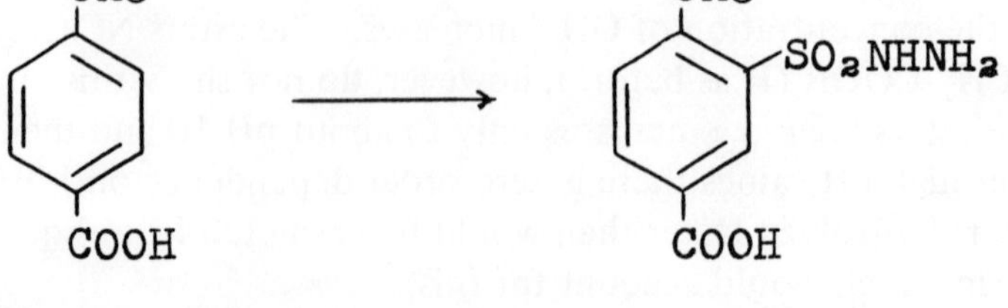

[structure: OMe, COOH] ⟶ [structure: OMe, SO_2NHNH_2, COOH]

Cremlyn, *J. Chem. Soc., C* **1968**, 11.

4.

i → ii

5.

stebisimine

Kametani, Kusama, and Fukumoto, *Chem. Commun.* **1967**, 1212.

6.

Russell and Lockhart, *Org. Syn.* **III,** 463.

7.

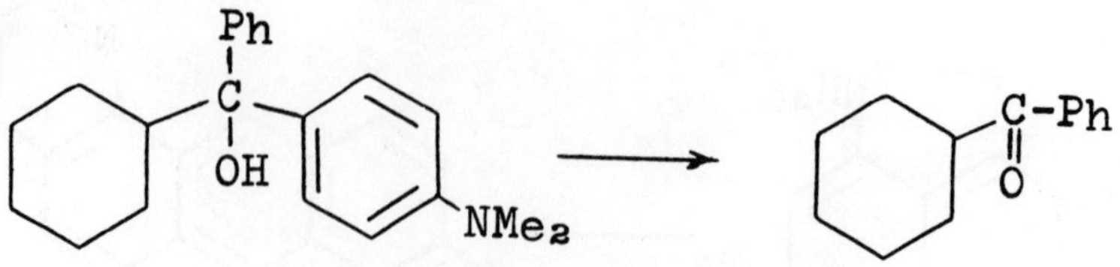

Ph, C, OH, NMe_2 → C–Ph, O

Sisti, Sawinski, and Stout, *J. Chem. Eng. Data* **9**, 108 (1964).

8.

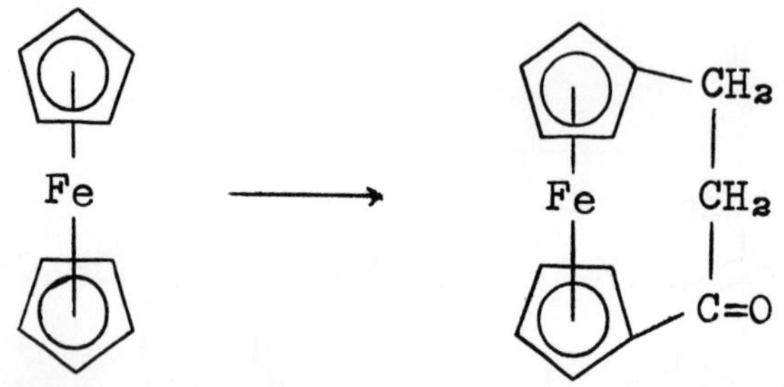

Fe → Fe, CH_2, CH_2, C=O

Rinehart, Curby, Gustafson, Harrison, Bozak, and Bublitz, *J. Amer. Chem. Soc.* **84**, 3263 (1962).

9.

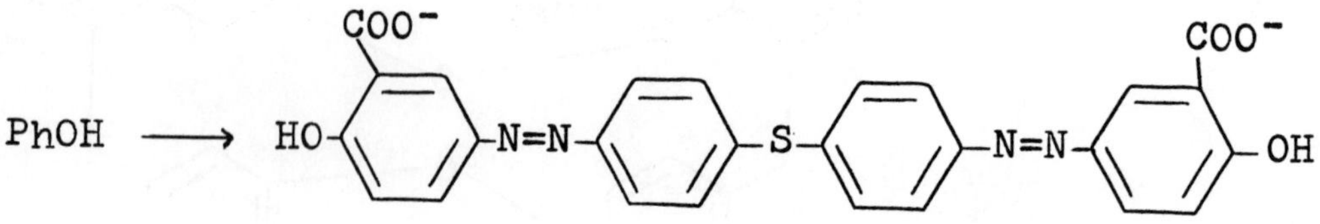

PhOH → COO^-, HO, N=N, S, N=N, COO^-, OH

i

anthracene yellow C

10.

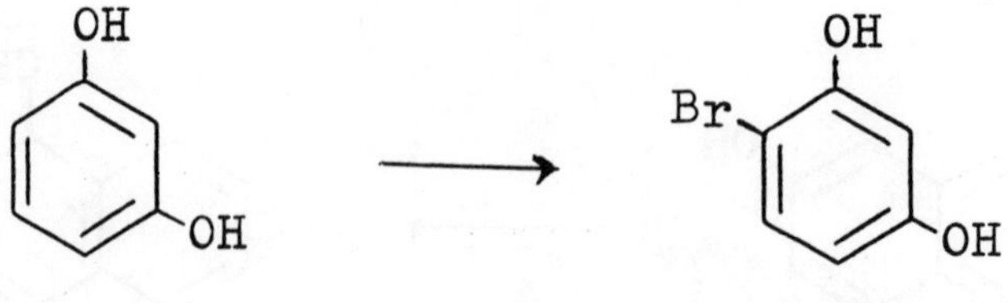

OH, OH → OH, Br, OH

i ii

11.

HO, O → MeO, MeO, OMe, O, O, O

Fukui and Nakayama, *Bull. Chem. Soc. Jap.* **41**, 1385 (1968).

12.

PhOH → OH, Br

Huston and Ballard, *Org. Syn.* **II**, 97.

13.

Me, Me, i → Me, Me, NH_2, Me, ii; Me, Me, NH_2, Me, iii

14.

NH_2, SO_2NH_2 → NH_2, Cl, Cl

Seikel, *Org. Syn.* **III**, 262.

15.

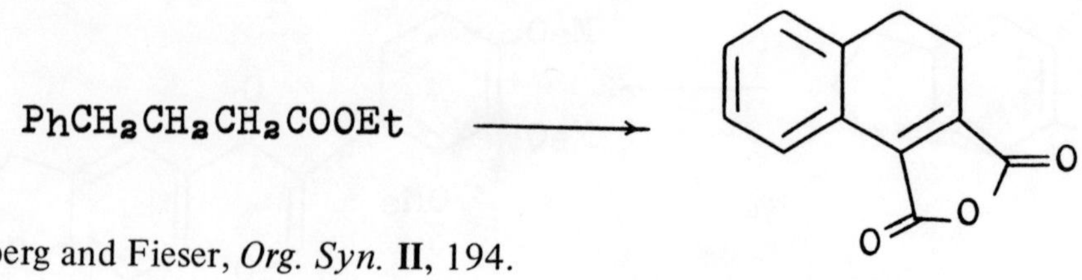

$PhCH_2CH_2CH_2COOEt$ ⟶

Hershberg and Fieser, *Org. Syn.* **II**, 194.

16.

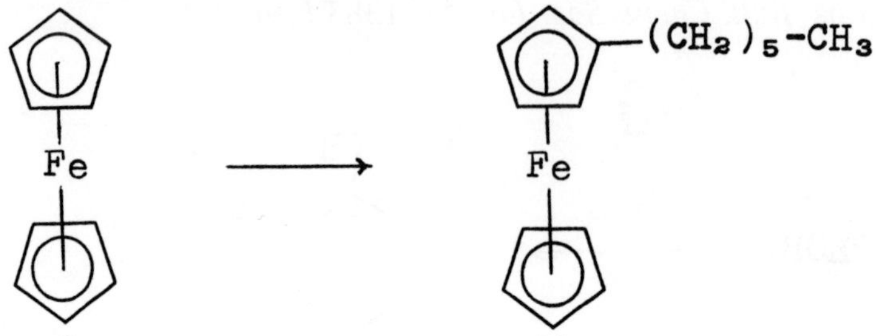

⟶ —$(CH_2)_5$—CH_3

Fe Fe

i ii

17.

C_6H_6 ⟶

NO_2

$COCH_3$

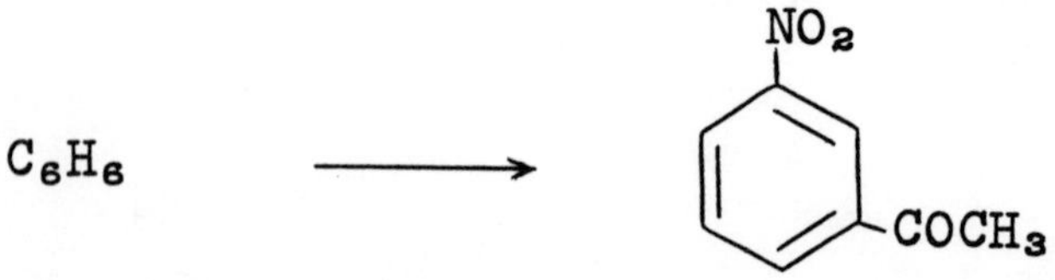

Adams and Noller, *Org. Syn.* **I**, 109; Corson and Hazen, *Org. Syn.* **II**, 434.

18.

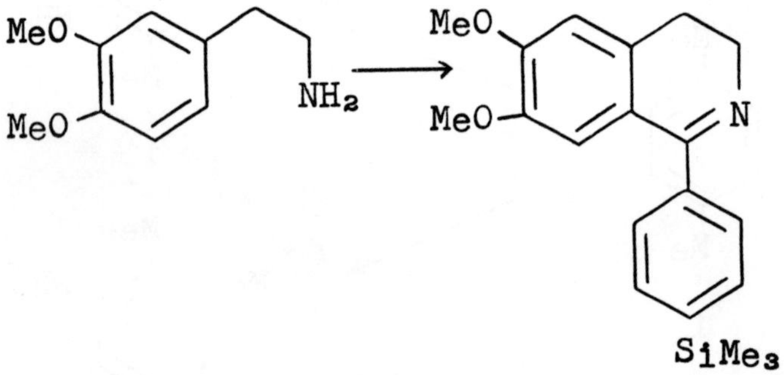

MeO

MeO

NH_2 ⟶

MeO

MeO

N

$SiMe_3$

Belsky, Gertner, and Zilkha, *Can. J. Chem.* **46**, 1921 (1968).

19.

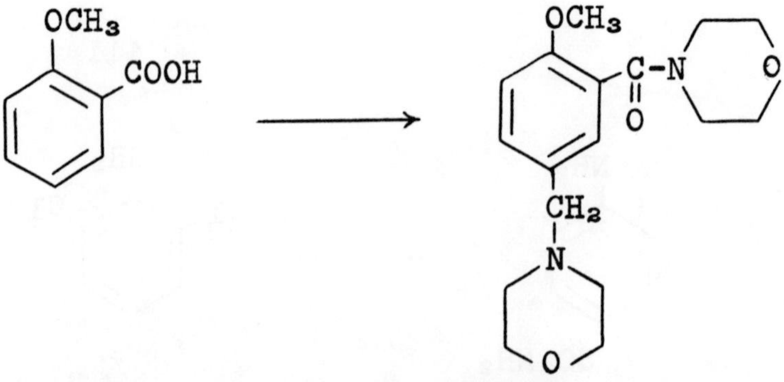

OCH_3

COOH ⟶

OCH_3

C—N O

O

CH_2

N

O

[1] Quelet, Dran, and Lallouz, *Bull. Soc. Chim. Fr.* **1969**, 1698.

20.

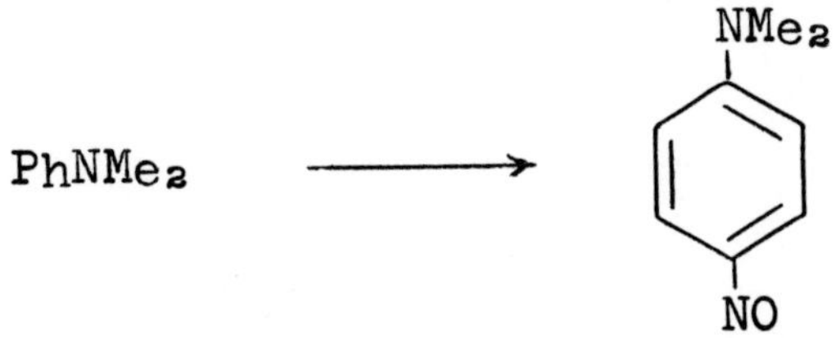

PhNMe2 ⟶ (p-nitroso-N,N-dimethylaniline)

NMe2

NO

Adams and Coleman, *Org. Syn.* **I**, 214; Bennett and Bell, *Org. Syn.* **II**, 223.

21.

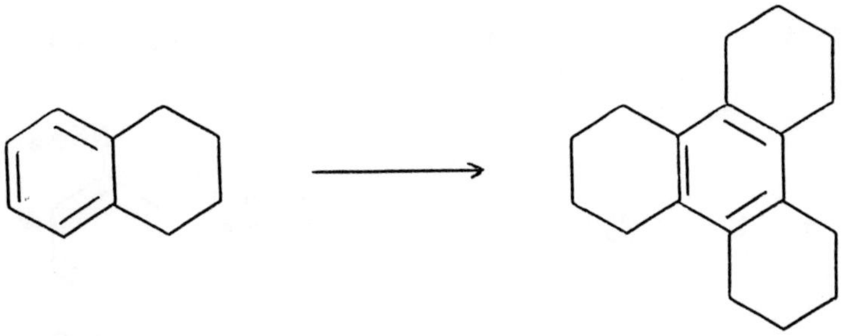

Hausigk and Kölling, *Chem. Ber.* **101**, 469 (1968).

22.

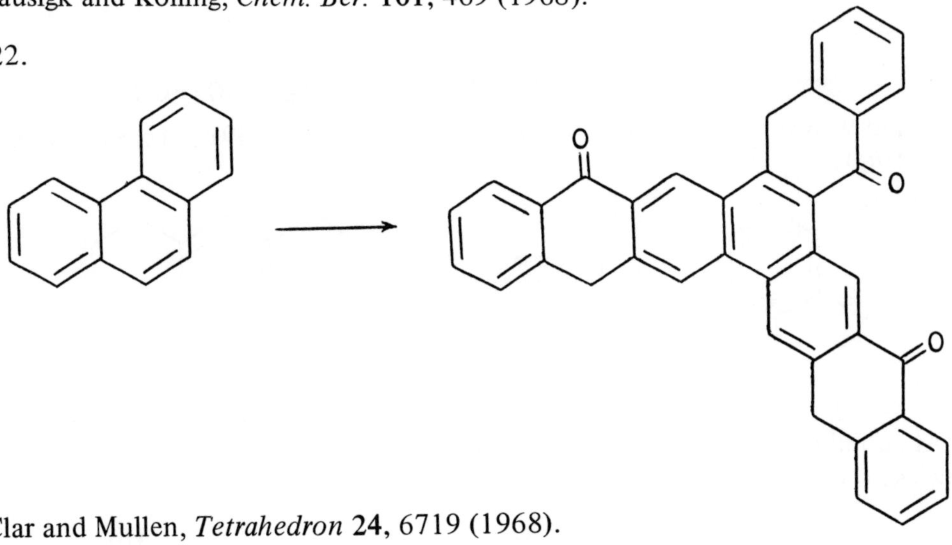

O

O

O

Clar and Mullen, *Tetrahedron* **24**, 6719 (1968).

23.

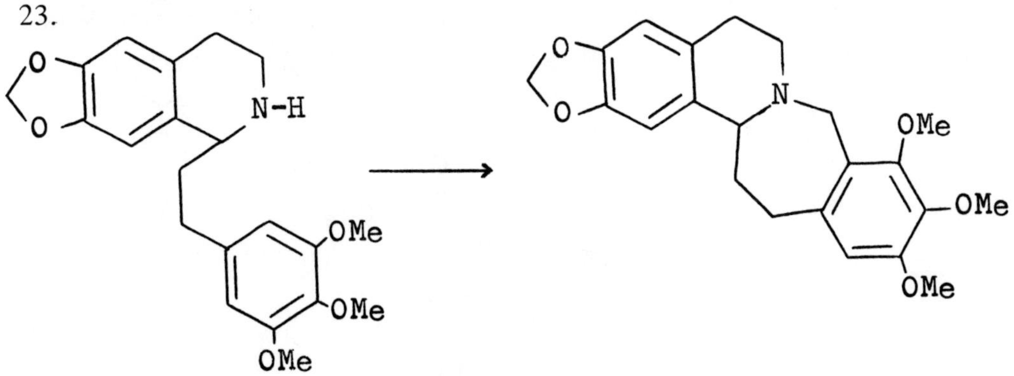

O

O

N-H

OMe

OMe

OMe

O

O

N

OMe

OMe

OMe

Brossi and Teitel, *Helv. Chim. Acta* **52**, 1228 (1969).

24.

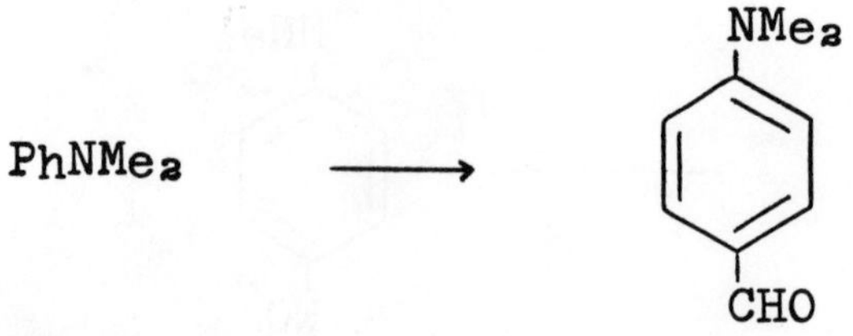

$$PhNMe_2 \longrightarrow$$ (4-dimethylaminobenzaldehyde: NMe_2 / benzene ring / CHO)

Campaigne and Archer, *Org. Syn.* **IV**, 331.

The following conversions (problems 25 to 32) have been reported in the literature. In each case suggest a reasonable mechanism.

25.

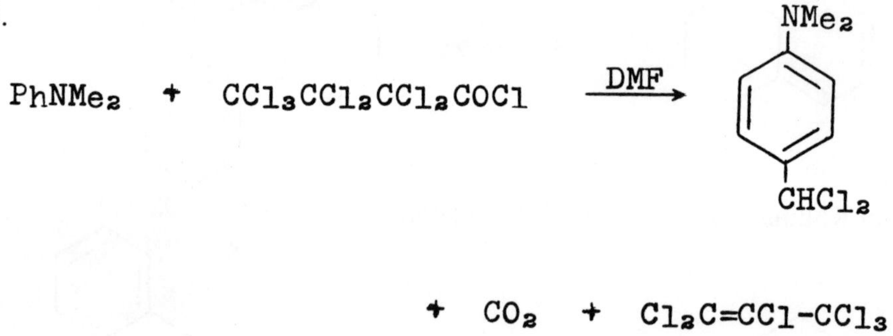

$$PhNMe_2 + CCl_3CCl_2CCl_2COCl \xrightarrow{DMF}$$ (NMe_2 / benzene ring / $CHCl_2$)

$$+ CO_2 + Cl_2C{=}CCl{-}CCl_3$$

Roedig and Wenzel, *Angew. Chem. Intern. Ed. Engl.* 8, 71 (1969); (*Angew. Chem.* **81**, 36 (1969)).

26.

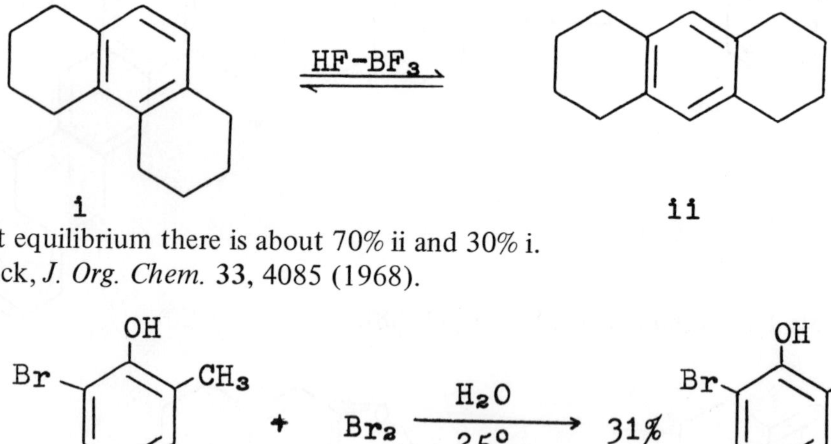

i $\underset{}{\overset{HF-BF_3}{\rightleftharpoons}}$ ii

At equilibrium there is about 70% ii and 30% i.

Bushwick, *J. Org. Chem.* 33, 4085 (1968).

27.

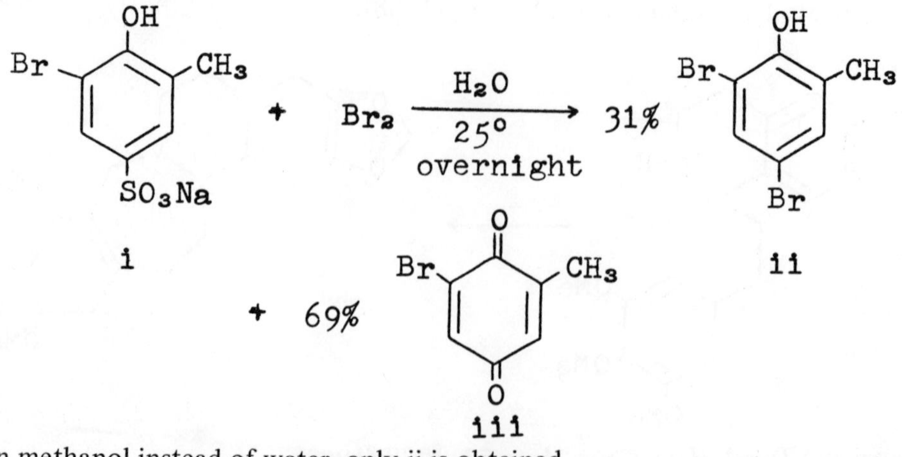

i (OH, Br, CH_3, SO_3Na) $+ Br_2 \xrightarrow[25^\circ \text{ overnight}]{H_2O}$ 31% ii (OH, Br, CH_3, Br)

+ 69% iii (O, Br, CH_3, O)

Note: in methanol instead of water, only ii is obtained.

28.

C_6H_6 + 1,1-dichloro-2,2-dimethylcyclopropane $\xrightarrow[\text{20-25°, 10-20 min}]{AlCl_3}$ 59% 1,2-dimethylindene + 7% 1,1-dimethyl-2-phenylindane

Buddrus, *Chem. Ber.* **101**, 4152 (1968).

29.

MeO, OMe, OMe (i) + PhCOOOH (excess) $\xrightarrow[\text{6 hr}]{CHCl_3}$ 27% MeO, OMe, O, O (ii)

Note: When other methoxybenzenes were treated with perbenzoic acid, it was noted that the reactivity increased with increasing number of methoxy groups.

30.

$\xrightarrow[\text{1 min}]{HNO_3\text{-}HOAc}$ NO_2

Allinger, Da Rooge, and Hermann, *J. Amer. Chem. Soc.* **83**, 1974 (1961).

31.

2-methylindole (CH_3, N–H) + diphenylcyclopropenone (Ph, Ph, O) $\xrightarrow[\text{EtOH, 3-12 hr}]{HCl}$ 87% (Ph, Ph, CH_3, $N^{\oplus}$–H)

Hill and Battiste, *Tetrahedron Lett.* **1968**, 5537; Battiste and Hill, *Tetrahedron Lett.* **1968**, 5541.

32.

$$PhCH_2\text{-}^{14}CH_2Cl \;+\; PhOMe \xrightarrow{AlCl_3} Ph\overset{1}{C}H_2\overset{2}{C}H_2\text{-}C_6H_4\text{-}OMe$$

i ii

Note: In ii the label was found to be equally distributed between C-1 and C-2.

In the following problems (33 to 37) give the products and suggest reasonable mechanisms.

33.

(benzo[b]naphtho[2,1-d]thiophene) $\xrightarrow[Cl_2CHOBu,\ 1\ hr]{SnCl_4}$

Buu-Hoï, Croisy, and Jacquignon, *J. Chem. Soc., C* **1969**, 339.

34.

(2-bromophenol: OH, Br) $\xrightarrow[pyridine]{EtCOCl}$ i ($C_9H_9BrO_2$) $\xrightarrow{AlCl_3}$

ii ($C_9H_9BrO_2$) $\xrightarrow{Br_2}$ (OH, Br, Br, COEt)

35.

(HO, Et, OH benzene) + F_3CCN $\xrightarrow{ZnCl_2}$

Whalley, *J. Chem. Soc.* **1951**, 665.

36.

(OMe, NH_2 benzene) + $HOOCCH_2COOH$ $\xrightarrow{POCl_3}$ $C_9H_7NO_2$

Kappe and Ziegler, *Tetrahedron Lett.* **1968**, 1947.

37.

Ph-N-NO_2 (para-NO_2 phenyl) $\xrightarrow{H_2SO_4}$ i

38.

	Et, Et, Me, Et	Et, Et, Me, Me, Et	Et, Et, Br, Et	Et, Et, Br, Br, Et
k_D/k_H	0.95	0.86	0.80	0.60

Explain the isotope effects observed when the compounds shown were brominated with Br_2 in DMF.

Nilsson and Olsson, *Acta Chem. Scand.* 23, 7 (1969).

39.

$PhNH_2 \xrightarrow{H_2SO_4} PhNH\text{-}SO_3H$ (i) $\longrightarrow$ p-$NH_2C_6H_4SO_3H$

It has been proposed that in the sulfonation of aniline, i is an intermediate. Devise an experiment to test this possibility.

Vrba and Allan, *Collect. Czech. Chem. Commun.* 33, 2502 (1968).

40.

$PhCH_2$-cyclopentane (i; positions 1, 2) $\xrightarrow[C_6H_6,\ 80^\circ\ 15\ min]{AlCl_3}$ Ph-cyclohexane (ii)

The conversion of i to ii, under the conditions given above, was irreversible: ii gave no i under these conditions. When i was labeled in the 1-position with ^{14}C, the label was found equally distributed in the cyclohexane ring: 1/6 at each position. Suggest a reasonable mechanism.

41.

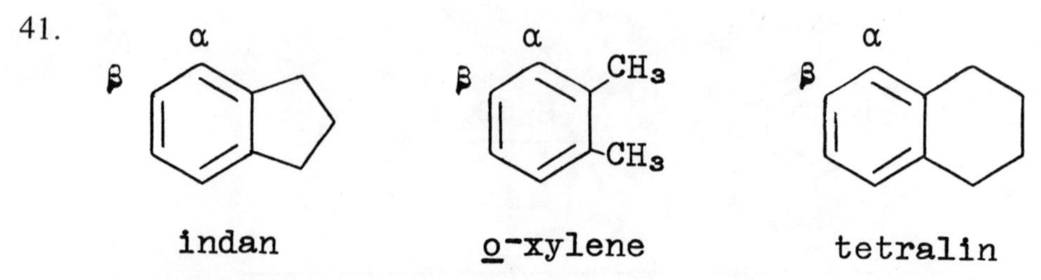

In bromination of these three compounds, the ratio of β to α substitution decreased in the order indan, *o*-xylene, tetralin. Explain.

Vaughan, Welch, and Wright, *Tetrahedron* **21**, 1665 (1965); see also Bassindale, Eaborn, and Walton, *J. Chem. Soc., B* **1969**, 12.

42. The rates of nitration of benzene, toluene, *p*-xylene, and mesitylene with anhydrous HNO_3 in CCl_4 were very similar and were all zero order in the aromatic compound. In two cases studied in detail (mesitylene and *p*-xylene) the reaction was sixth order in HNO_3. Nitration of mesitylene was substantially faster at 0° than at 40°, and the order was approximately 6 at both temperatures. Suggest an explanation.

Note: It is known that the equilibrium $2\ HNO_3 \rightleftharpoons NO_2^+ + NO_3^- + H_2O$ is shifted to the right with decreasing temperature.

Bonner, Hancock, and Rolle, *Tetrahedron Lett.* **1968**, 1665; Bonner, Hancock, Rolle, and Yousif, *J. Chem. Soc., B* **1970**, 314.

43. Treatment of i with BF_3 gave ii with the tritium distributed as follows:

T (~47%)

T (~53%)

i   ii

Explain.

Jackson, Naidoo, and Smith, *Tetrahedron* **24**, 6119 (1968).

44.

H_2SO_4

When a series of N-nitroso-N-methylanilines was treated with H_2SO_4, the Fischer-Hepp rearrangement (shown above) took place. When excess urea was added to the system, no C-nitroso product was detected when X = Cl, Br, COOH, or NO_2; but when X = H, Me, OMe, or OH at least some C-nitroso product could still be isolated, even in the presence of excess urea. Explain.

45. The partial rate factors for the bromination of toluene are: o_f^{Me} 600; m_f^{Me} 5.5; p_f^{Me} 2420. Calculate the overall rates of bromination (relative to benzene) for a. *o*-xylene; b. *p*-xylene; c. hemimellitene (1,2,3-trimethylbenzene); d. isodurene (1,2,3,5-tetramethylbenzene); and e. pentamethylbenzene. Experimental values are (benzene = 1.00) a. 5.32×10^3; b. 2.52×10^3; c. 1.67×10^6; d. 0.42×10^9; and e. 0.81×10^9.

46. The rate of nitration of toluene at 25° with HNO_3 in HOAc was 28.8 times that of benzene. Isomer distribution of the nitrotoluenes produced was: ortho 56.9%; meta 2.8%; para 40.3%. Calculate the three partial rate factors.

47. Given the partial rate factors for the nitration of toluene in HOAc at 25°, calculated in the previous problem (see the answer section) calculate the relative ratios expected for the mononitration of a. *m*-xylene and b. *o*-xylene. Experimental results were a. 11% 2-nitro, 89% 4-nitro, and 0% 5-nitro-1,3-dimethylbenzene; b. 73% 3-nitro and 27% 4-nitro-1,2-dimethylbenzene.

48.

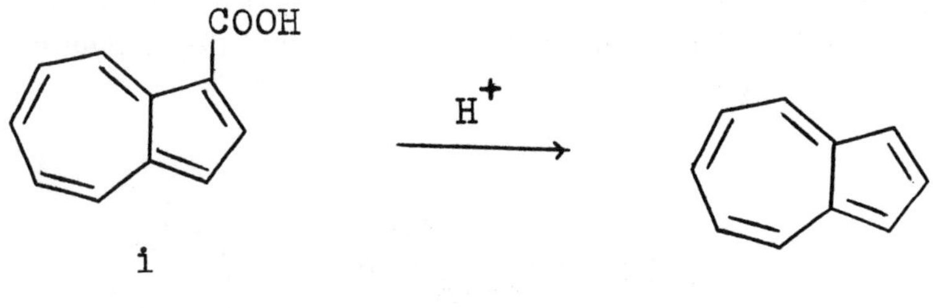

COOH

$\xrightarrow{H^+}$

i

For the decarboxylation of i, there is a first-order dependence on the acidity at low concentrations of H^+, but an independence of acidity at high concentrations. Additionally, it was determined that at low H^+ concentrations (below 0.01 *M*) there is virtually no ^{13}C isotope effect, while at high concentrations (e.g. 0.3 *M*) there is a ^{13}C isotope effect of 1.035 to 1.043 (the carbon whose isotope effect was measured is the COOH carbon, not any of the ring carbons). At very high acid concentrations, decarboxylation does not take place. Suggest a mechanism to account for these observations.

Huang and Long, *J. Amer. Chem. Soc.* **91**, 2872 (1969).

49.

CH3 CH3 (i: 1,8-dimethylnaphthalene, positions 2 and 4 labelled) — 1, 2, 3, 5 labelled on ii (acenaphthene)

i ii

Explain why nitration of i (in Ac_2O at 0°) gave 92.7% 2-nitro and 7.3% 4-nitro-substitution, while similar nitration of ii gave chiefly 5- nitro (64.1%) and much less (35.9%) 3-nitro substitution.

Davies and Warren, *J. Chem. Soc., B* **1969**, 873.

50.

Me, CHO, Me, Me (i) $\xrightarrow[H_2SO_4]{100\%}$ Me, Me, Me + CO

i

Isotope effects were measured for this reaction in two different ways. The rate of decarboxylation of i was compared with that of ArCDO (an isotope effect of 1.8 was found); and the rate of the reaction in H_2SO_4 was compared with that in D_2SO_4 (in this case k_H/k_D was 2.4). In the latter case the isotope effect decreased with decreasing concentration of H_2SO_4 in H_2O. Explain.

Schubert and Burkett, *J. Amer. Chem. Soc.* **78**, 64 (1956).

51. Halogens (Br or I) can be removed from aromatic rings if they are ortho to another halogen (Br, I, or Cl), or better, ortho to two halogens, by treatment with excess *t*-BuOK in 50% *t*-BuOH-50% Me_2SO. The halogen removed appears as halide ion, e.g.:

Cl, Br, Cl + t-BuOK $\xrightarrow[Me_2SO]{t\text{-BuOH}}$ Cl, Cl + Br^-

Relative reactivities of some compounds were (halogen removed is underlined):

51 Cont.

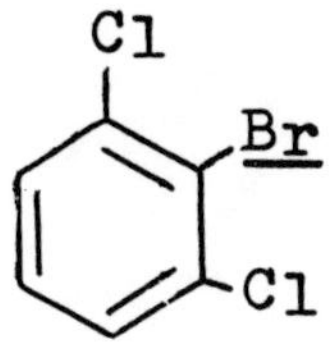

1 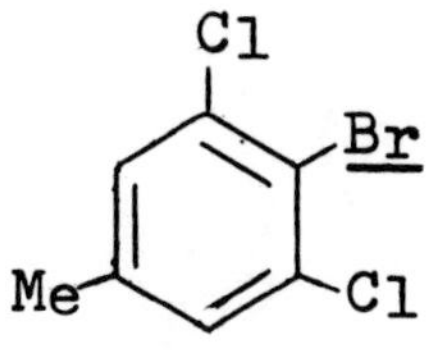 0.2 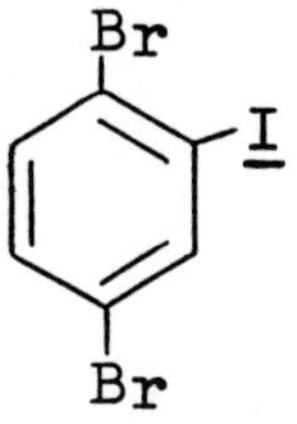 1.4 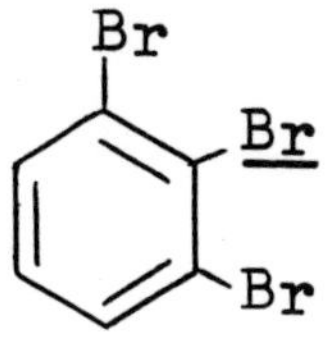 4.6

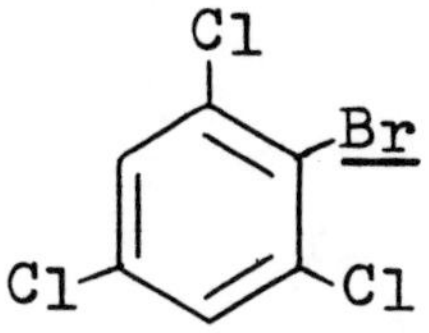

33 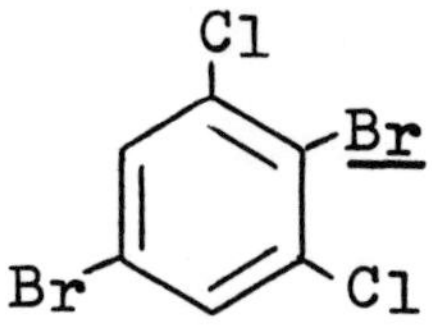 34 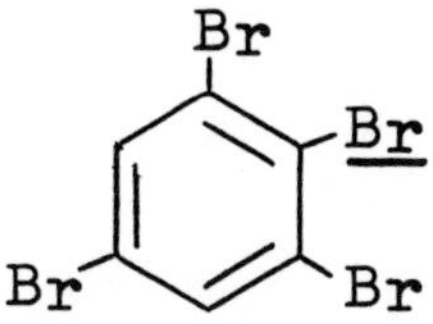 80 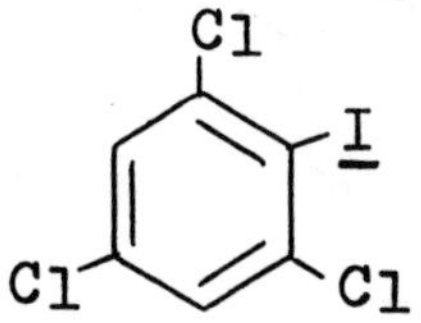 900

Certain other dipolar aprotic solvents could substitute for Me_2SO, e.g., $AcNMe_2$, N-methylpyrrolidone, sulfolane, though reactivities were lower, but the reaction failed in *t*-BuOH alone, without cosolvent, or when the cosolvent was Ph_2SO or DMF. A free radical mechanism was ruled out by demonstration that similar substrates exhibited different reactivity patterns in known free radical reactions; and by a failure to isolate biaryls, which are known to form when aryl free radicals are involved. Suggest a mechanism to account for these facts.

Bunnett and Victor, *J. Amer. Chem. Soc.* **90,** 810 (1968).

52. Compare the reactivity towards electrophilic substitution of furan, thiophene, benzene, pyrrole.

53. When 1-phenyl-5-chloropentane (i) is treated with $AlCl_3$, the product is 1-methyltetralin (ii). Devise an experiment to distinguish between two possible pathways, a and b:

53 Cont.

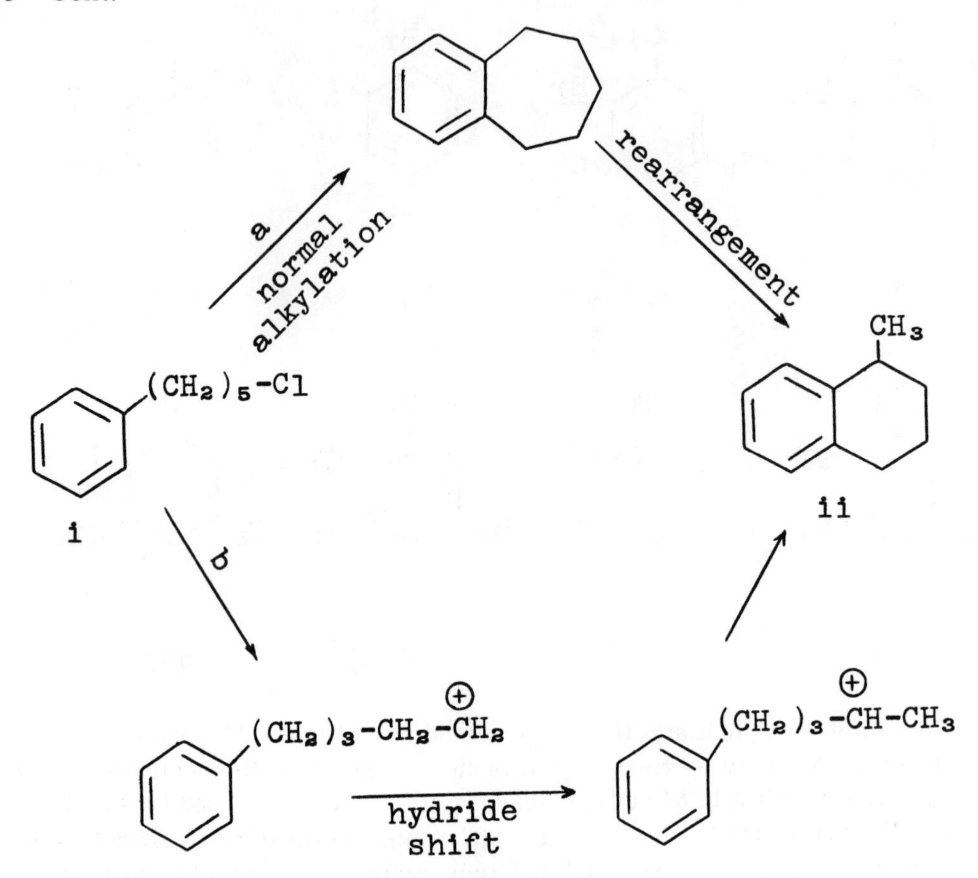

Barclay and Sanford, *Can. J. Chem.* **46**, 3315 (1968).

54. In each case predict the principal position of electrophilic attack:

a. $PhNHCOCH_3$ (Friedel-Crafts acylation)

b. $Ph\overset{+}{A}sMe_3$ (nitration)

c. $PhCH_2\overset{+}{A}sMe_3$ (nitration)

d. $PhSO_3H$ (Friedel-Crafts alkylation)

e. $Ph{-}CH{=}CH{-}CF_3$ (nitration)

f. g. h.

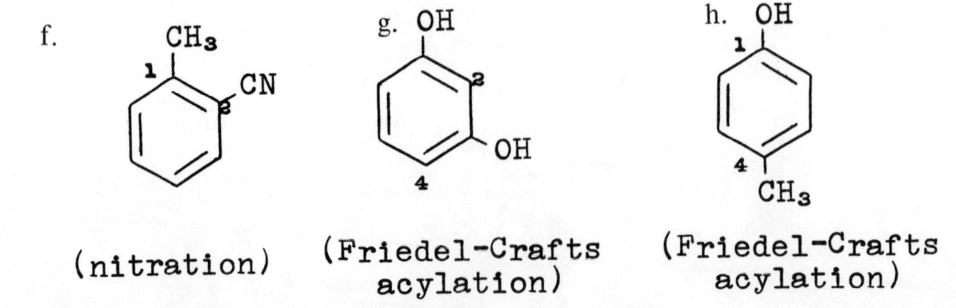

54 Cont.

i. CH_3 1, 4 t-Bu

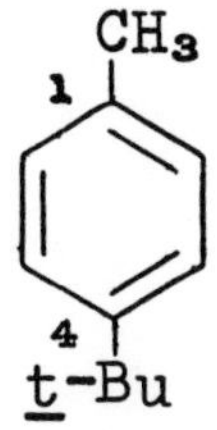

(Friedel-Crafts acylation)

j. CH_3 NO_2 3, 4, 6 CH_3

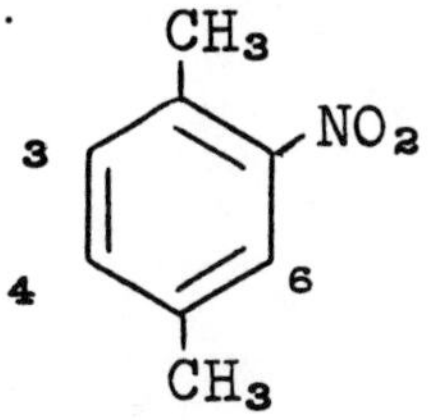

(nitration)

k. CH_3 OH 5, 4 NO_2

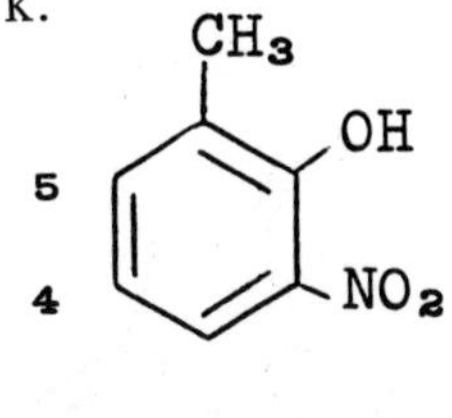

(Friedel-Crafts acylation)

l. Cl 2' O 3, 4

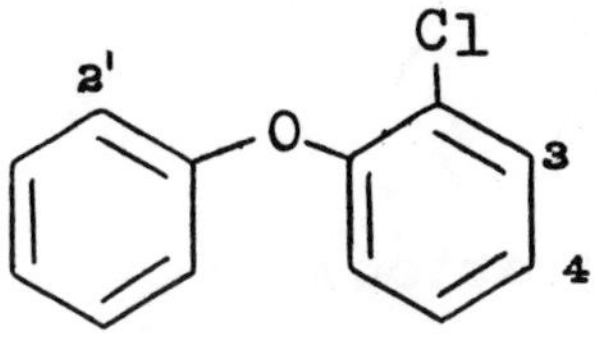

(Friedel-Crafts acylation)

m. 1, 2', 2

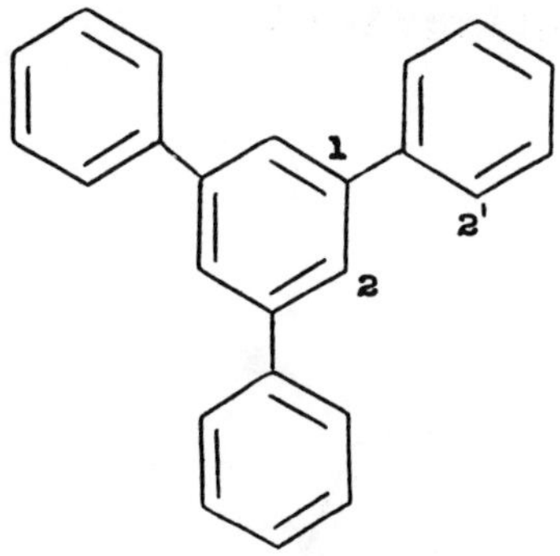

(nitration)

n. CH_3

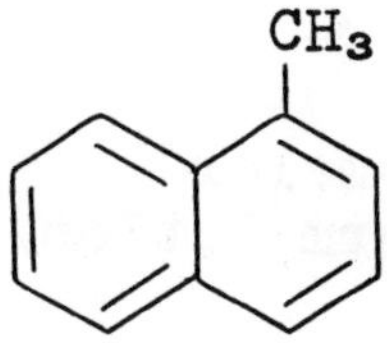

(detritiation)

o. OMe

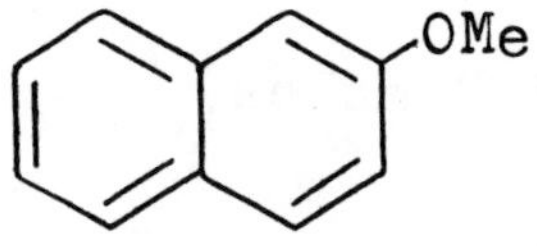

(detritiation)

p.

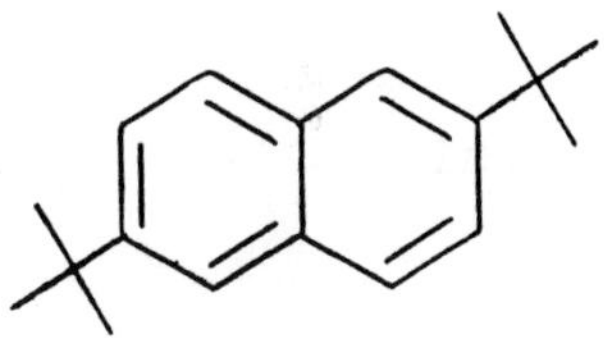

(bromination)

q. 9 1 2

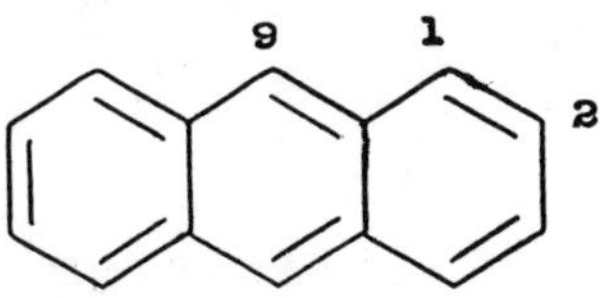

(Vilsmeier formylation)

r. 1, 3, 4

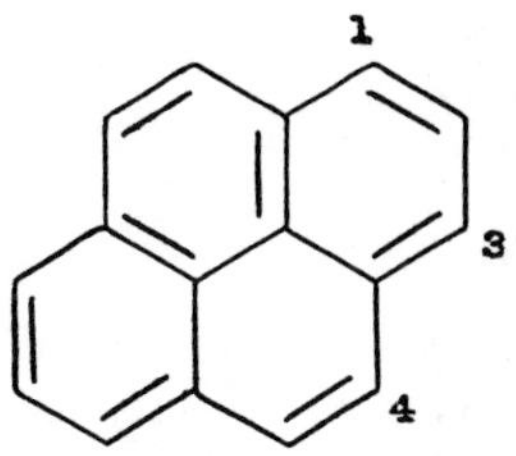

1. i-PrCl + $AlCl_3$

2. t-BuCl + $AlCl_3$

54 Cont.

s.

(nitration)

t.

(nitration)

u.

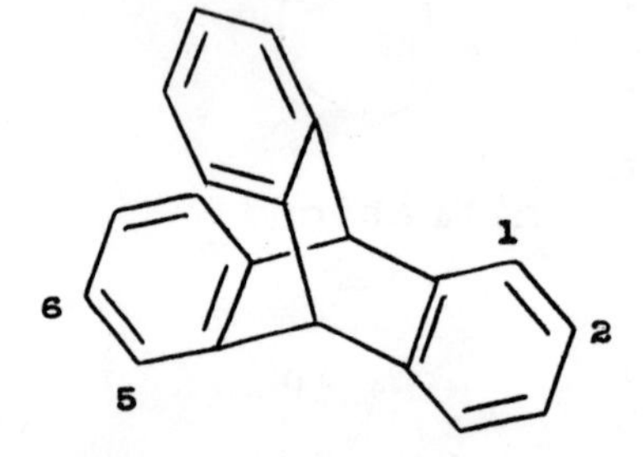

(nitration: 1 and 2 moles)

v.

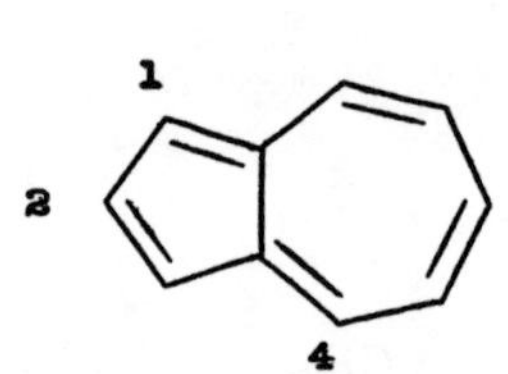

(Vilsmeier formylation)

w.

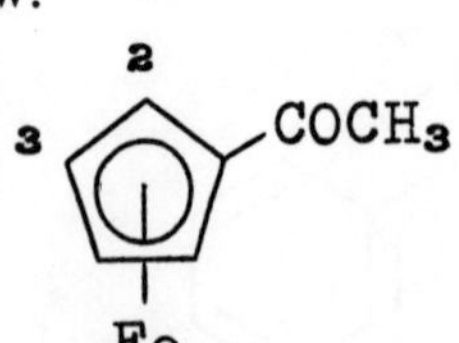

(Friedel-Crafts acylation)

x.

(Vilsmeier formylation)

y.

(Vilsmeier formylation)

54 Cont.

z.

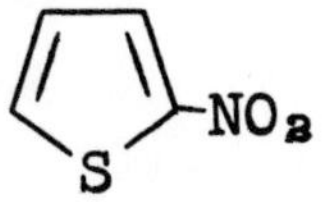

S

NO_2

(nitration)

aa.

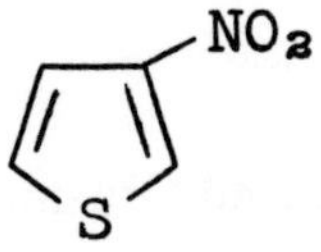

NO_2

S

(nitration)

bb.

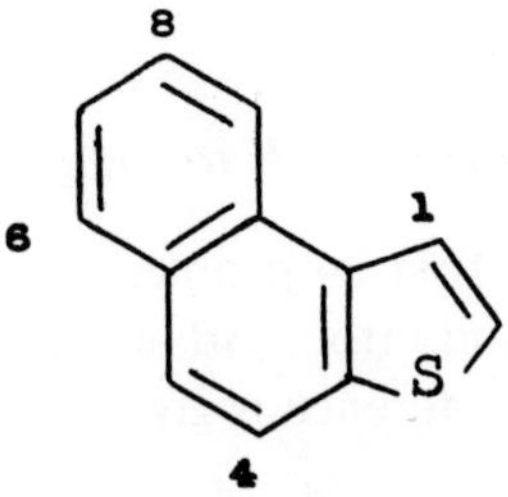

8

1

6

S

4

(nitration)

cc.

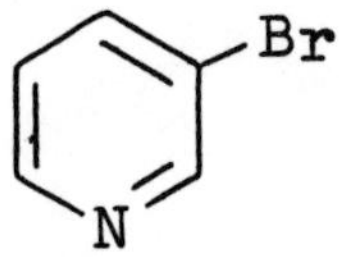

Br

N

(nitration)

dd.

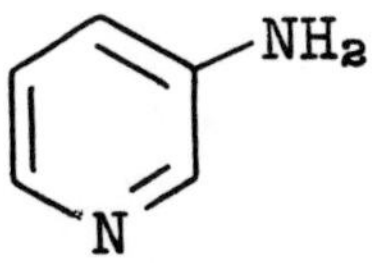

NH_2

N

(chlorination)

ee.

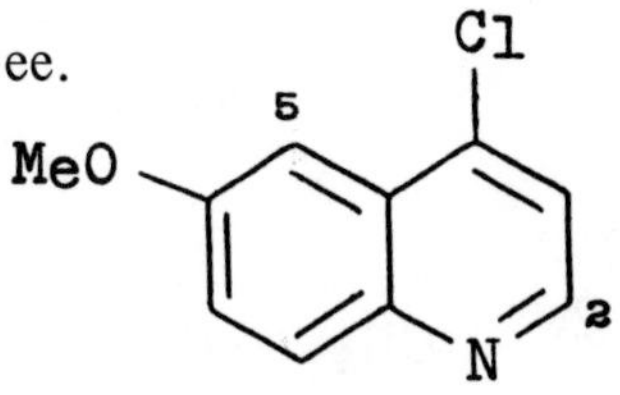

Cl

5

MeO

N

2

(nitration)

ff.

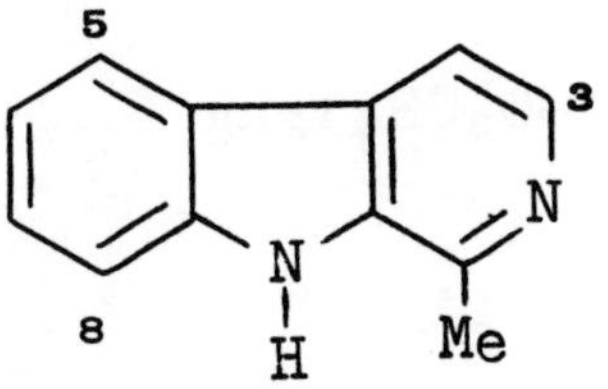

5

3

N

N

H

8

Me

(bromination)

gg.

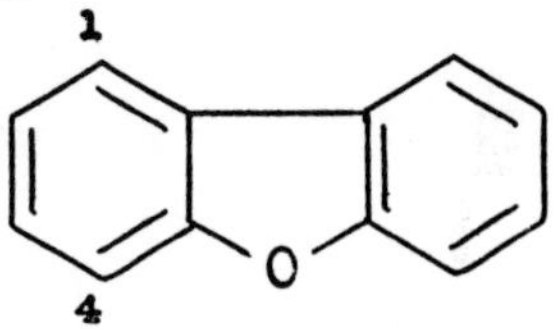

1

O

4

(Friedel-Crafts acylation)

PROBLEM SET 6

ALIPHATIC ELECTROPHILIC SUBSTITUTION

Corresponds to Chapter 12 of Advanced Organic Chemistry.

The first 24 problems are synthetic. In each case you are asked to convert one compound into another. In every case the conversion has actually been carried out, and the reference is given.

1.

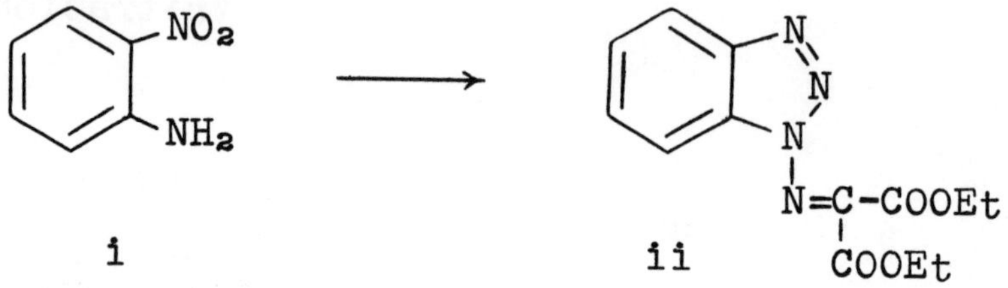

NO_2

NH_2

i

N

N

N

N=C-COOEt

COOEt

ii

2.

CH_3 CH_3

CH_3

$N-CH_2Ph$

CH_3 CH_3

CH_3

N-Cl

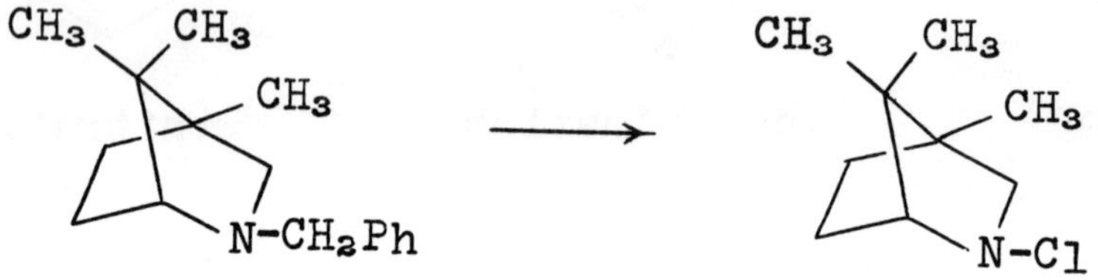

Gassman and Cryberg, *J. Amer. Chem. Soc.* **91**, 2047 (1969).

3. $PhSO_2CH_3$ → $(PhSO_2CH_2)_2Hg$

i ii

4.

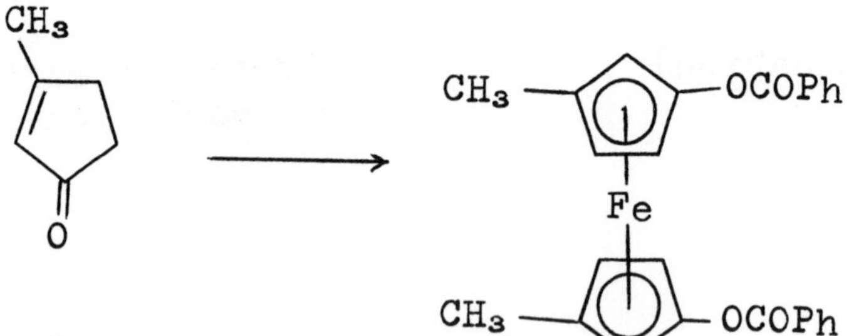

CH_3

O

i

CH_3 OCOPh

Fe

CH_3 OCOPh

ii

5.

3-thiophene-$CH(OEt)_2$ ⟶ 2-iodo-3-thiophenecarboxaldehyde (CHO, I)

Guilard, Fournari, and Person, *Bull. Soc. Chim. Fr.* **1967**, 4121.

6.

AcO-substituted methylcyclohexene ⟶ HO-substituted methylcyclohexene with COOMe

Bruck, Clark, Davidson, Günther, Littlewood, and Lythgoe, *J. Chem. Soc., C* **1967**, 2529.

7.

$CH_2(COOEt)_2$ ⟶ $(CH_3)_2CH\text{-}CH(Br)\text{-}COOH$

Marvel and du Vigneaud, *Org. Syn.* **II**, 93.

8.

MeOOC-furan (isopropyl) ⟶ 2-acyl-2-methylcyclopentanone furyl ketone (O, O)

Büchi and Wüest, *Tetrahedron* **24**, 2049 (1968).

9.

$PhCH_2Cl$ ⟶ $PhCH_2CD_2COOEt$

The only source of deuterium is D_2O.
Barclay and Sanford, *Can. J. Chem.* **46**, 3315 (1968).

10.

$PhCH_2CN$ ⟶ Ph_2CHCN

Robb and Schultz, *Org. Syn.* **III**, 347.

11.

2-methylcyclopentanone (CH_3, O) **i** ⟶ $CH_3\text{-}CH(CH_3)\text{-}CH_2\text{-}CH_2\text{-}C(CH_3)_2\text{-}CONH_2$ **ii**

12.

Cyclopropyl–Br ⟶ Cyclopropyl–O-t-Bu

Note: S_N reactions on cyclopropyl halides invariably lead to ring opening.

Longone and Miller, *Tetrahedron Lett.* **1967**, 4941.

13.

Cyclooctyl–Br ⟶ Cyclooctyl–HgBr

Cope and Engelhart, *J. Amer. Chem. Soc.* **90**, 7092 (1968).

14.

$NH_2-C(=O)-CH_2-CN$ ⟶ (1,2,4-triazine ring: H–N, C=O, C–CN, =N, N–Ph, C=O)

O

CN

H-N

N

O

N

Ph

Slouka, *Monatsh. Chem.* **99**, 1808 (1968).

15.

Fe (ferrocene) ⟶ Fe (ferrocene)–SiEt$_3$

i ii

16. $(CH_3)_2CHCH_2COOH$ ⟶ $(CH_3)_2CHCH(NH_2)COOH$

valine

Marvel, *Org. Syn.* **III**, 848.

17.

$EtOOC-(CH_2)_4-COOEt$ ⟶ $N_3-C(=O)-(CH_2)_4-C(=O)-N_3$

Smith, *Org. Syn.* **IV**, 819.

18.

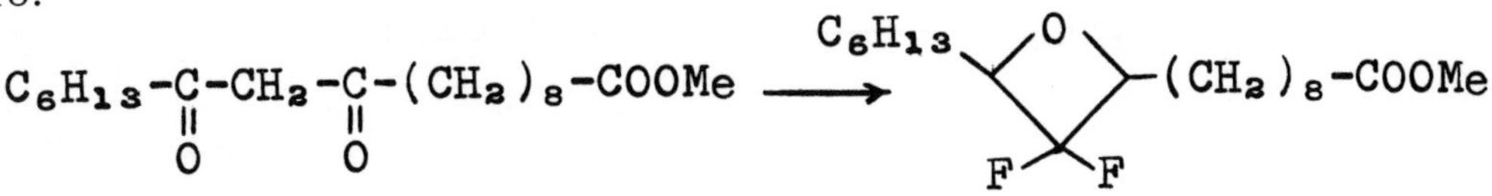

C_6H_{13}-C(=O)-CH_2-C(=O)-$(CH_2)_8$-COOMe → (oxetane: C_6H_{13}, O, F, F, $(CH_2)_8$-COOMe)

Stein and Haack, *J. Org. Chem.* **33**, 3784 (1968).

19.

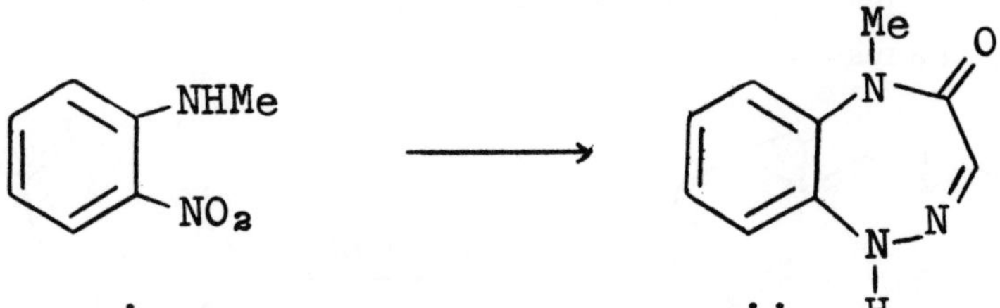

NHMe, NO_2 → Me, N, O, N, N, H

i ii

20.

Et, N, HOOC → Et, N, CH_3CO

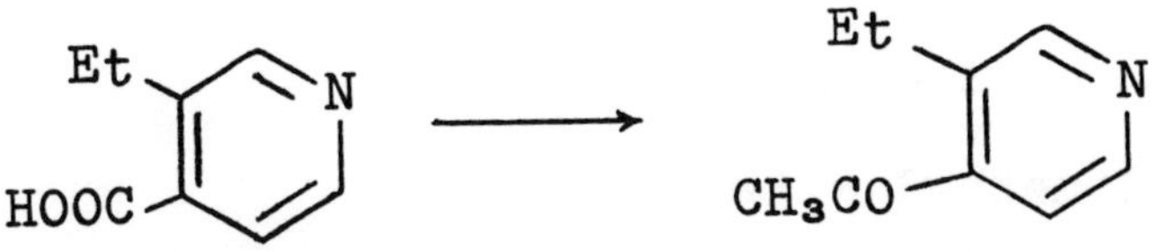

Jackson, Gaskell, Wilson, and Joule, *Chem. Commun.* **1968**, 364.

21.

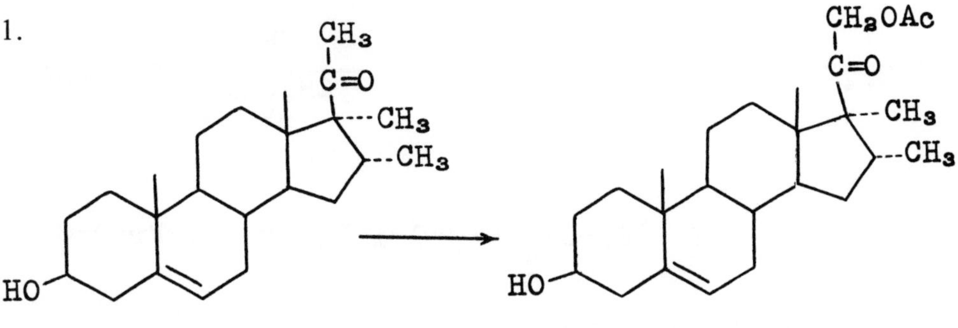

CH_3, C=O, CH_3, CH_3, HO → CH_2OAc, C=O, CH_3, CH_3, HO

i ii

22.

Me, Me, Br, Br → Me, Me, H, OH

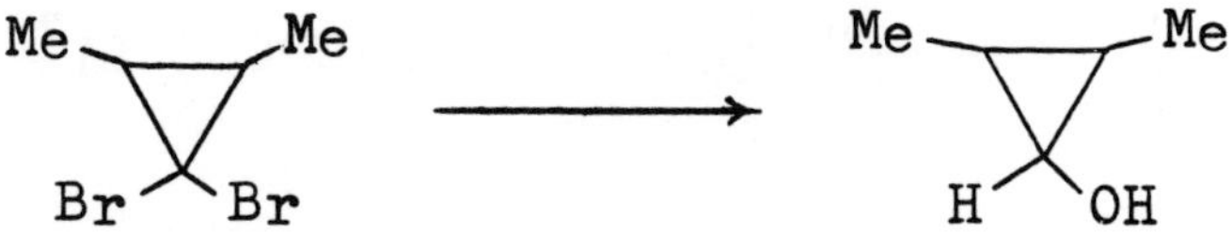

Note: S_N reactions on cyclopropyl halides invariably lead to ring opening.

Longone and Wright, *Tetrahedron Lett.* **1969**, 2859.

23.

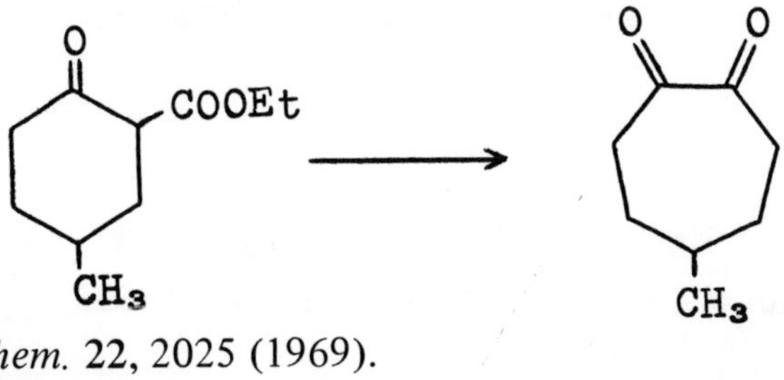

O, COOEt, CH_3 → O, O, CH_3

Saraf, *Aust. J. Chem.* **22**, 2025 (1969).

24.

$$PhCH_2NH_2 \longrightarrow PhCH_2\text{-}N(NO)\text{-}COOEt$$

Gutsche and Johnson, *Org. Syn.* **IV**, 780.

The following conversions (problems 25 to 44) have been reported in the literature. In each case suggest a reasonable mechanism.

25.

i → ii (aq. KOH, MeOH, warming, 0.3 hr; 47%)

i: radicicol dimethyl ether (MeO, Cl, OMe, O)

ii (MeO, Cl, OMe, O)

26.

$$PhN_2^+\ BF_4^- + F\text{-}NH\text{-}C(=O)\text{-}O\text{-}i\text{-}Pr \xrightarrow[CH_2Cl_2,\ 0^\circ]{KF} 71\%\ PhN_3$$

$$+\ 18\%\ i\text{-}Pr\text{-}O\text{-}C(=O)\text{-}N(F)\text{-}C(=O)\text{-}O\text{-}i\text{-}Pr + i\text{-}Pr\text{-}O\text{-}C(=O)\text{-}F$$

Baum, *J. Org. Chem.* **33**, 4333 (1968).

27.

4-Ph-1,2,3-thiadiazole (N, N, S, Ph) → PhC≡CS⁻ Li⁺ (BuLi, hexane-THF, -60°)

Raap and Micetich, *Can. J. Chem.* **46**, 1057 (1968).

28.

$ClCH_2$-furan (i) → CH_3-furan-CN (ii) (aq. NaCN)

29. $$EtAlCl_2 + EtCl \xrightarrow{80^\circ} C_2H_6 + C_2H_4 + AlCl_3$$

Pasynkiewicz and Kuran, *J. Organometal. Chem.* **16**, 43 (1969).

30.

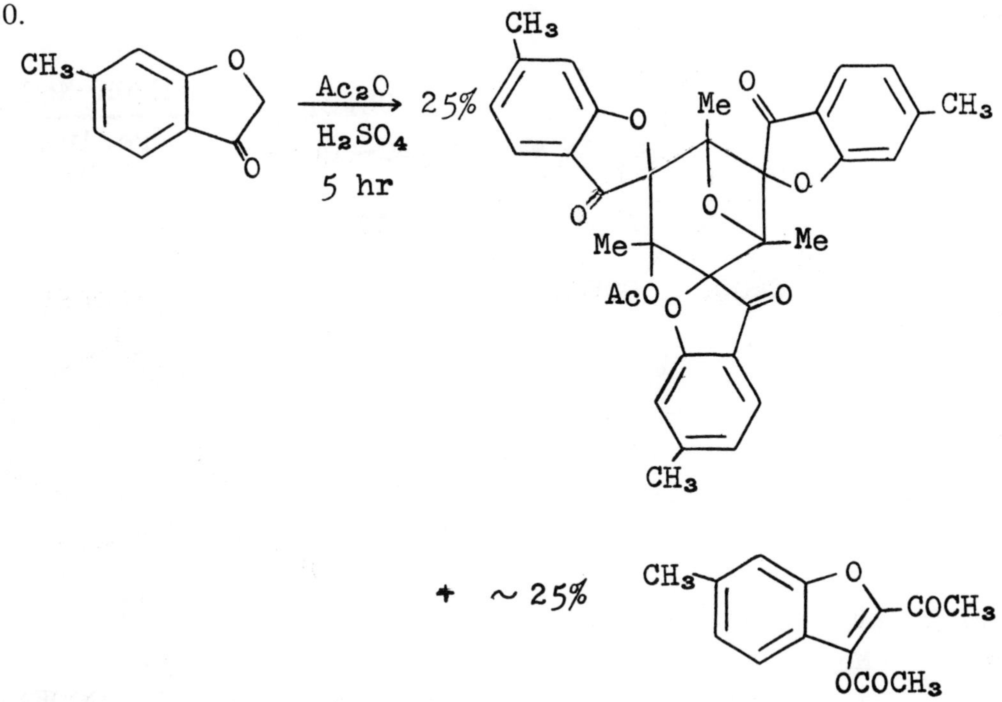

Ac_2O, H_2SO_4, 5 hr → 25% + ~25%

Bohlmann and Zdero, *Chem. Ber.* **101**, 3961 (1968).

31.

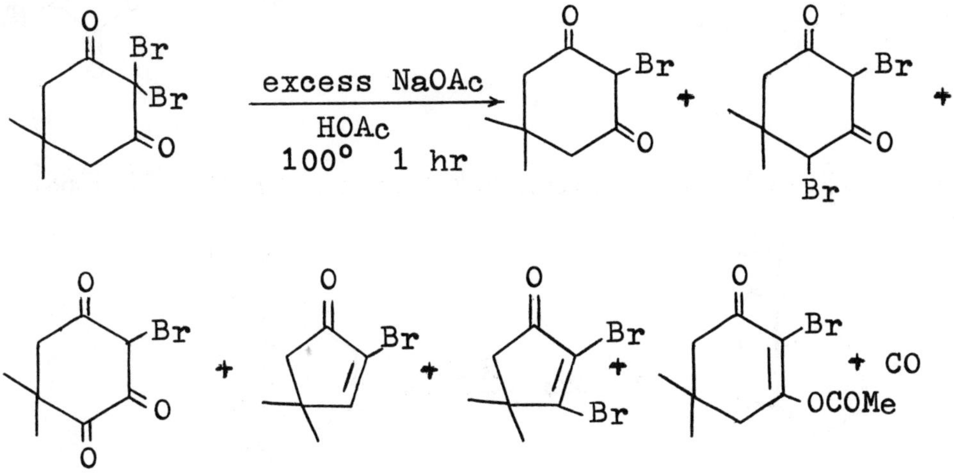

excess NaOAc, HOAc, 100° 1 hr → + CO

De Pooter and Schamp, *Bull. Soc. Chim. Belges* **77**, 377 (1968).

32.

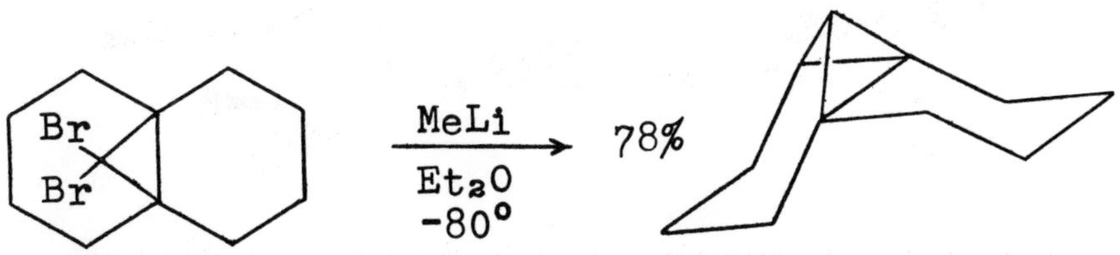

MeLi, Et_2O, −80° → 78%

Vaidyanathaswamy and Devaprabhakara, *Chem. Ind.* (London) **1968**, 515.

33.

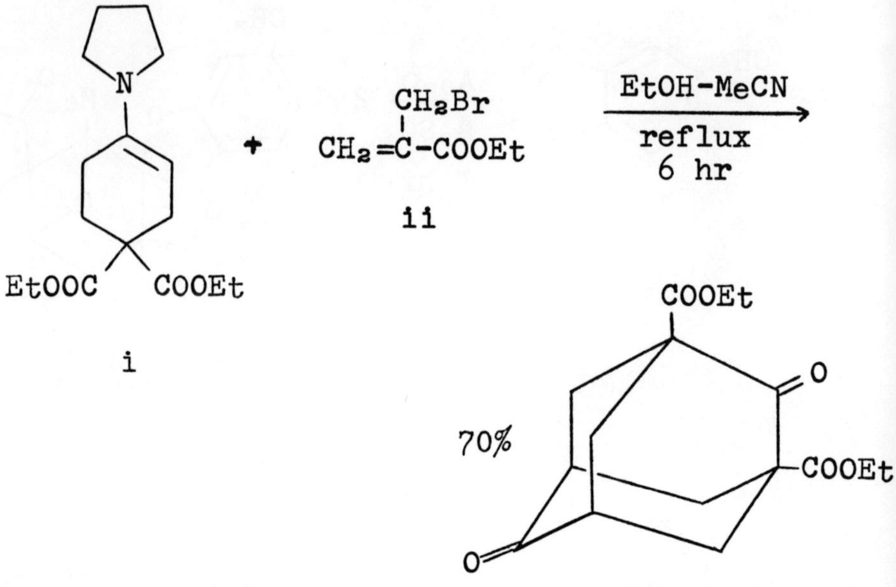

EtOOC COOEt

i

+ $CH_2=C(CH_2Br)-COOEt$

ii

EtOH-MeCN, reflux, 6 hr →

70%

COOEt, O, COOEt, O

iii

34.

Me, Me, COOEt, O, COOH, O + HONO

H_2O-HOAc, 0° 30 min →

Me, Me, COOEt, O, NHOH, O

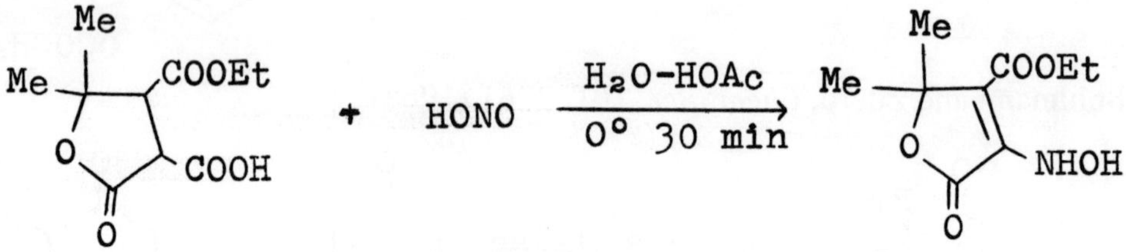

Torii, Endo, Oka, Kariya, and Takeda, *Bull. Chem. Soc. Jap.* **41**, 2707 (1968).

35.

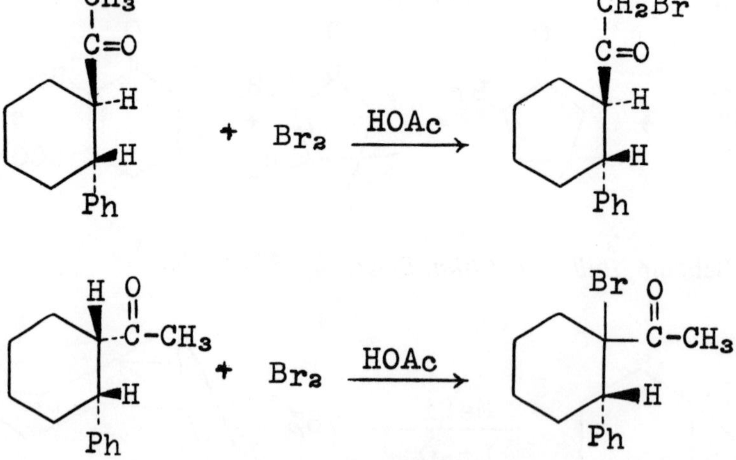

CH_3, C=O, H, H, Ph + Br_2 —HOAc→ CH_2Br, C=O, H, H, Ph

H, O, C–CH_3, H, Ph + Br_2 —HOAc→ Br, O, C–CH_3, H, Ph

Explain why the two isomers are brominated in opposite directions.
Zimmerman, *J. Amer. Chem. Soc.* **79**, 6554 (1957).

36. (cyclooctane diazo compound, $=N_2$) $\xrightarrow{160°}$ 46% (bicyclo[3.3.0]octane) + 45% (cyclooctene) + 9% (bicyclo[5.1.0]octane)

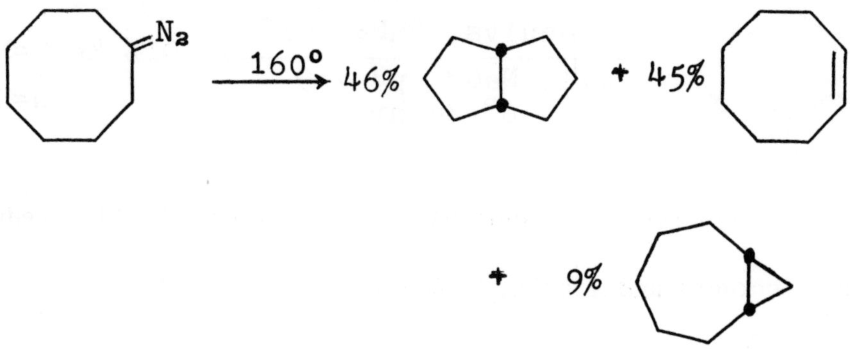

Friedman and Shechter, *J. Amer. Chem. Soc.* **83**, 3159 (1961).

37. $PhNH_2$ + $Cl_3C\text{-}C(=O)\text{-}CCl_3$ ⟶ $PhNH\text{-}C(=O)\text{-}CCl_3$

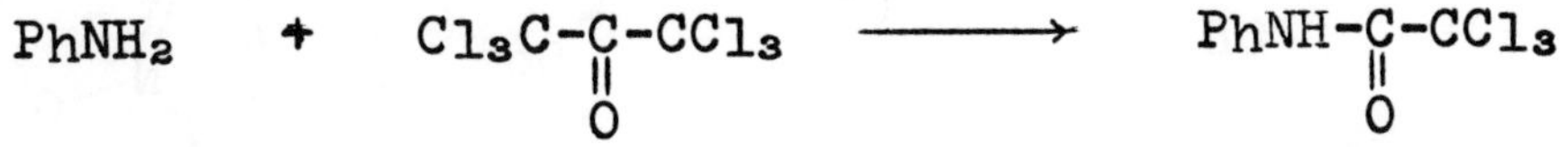

i ii iii

38. (tetraphenyl bi-imidazolidinylidene) + (5,5-dimethyl-2-diazonio-cyclohexenone-olate, Me, Me, $O^{\ominus}$, $N_2^{\oplus}$, O) $\xrightarrow{PhMe}$

6.5% (1,3-diphenylimidazolidin-2-ylidene =N-N= 5,5-dimethylcyclohexane-1,3-dione, Ph, Ph, O, O, Me, Me)

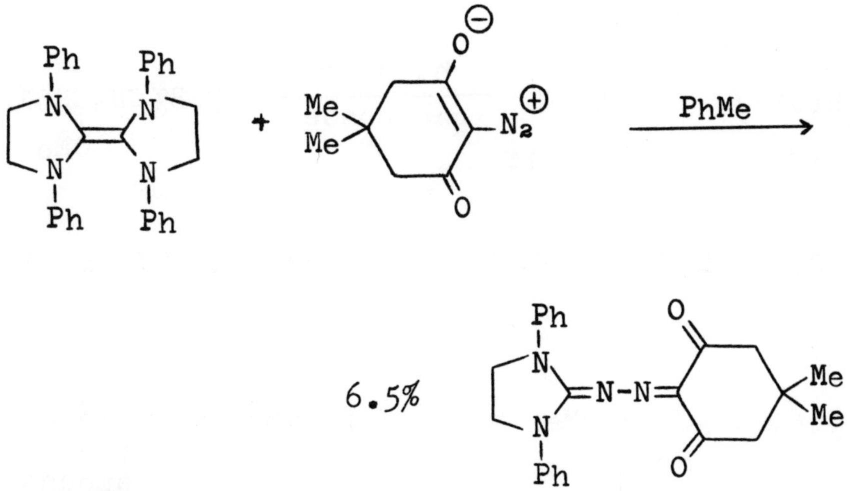

Regitz, Liedhegener, and Stadler, *Justus Liebigs Ann. Chem.* **713**, 101 (1968).

39. (1-methyl-2-methyl-2-phenylcyclopentanol: CH_3, OH, Ph, CH_3) $\xrightarrow[Me_2SO,\ 25°]{MeSOCH_2^-}$ (CH_3, OH, Ph, CH_3)

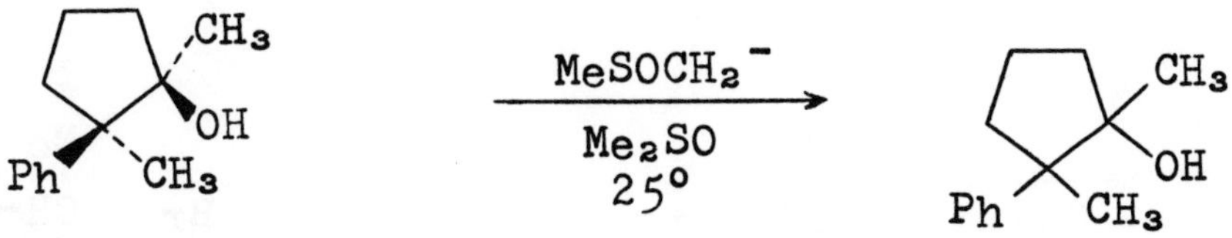

(-)-(1R,2R)-cis isomer

completely equilibrated and racemized

Note: Both (-)-(1R,2R)-*cis*-1,2-dimethyl-2-phenylcyclohexanol (shown above) and the (-)-(1S,2R)-*trans* isomer are completely equilibrated and racemized when treated with Me_2SO-$MeSOCH_2Na$ at 25°. That is, an equilibrium mixture of all four diastereomers is formed. Explain.

Hoffman and Cram, *J. Amer. Chem. Soc.* **91**, 1000 (1969).

40. [2,4-dimethylcyclohexane-1,3-dione] $\xrightarrow[\text{dil. NaOH pH 7-9, } 0^\circ \text{ 24 hr}]{\text{2 equivs. } PhN_2^+ \ Cl^-}$ 91% Me–C(N=NPh)=N–NHPh

Note: the same product was also obtained from 2,4,6-trimethyl-1,3-hexanedione.

Hargreaves, Hickmott, and Hopkins, *J. Chem. Soc., C* **1968**, 2599.

41. Me–CH=C(Me)–C(=N–OH)–Ph (i) $\xrightarrow[0^\circ]{\text{HONO, HOAc-}H_2O}$ 96% iii

Ph–CH=C(Me)–C(=N–OH)–Me (ii) $\xrightarrow[0^\circ]{\text{HONO, HOAc-}H_2O}$ 96% iii

iii: pyrazole N-oxide bearing Me, Me, Ph substituents, $N^{\oplus}$–$O^{\ominus}$, N–OH

42. $CH_3CH(OMe)_2$ (120 torr) + CH_2N_2 (40 torr) $\xrightarrow[\text{gas phase, } 40^\circ]{h\nu \text{ 3 hr}}$ 57.8% $CH_3CH(OMe)OEt$ +

22.2% $CH_3CH_2CH(OMe)_2$ + 11.5% $Me_2C(OMe)_2$ +

4.6% $CH_3CH(OMe)CH_2OMe$ + 3.9% $MeOCH=CH_2$ + Me_2O small amount

Kirmse and Buschhoff, *Chem. Ber.* **102**, 1087 (1969).

43. i [steroid: 3-AcO, 5-H, 9-Br, 11-oxo, 17-CH(Me)OAc] $\xrightarrow[15^\circ \text{ 2.5 days}]{\text{HOAc-HBr}}$ ii [steroid: 3-AcO, 5-H, 11-oxo, 12-Br, 17-CH(Me)OAc]

44. (2-chloro-4-nitro-5-oxido-benzenediazonium: O_2N, Cl, $\overset{\oplus}{N_2}$, $O^{\ominus}$) + $PhCH_2CHCOOEt$ (with $COCH_3$ on CH) $\xrightarrow[EtOH,\ 0°]{H_2O-OH^-}$

90% (Cl, O_2N, OH substituted ring)–$NH-N=C(CH_2Ph)-COOEt$

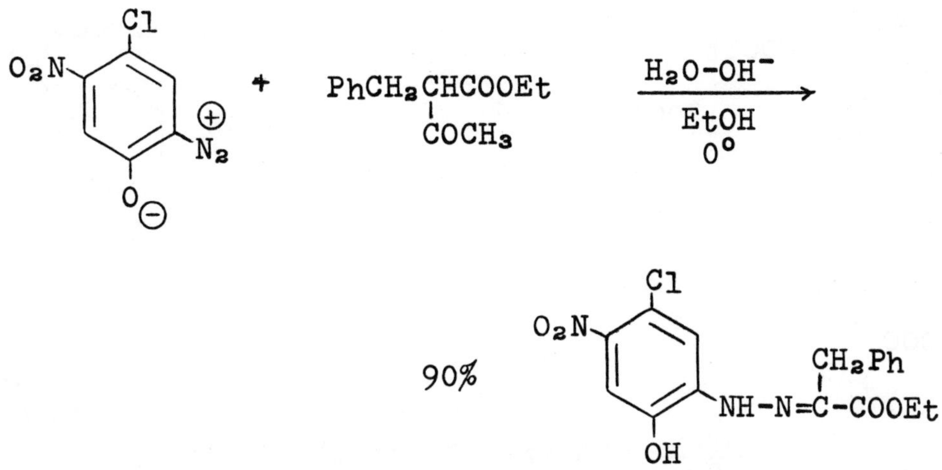

Ried and Kleeman, *Justus Liebigs Ann. Chem.* **713**, 127 (1968).

45. Dialkylmagnesium compounds (R_2Mg) are often prepared by treatment of ordinary ethereal Grignard solutions with dioxane, which precipitates all of the magnesium halide and leaves R_2Mg in ether-dioxane. However, if one desires R_2Mg with no ether present, it is usually difficult to remove all of the ether from such solutions. Suggest another way of preparing ether-free dialkylmagnesiums.

Ashby and Arnott, *J. Organometal. Chem.* **14**, 1 (1968).

46. Why is i a particularly stable diazonium ion?

(indolizine with $\overset{\oplus}{N_2}$ substituent; ring N labelled N)

Tedder, Todd, and Gibson, *J. Chem. Soc., C* **1969**, 1279.

47. $Me_3Sn-CH=C=CH_2$ + SO_2 ⟶ $Me_3SnO-S(=O)-CH_2-C\equiv CH$

$Me_2Sn(-CH=C=CH_2)(-CH=C=CH_2)$ + SO_2 ⟶ $Me_2Sn(-S(=O)-CH_2-C\equiv CH)(-O-S(=O)-CH_2-C\equiv CH)$

$Ph_3Sn-CH_2-C\equiv CH$ + SO_2 ⟶ $Ph_3SnO-S(=O)-CH=C=CH_2$

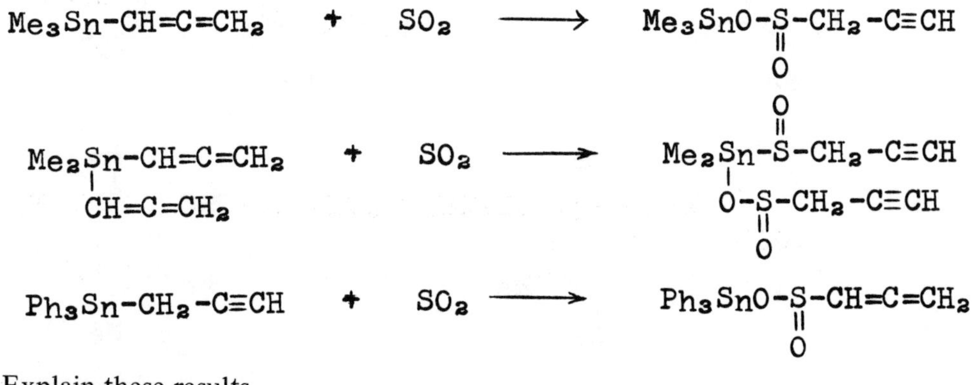

Explain these results.

Kitching, Fong, and Smith, *J. Amer. Chem. Soc.* **91**, 767 (1969).

48. 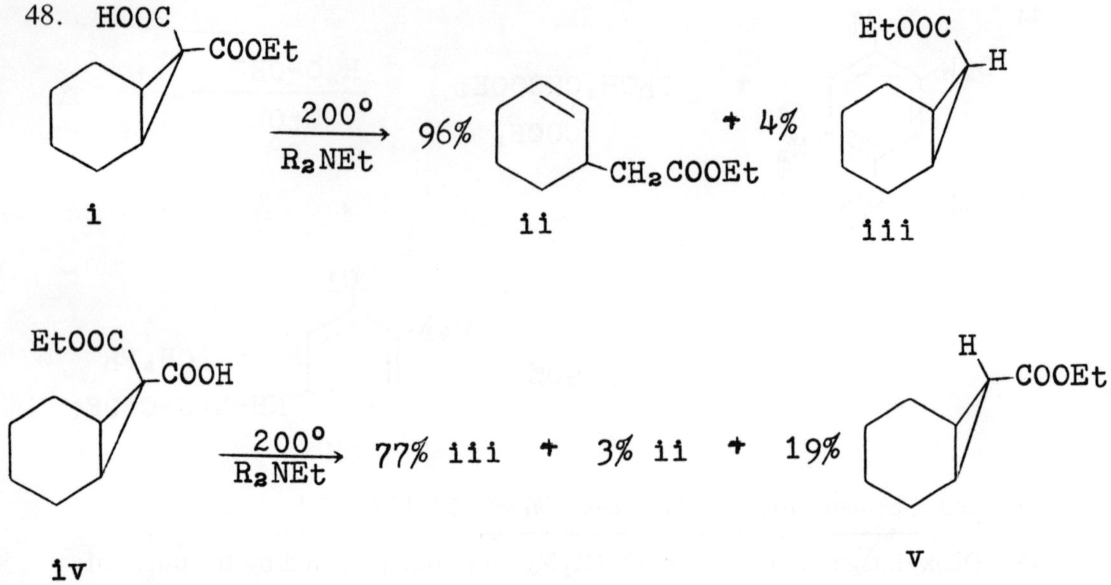 HOOC —COOEt (i) $\xrightarrow[R_2NEt]{200°}$ 96% (ii) —CH_2COOEt + 4% (iii) EtOOC —H

i ii iii

EtOOC —COOH (iv) $\xrightarrow[R_2NEt]{200°}$ 77% iii + 3% ii + 19% (v) H —COOEt

iv v

When heated at 200° in dicyclohexylethylamine, the isomers i and iv gave different results, as shown. Explain.

49.

$PhCH_2CH_2C{\equiv}C{-}C{\equiv}CCH_2CH_2Ph \xrightarrow[t\text{-BuOH},\ 75°]{t\text{-BuOK}}$

PhCH=CH-CH=CH-CH=CH-CH=CHPh

A diaryl conjugated tetraene was also obtained when the benzene rings carried ortho, meta, or para methyl substituents, but when the rings carried four substituents, a different result was obtained:

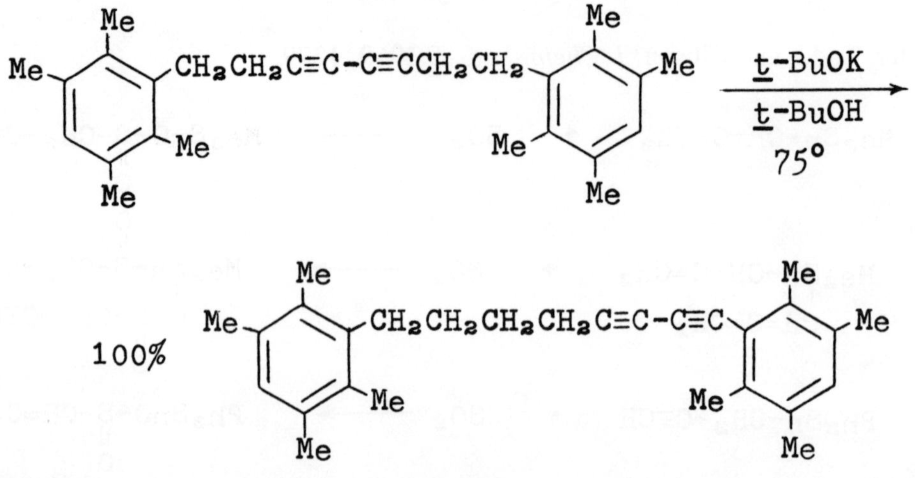

Me, Me, Me, Me (tetramethylphenyl) —$CH_2CH_2C{\equiv}C{-}C{\equiv}CCH_2CH_2$— (tetramethylphenyl) Me, Me, Me, Me $\xrightarrow[t\text{-BuOH},\ 75°]{t\text{-BuOK}}$

100% Me, Me, Me, Me (tetramethylphenyl) —$CH_2CH_2CH_2CH_2C{\equiv}C{-}C{\equiv}C$— (tetramethylphenyl) Me, Me, Me, Me

Explain.

Hubert and Anciaux, *Bull. Soc. Chim. Belges* 77, 513 (1968).

50. $CH_3CH_2CH(CH_3)-HgBr$ (i) + Mg ⟶ $CH_3CH_2CH(CH_3)-Hg-CH(CH_3)CH_2CH_3$ (ii)

i: $[\alpha]_D^{25}$ $-5.05°$ ii: $[\alpha]_D^{25}$ $-7.57°$

Assuming that i is optically pure, devise experiments to determine: a. if this reaction proceeds with retention or inversion of configuration, and b. whether any racemization occurred.

51. 2-Me-2-Ph-cyclopentanone $\xrightarrow{t\text{-BuOK}}$ $PhCH(CH_3)CH_2CH_2CH_2COO^-$

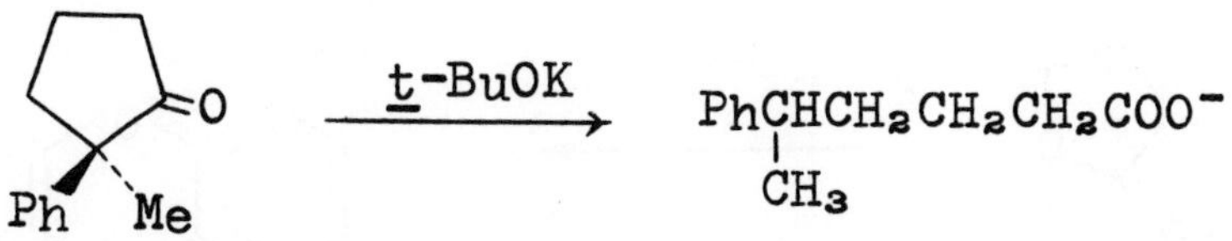

When this reaction was carried out in the solvent *t*-BuOH, the product was formed with about 60% net retention of configuration; however, when a solution of 2.1 *M t*-BuOH in Me_2SO was used as solvent, there was practically complete racemization (about 1% net retention). Explain.

Note: in the latter solvent, unreacted starting material was found not to have racemized.

Hoffman and Cram, *J. Amer. Chem. Soc.* **91**, 1009 (1969).

PROBLEM SET 7
AROMATIC NUCLEOPHILIC SUBSTITUTION

Corresponds to Chapter 13 of Advanced Organic Chemistry.

The first 15 problems are synthetic. In each one you are asked to convert one compound into another. In every case the conversion has actually been carried out, and the reference is given.

1.

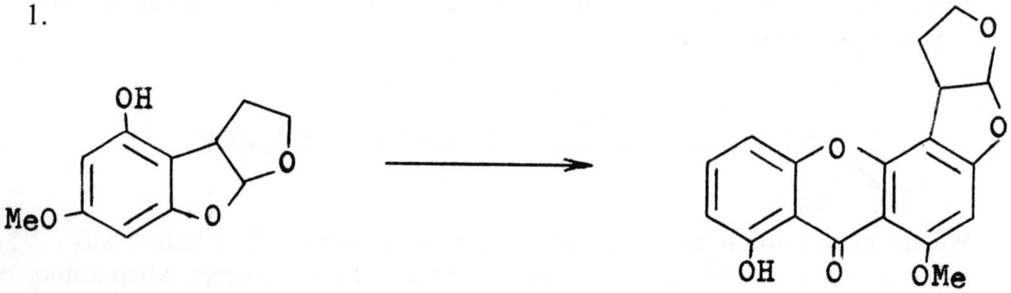

Rance and Roberts, *Tetrahedron Lett.* **1969**, 277.

2.

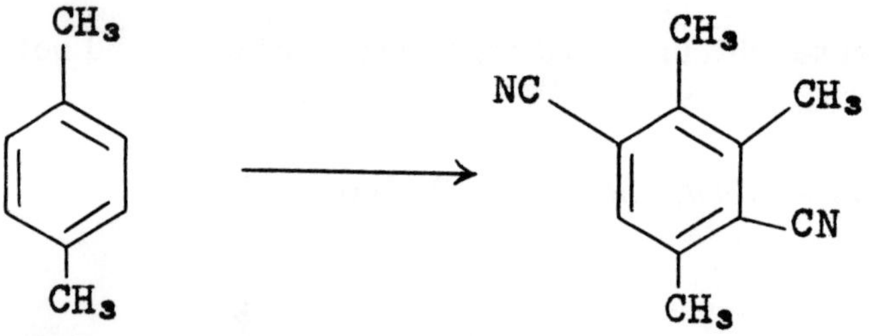

Wals and Nauta, *Rec. Trav. Chim.* **87**, 65 (1968).

3.

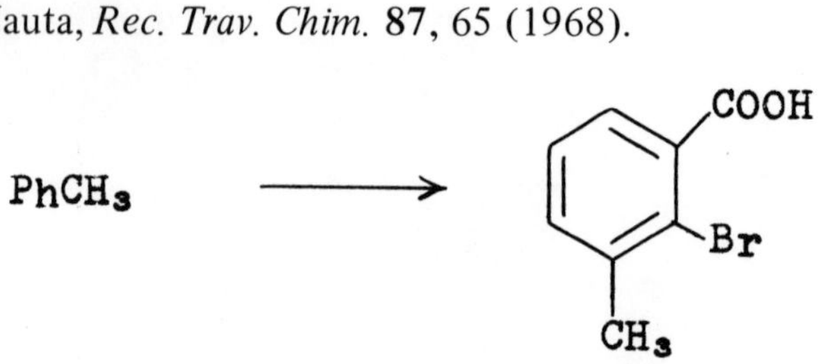

$PhCH_3$ ⟶

4.

PhCOOH ⟶

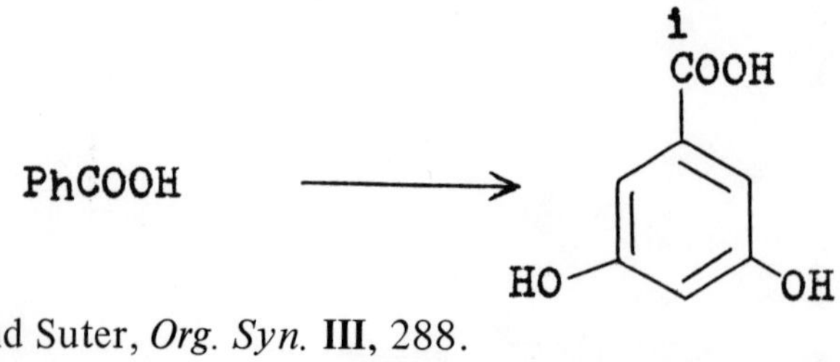

Weston and Suter, *Org. Syn.* **III**, 288.

5.

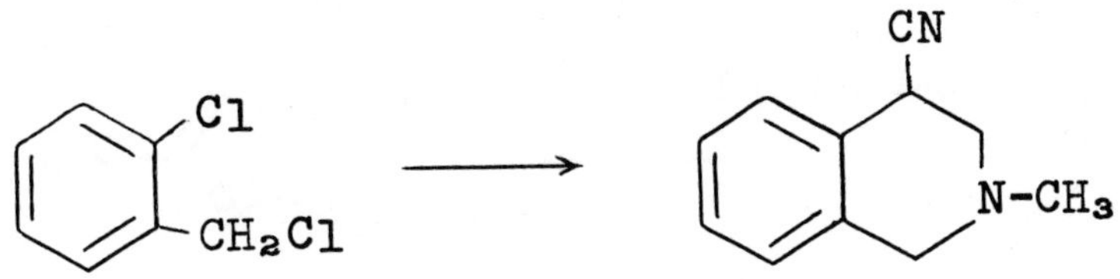

Cl
CH_2Cl
CN
$N\text{-}CH_3$

Julia, Igolen, and Le Goffic, *Bull. Soc. Chim. Fr.* **1968**, 310.

6.

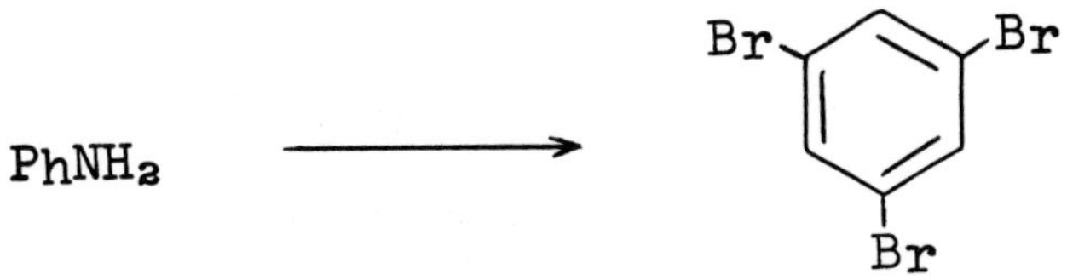

$PhNH_2$
Br
Br
Br

Coleman and Talbot, *Org. Syn.* **II**, 592.

7. $PhNH_2$

N
NH_2

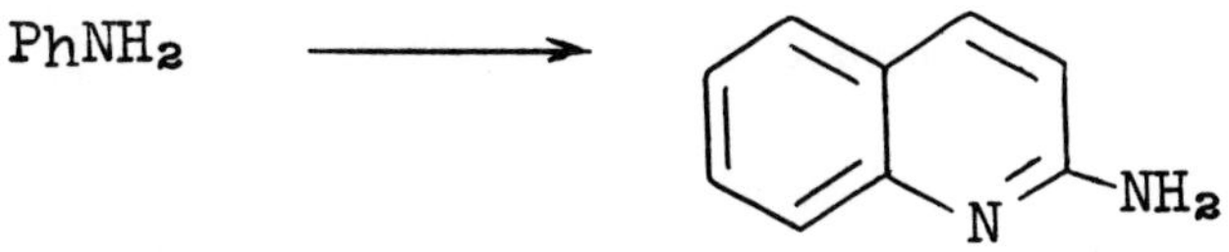

i

8.

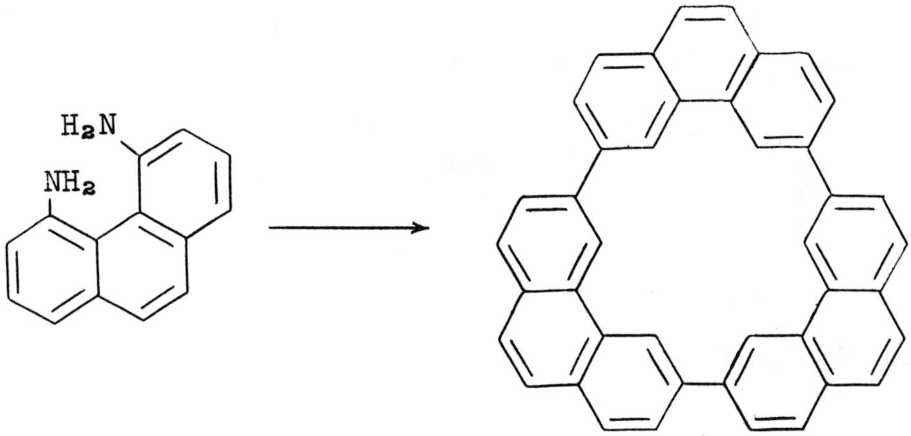

H_2N
NH_2

Staab, Braünling, and Schneider, *Chem. Ber.* **101**, 879 (1968).

9. NO_2
Me
NO_2

MeO Me Me OMe

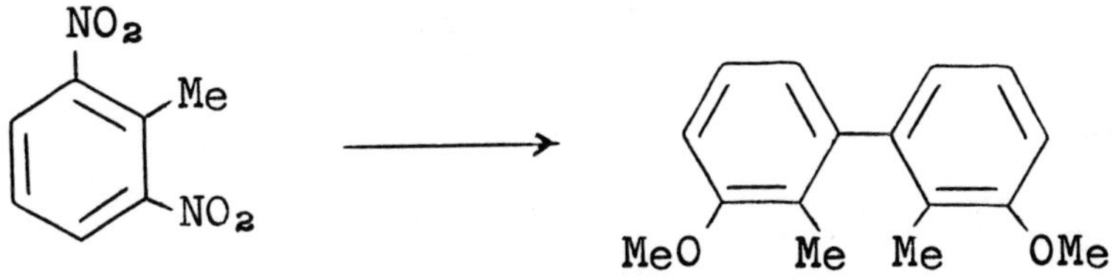

Lounasmaa, *Acta Chem. Scand.* **22**, 2388 (1968).

10. [Structure i: pentafluorophenyl-CH_2COCl (F at each ring position, CH_2COCl substituent) → Structure ii: 4,5,6,7-tetrafluoro-2-methylbenzofuran (CH_3, O)]

i → ii

11. $\overset{\oplus}{PhCH_2NMe_3}$ → hexamethylbenzene (six CH_3 groups)

Kantor and Hauser, *J. Amer. Chem. Soc.* **73**, 4122 (1951).

12. 2,6-dimethoxyphenol (OH, MeO, OMe) → 1,3-dimethoxybenzene (MeO, OMe)

Pirkle and Zabriskie, *J. Org. Chem.* **29**, 3124 (1964).

13. 4-chlorobenzenesulfonic acid (Cl, SO_3H) → 2,6-dinitroaniline (NH_2, O_2N, NO_2)

Schultz, *Org. Syn.* **IV**, 364.

14. $PhCH_3$ → 3-bromotoluene (CH_3, Br)

i

15. 2-chlorobenzonitrile (Cl, CN) → 2-fluoro-5-nitrobenzonitrile (F, CN, NO_2)

Wilshire, *Aust. J. Chem.* **20**, 1663 (1967).

The following conversions (problems 16 to 25) have been reported in the literature. In each case suggest a reasonable mechanism.

16. OH, HO, OH + Me_2NH $\xrightarrow{130-50°}$ 68% OH, Me_2N, NMe_2

+ 22% NMe_2, Me_2N, NMe_2

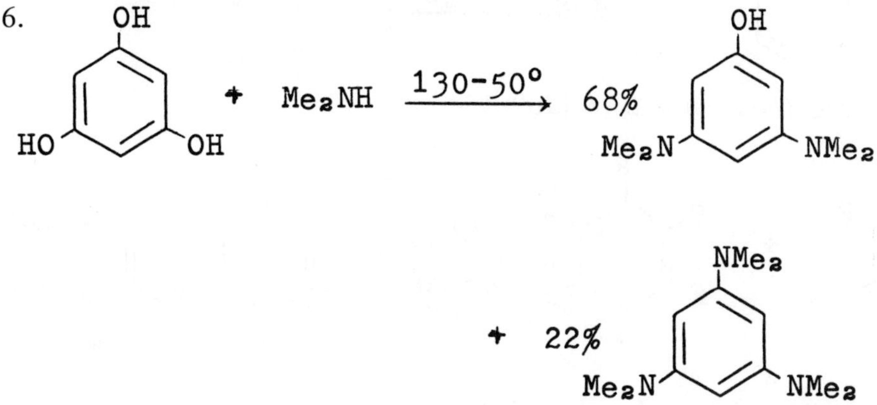

Note: 1,3,5-trimethoxybenzene does not react with Me_2NH, even at 200°.

Effenberger and Niess, *Chem. Ber.* **101**, 3787 (1968).

17. Cl, N, Ph, N (i) + KNH_2 $\xrightarrow[\text{liq. } NH_3 \text{ } -33° \text{ 4 hr}]{\text{abs. } Et_2O}$ 39% CH_3, N, N, Ph, N (ii)

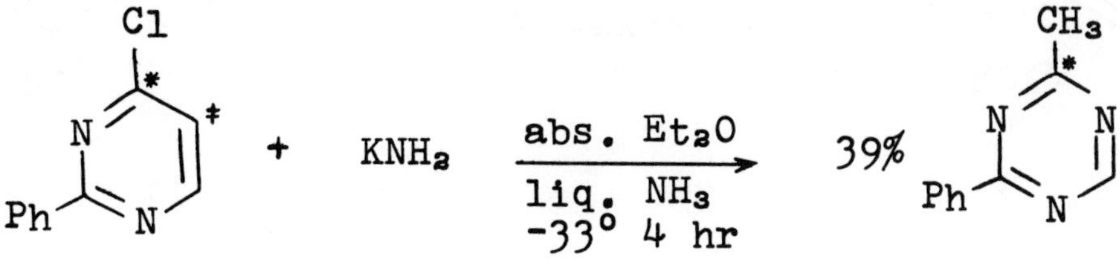

Note: The reaction was run with i labeled with ^{14}C. The position of the label in the starting material and the product is shown by *. The carbon marked ‡ appears as the CH_3 group in the product.

van Meeteren and van der Plas, *Rec. Trav. Chim.* **86**, 15 (1967); See also van der Plas, Zuurdeeg, and van Meeteren, *Rec. Trav. Chim.* **88**, 1156 (1969).

18. SO_2NMe_2, NH–C(=O), OH, Br $\xrightarrow[Et_2O\text{–(N-methylpyrrolidone)}]{NaH}$ 70% H, N–C(=O), O, SO_2NMe_2

+ 3.4% O–C(=O), SO_2NMe_2, N–H

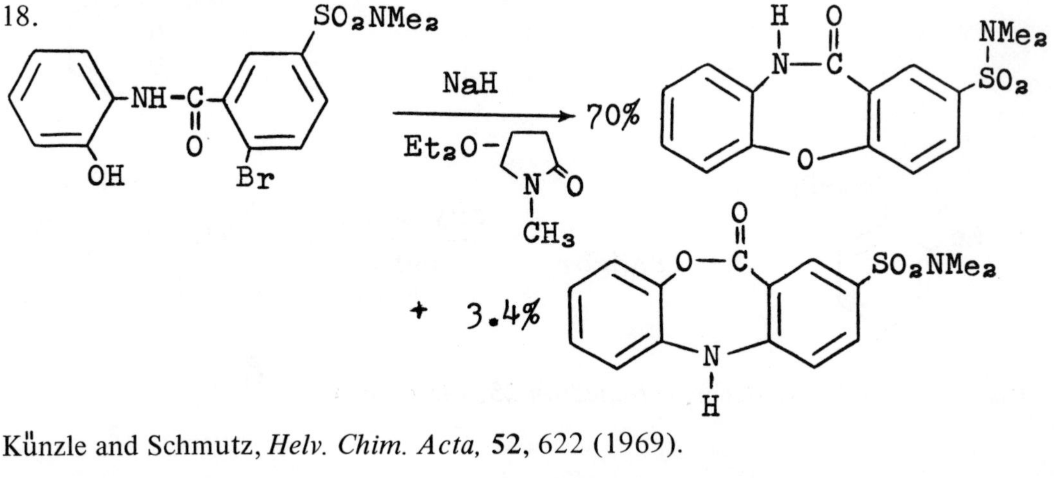

Künzle and Schmutz, *Helv. Chim. Acta,* **52**, 622 (1969).

19. 4-aminoquinolizinium + HONO $\xrightarrow{H_2O}$ pyrazolo/triazolopyridine with CH=CH-CHO

NH$_2$ + HONO $\xrightarrow{H_2O}$ CH=CH-CHO, N, N-N

Davies and Jones, *Tetrahedron Lett.* **1969**, 1549.

20. F, F, F, F, Li + F, F, F, F, S$^-$ $\xrightarrow[\text{hexane, } 0^\circ]{Et_2O}$ F, F, F, F, S, F, F, F, F

Chambers and Spring, *Tetrahedron Lett.* **1969**, 2481.

21. $\oplus$SMe$_2$ Br$^-$ $\xrightarrow[\text{steel bomb, 2 weeks}]{\text{liq. } NH_3}$ 76.4% CH$_2$SMe

i → ii

22. N=N-Ph, MeO + PhMgBr $\xrightarrow[\text{reflux, 5 hr}]{\text{dry } C_6H_6\text{-}Et_2O}$ Ph, Ph, N=N, MeO

N=N-Ph, Me + PhMgBr $\xrightarrow[\text{reflux, 5 hr}]{\text{dry } C_6H_6\text{-}Et_2O}$ N=N-Ph, Me, Ph

Risaliti, Bozzini, and Stener, *Tetrahedron* **25**, 143 (1969).

23.

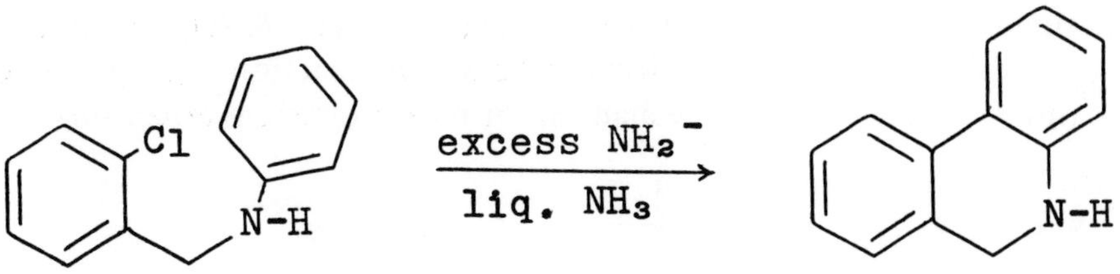

Cl, N-H

excess NH_2^-

liq. NH_3

N-H

i

Note: Similar treatment of the N-methyl derivative of i gave only a complex mixture.

Kessar, Gopal, and Singh, *Tetrahedron Lett.* **1969**, 71.

24.

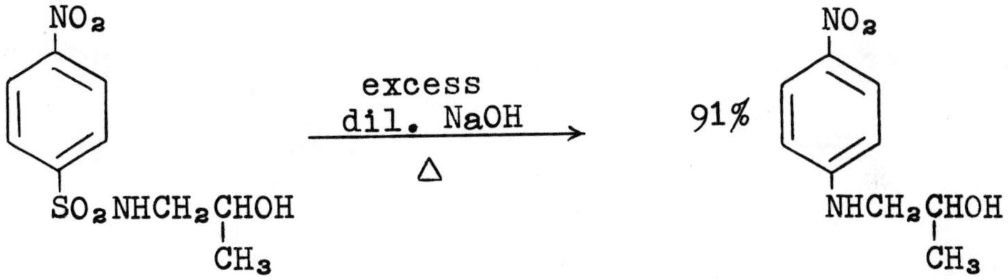

NO_2

SO_2NHCH_2CHOH
CH_3

i

excess
dil. NaOH
Δ

91%

NO_2

$NHCH_2CHOH$
CH_3

ii

25.

O

CH_2 OMe OMe
CN

NaOMe
Me_2SO
140° 10 min

OH
OMe
CN

almost 100% yield

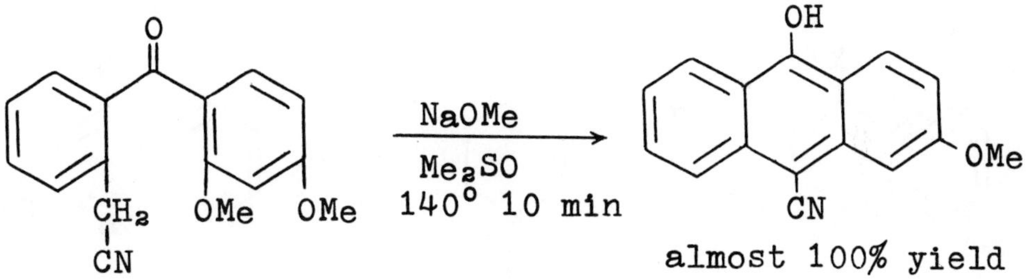

Davies, Davies, and Hassall, *Chem. Commun.* **1968**, 1555.

26. Predict the product:

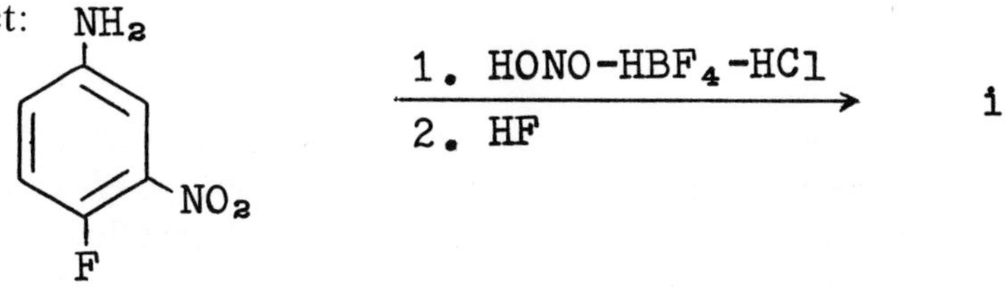

NH_2
NO_2
F

1. HONO-HBF_4-HCl
2. HF

i

27. Predict the product:

F, F, F, F, F, N

+ $CF_3CF{=}CF_2$

KF-sulfolan
130° 12 hr

94% $C_8F_{11}N$

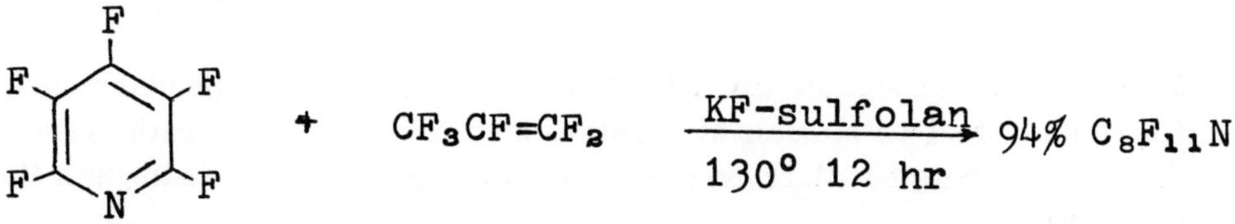

Chambers, Jackson, Musgrave, and Storey, *J. Chem. Soc., C* **1968**, 2221.

28. The pyrimidine i, upon treatment for 5 min at −75° with KNH_2 in liquid NH_3, gave ii. Devise an experiment to decide whether the reaction takes place by path a (intermediate complex mechanism) or path b (aryne mechanism):

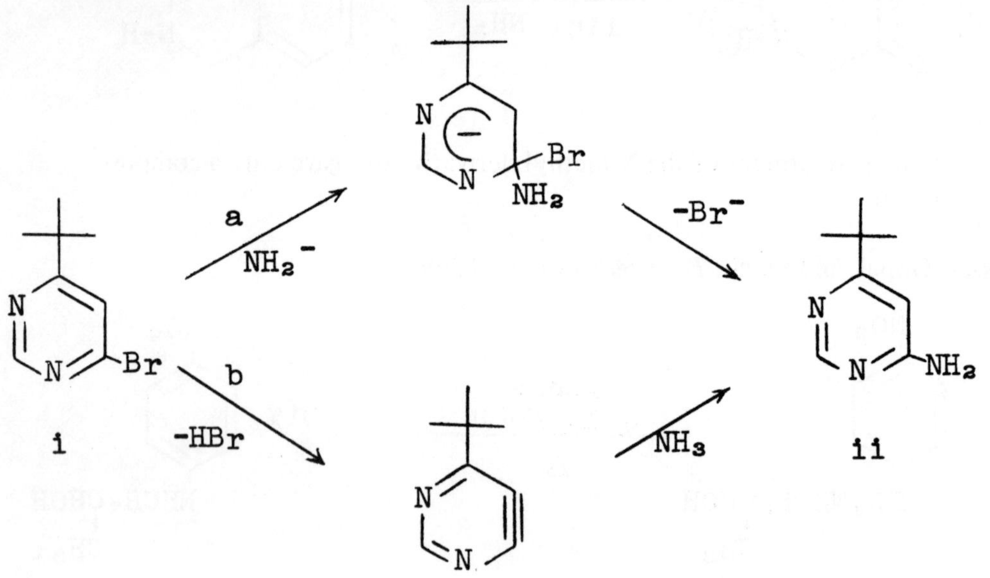

a NH_2^-
$-Br^-$
Br
NH_2
N
N
Br
b
$-HBr$
NH_3
i
ii
N
N
NH_2

29.

X
N
OMe
+
N
D
i
N
N
OMe

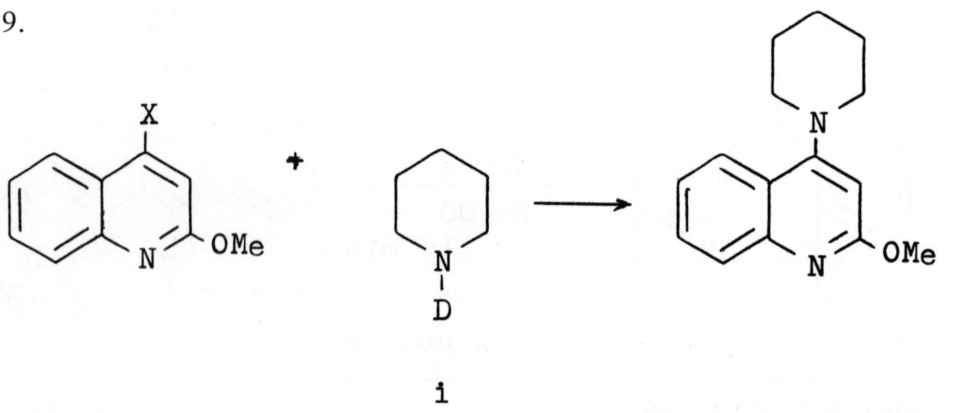

In the above reaction, when the rate with undeuterated piperidine was compared with that of i, a small but definite isotope effect ($k_H/k_D \approx 1.25$) was noted when X was F, but there was no isotope effect when X was Cl. Explain.
Giardi, Illuminati, and Sleiter, *Tetrahedron Lett.* **1968**, 5505.

30. When the hydrazones i are treated with polyphosphoric acid, the heterocycles ii are obtained. Two possible mechanisms involve nucleophilic (path a) and electrophilic (path b) attack on ring A. Devise experiments to distinguish these mechanisms.

30 Cont.

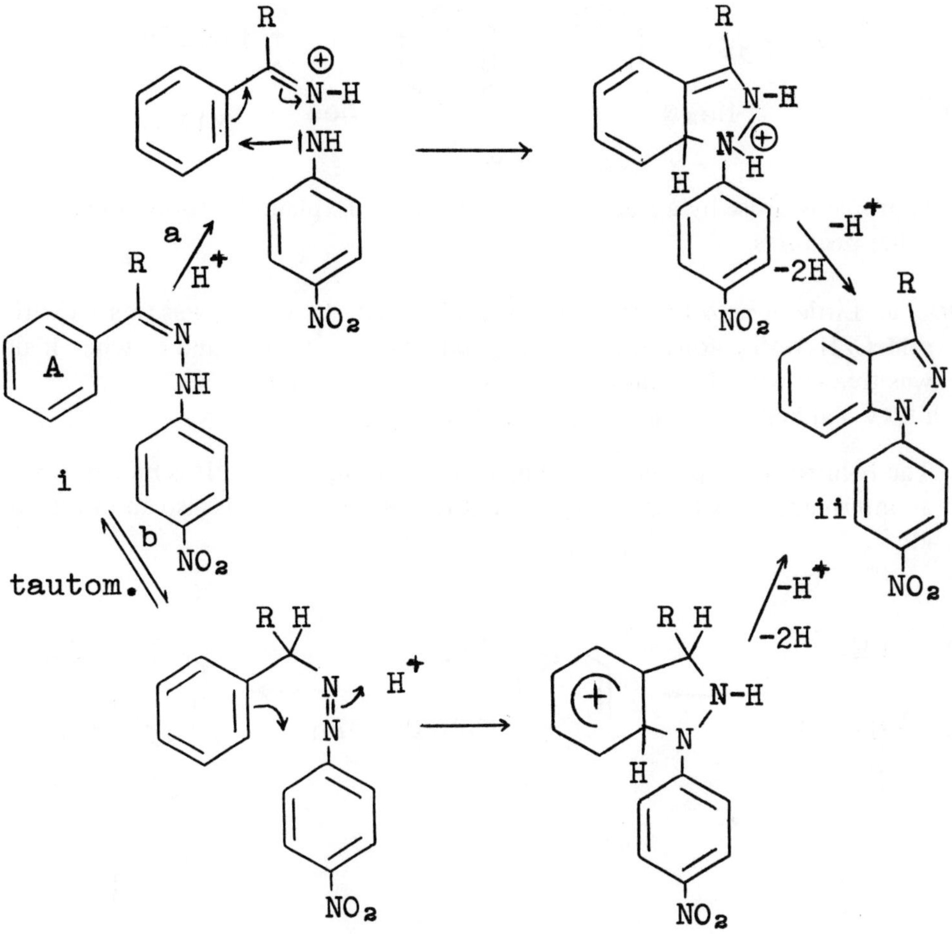

R
⊕
N-H
NH
a
H^+
R
N
NH
A
i
NO_2
b
tautom.
R
N-H
N⊕
H
H
NO_2
$-H^+$
$-2H$
R
N
N
ii
NO_2
R H
N
N
H^+
NO_2
R H
⊕
N-H
N
H
NO_2
$-H^+$
$-2H$
NO_2

Robinson, *Tetrahedron Lett.* **1967**, 5085.

31.

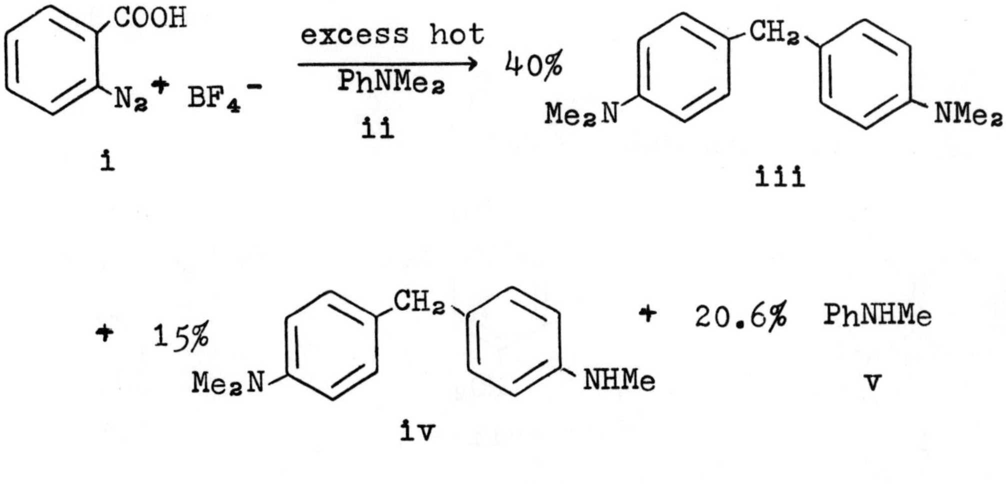

COOH
N_2^+ BF_4^-
i
excess hot
$PhNMe_2$
ii
40%
CH_2
Me_2N
NMe_2
iii

+ 15%
CH_2
Me_2N
NHMe
iv
+ 20.6% PhNHMe
v

31 Cont.

+ 13% $Me_2N-C_6H_4-C_6H_4-NMe_2$ + PhCOOH

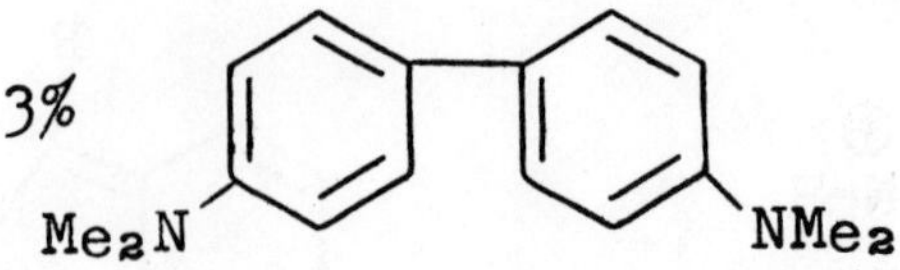

vi vii

vi probably arose by a free radical mechanism. Explain the formation of the other products.

Notes: a. Little or no o-FC_6H_4COOH was obtained; b. when i was treated with p-$MeC_6H_4NMe_2$ none of the above products was formed; and c. when ii alone was treated with BF_3, these products were not obtained.
Suschitzky and Sellers, *Tetrahedron Lett.* **1969**, 1105.

32. The N-nitroso compound i spontaneously rearranges to iv. It is known that ii is an intermediate. Devise an experiment to determine if iii is also an intermediate.

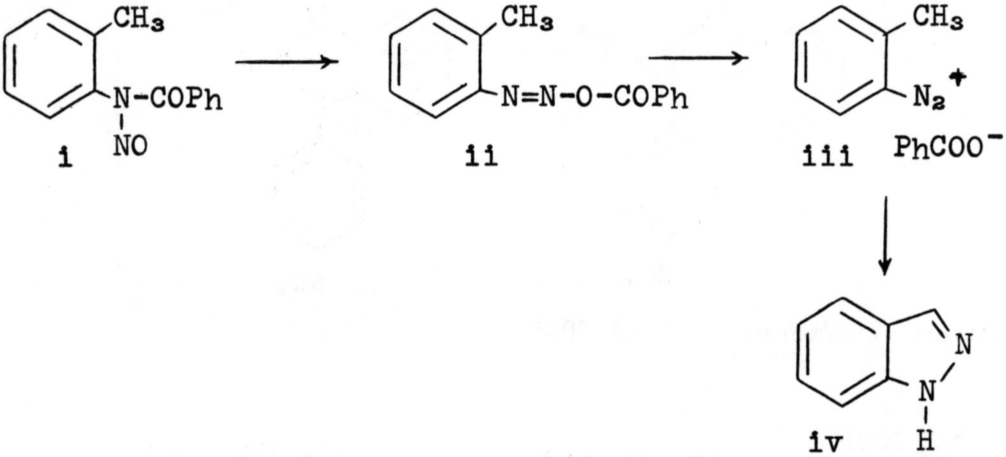

CH3, N–COPh, NO (i) → CH3, N=N–O–COPh (ii) → CH3, N_2^+ PhCOO⁻ (iii) → N, N–H (iv)

33. Arrange in order of reactivity to OMe^-.

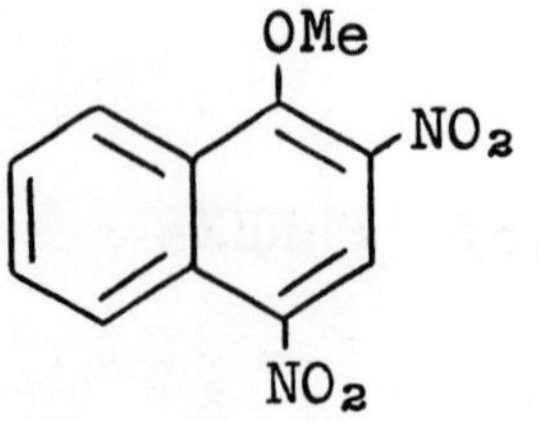

OMe, NO2, NO2 (i) OMe, NO2, NO2 (ii) OMe, O2N, NO2, NO2 (iii)

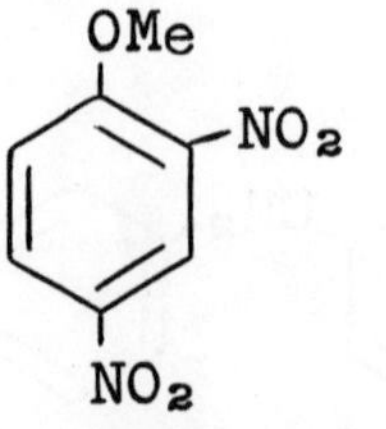

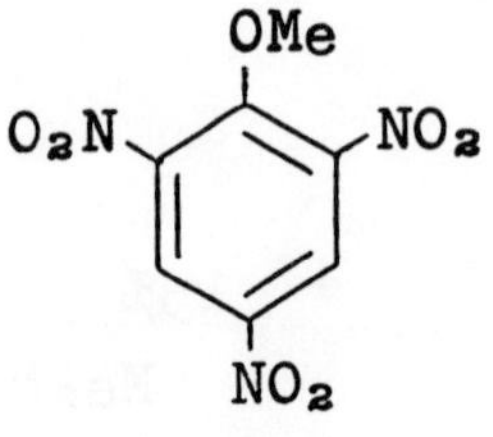

34. Arrange in order of reactivity to *p*-nitrophenoxide ion.

PhCl

i ii 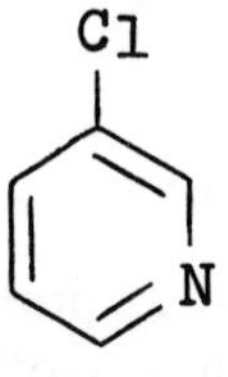 iii

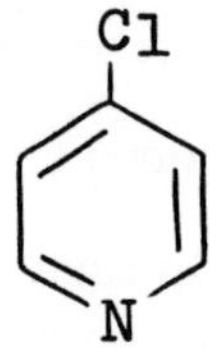

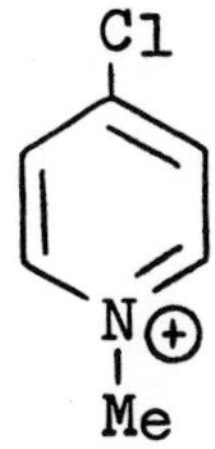

iv 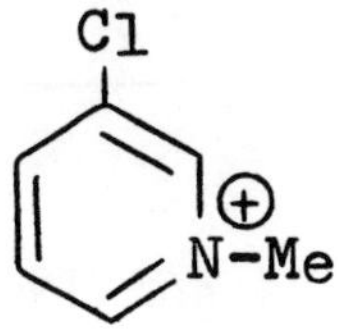 v 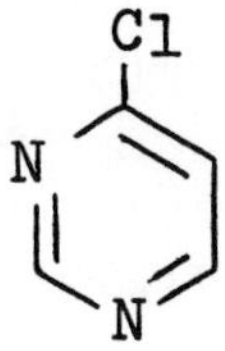 vi

35. In each case, predict the principal product when the compound is treated with NH_2^-.

a. b.  c. 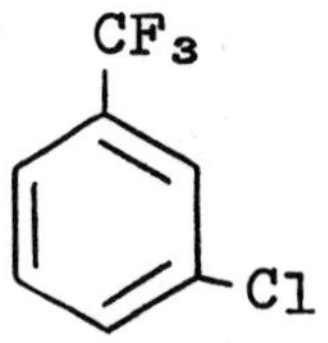 d.

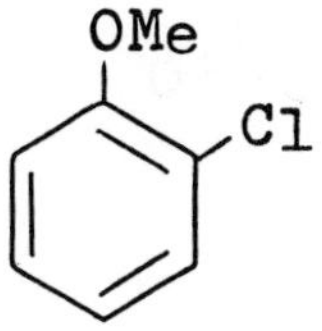

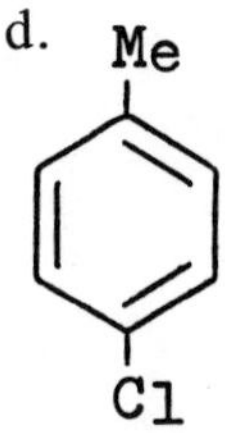

e. 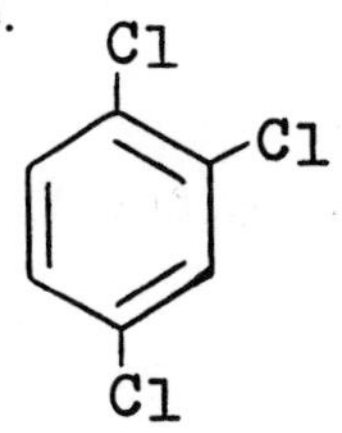 f. 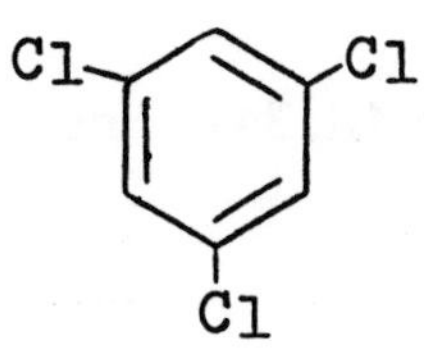 g.

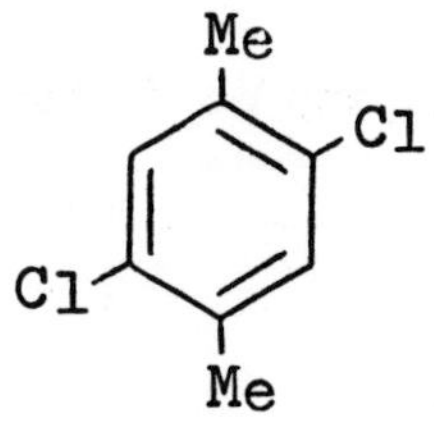

PROBLEM SET 8
FREE-RADICAL SUBSTITUTION

Corresponds to Chapter 14 of Advanced Organic Chemistry.

The first 22 problems are synthetic. In each one you are asked to convert one compound into another. In every case the conversion has actually been carried out, and the reference is given.

1.

Moon and Ganz, *J. Org. Chem.* **34**, 465 (1969).

2.

COOMe COOH COOH COOMe → COOH Br Br COOH

Blight, Coppen, and Grove, *J. Chem. Soc., C* **1969**, 552.

3.

i → ii

4.

NO_2 → NO_2 NO_2

Hodgson, Mahadevan, and Ward, *Org. Syn.* **III**, 341; Price and Voong, *Org. Syn.* **III**, 664.

5.

CH_2 → O-C-Ph O CH_2

Cross and Whitham, *J. Chem. Soc.* **1961**, 1950.

6.

Corey and Sauers, *J. Amer. Chem. Soc.* **81**, 1739 (1959).

7.

```
     Me                         Me              Me
     |                          |               |
HC≡C-C-Cl   ──→   MeOOC-CH2-C-C≡C-C≡C-C-CH2-COOMe
     |                          |               |
     Me                         Me              Me
```

Cordes, Prelog, Troxler, and Westen, *Helv. Chim. Acta* **51**, 1663 (1968).

8.

i ──→ ii

9.

i ──→ ii

10.

```
Et-C-Pr   ──→   Either R or S    Et
   ||                            |
   O                          Bu-C-Pr
                                 |
                                 (CH2)5-Me

                 (one enantiomer only)
```

Wynberg, Hekkert, Houbiers, and Bosch, *J. Amer. Chem. Soc.* **87**, 2635 (1965).

11.

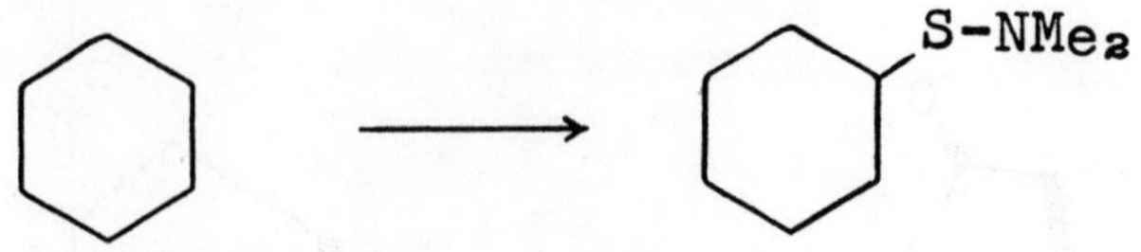

Cyclohexane ⟶ cyclohexyl–S–NMe_2

Müller and Schmidt, *Chem. Ber.* **96**, 3050 (1963).

12.

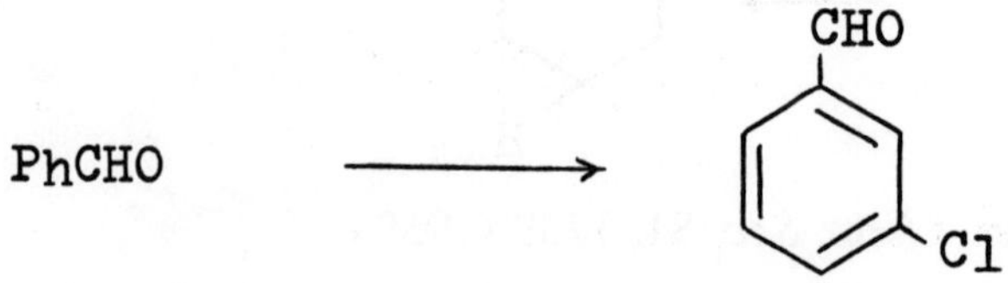

PhCHO ⟶ m-chlorobenzaldehyde (CHO, Cl)

Buck and Ide, *Org. Syn.* **II**, 130; Icke, Redemann, Wisegarver, and Alles, *Org. Syn.* **III**, 644.

13.

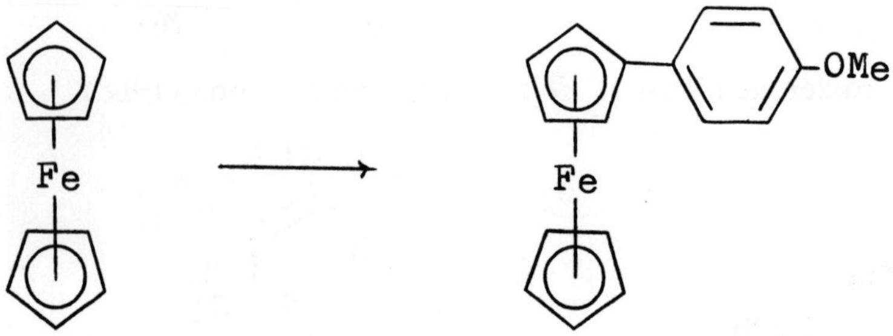

Ferrocene (Fe) ⟶ (p-methoxyphenyl)ferrocene (OMe, Fe)

Broadhead and Pauson, *J. Chem. Soc.* **1955**, 367.

14.

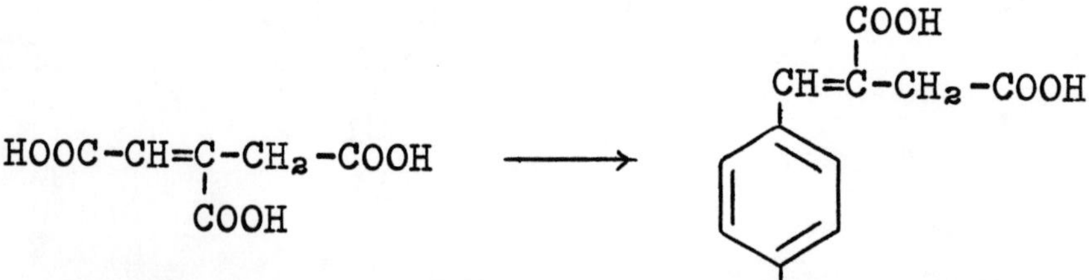

HOOC–CH=C(COOH)–CH_2–COOH ⟶ p-ClC_6H_4–CH=C(COOH)–CH_2–COOH

aconitic acid

Mathur, Krishnamurti, and Pandit, *J. Amer. Chem. Soc.* **75**, 3240 (1953).

15.

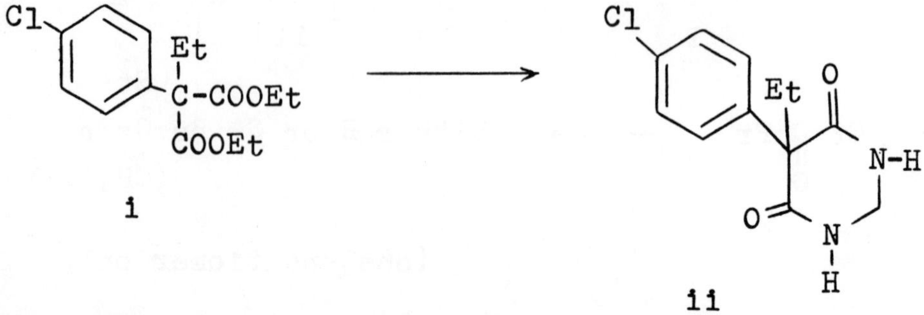

p-ClC_6H_4–C(Et)(COOEt)–COOEt ⟶ 5-(p-chlorophenyl)-5-ethyl-hexahydropyrimidine-4,6-dione (Cl, Et, O, N–H, O, N–H)

i ⟶ ii

16.

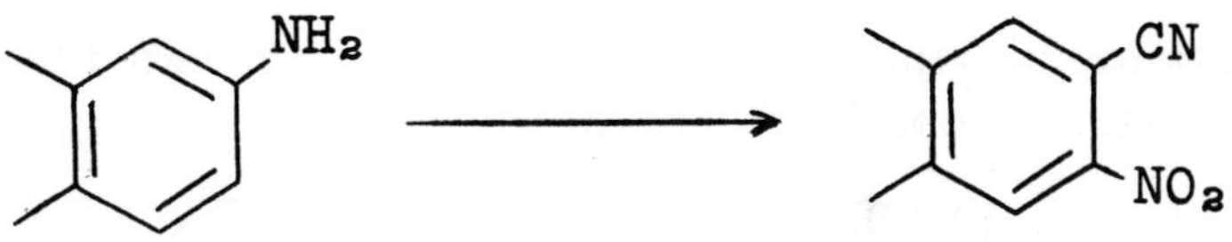

NH_2 → CN, NO_2

Brändström and Carlsson, *Acta Chem. Scand.* **21**, 983 (1967).

17.

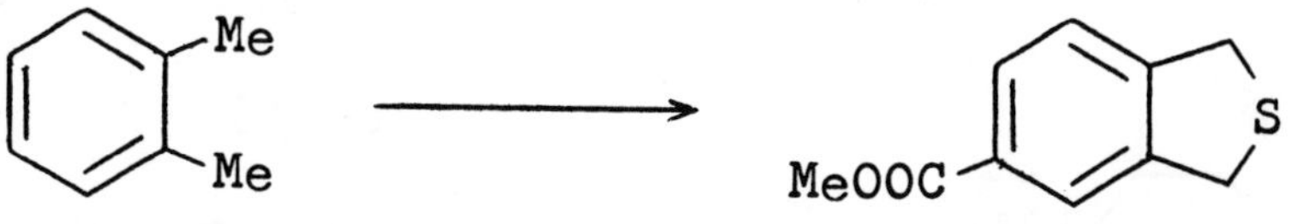

Me, Me → MeOOC, S

i ii

18.

CH_3, $CH(CH_3)_2$, CH_2, $CON(CH_3)_2$ → CH_3, $CH(CH_3)_2$, CH_3

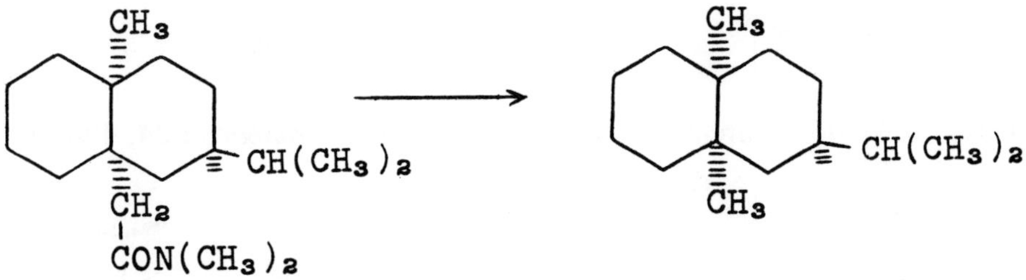

Dawson and Ireland, *Tetrahedron Lett.* **1968**, 1899.

19. CH_3COCH_2COOEt → $CH_3CH_2CH_2CH(NO_2)COOEt$

Emmons and Freeman, *J. Amer. Chem. Soc.* **77**, 4391 (1955).

20. $CH_3(CH_2)_7CH(CH_3)COOH$ → $CH_3(CH_2)_7CH(CH_3)(CH_2)_8COOH$

i ii

21.

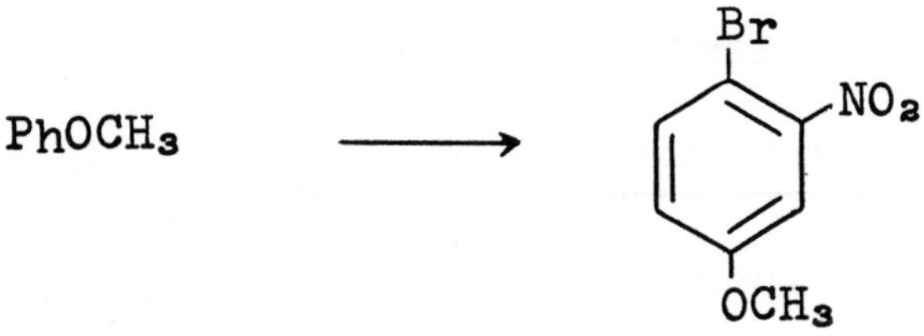

$PhOCH_3$ → Br, NO_2, OCH_3

i

22.

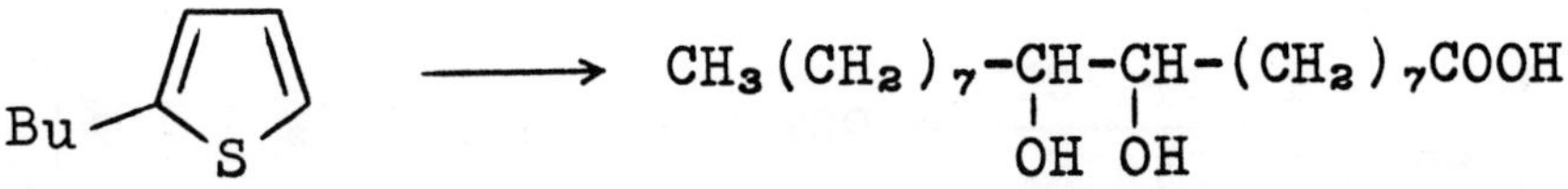

Bu, S → $CH_3(CH_2)_7$-CH(OH)-CH(OH)-$(CH_2)_7COOH$

McGhie, Ross, Laney, and Barker, *J. Chem. Soc., C* **1968**, 1; Grey, McGhie, and Ross, *J. Chem. Soc.* **1960**, 1502.

The following conversions (problems 23 to 37) have been reported in the literature. In each case suggest a reasonable mechanism.

23. Cycloheptanol $\xrightarrow[\text{boiling } C_6H_6]{Pb(OAc)_4,\ K_2CO_3-CaCO_3}$ 8-oxabicyclo[3.2.1]octane (15.3%) + cycloheptanone (7.9%) + cycloheptyl acetate (38.2%) + [$AcO(CH_2)_6CHO$ + $CH_3CH(OAc)(CH_2)_5CHO$] (3%)

Mihailović, Čeković, Andrejević, Matić, and Jeremić, *Tetrahedron* **24**, 4947 (1968).

24. (+)-β-himachalene $\xrightarrow[\text{in air}]{480\text{-}490°}$ 35-40% (+)-cuparene

Subba, Rao, Damodaran, and Dev, *Tetrahedron Lett.* **1968**, 2213.

25. 1,8-Bis(phenylethynyl)naphthalene (Ph–C≡C, Ph–C≡C) $\xrightarrow{\text{warming to 100° or pyrolyzed in pentane or exposed to } AlCl_3}$ Ph-substituted benzo-fused fluoranthene

Bossenbroek and Shechter, *J. Amer. Chem. Soc.* **89**, 7111 (1967).

26. 1-Hydroxycyclohexyl-CH_2COOAg (OH) $\xrightarrow[\text{2. } LiAlH_4]{\text{1. } Br_2-CCl_4}$ 90% $CH_3(CH_2)_4CHCH_3$ (OH) + 10% 1-methylcyclohexanol (OH, CH_3)

Kundu and Sisti, *J. Org. Chem.* **34**, 229 (1969).

27.

O

C-COOH

O

i

$HgO\text{-}I_2$

CCl_4

$h\nu$

30%

I

ii

28.

Me Me

Me OH

Me OMe

air

hexane

24 hr

97% HOO-C—C-COOMe

Me Me

Me Me

Gibson and DePuy, *Tetrahedron Lett.* **1969**, 2203.

29.

$CH_2CH_2CH_2COOH$

N_2^+ BF_4^-

i

+ Ti^{III} → ~30% $PhCH_2CH_2CH_2COOH$

ii

+ ~15% $PhCH_2CH_2CHCOOH$

$PhCH_2CH_2CHCOOH$

iii

30.

+ $Cl_2C{=}CClH$ $\xrightarrow{250^\circ\text{-}360^\circ}$ $CH{=}CCl_2$

Hardwick, *Intl. J. Chem. Kinetics* 1, 325 (1969).

31.

Me

S-(+) Et-CH-CH_2CH_2COOH

alk. MnO_4^-

Me

Et

O

O

Note: At pH = 13 there was 35% retention of configuration.

Wiberg and Fox, *J. Amer. Chem. Soc.* **85**, 3487 (1963).

32.

i + H_2O_2 $\xrightarrow[\text{aq. HCl}]{NaNO_2}$ ii (OH) + iii (NO_2) + iv (ONO_2)

33.

OMe, N-Et, BF_4^- N_2^+, O $\xrightarrow[\text{acetone, 10 min}]{\text{Cu powder}}$ 22% MeO, N-Et, O

+ 2.4% OMe, N-H, O (i) + 5% OMe, OMe, Et, N, O, Et, N, O

Note: Product i is lacking the N-ethyl group.

Hey, Rees, and Todd, *J. Chem. Soc., C* **1967**, 1518; Brown, Hey, and Rees, *J. Chem. Soc.* **1961**, 3873.

34.

OMe, MeOOC $\xrightarrow[\text{2. } H_2O]{\text{1. } Br_2\text{-}Et_2O}$ 23% OMe, Br, O, MeOOC

Note: The addition of small amounts of Bz_2O_2 increased the yield to 38%.

Davis and Watkins, *Aust. J. Chem.* **21**, 2769 (1968).

35.

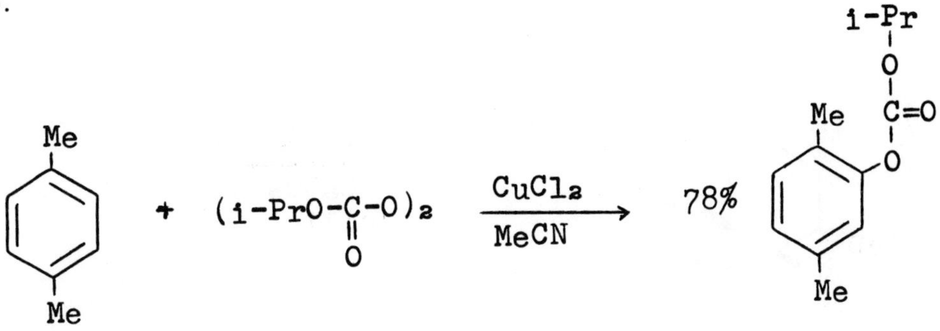

Me / Me + (i-PrO-C(=O)-O)$_2$ $\xrightarrow[MeCN]{CuCl_2}$ 78% Me / Me / O-C(=O)-O-i-Pr

Kovacic, Reid, and Kurz, *J. Org. Chem.* **34**, 3302 (1969).

36.

PhCHO + PhC(=O)-O-O-CMe$_3$ (i) ⟶ PhCOOCH(Ph)CH(Ph)OCOPh (ii) + PhCOOH + CO_2

37.

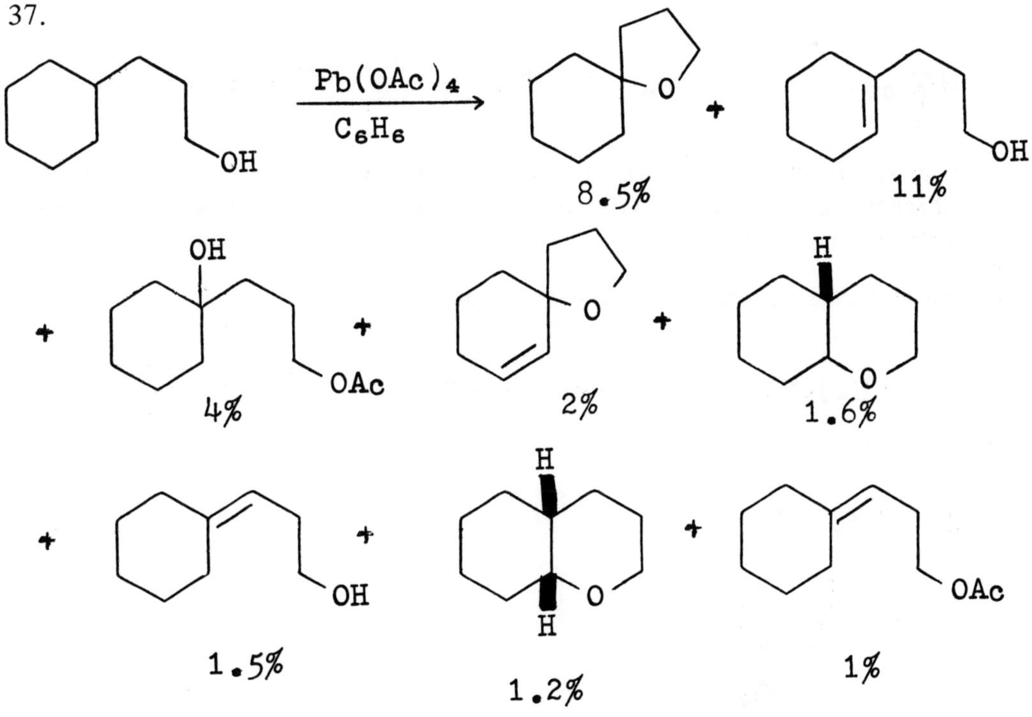

OH $\xrightarrow[C_6H_6]{Pb(OAc)_4}$ O (8.5%) + OH (11%) + OH / OAc (4%) + O (2%) + H / O (1.6%) + OH (1.5%) + H / H / O (1.2%) + OAc (1%)

+ 9 other products

Explain the formation of the products shown.
Mihailović, Konstantinović, Milovanović, Janković, Čeković, and Jeremić, *Chem. Commun.* **1969**, 236.

---

38. Predict the product:

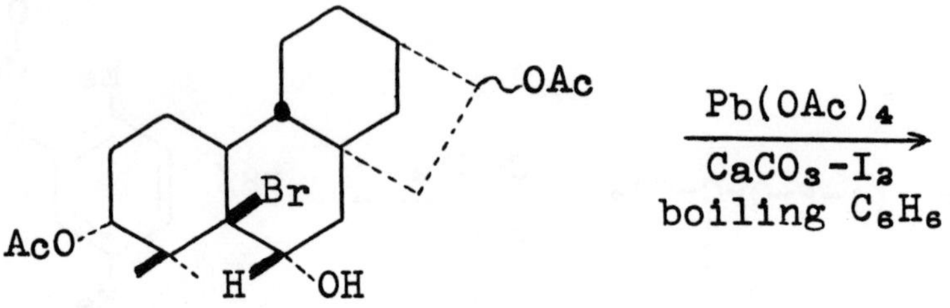

AcO, H, OH, Br, OAc

$\xrightarrow[\text{CaCO}_3\text{-I}_2\text{, boiling } C_6H_6]{Pb(OAc)_4}$

Ogawa, Mori, and Matsui, *Tetrahedron Lett.* **1968**, 125.

39.

$$CH_3\text{-}CH(CH_3)\text{-}CH(CH_3)\text{-}CH_3 + SO_2 + Cl_2 \xrightarrow{h\nu} CH_3\text{-}CH(CH_3)\text{-}CH(CH_3)\text{-}CH_2SO_2Cl$$

$$(CH_3)_3CH + SO_2 + Cl_2 \xrightarrow{h\nu} (CH_3)_2CH\text{-}CH_2Cl$$

In sharp contrast to chlorination or bromination, in this reaction *only* primary (or secondary, if present) hydrogen is replaced: no tertiary product is formed at all. Explain why.

40. List in decreasing order of reactivity towards Cl.

a. $(CH_3)_3CCN$ b. $(CH_3)_3CPh$ c. $(CH_3)_3C\text{-}C(=O)\text{-}Ph$

d. 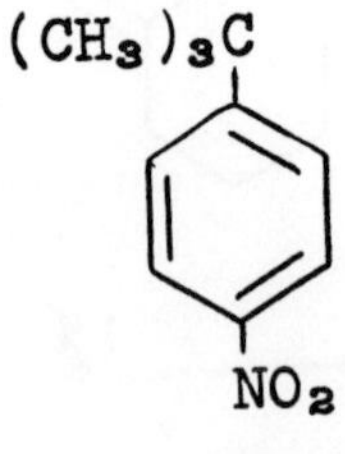 $(CH_3)_3C\text{-}C_6H_4\text{-}NO_2$ (para)

41. Predict which position is most likely to be attacked by Cl• .

a.

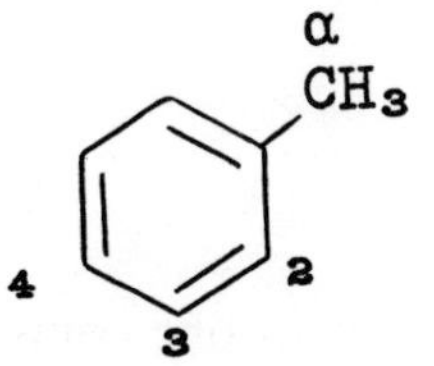

b.

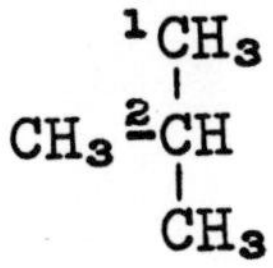

c. 4 3 1
$CH_3CH_2CF_2CH_3$

d. 4 3 2
$CH_3CH_2CH_2CN$

e. α 1 2
$CH_3COOCH_2CH_3$

f. 1 2 3
$CCl_3COOCH_2CH_2CH_3$

PROBLEM SET 9

ADDITION TO CARBON-CARBON MULTIPLE BONDS

Corresponds to Chapter 15 of Advanced Organic Chemistry.

The first 24 problems are synthetic. In each one you are asked to convert one compound into another. In every case the conversion has actually been carried out, and the reference is given.

1\. OH → CH_2COOH CH_2COOH

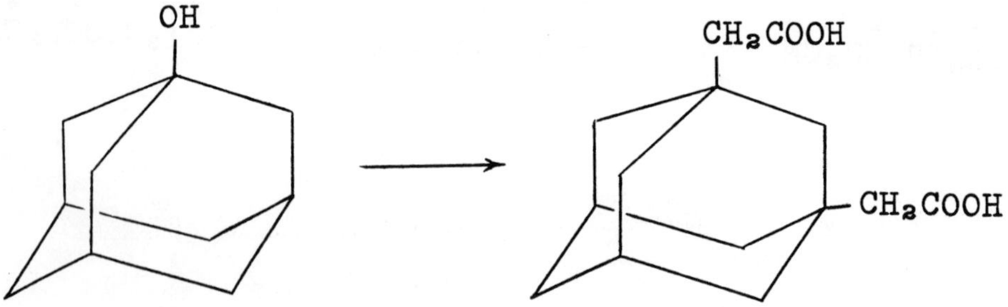

Bott, *Chem. Ber.* **101**, 564 (1968).

2\. F F Br F F F → F F F F

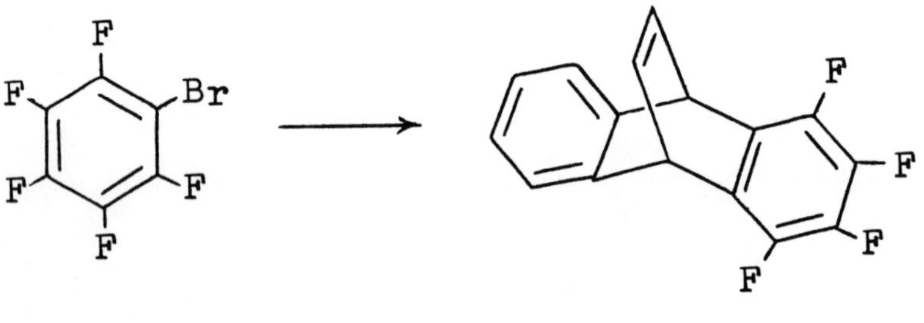

Brewer, Eckhard, Heaney, and Marples, *J. Chem. Soc., C* **1968**, 664.

3\. COOH → O

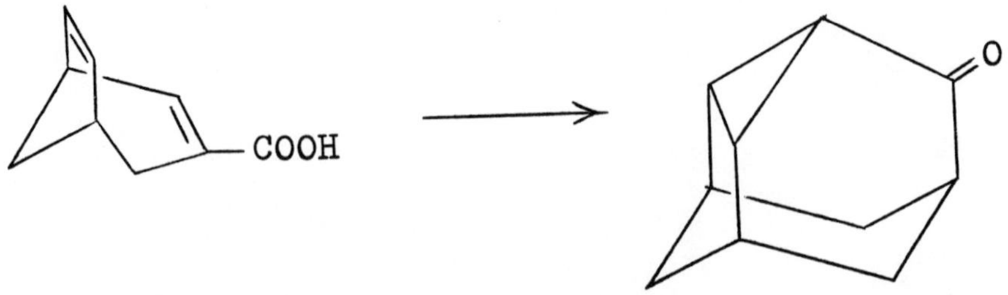

Baldwin and Foglesong, *J. Amer. Chem. Soc.* **90**, 4303 (1968).

4\. CH_3 i COOH → CH_3 ii COOH

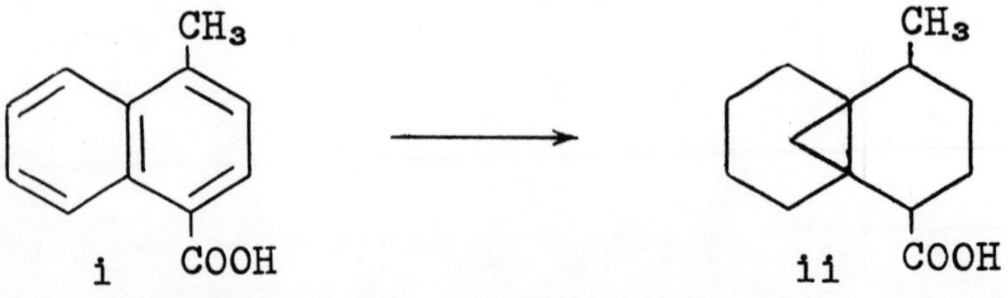

5.

HO Ph Ph

Zimmerman, Crumrine, Döpp, and Huyffer, *J. Amer. Chem. Soc.* **91**, 434 (1969).

6.

MeO COOH → Br O O Me C O

Bachi, Epstein, Herzberg-Minzly, and Loewenthal, *J. Org. Chem.* **34**, 126 (1969).

7.

$MeO(CH_2)_4-C(=CH(CH_2)_3OMe)-(CH_2)_4-OMe$ → CH_2-CH_2-CH … CH_2 (cyclophane)

i ii

8.

$CH_2=C(CH_3)CH_2CH_2C(CH_3)=CH_2$ → CH_3 CH_3

Skatterbøl, *J. Org. Chem.* **31**, 2789 (1966).

9. $ClCH_2CH=CHCH_2OH$ ⟶ $MeOOCCH=CHCH_2CH=CHCH_2CH=CHCOOMe$

i ii

10.

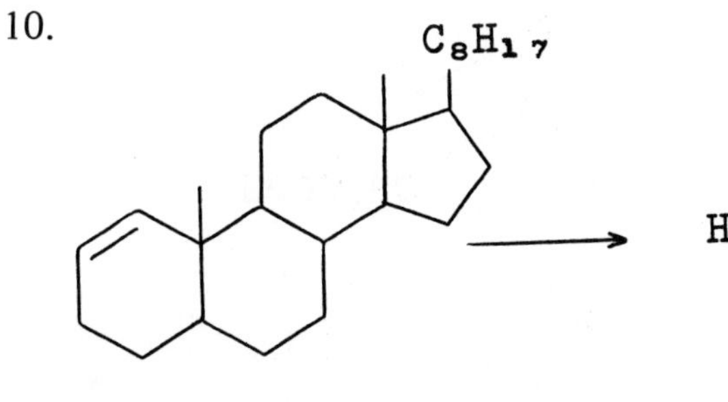

→ HO C_8H_{17}

Sondheimer and Nussim, *J. Org. Chem.* **26**, 630 (1961).

11.

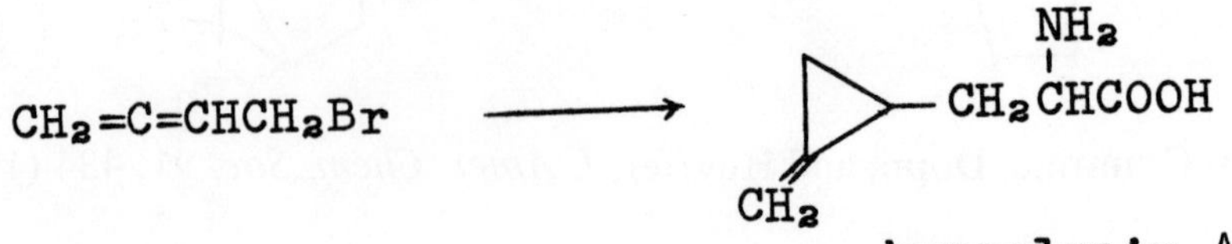

$CH_2{=}C{=}CHCH_2Br$ ⟶ [cyclopropane with $=CH_2$ and $CH_2CH(NH_2)COOH$]

hypoglycin A

Black and Landor, *J. Chem. Soc., C* **1968**, 283, 288.

12.

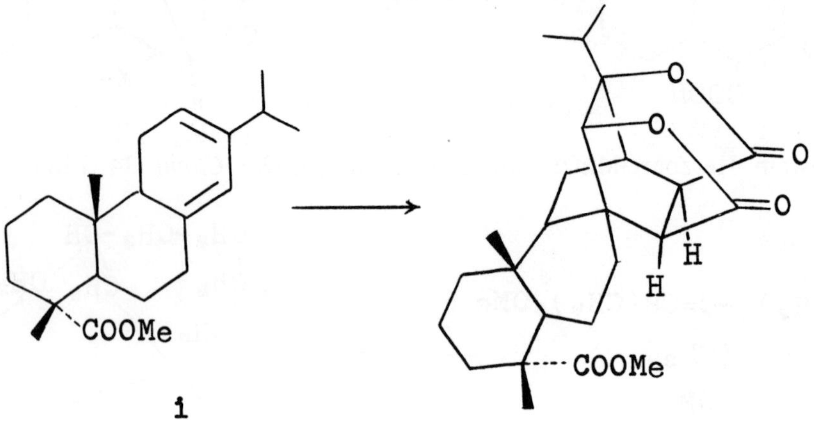

i ⟶ ii

COOMe

H

H

O

O

O

O

COOMe

13.

$^{14}CHCl_3$ ⟶ [naphthalene with ^{14}C–Cl]

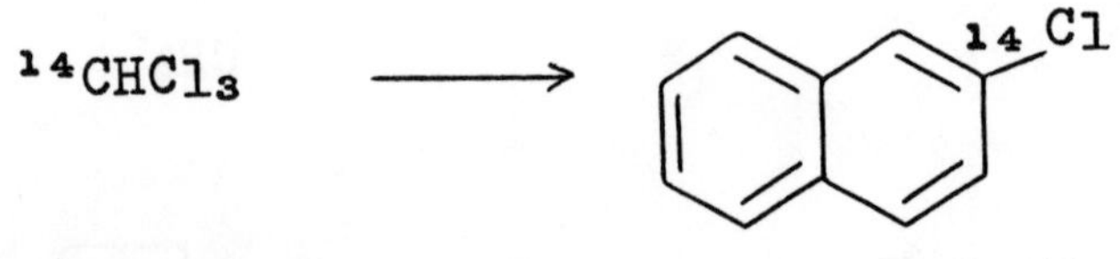

van der Jagt, den Hollander, and van Zanten, *Rec. Trav. Chim.* **87**, 1148 (1968).

14.

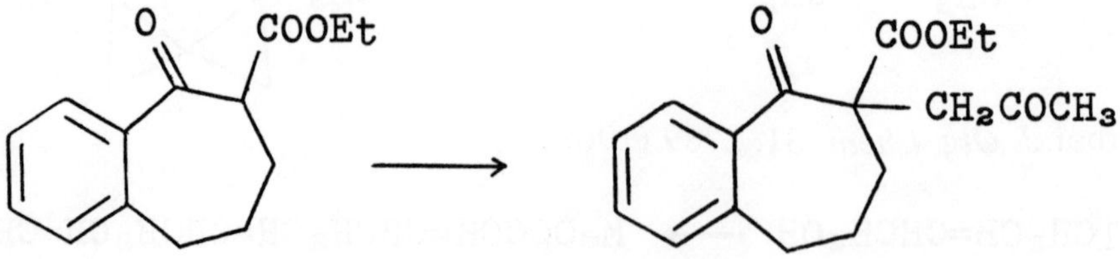

O COOEt ⟶ O COOEt CH_2COCH_3

Kloster-Jensen, Kováts, Eschenmoser, and Heilbronner, *Helv. Chim. Acta* **39**, 1051 (1956).

15.

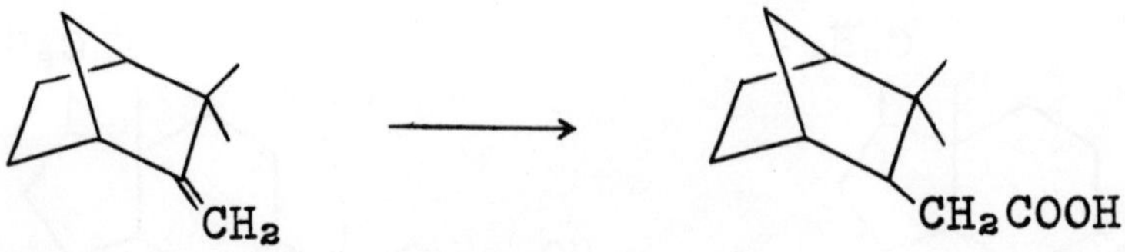

CH_2 ⟶ CH_2COOH

Suga and Watanabe, *Aust. J. Chem.* **20**, 2033 (1967).

16.

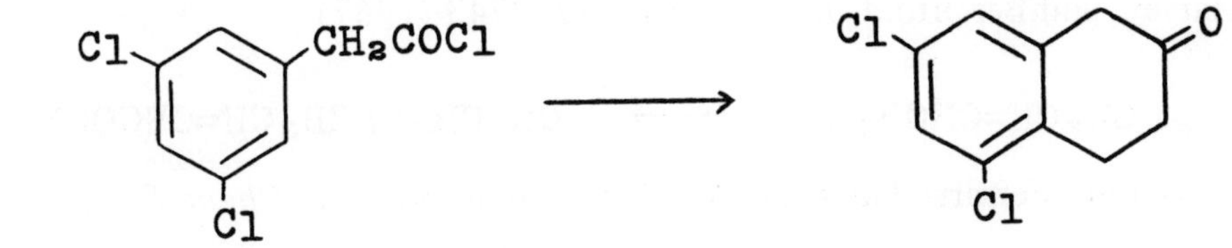

Cl CH_2COCl Cl → Cl O Cl

Rosowsky, Battaglia, Chen, and Modest, *J. Org. Chem.* **33**, 4288 (1968).

17.

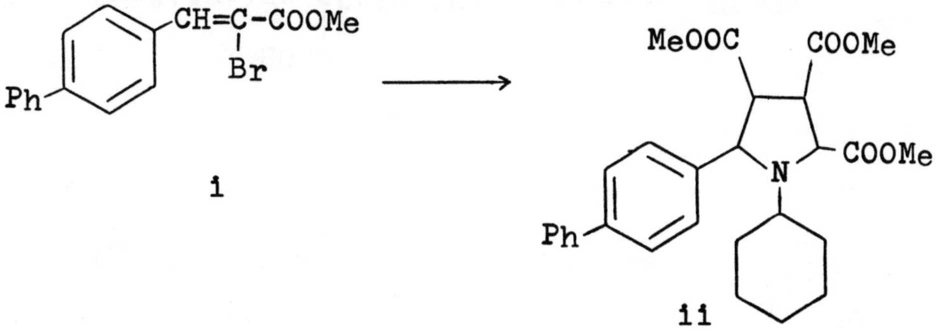

CH=C(Br)-COOMe Ph i → MeOOC COOMe COOMe N Ph ii

18.

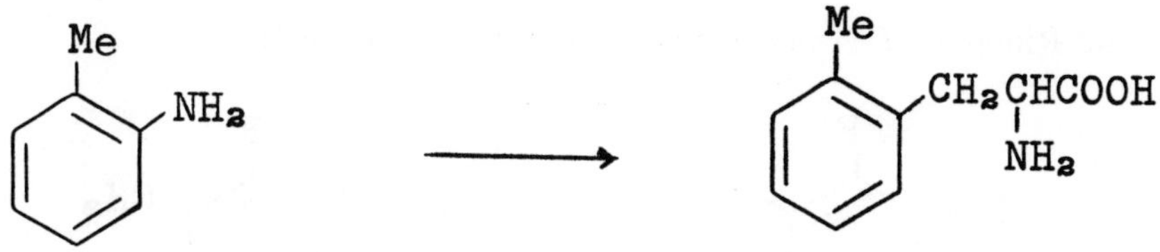

Me NH_2 → Me $CH_2CHCOOH$ NH_2

Cleland, *J. Org. Chem.* **34**, 744 (1969).

19.

Ph-C(Me)=CH_2 → Ph-CH(Me)CH_2CH_2CH(Me)-Ph

Richards and Scilly, *J. Chem. Soc., C* **1969**, 55.

20.

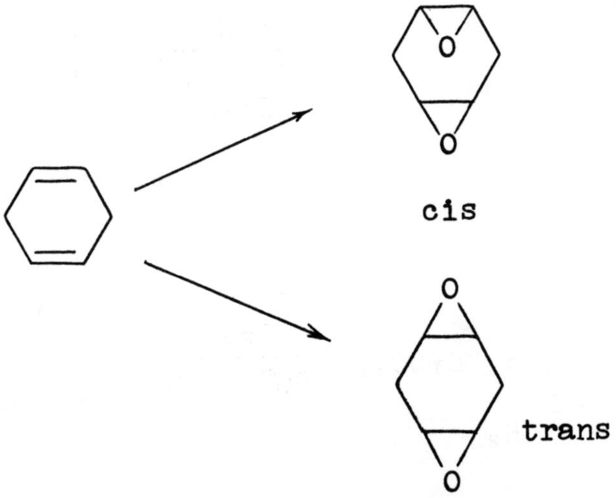

O O cis

O O trans

Craig, Harvey, and Berchtold, *J. Org. Chem.* **32**, 3743 (1967).

21.

$$CH_3CH{=}CHCH_2Cl \longrightarrow CH_3CH{=}CHCH_2CH{=}CHCOOMe$$

Chiusoli, Dubini, Ferraris, Guerrieri, Merzoni, and Mondelli, *J. Chem. Soc., C* **1968**, 2889.

22.

$$HC{\equiv}CH \longrightarrow CH_3(CH_2)_3\underset{\displaystyle COOH}{\underset{|}{C}}{=}CH(CH_2)_3CH_3$$

i

23.

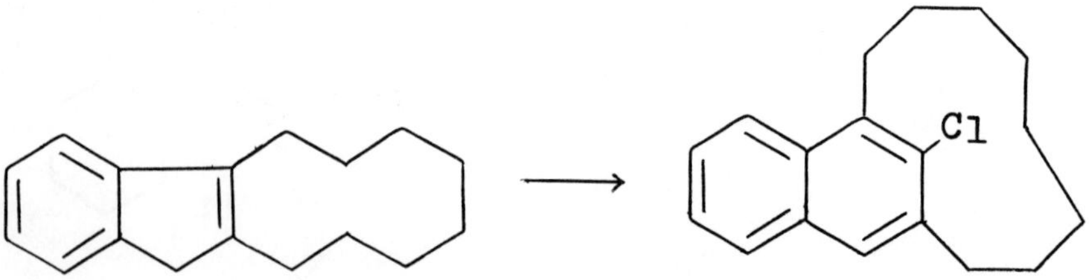

Parham and Rinehart, *J. Amer. Chem. Soc.* **89**, 5668 (1967).

24.

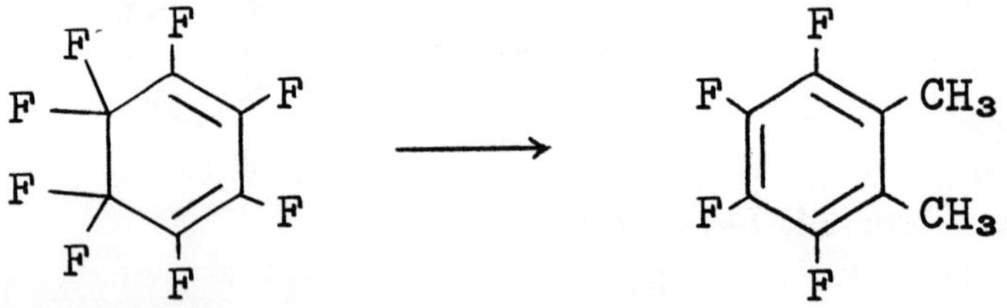

Anderson, Feast, and Musgrave, *J. Chem. Soc., C* **1969**, 211.

The following conversions (problems 25 to 58) have been reported in the literature. In each case suggest a reasonable mechanism.

25.

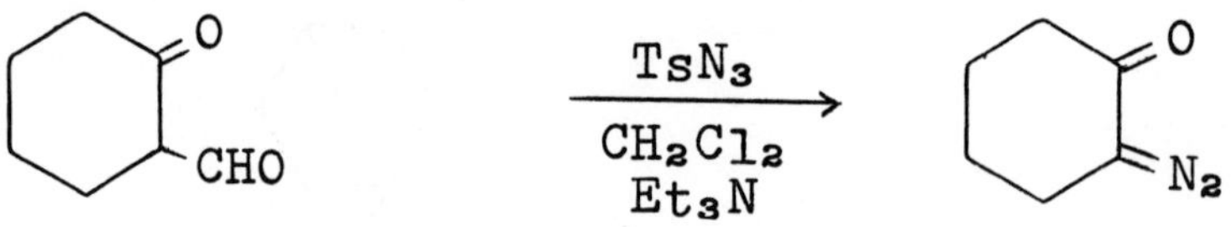

i $\xrightarrow[CH_2Cl_2,\ Et_3N]{TsN_3}$ ii

26.

$+ \ BrCCl_3 \xrightarrow[80^\circ\ 24\ hr]{Bz_2O_2}$

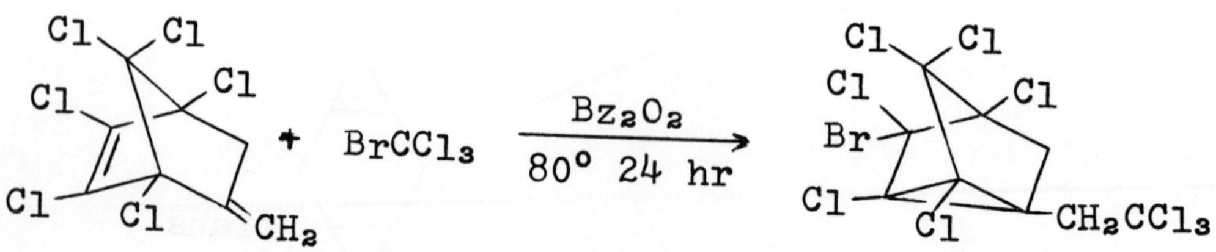

Alden and Davies, *J. Chem. Soc., C* **1967**, 1017.

27.

COO^- N_2^+ + SO_2 $\xrightarrow{\Delta}$ 9%

i ii iii

28.

O_2N S N Br + N-H $\xrightarrow[\text{overnight stirring}]{Me_2SO}$

44% NO_2 N-CH=C-SCN + 11% O_2N S N N

Ilvespää, *Helv. Chim. Acta* **51**, 1723 (1968).

29.

$(MeO)_2C{=}C(OMe)_2$ + $NCCH_2CN$ $\xrightarrow{\text{no base}}$ $(MeO)_2CH{-}C(OMe){=}C(CN){-}CN$

Hoffmann, *Angew. Chem. Intern. Ed.* **7**, 754 (1968), p. 759 [*Angew. Chem.* **80**, 823 (1968)].

30.

O OEt $\xrightarrow[2.\ H_2O]{1.\ EtMgBr}$ 42% Et-CH(OH) OEt

Quelet and d'Angelo, *Bull. Soc. Chim. Fr.* **1967**, 3390; *C. R. Acad. Sci. Paris, Ser. C* **264**, 216 (1967).

31.

AgOAc-I_2 (1:4.4)

HOAc

OAc

$CONH_2$

OH

3-chloroperbenzoic acid

CH_2Cl_2 14 days

Dobson, Davis, Hartung, and Manson, *Can. J. Chem.* **46**, 2843 (1968).

32.

EtOOC, COOEt, C=C, H, H

+

200°

40 hr

EtOOC COOEt

59%

cis and trans

Notes: The same mixture of isomers (mostly trans) was obtained from either diethyl maleate (shown) or diethyl fumarate. The two olefins do not interconvert under these conditions.

Reich and Cram, *J. Amer. Chem. Soc.* **91**, 3517 (1969).

33.

Br$^-$ + $LiAlH_4$ $\xrightarrow[\text{3 hr}]{\text{THF}}$ $CH{=}CHCH{=}CH_2$

Miyadera and Kishida, *Tetrahedron* **25**, 397 (1969).

34.

COOMe $\xrightarrow[h\nu\text{-sensitizer}]{O_2}$ COOMe

i 40–50% ii

+ COOMe + COOMe OOH

20–30% iii 5–10% iv

35.

$PhC{\equiv}C{-}C({=}O){-}Ph$ + $EtOOC{-}CH(Me){-}COOEt$ $\xrightarrow[Et_2O]{\text{1 equiv. NaOEt}}$

$EtOOC{-}C(Me)(COOEt){-}C(Ph){=}CH{-}C({=}O){-}Ph$ + (Ph, EtOOC, Ph, Me pyranone) + (Ph, Ph, Me pyranone)

Buggle, Hughes, and Philbin, *Chem. Ind.* (London) **1969**, 77.

36.

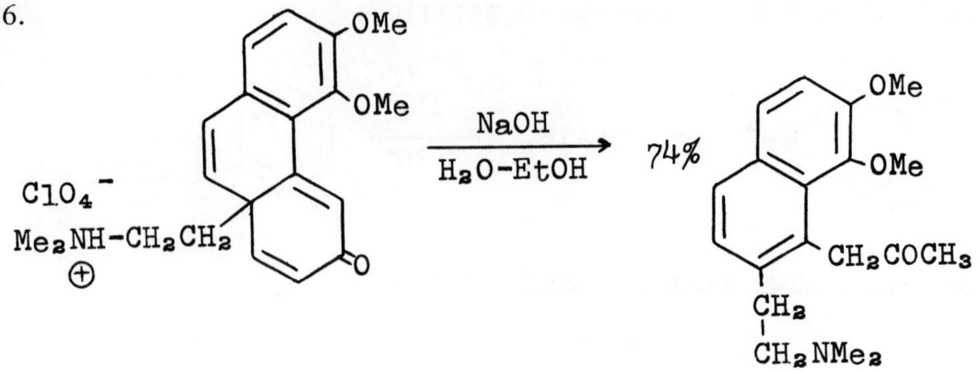
OMe
OMe
ClO4⁻
Me2NH-CH2CH2
⊕
O
NaOH
H2O-EtOH
74%
OMe
OMe
CH2COCH3
CH2
CH2NMe2

Fleischhacker and Vievöck, *Monatsh. Chem.* **100**, 163 (1968).

37.

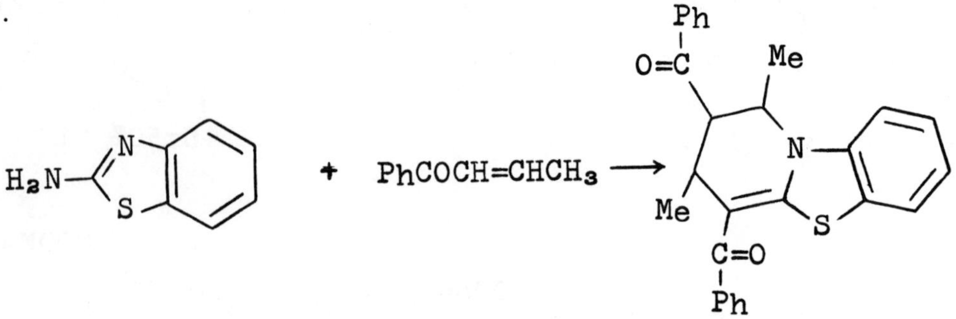
H2N
N
S
+
PhCOCH=CHCH3
Ph
O=C
Me
N
Me
S
C=O
Ph

Findlay, Langler, Podesva, and Vagi, *Can. J. Chem.* **46**, 3659 (1968).

38.

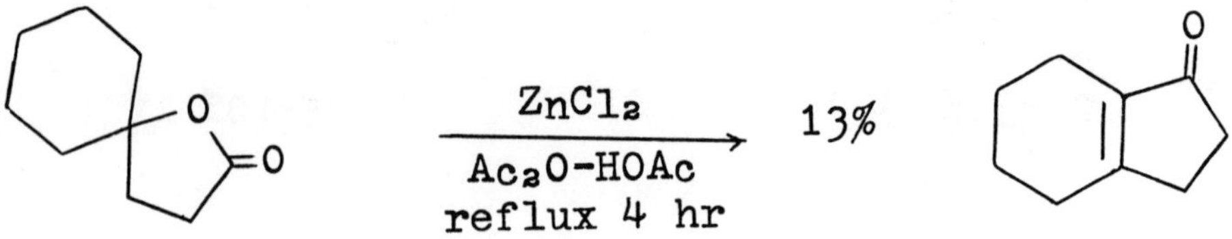
O
O
ZnCl2
Ac2O-HOAc
reflux 4 hr
13%
O

Mathieson, *J. Chem. Soc.* **1951**, 177.

39.

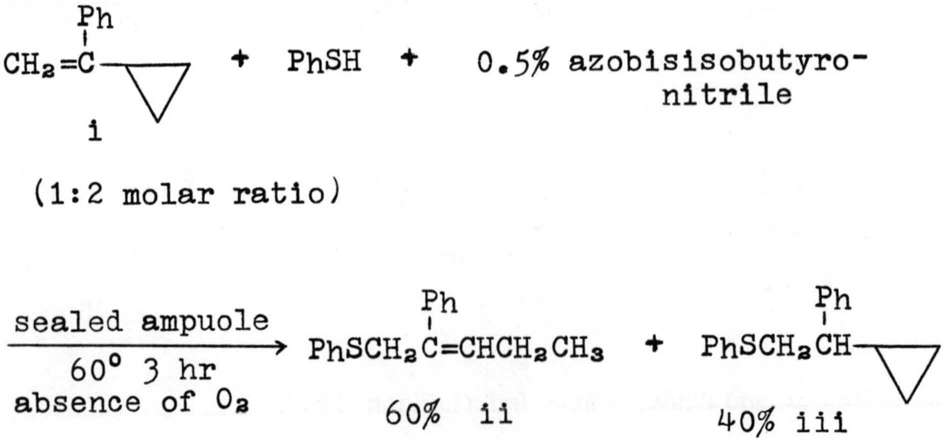
Ph
CH2=C
i
+
PhSH
+
0.5% azobisisobutyro-
nitrile
(1:2 molar ratio)
sealed ampuole
60° 3 hr
absence of O2
Ph
PhSCH2C=CHCH2CH3
60% ii
+
Ph
PhSCH2CH
40% iii

40.

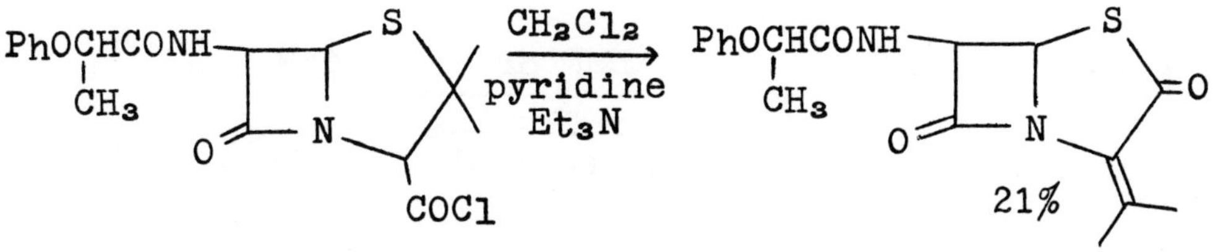

PhOCHCONH (CH3) ... S ... N ... O ... COCl → (CH_2Cl_2, pyridine, Et_3N) PhOCHCONH (CH3) ... S ... =O ... N ... O ... 21%

α-phenoxyethylpenicillin

Wolfe, Godfrey, Holdrege, and Perron, *Can. J. Chem.* **46**, 2549 (1968).

41.

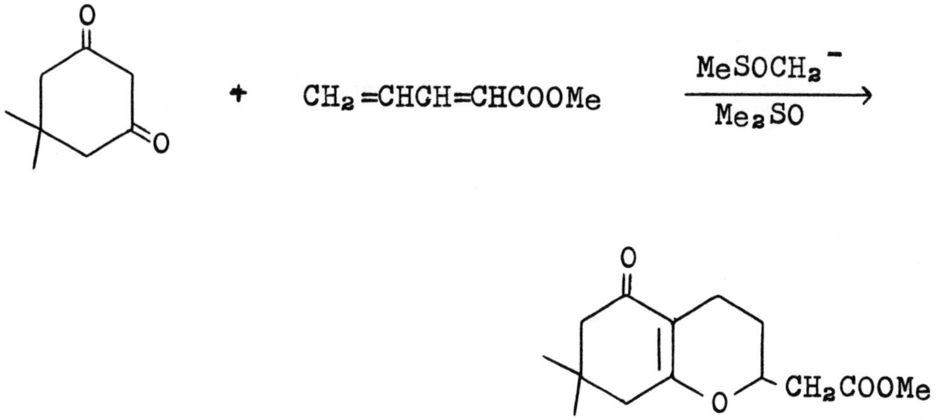

O ... O + CH_2=CHCH=CHCOOMe → ($MeSOCH_2^-$ / Me_2SO)

O ... O ... CH_2COOMe

Danishefsky, Koppel, and Levine, *Tetrahedron Lett.* **1968**, 2257.

42.

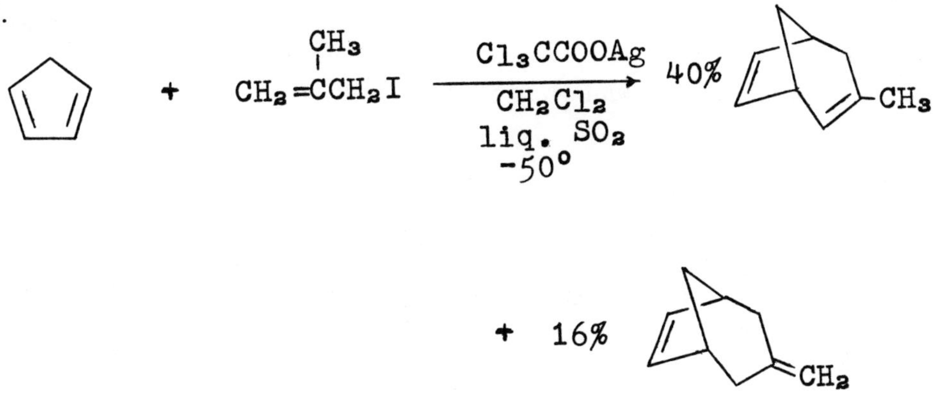

+ CH_2=C(CH_3)CH_2I → ($Cl_3CCOOAg$ / CH_2Cl_2, liq. SO_2, -50°) 40% ... CH_3

+ 16% ... CH_2

Hoffmann, Joy, and Suter, *J. Chem. Soc., B* **1968**, 57.

43.

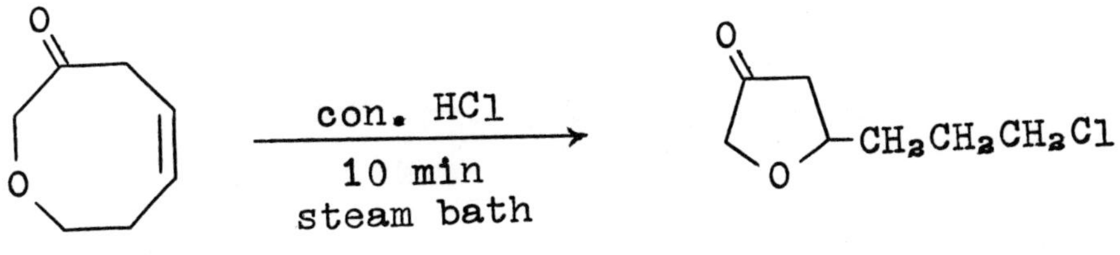

O ... O → (con. HCl / 10 min, steam bath) O ... O ... $CH_2CH_2CH_2Cl$

i ii

44.

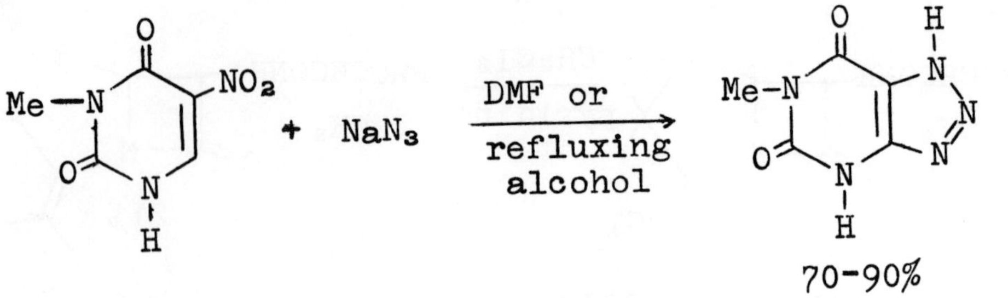

$$\text{5-nitro-3-methyluracil} + NaN_3 \xrightarrow[\text{refluxing alcohol}]{\text{DMF or}} \text{triazolopyrimidinedione (Me-N)} \quad 70\text{-}90\%$$

Blank and Fox, *J. Amer. Chem. Soc.* **90**, 7175 (1968).

45.

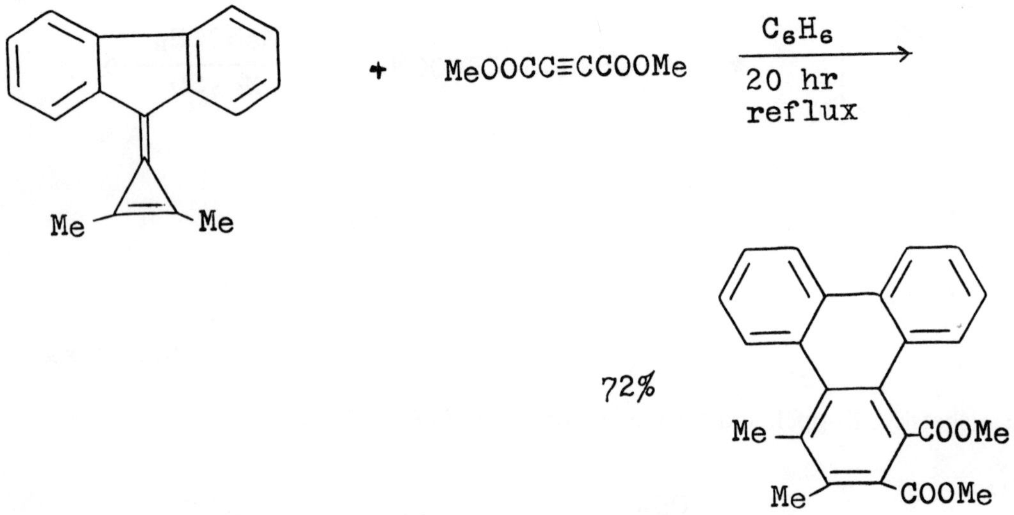

$$\text{9-(2,3-dimethylcyclopropenylidene)fluorene} + MeOOCC{\equiv}CCOOMe \xrightarrow[\text{20 hr reflux}]{C_6H_6} 72\% \ \text{dimethyl 3,4-dimethyltriphenylene-1,2-dicarboxylate (Me, Me, COOMe, COOMe)}$$

Prinzbach and Fischer, *Helv. Chim. Acta* **50**, 1692 (1967).

46.

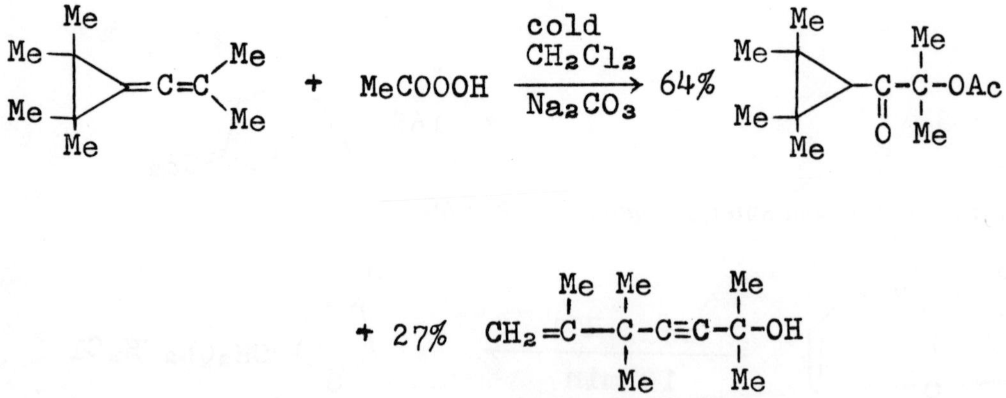

$$\text{tetramethylcyclopropylidene}{=}C{=}C(Me)_2 + MeCOOOH \xrightarrow[Na_2CO_3]{\text{cold } CH_2Cl_2} 64\% \ \text{(tetramethylcyclopropyl)}{-}\underset{\underset{O}{\|}}{C}{-}C(Me)_2{-}OAc$$

$$+ 27\% \quad CH_2{=}C(Me){-}C(Me)_2{-}C{\equiv}C{-}C(Me)_2{-}OH$$

Crandall, Paulson, and Bunnell, *Tetrahedron Lett.* **1968**, 5063.

47. i + CH_2N_2 $\xrightarrow[\text{3 days}]{Et_2O}$ ~44% $CH_2{=}C(Ph){-}CH(Me){-}N_3$ (ii)

+ ~28% iii + ~28% iv

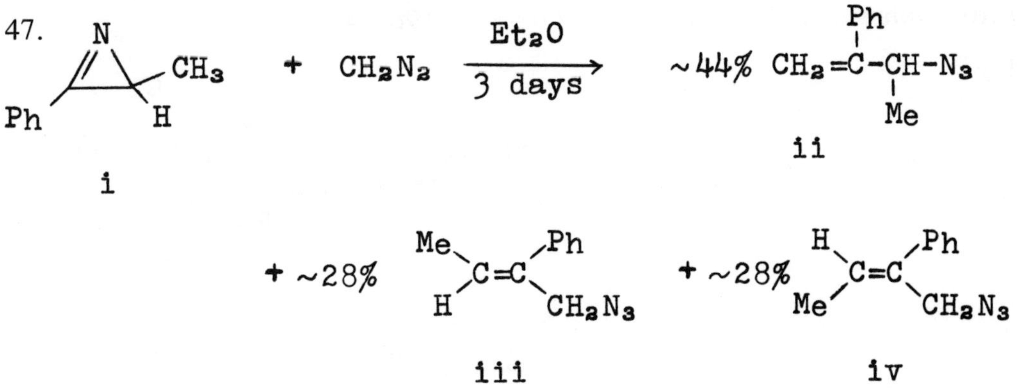

48. $\xrightarrow[160°]{\text{1,2,4-trichlorobenzene}}$ 66%

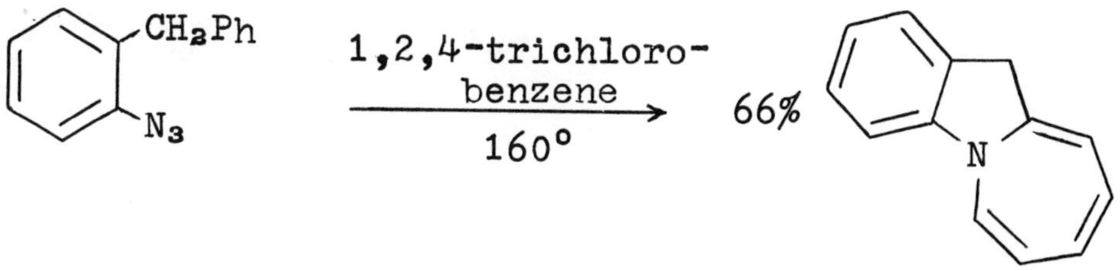

Krbechek and Takimoto, *J. Org. Chem.* **33**, 4286 (1968).

49. + Br_2 $\xrightarrow[5°\ 2\ hr]{CHCl_3}$ 77%

+ 6%

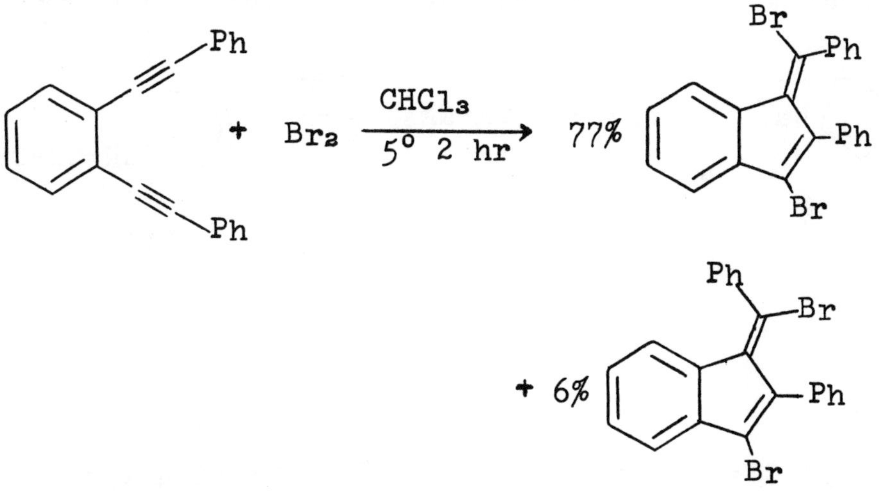

Whitlock, Sandvick, Overman, and Reichardt, *J. Org. Chem.* **34**, 879 (1969).

50. + $Cl{-}C(Me)_2{-}C{\equiv}CH$ $\xrightarrow[\text{anhyd. } K_2CO_3]{\text{acetone}}$

30%

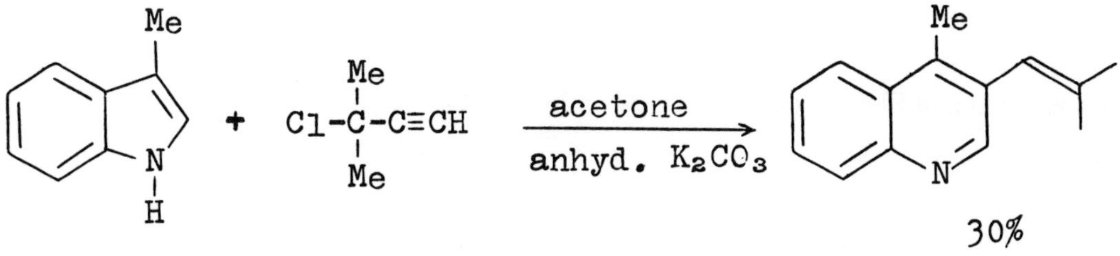

Bycroft, Johnson, and Landon, *Chem. Commun.* **1969**, 463.

51.

PhC≡CEt + $Cl_3CCOONa$ —(glyme, 85° 15 hr)→ 8% iii (2,2-dichloro-3-ethyl-4-phenylcyclobut-3-enone; ring substituents O, Cl, Cl, Ph, Et)

i ii iii

+ 7% iv (4,4-dichloro-2-ethyl-3-phenylcyclobut-2-enone; ring substituents Cl, Cl, O, Ph, Et) + v (phenylethylcyclopropenone; O, Ph, Et)

iv v

small amount

Note: i, treated with $Cl_2CHCOCl$ and Et_3N gave no iii and only about 1% iv.

52. (cyclopentadienide $^{-}$) Na^+ + Ph-C(=O)-CH=CH-NMe_2 —(THF, 2 hr)→ (cyclopentadienide $^{-}$)-CH=CH-C(=O)-Ph

i ii iii

53.

$CH_3(CH_2)_9CH=CH_2$ + (1,3,5-trioxane) —(Me_2CO, $h\nu$)→ $CH_3(CH_2)_{11}$-(1,3,5-trioxan-2-yl)

24%

Rosenthal and Elad, *J. Org. Chem.* **33**, 805 (1968).

54.

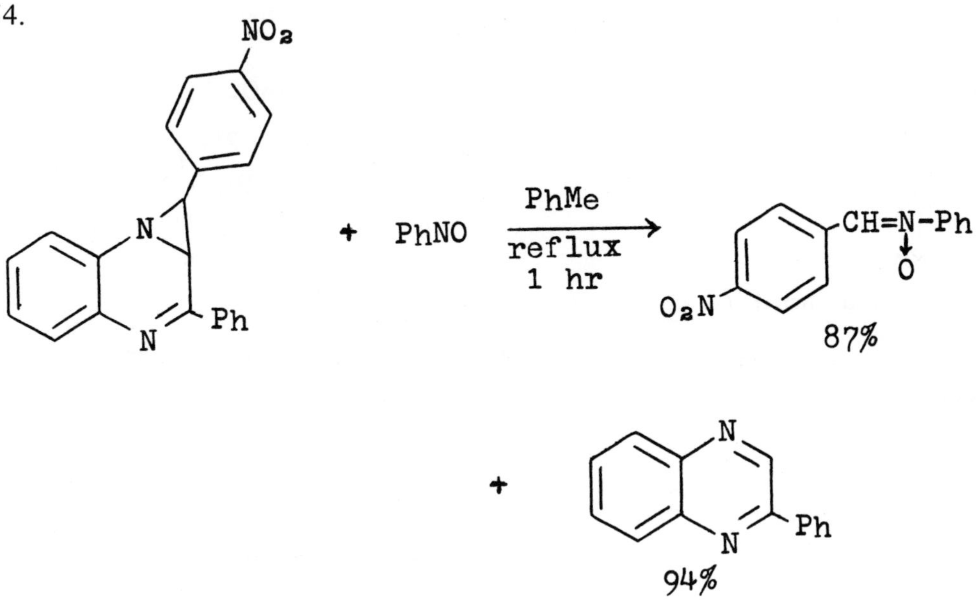

NO_2

+ PhNO $\xrightarrow[\text{reflux, 1 hr}]{\text{PhMe}}$ O_2N–C_6H_4–CH=N(→O)–Ph 87%

+ 2-phenylquinoxaline (Ph) 94%

Heine and Henzel, *J. Org. Chem.* **34**, 171 (1969).

55.

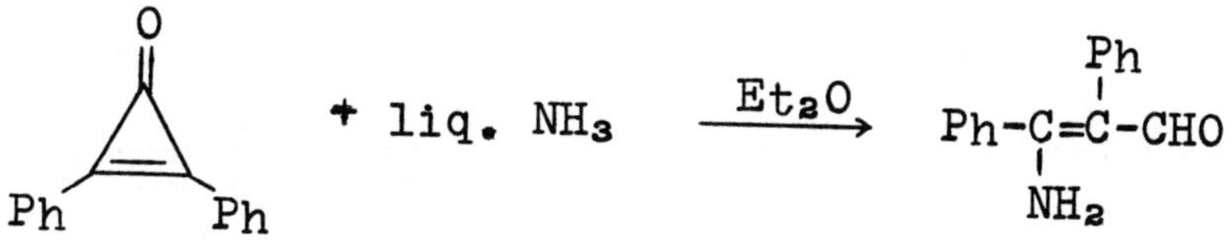

Diphenylcyclopropenone (Ph, Ph) + liq. NH_3 $\xrightarrow{Et_2O}$ Ph–C(NH_2)=C(Ph)–CHO

Toda, Mitote, and Akagi, *Chem. Commun.* **1969**, 228.

56.

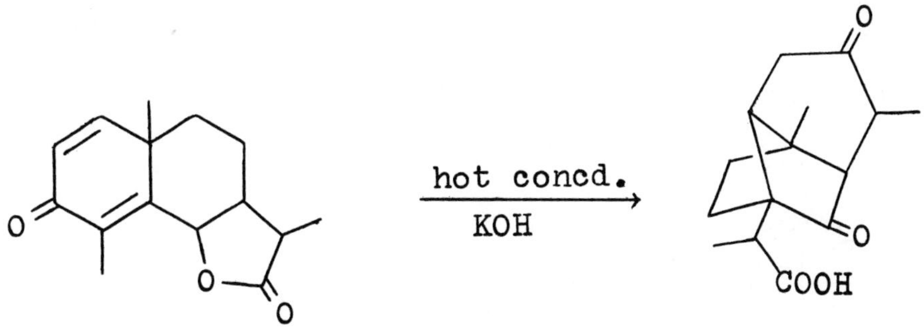

i
santonin

$\xrightarrow[\text{KOH}]{\text{hot concd.}}$

ii
santonic
acid

(COOH)

57.

$(Me_2N)_2C{=}C(NMe_2)_2$ $\xrightarrow[H_2O]{HCl}$ 47% Me_2N–C(=O)–CHO

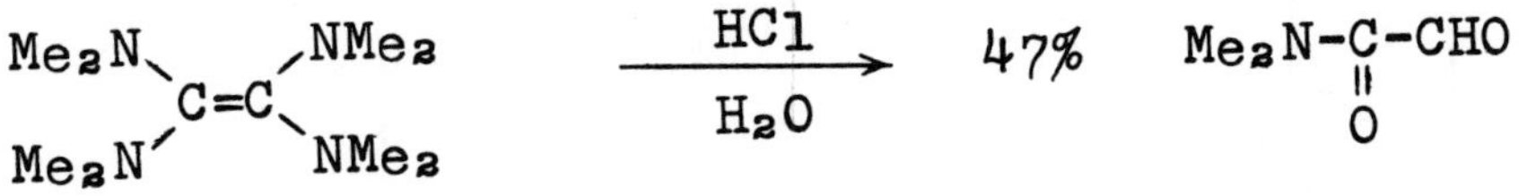

Wiberg and Buchler, *Z. Naturforsch; B* **19**, 9 (1964).

58.

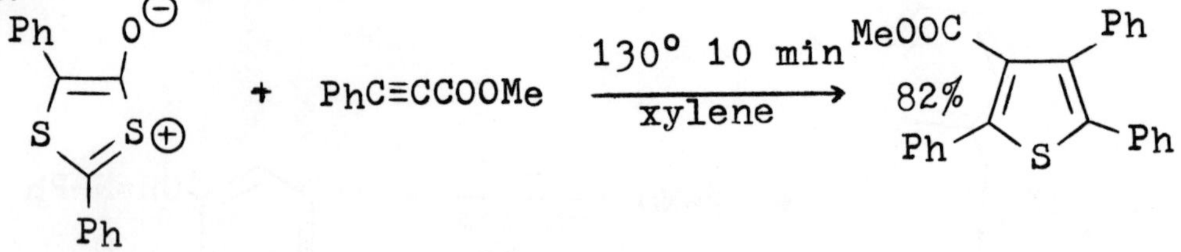

Ph O⊖ ; S S⊕ ; Ph + PhC≡CCOOMe $\xrightarrow[\text{xylene}]{130^\circ\ 10\ \text{min}}$ 82% MeOOC, Ph, Ph, S, Ph

Gotthardt and Christl, *Tetrahedron Lett.* **1968**, 4747, 4751.

In the following problems (59 to 69) give the products and suggest reasonable mechanisms.

59.

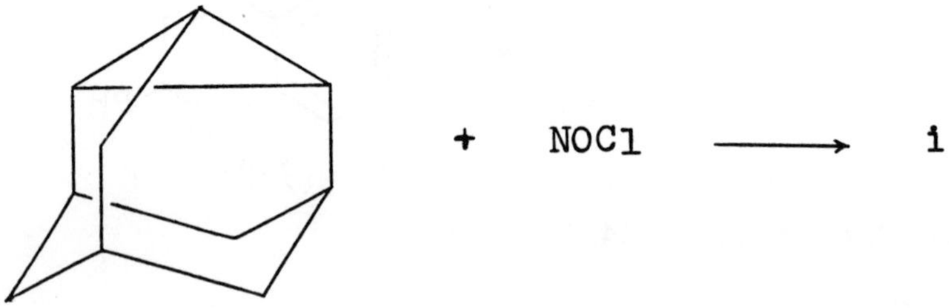

+ NOCl ⟶ i

60.

$CF_2ClCF_2CCl_3$ + $Me_2C{=}CH_2$ $\xrightarrow[100^\circ\ \ 4\ \text{hr}]{Bz_2O_2}$

25% i ($C_7H_7Cl_3F_4$) + 13% ii ($C_7H_8Cl_4F_4$)

Tarrant and Tandon, *J. Org. Chem.* **34**, 864 (1969).

61.

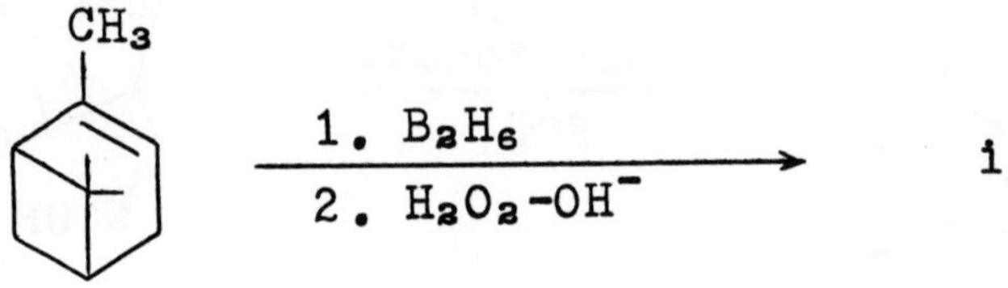

CH_3 $\xrightarrow[2.\ H_2O_2\text{-}OH^-]{1.\ B_2H_6}$ i

Give the configuration of the product.

62.

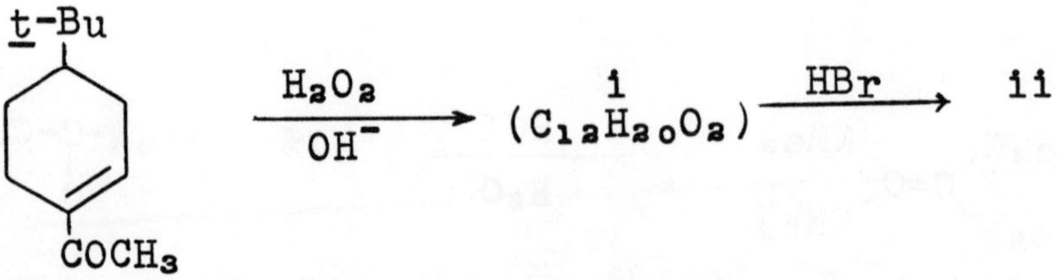

t-Bu ; $COCH_3$ $\xrightarrow[OH^-]{H_2O_2}$ i ($C_{12}H_{20}O_2$) $\xrightarrow{HBr}$ ii

Give the configurations of i and ii.

63.

(ferrocene: one Cp ring bearing $COCH_3$, the other bearing $\underset{\underset{O}{\|}}{C}-CH{=}CH-Ph$)

$$\xrightarrow[\text{EtOH, 12 hr}]{\text{NaOH}}$$

Barr and Watts, *Tetrahedron* **24**, 3219 (1968).

64.

$$NCCH_2COOEt \xrightarrow[2.\ CH_2{=}C(Cl){-}CN]{1.\ NaH} 55\%\ \ \underset{(C_8H_8O_2N_2)}{i}$$

Madsen, Preston, and Lawesson, *Ark. Kemi* **28**, 395 (1968).

65.

AcO

β-ergosteryl acetate

$$+\ \ NC-\underset{NC}{C}{=}\underset{CN}{C}-CN$$

$$\xrightarrow{C_6H_6} \underset{(C_{36}H_{46}N_4O_2)}{i}$$

Note: i has three nonconjugated double bonds.

Lautzenheiser and LeQuesne, *Tetrahedron Lett.* **1969**, 207.

66.

Ph(H)C=C(H)Ph + Ph(Me)(H)C–C(=O)–OOH $\xrightarrow[0^\circ]{CHCl_3}$

trans-stilbene S-(+)-2-phenylperoxypropionic acid

Give the configuration of the product.

Montanari, Moretti, and Torre, *Chem. Commun.* **1969**, 135.

67.

$$Me_2CH-NH-Cl\ \ +\ \ CH_2{=}CHCl \xrightarrow[h\nu\ 25^\circ]{H_2SO_4-HOAc}$$

Neale and Marcus, *J. Org. Chem.* **33**, 3457 (1968).

68.

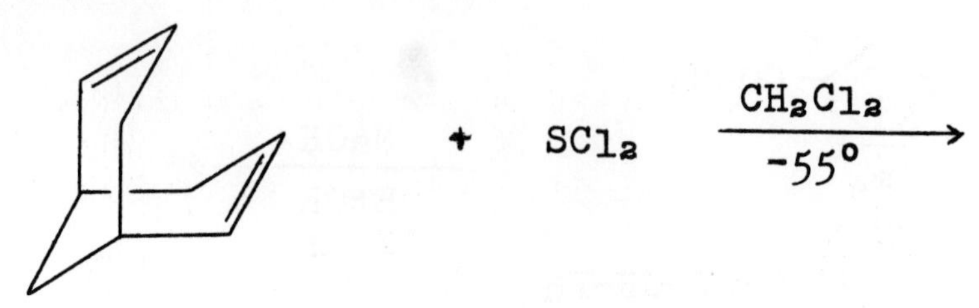

Stetter and Schwartz, *Chem. Ber.* **101**, 2464 (1968).

69. a. $CH_2=CH-CH=CH-SiMe_3$ + EtS^- (molar ratio 1:1) $\xrightarrow[\text{100° 67 hr}]{\text{sealed tube}}$

b. $CH_2=CH-CH=CH-SiMe_3$ + EtSH (molar ratio 1:3) $\xrightarrow[(t\text{-Bu})_2O_2\ \ 100°\ 4\ \text{hr}]{\text{sealed tube}}$

---

70. $CH_2=C(CH_3)-COOMe$ + $CH_3CH(Cl)COOR$ $\xrightarrow[C_6H_6\ 2\ \text{hr}]{\text{NaH}}$ cyclopropane (Me, MeOOC on one carbon; Me, COOR on the other)

R = (−)-menthyl

a. Suggest a reasonable mechanism.

b. When the solvent was varied over a range from 100% benzene through 1:1 benzene-DMF to 100% DMF the cis-trans ratio of the product changed from 100:0 to 1:0.43, while the RR/SS ratio changed from 1.02 to 0.70. Explain.

Inouye, Inamasu, Horiike, Ohno, and Walborsky, *Tetrahedron* **24**, 2907 (1968).

71. β-Hydroxyalkynes can be reduced to the corresponding alkenes with $LiAlH_4$. For i and other β-hydroxy *terminal* alkynes, the rate of the reaction was increased by the addition of NaOMe, but decreased by the addition of *t*-BuOK. Explain.

(i: cyclohexane bearing C≡CH and OH on the same carbon)

Molloy and Hauser, *Chem. Commun.* **1968**, 1017.

72.

$Ar-CH=CH-CHN_2$ $\xrightarrow[\text{room temp.}]{\text{spontaneous}}$ ~ 100% (3-Ar-pyrazole, N–H)

72. Cont.

In this reaction the relative rates for four aryl groups were as follows: $m\text{-}O_2NC_6H_4$ 19.3; $p\text{-}ClC_6H_4$ 31.2; Ph, 36.4; and $p\text{-}MeC_6H_4$ 44.3. Suggest a mechanism which would account for the relatively small effect of the ring substitutents.

Brewbaker and Hart, *J. Amer. Chem. Soc.* **91**, 711 (1969).

73.

2 (i, N-COOEt azepine) —Δ→ ii (COOEt, N, N, COOEt)

Is this thermal reaction allowed by the Woodward-Hoffmann principle of conservation of orbital symmetry?

74.

i (O_2, S, N-H, N) + Cl_2 —dry CH_2Cl_2→ ii + N_2 —H_2O→ iii (O, S, O, OH)

What is the structure of ii, and explain the reactions.

75.

benzyne + (CH_3, H, C=C, H, H, C=C, H, CH_3) —25°→ 60% i (Me, Me) + 6% ii

Benzyne reacts with *trans-trans*-2,4-hexadiene (shown above) to give 60% of the normal Diels-Alder product (i) and only 6% of the benzyne dimerization product biphenylene (ii). In contrast, the reaction of benzyne with *cis-trans*-2,4-hexadiene gives 70% ii and not more than 8% of i. Explain.

Atkin and Rees, *Chem. Commun.* **1969**, 152.

76.

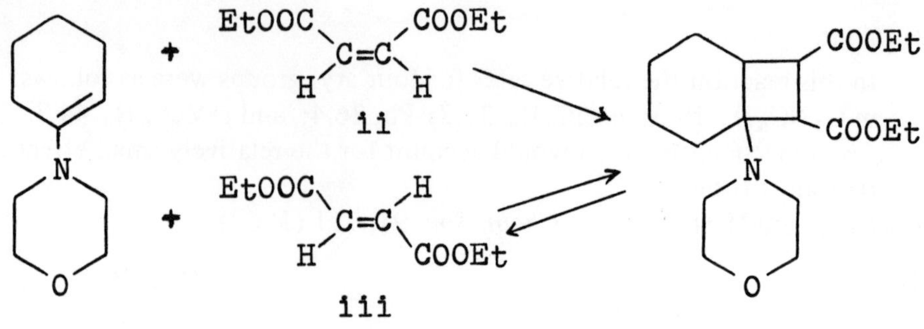

EtOOC COOEt C=C H H ii

N O i

EtOOC H C=C H COOEt iii

COOEt COOEt N O iv

1-Morpholinocyclohexene (i) gave the same isomer (iv) with either diethyl maleate (ii) or diethyl fumarate (iii). When i was treated with a large excess of diethyl maleate, the unchanged olefin was found to have been converted to diethyl fumarate. However, under the reaction conditions diethyl maleate alone, without i, did not isomerize to diethyl fumarate. iv when heated gave only diethyl fumarate (and i) but no diethyl maleate. Suggest a mechanism consistent with these results.

Risaliti, Valentin, and Forchiassin, *Chem. Commun.* **1969**, 233.

77.

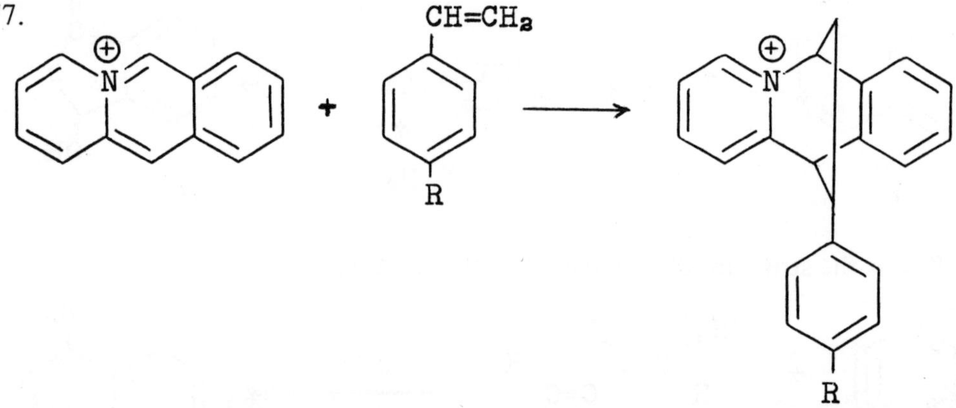

⊕ N + $CH=CH_2$ R → ⊕ N R

In the Diels-Alder reaction between *p*-substituted styrenes and acridizinium ion, rates were as follows ($k \times 10^{-3}$): R = NO_2 2.0; R = H 5.8; R = CH_3 9.8; and R = CH_3O 25. What conclusion can be drawn about the mechanism?

Bradsher and Stone, *J. Org. Chem.* **33**, 519 (1968).

78.

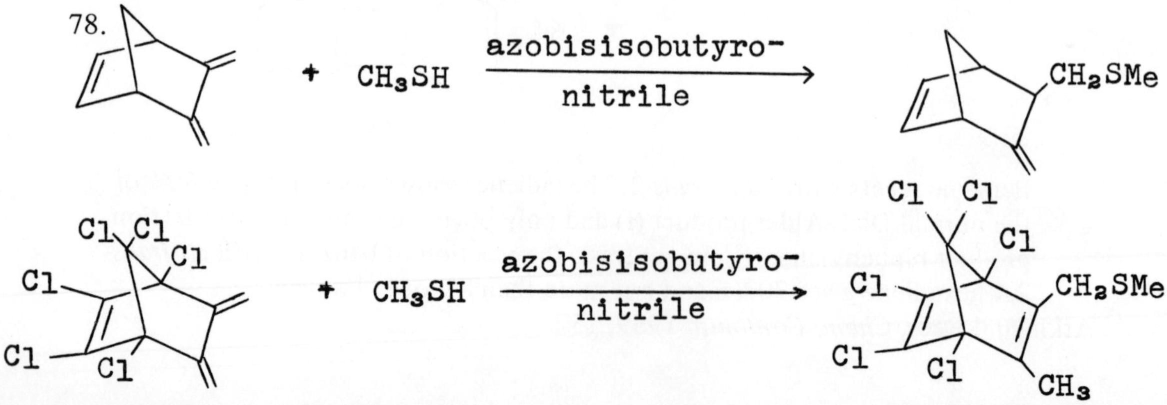

+ CH_3SH azobisisobutyronitrile → CH_2SMe

Cl Cl Cl Cl Cl Cl + CH_3SH azobisisobutyronitrile → Cl Cl Cl Cl Cl Cl CH_2SMe CH_3

78. Cont.

Explain the difference in products.

Alden, Claisse, and Davies, *J. Chem. Soc.* **1968**, 1228 (1968).

79.

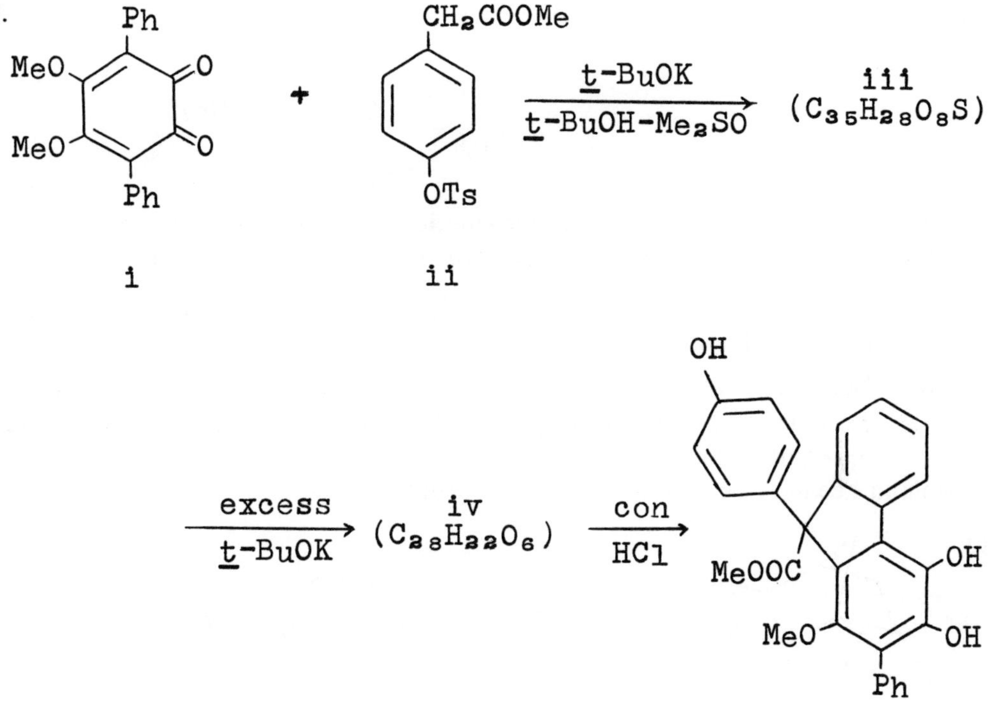

Ph, MeO, MeO, O, O, Ph (i) + CH_2COOMe, OTs (ii) $\xrightarrow[\underline{t}\text{-BuOH-}Me_2SO]{\underline{t}\text{-BuOK}}$ iii ($C_{35}H_{28}O_8S$)

$\xrightarrow[\underline{t}\text{-BuOK}]{\text{excess}}$ iv ($C_{28}H_{22}O_6$) $\xrightarrow[\text{HCl}]{\text{con}}$ OH, MeOOC, OH, OH, MeO, Ph (v)

Write structures for iii and iv and mechanisms for all steps.

80.

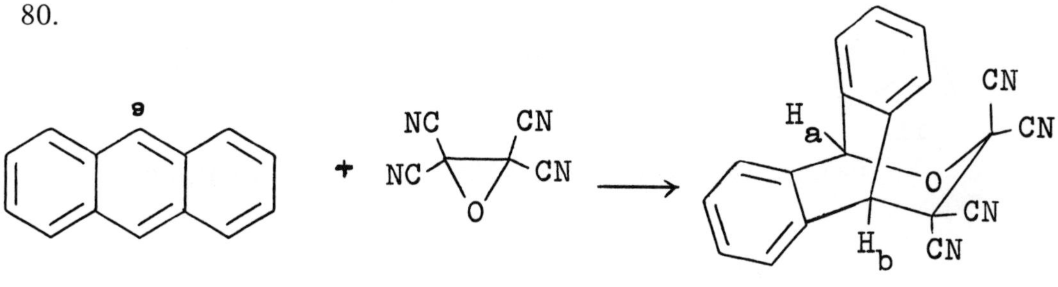

9 (i) + NC, CN, NC, CN, O $\longrightarrow$ H_a, CN, CN, O, CN, H_b, CN (ii)

In the above reaction, when 9-deuteroanthracene was used instead of i, there was formed 19% more ii with H_a = D, H_b = H (iii) than ii with H_a = H, H_b = D (iv); that is, iii was formed 19% faster than iv. Explain this result.

Brown and Cookson, *Tetrahedron* **24**, 2551 (1968).

81. The reaction between dimethylarsine and hexafluoro-2-butyne gives mostly trans adduct:

81. Cont.

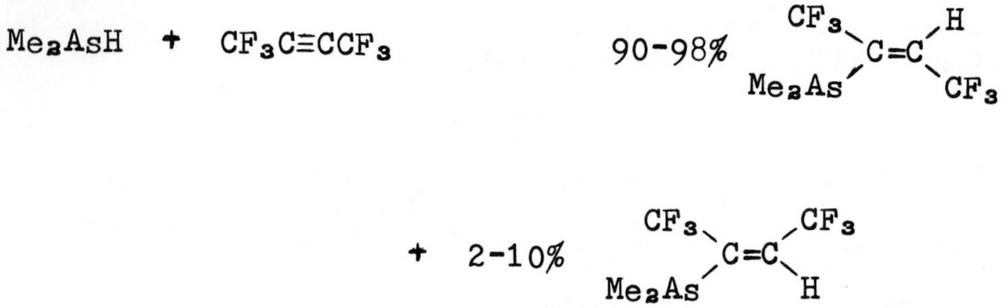

Me_2AsH + $CF_3C{\equiv}CCF_3$ → 90-98% $(CF_3)(Me_2As)C{=}C(H)(CF_3)$

+ 2-10% $(CF_3)(Me_2As)C{=}C(CF_3)(H)$

The reaction followed second order kinetics: first order in each reactant, and the rate was not affected by the addition of free radical inhibitors.

a. From your general knowledge of addition mechanisms, what is the most likely mechanism?

b. Devise an experiment to test whether the addition is intra or intermolecular, that is, whether the Me_2 As and H moieties are added from the same or different molecules of Me_2 AsH.

Cullen and Leeder, *Can. J. Chem.* **47**, 2137 (1969).

82.

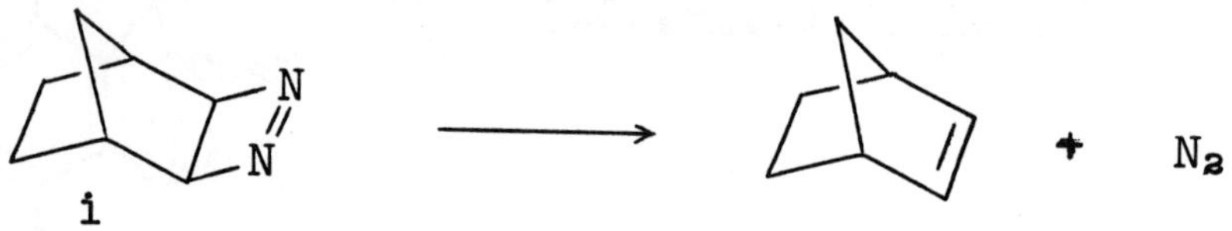

i → + N_2

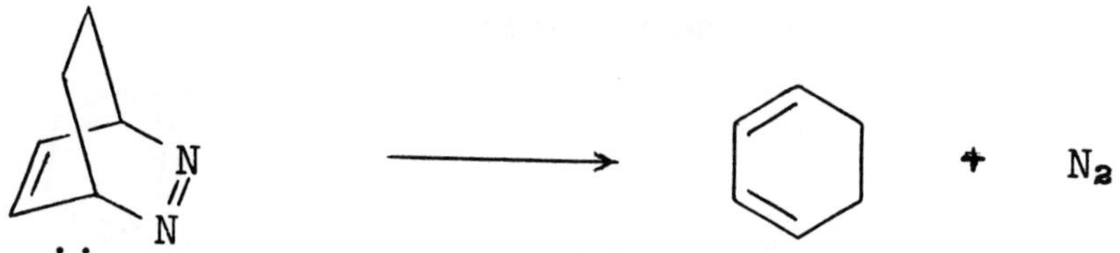

ii → + N_2

At −78°, thermal decomposition of ii is 10^{22} times faster than that of i. Explain.

83.

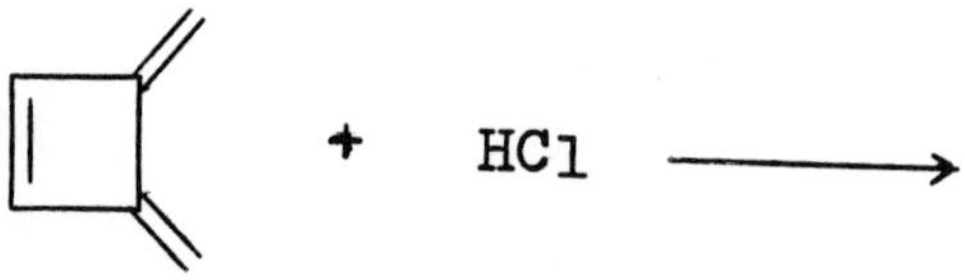

+ HCl →

In this reaction three products are formed, each of which has added two moles of HCl. Write the structures.

84.

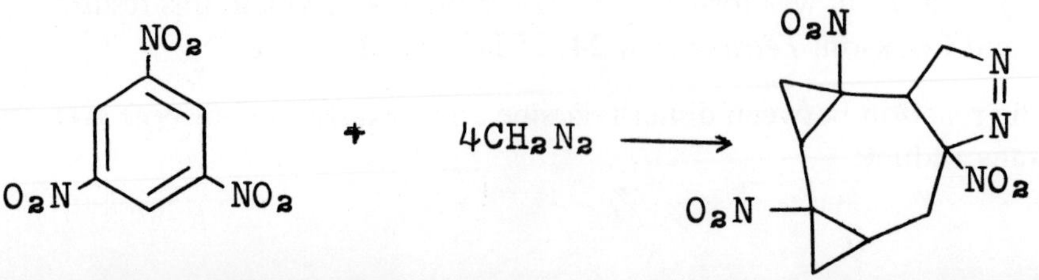

+ $4CH_2N_2$ →

84. Cont.

a. Suggest a reasonable mechanism.

b. Why is the product not i?

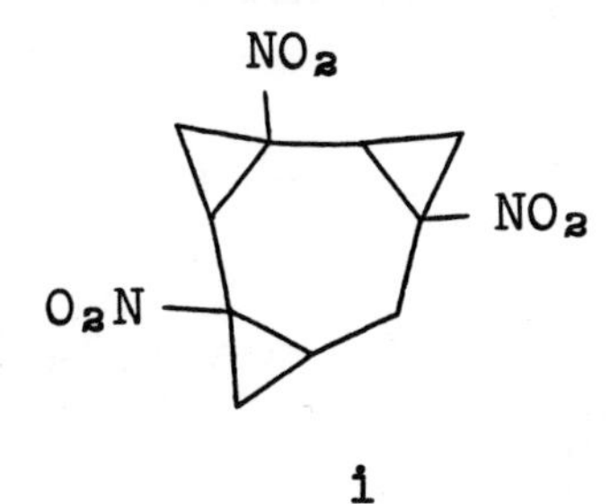

i

Louwrier-de Wal and de Boer, *Rec. Trav. Chim.* **87**, 699 (1968).

85.

a. MeLi + CH_2Cl_2 + → 41% —Cl

cis and trans

i

+ 2% — CH_3

cis and trans

ii

b. MeLi + CH_2Cl_2 + → 4-6% i

+ 30-31% ii + 1%

iii

MeLi was prepared from CH_3Cl and Li in (a) and from CH_3I and Li in (b). In each case the total MeLi solution (including the halide ion) was added to a mixture of CH_2Cl_2 and cyclohexene. Explain the differences in the ratio of i to ii, and also why a small amount of norcarane (iii) was formed in b, but not in a.

Note: i treated with MeLi does not give ii.

Dilling and Edamura, *J. Org. Chem.* **32**, 3492 (1967).

86. In each case predict the principal product or products.

a. $CH_3-C(CH_3)=CH-CH_3 + H_2O \xrightarrow{H^+}$

b. $CH_2=CH-CHO + HCl \longrightarrow$

c. $CF_3-C\equiv C-CH_2-CH=C(CH_3)-CH_3$ + one mole $Br_2 \longrightarrow$

d. $Me_3Si-CH=CH_2 + HI \longrightarrow$

e. $\underline{p}$-$Me_3SiC_6H_4CH=CH_2 + HBr \longrightarrow$

f. $\underline{p}$-$Cl_3SiC_6H_4CH=CH_2 + HBr \longrightarrow$

g. $CH_2=C(OEt)-CHO + EtOH \xrightarrow{H^+}$

h. $CF_3-C\equiv CH$ + one mole $HBr \longrightarrow$

i. $\underline{p}$-$MeC_6H_4CH=CH_2 + HSCH_2COOH \xrightarrow{Bz_2O_2}$

j. $\underline{p}$-$CH_3CONHC_6H_4C\equiv CH$ + one mole $HBr \longrightarrow$

k. p-$O_2NC_6H_4CH=CHPh + BrCCl_3 \xrightarrow{(\underline{t}\text{-}Bu)_2O_2}$

l. $F_2C=CH-CH=CH_2 + CF_2=CCl_2$ (2 + 2 cyclo-addition) $\longrightarrow$

m. $CH_2=C(Cl)-CH=CH_2 + CF_2=CCl_2$ (2 + 2 cyclo-addition) $\longrightarrow$

n. 2,4,6-$Me_3C_6H_2$-C(=O)-CH=CH-C_6H_2-2,4,6-Me_3 + KCN $\xrightarrow{HOAc}$

PROBLEM SET 10

ADDITION TO CARBON-HETERO MULTIPLE BONDS

Corresponds to Chapter 16 of Advanced Organic Chemistry.

The first 45 problems are synthetic. In each one you are asked to convert one compound into another. In every case the conversion has actually been carried out, and the reference is given.

1.

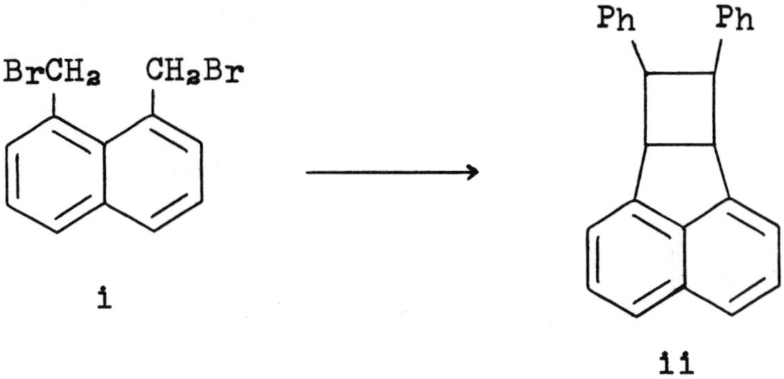

BrCH$_2$ CH$_2$Br

i

Ph Ph

ii

2.

$MeOOCCH_2\underset{\underset{O}{\|}}{C}CH_2COOMe$

O

O

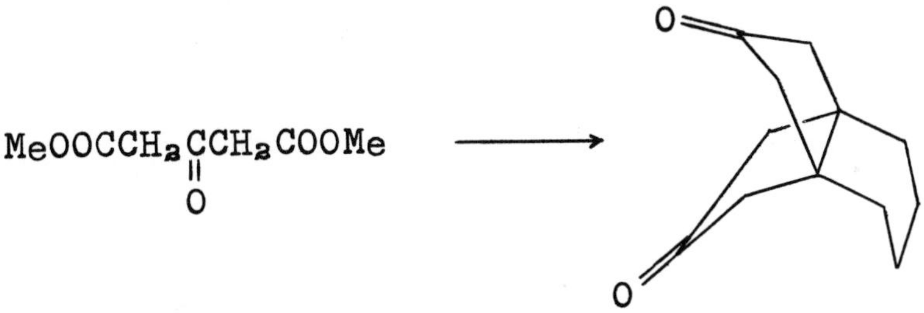

Weiss and Edwards, *Tetrahedron Lett.* **1968**, 4885.

3.

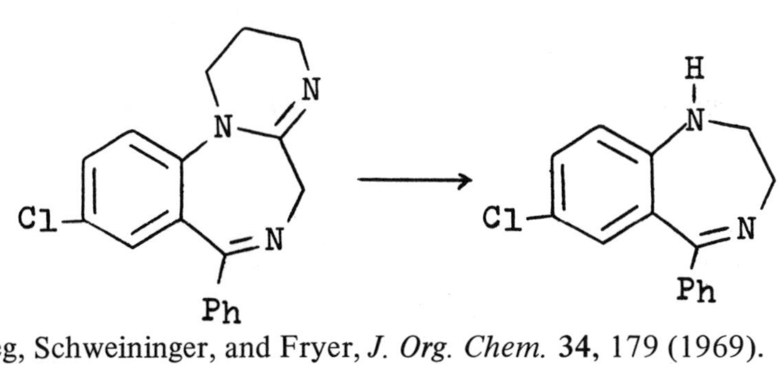

N N Cl N Ph

H N Cl N Ph

Derieg, Schweininger, and Fryer, *J. Org. Chem.* **34**, 179 (1969).

4.

CHO OH

O

4. Cont.

Schweizer, Berninger, Crouse, Davis, and Logothetis, *J. Org. Chem.* **34**, 207 (1969).

5.

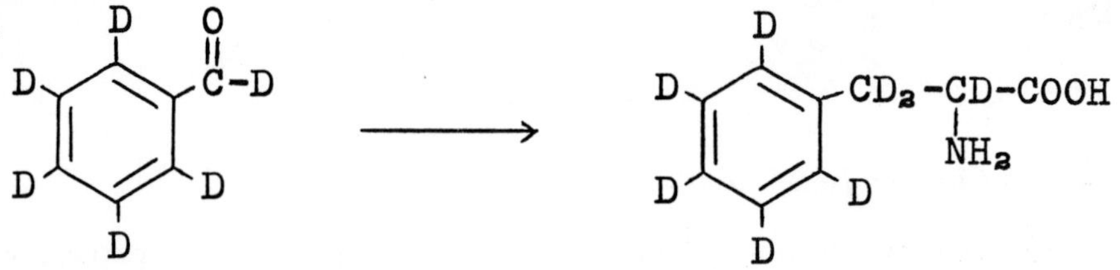

i

optically active
deuterated L-phenylalanine

ii

6.

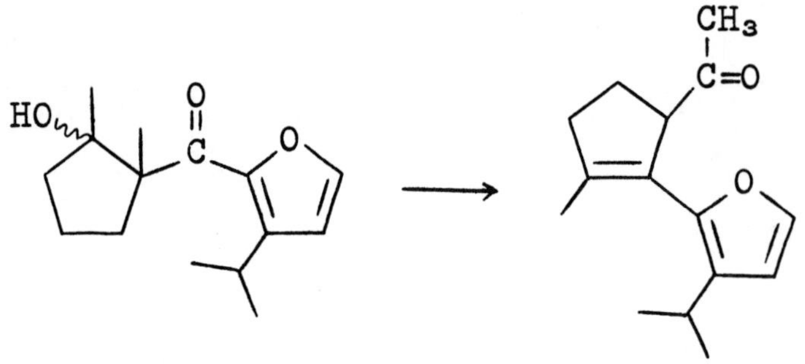

Büchi and Wüest, *Tetrahedron* **24**, 2049 (1968).

7.

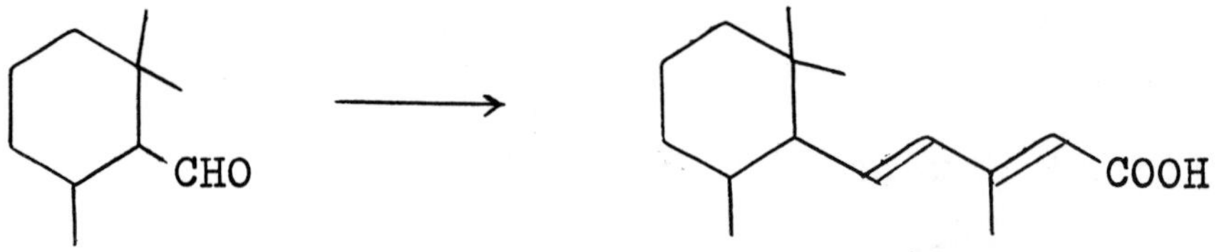

Blatz, Balasubramaniyan, and Balasubramaniyan, *J. Amer. Chem. Soc.* **90**, 3282 (1968).

8.

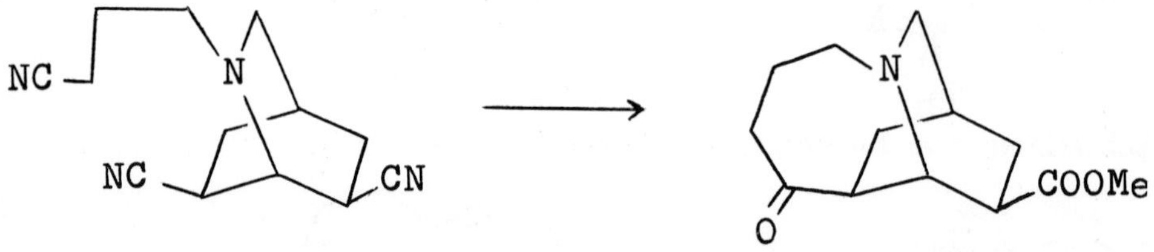

Ikezaki, Wakamatsu, and Ban, *Chem. Commun.* **1969**, 88.

9.

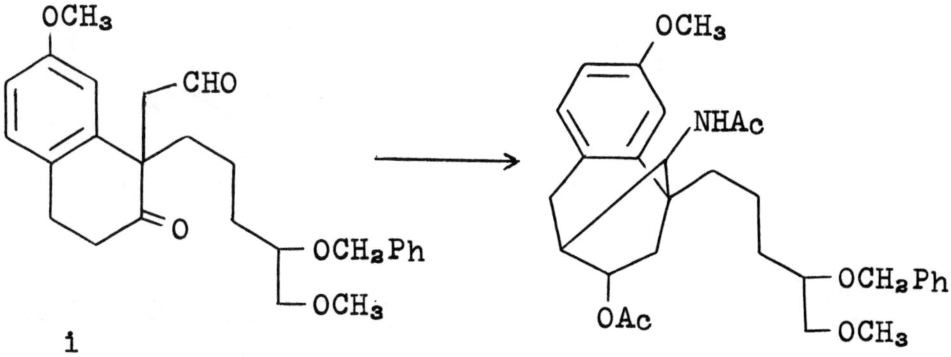

i

ii

10.

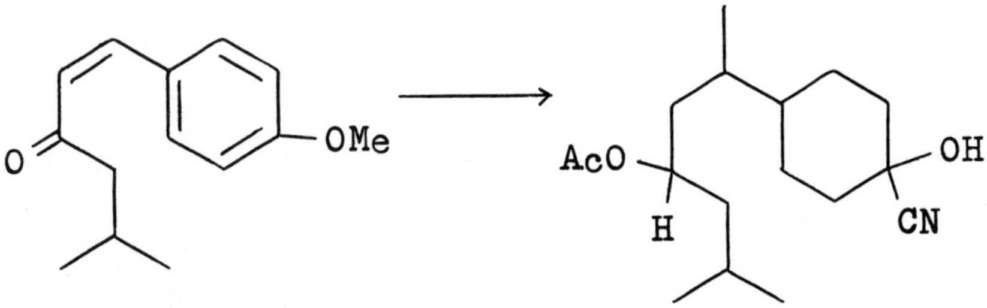

Ayyar and Krishna Rao, *Can. J. Chem.* **46**, 1467 (1968).

11.

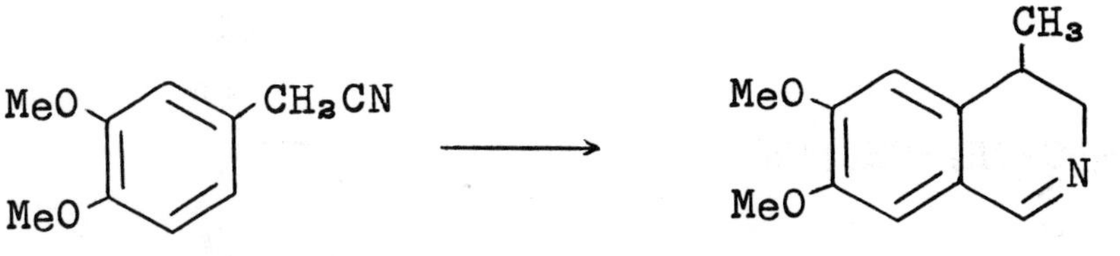

Wenkert and Haugwitz, *Can. J. Chem.* **46**, 1160 (1968).

12.

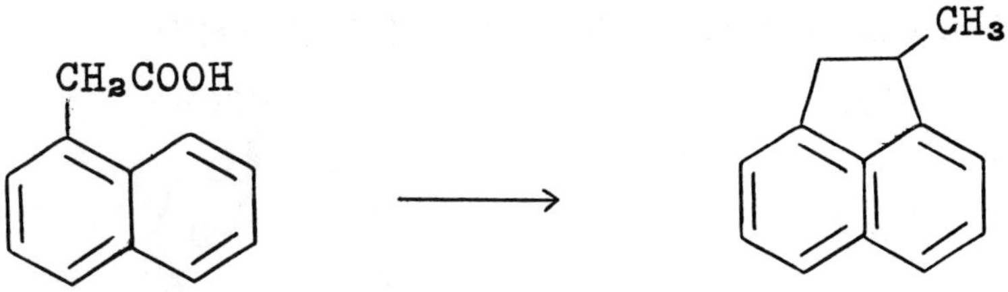

Bosch and Brown, *Can. J. Chem.* **46**, 715 (1968).

13.

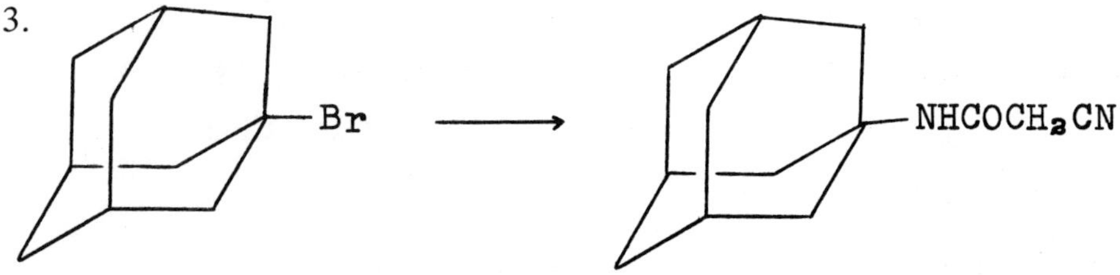

i

ii

14.

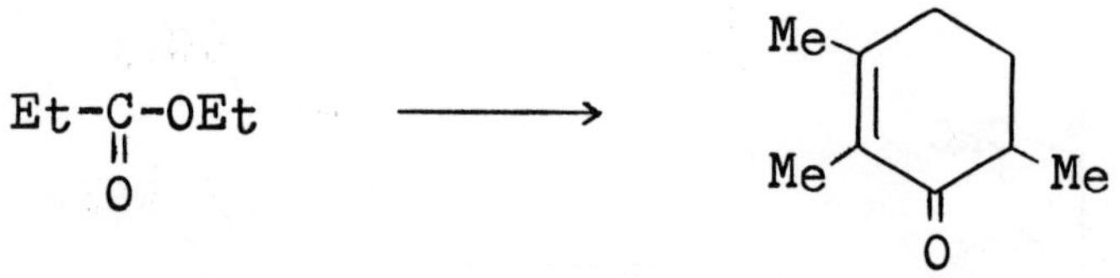

$Et-\underset{\underset{O}{\|}}{C}-OEt \longrightarrow$

Ichikawa, Owatari, and Kato, *Bull. Chem. Soc. Jap.* **41**, 1228 (1968).

15.

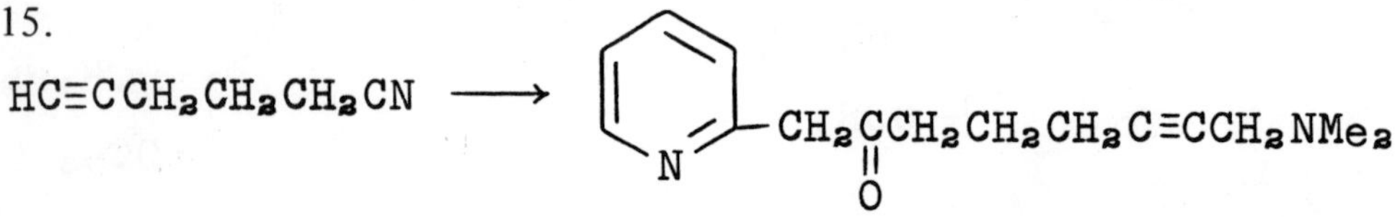

$HC{\equiv}CCH_2CH_2CH_2CN \longrightarrow$ (2-pyridyl)$-CH_2\underset{\underset{O}{\|}}{C}CH_2CH_2CH_2C{\equiv}CCH_2NMe_2$

Compagnon, Miocque, and Gautier, *Bull. Soc. Chim. Fr.* **1968**, 4127, 4136.

16.

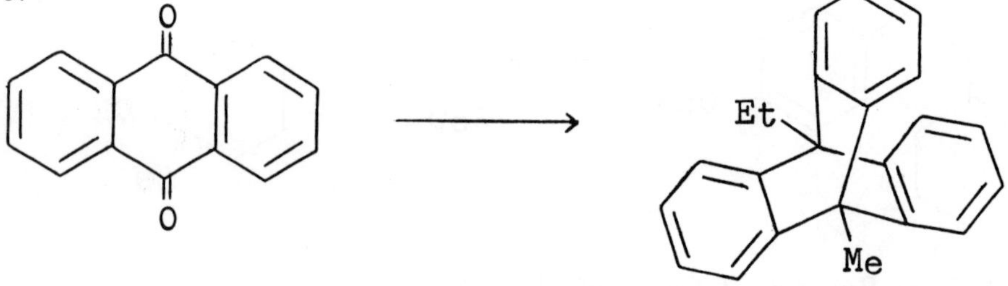

17.

$Ph-\underset{\underset{O}{\|}}{C}-CH_2-CH_2-CN \longrightarrow$

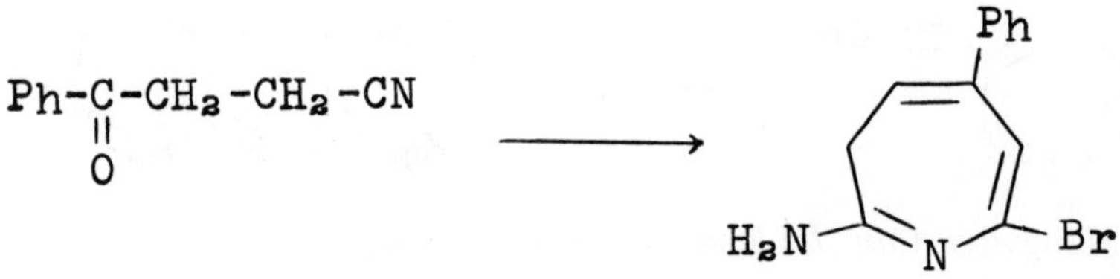

Nasutavicus, Tobey, and Johnson, *J. Org. Chem.* **32**, 3325 (1967).

18.

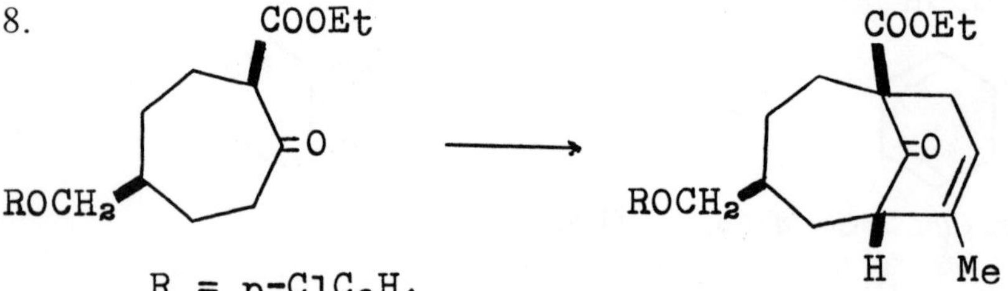

R = $\underline{p}$-ClC_6H_4

Marshall and Partridge, *J. Amer. Chem. Soc.* **90**, 1090 (1968).

19.

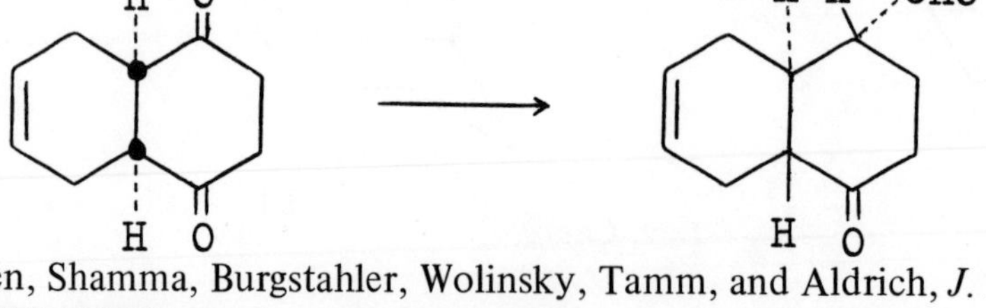

van Tamelen, Shamma, Burgstahler, Wolinsky, Tamm, and Aldrich, *J. Amer. Chem. Soc.* **91**, 7315 (1969).

20. CHO, OCH_3 (i) ⟶ CH_3O–C_6H_4–CH_2–C(=O)–C_6H_4–OCH_3 (ii)

i ii

21. $CH_3COCH_2CH_2COOH$ ⟶

levulinic acid

Frank, Schmitz, and Zeidman, *Org. Syn.* **III**, 328.

22.

Rioult and Vialle, *Bull. Soc. Chim. Fr.* **1968**, 4477, 4483.

23.

Taylor and Wibberley, *J. Chem. Soc., C* **1968**, 2693.

24.

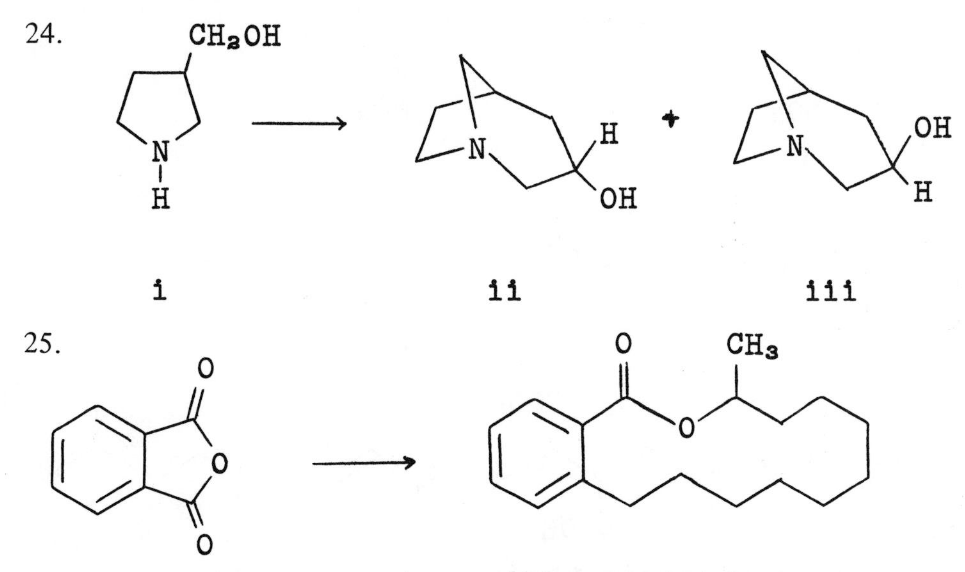

25.

dideoxyzearalane

Wehrmeister and Robertson, *J. Org. Chem.* **33**, 4173 (1968).

26.

$$PhCH_2{-}\underset{\underset{O}{\|}}{C}{-}NHPh \longrightarrow Ph{-}\underset{\underset{HO}{|}}{\overset{\overset{Ph}{|}}{C}}{-}\underset{\underset{Ph}{|}}{CH}{-}\underset{\underset{O}{\|}}{C}{-}NHPh$$

Kaiser, von Schriltz, and Hauser, *J. Org. Chem.* **33**, 4275 (1968).

27.

$$CF_2Cl{-}\underset{\underset{O}{\|}}{C}{-}CF_2Cl \longrightarrow CF_2Cl{-}\underset{\underset{OEt}{|}}{\overset{\overset{CF_2Cl}{|}}{C}}{-}{}^{18}O{-}\underset{\underset{O}{\|}}{C}{-}CH_3$$

Notes: Labeling need not be 100%. The source of ^{18}O must be $H_2{}^{18}O$.

Newallis, Lombardo, and McCarthy, *J. Org. Chem.* **33**, 4169 (1968).

28. Cyclopentanone ⟶ 3,3-difluoro-bicyclic compound bearing CH_3 (structures)

Lack and Roberts, *J. Amer. Chem. Soc.* **90**, 6997 (1968).

29. $CH_3CH_2CH_2NO_2 \longrightarrow CH_3CH_2{-}\underset{\underset{O}{\|}}{C}{-}CH_2CH_2COOH$

Kloetzel, *J. Amer. Chem. Soc.* **70**, 3571 (1948).

30.

COOH / NO_2 (i) ⟶ CN / NO_2 (ii)

⟶

COOH

COOH

β-carotene-14,15-dicarboxylic acid

Haeck and Kralt, *Rec. Trav. Chim.* **87**, 709 (1968).

32.

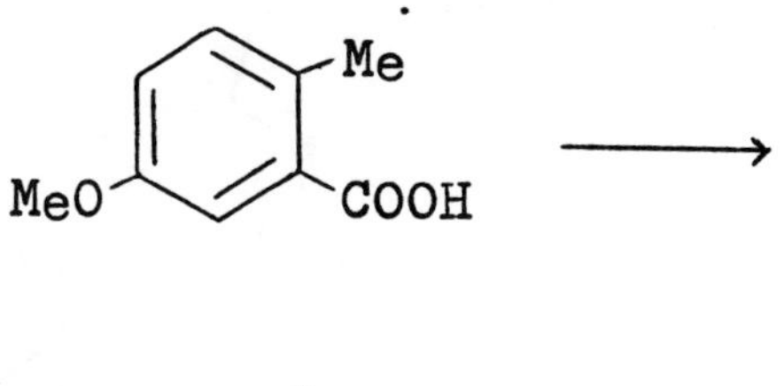

MeO, Me, COOH → Me, O, O

Bruck, Clark, Davidson, Günther, Littlewood, and Lythgoe, *J. Chem. Soc., C* **1967**, 2529.

33.

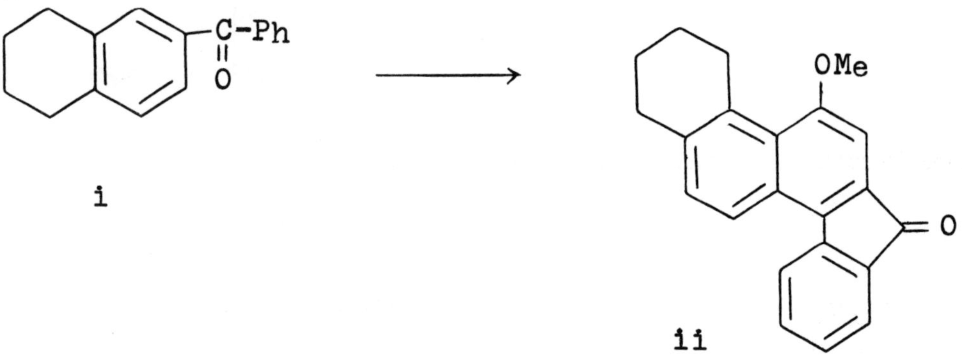

C-Ph, O → OMe, O

i

ii

34.

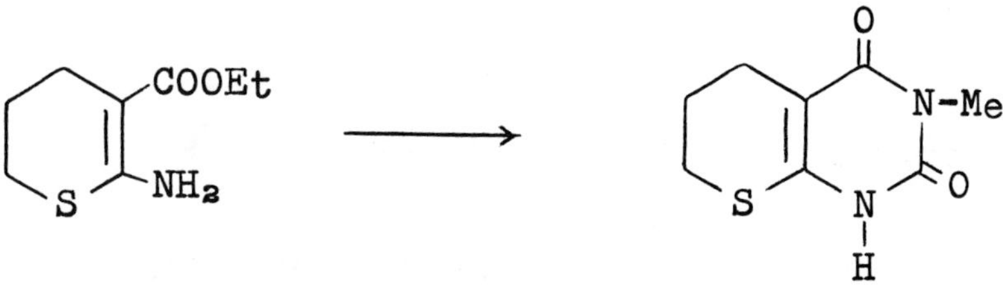

COOEt, NH_2, S → O, N-Me, O, S, N, H

Wamhoff, *Chem. Ber.* **101**, 3377 (1968).

35.

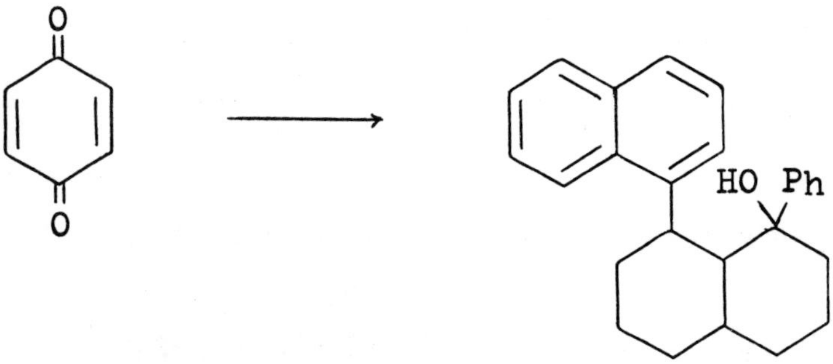

O, O → HO Ph

Bailey and Shuttleworth, *J. Chem. Soc., C* **1968**, 1115.

36.

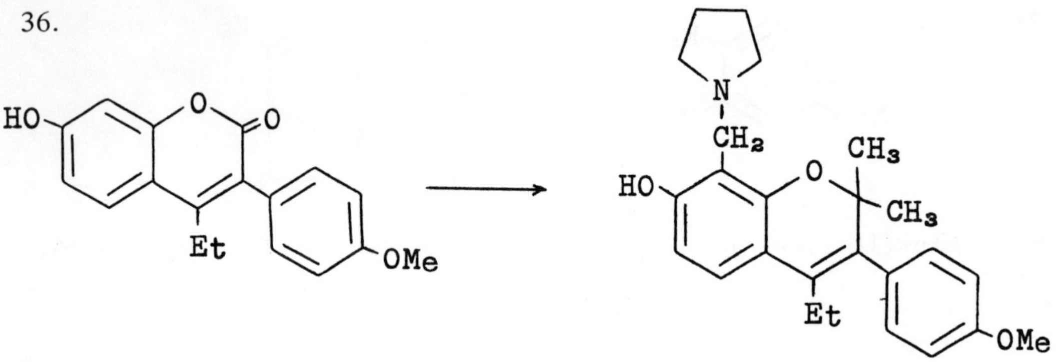

Cook and Wall, *J. Org. Chem.* **33**, 2998 (1968).

37.

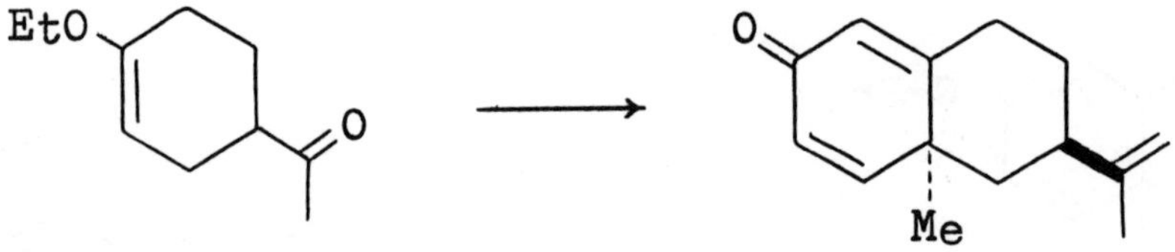

Pesaro, Bozzato, and Schudel, *Chem. Commun.* **1968**, 1152.

38.

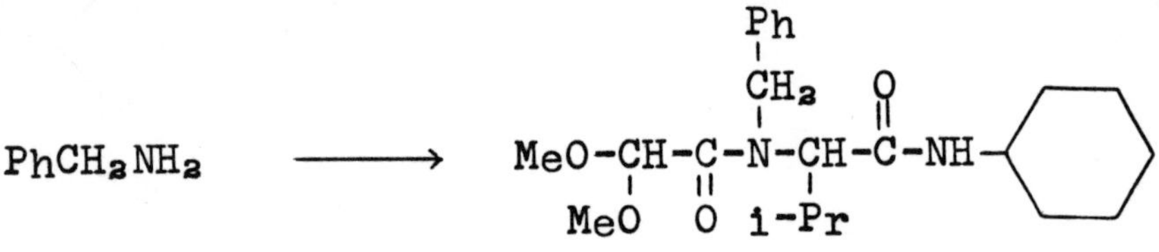

$PhCH_2NH_2$ (**i**) ⟶ $MeO-CH(OMe)-C(=O)-N(CH_2Ph)-CH(i\text{-}Pr)-C(=O)-NH-$cyclohexyl (**ii**)

39.

$Pr-C(=O)-Pr$ ⟶ $Pr-C(Pr)(Br)-(CH_2)_3-COOMe$

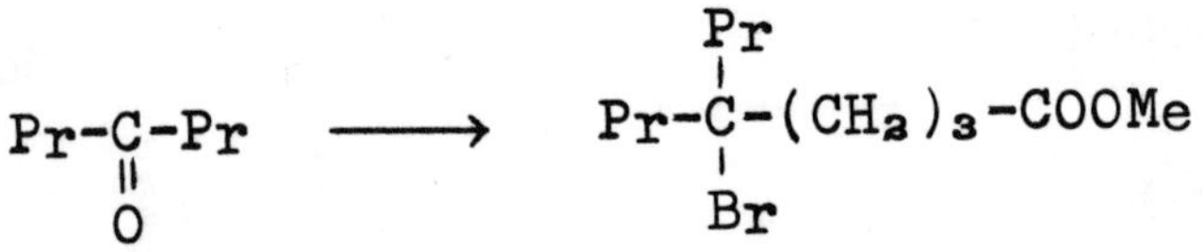

Eiter, Truscheit, and Boness, *Justus Liebigs Ann. Chem.* **709**, 29 (1967).

40.

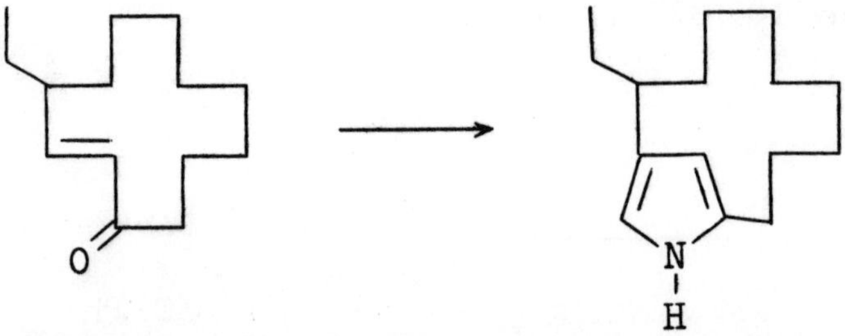

Wasserman, Keith, and Nadelson, *J. Amer. Chem. Soc.* **91**, 1264 (1968).

41.

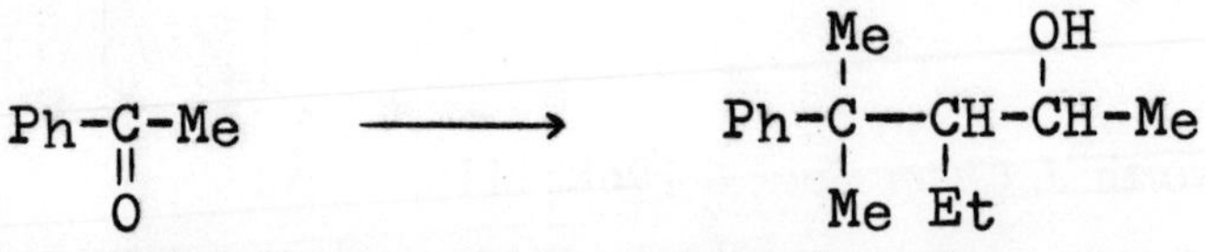

$Ph-C(=O)-Me$ ⟶ $Ph-C(Me)_2-CH(Et)-CH(OH)-Me$

Mosher, Berger, Foldi, Gardner, Kelly, and Nebel, *J. Chem. Soc., C* **1969**, 121.

42. NC-CH_2-COOH ⟶ (1,3-dithiolan-2-ylidene)C(NC)(COOEt)

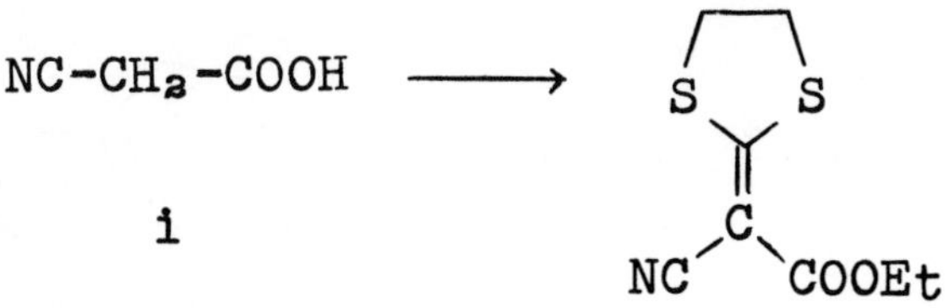

i ii

43. CH_3-CH=CH-CH_2OH ⟶ CH_2=CH-CH(CH_3)-NH_2

i ii

44. Rhodoxanthin was prepared by conversion of i to ii; reduction of ii to iii with $LiAlH_4$, and conversion of iii to iv. Show how to convert i to ii and iii to iv.

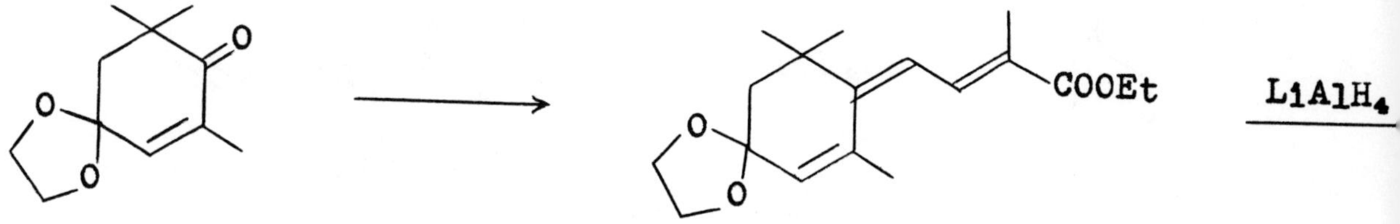

i ⟶ ii (COOEt) $\xrightarrow{LiAlH_4}$

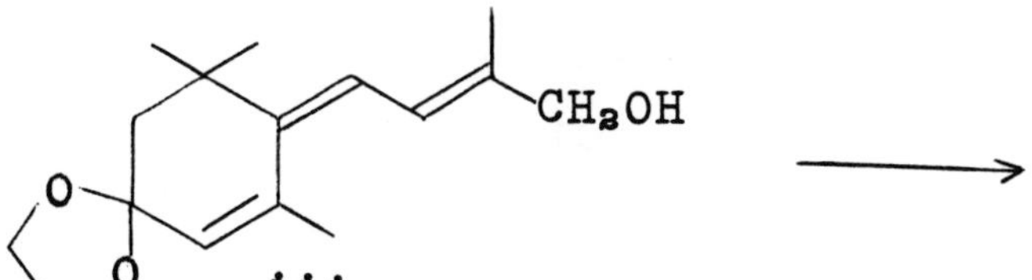

iii (CH_2OH) ⟶

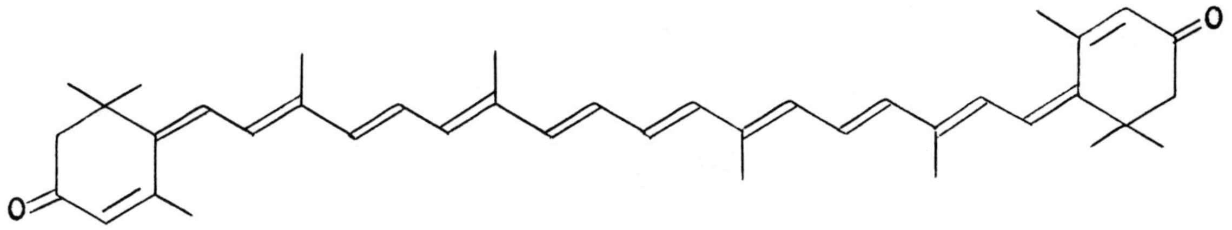

iv

rhodoxanthin

Mayer, Montavon, Rüegg, and Isler, *Helv. Chim. Acta* **50**, 1606 (1967).

45.

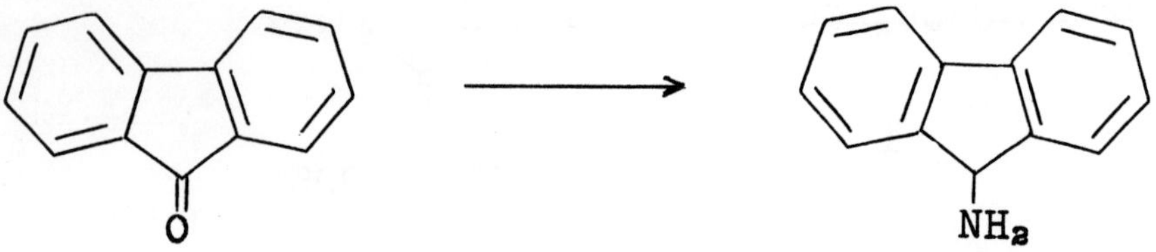

i → ii

The following conversions (problems 46 to 83) have been reported in the literature. In each case suggest a reasonable mechanism.

46.

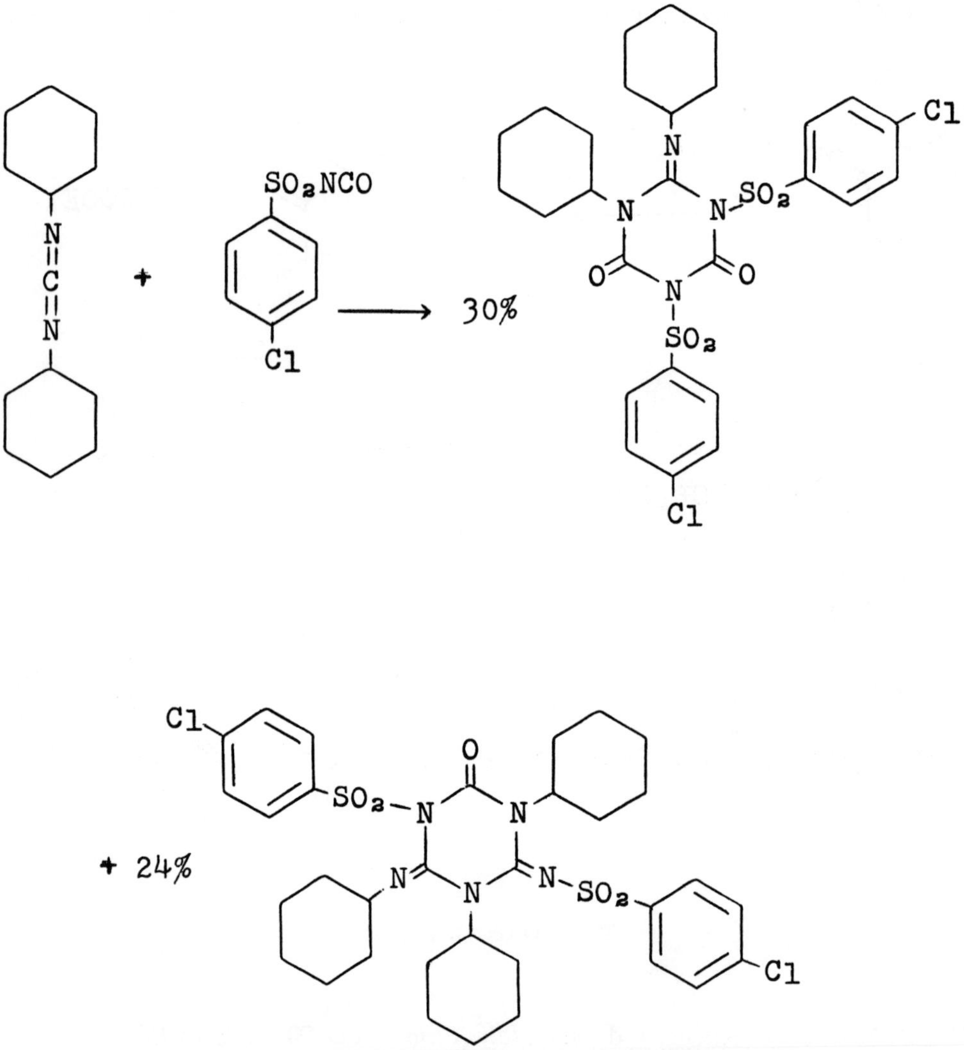

Ulrich, Tucker, Stuber, and Sayigh, *J. Org. Chem.* **34**, 2250 (1969).

47.

Cl ... C(=O)-Ph ... NH_2 + $\overset{\ominus}{C}H_2$-$\overset{\oplus}{S}Me_2$ ⟶ 72% Cl ... Ph ... N-H

Bravo, Gaudiano, and Umani-Ronchi, *Tetrahedron Lett.* **1969**, 679.

48.

Ph-C(=N-$\overset{\oplus}{N}Me_3$)-CH_3 + $CH_3SO\overset{\ominus}{C}H_2$ $\xrightarrow[\text{1 hr}]{Me_2SO}$ Ph ... Ph ... N-H + NMe_3

Sato, Kato, and Ohta, *Bull. Chem. Soc. Jap.* **40**, 2936 (1967).

49.

Cl ... N + $^{15}NH_3$ $\xrightarrow[\text{140° 36 hr}]{H_2O}$ $\overset{*}{N}H_2$ ... N^*

closed system

both nitrogens equally labeled

Wahren, *Tetrahedron* **24**, 441, 451 (1968).

50.

Ph ... S ... H ... (+) ... Cl^- ... N ... NH_2 ... Ph (i) $\xrightarrow[\text{2. HCl}]{\text{1. } PhNH_2 \text{ overnight}}$ Ph ... S ... N ... NHPh (ii)

Note: PhN=C(Ph)–SCH_2CN (iii) also gave ii on similar treatment.

51.

$$Me_2C{=}CHCH_2CH_2CN \xrightarrow[\text{20 hr}]{\text{cold } 98\%\ H_2SO_4}$$

Me, Me, O, N, N, Me, Me

23% i

Me, Me, H, N, O, N, H, O, Me, Me

4% ii

Note: ii could not be converted to i under the reaction conditions.

Ducker and Gunter, *Aust. J. Chem.* **21**, 2809 (1968).

52.

$N{-}CH_2CH_2NH_2$, Ph $\xrightarrow[\text{acetone}]{\text{excess}}$ 65%

Me, H, Me, N, N, Ph

Winn and Zaugg, *J. Org. Chem.* **34**, 249 (1969).

53.

$$ClCH{=}CH{-}\underset{\underset{O}{\|}}{C}{-}(CH_2)_{11}{-}\underset{\underset{O}{\|}}{C}{-}CH{=}CHCl \xrightarrow[\text{2. AcOH}]{\text{1. } NH_3 \text{ in abs EtOH, 16 hr}} 65\%$$

i

O, C, N, $(CH_2)_{11}$

ii

54.

$$CH_2{=}CH{-}\underset{\underset{CH_3}{|}}{C}{=}CH_2 \xrightarrow[\text{2. } H_2O{-}H^+]{\text{1. } ClSO_2{-}N{=}C{=}O \text{ abs } Et_2O}$$

isoprene

CH_3, O, O + CH_3, N, H, O

Haug, Lohse, Metzger, and Batzer, *Helv. Chim. Acta* **51**, 2069 (1968).

55.

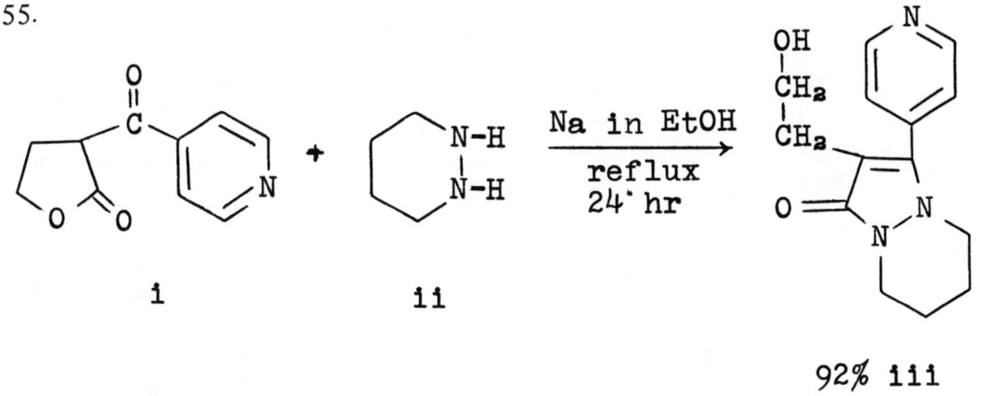

i + ii $\xrightarrow[\text{reflux, 24 hr}]{\text{Na in EtOH}}$ 92% iii

56.

$\xrightarrow[\text{ether, 0°}]{BF_3\text{-etherate}}$ 37%

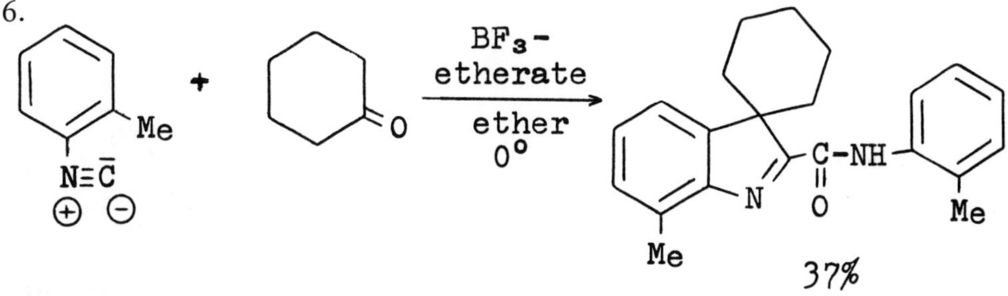

Zeeh, *Chem. Ber.* **101**, 1753 (1968).

57.

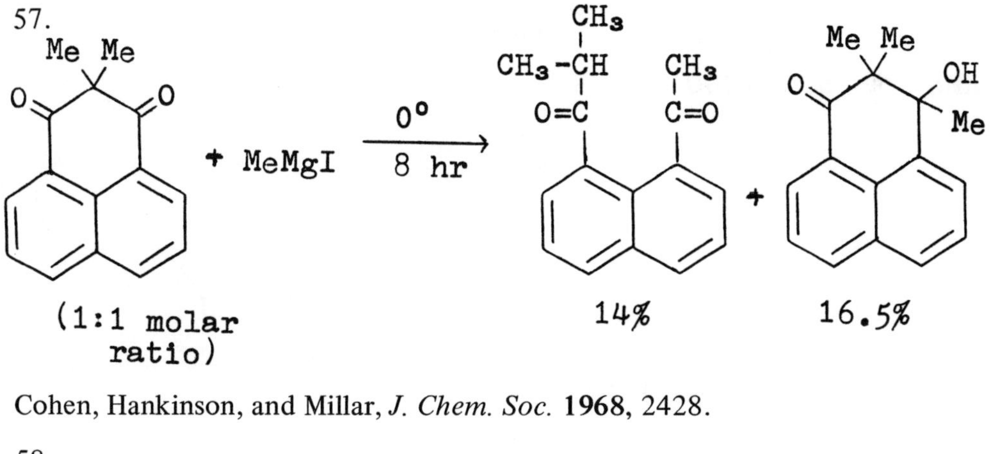

(1:1 molar ratio) + MeMgI $\xrightarrow[\text{8 hr}]{0°}$ 14% + 16.5%

Cohen, Hankinson, and Millar, *J. Chem. Soc.* **1968**, 2428.

58.

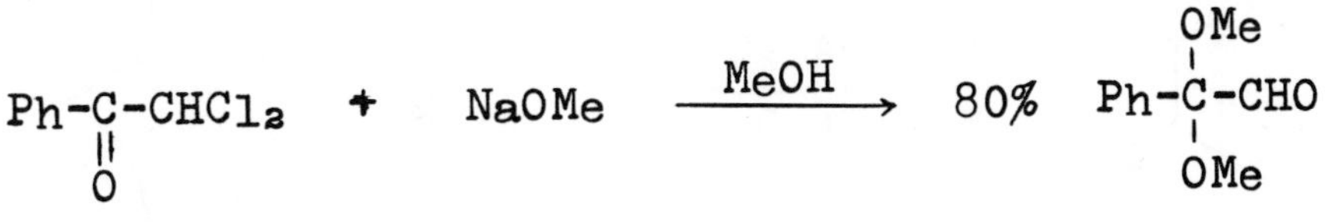

$Ph-C(=O)-CHCl_2$ + NaOMe $\xrightarrow{MeOH}$ 80% $Ph-C(OMe)_2-CHO$

Henery-Logan and Fridinger, *Chem. Commun.* **1968**, 130.

59.

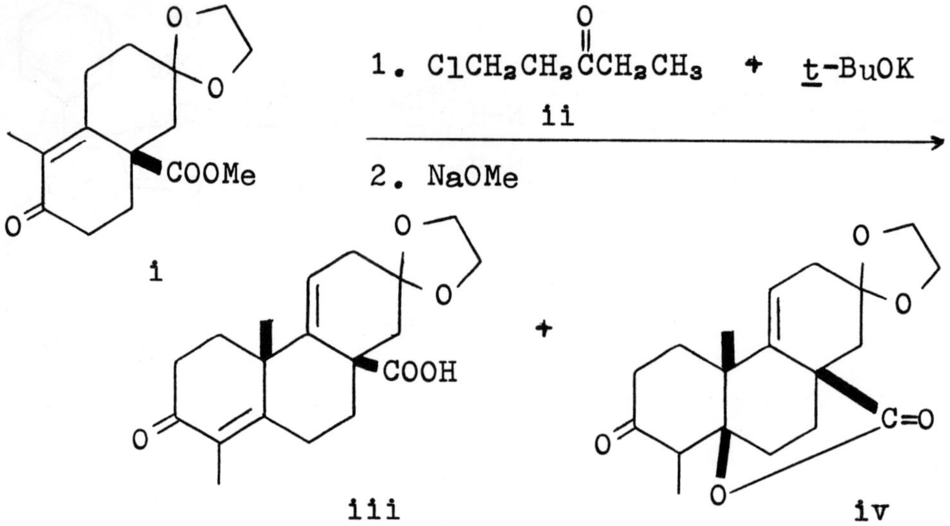

1. $ClCH_2CH_2\overset{O}{\overset{\|}{C}}CH_2CH_3$ + t-BuOK
ii

2. NaOMe

i → iii + iv

COOMe

COOH

C=O

60.

PhNHMe + $HOCH_2CHO$

room temp.

Me Ph N OH CH_2OH N Me

hot aq. EtOH

N Me

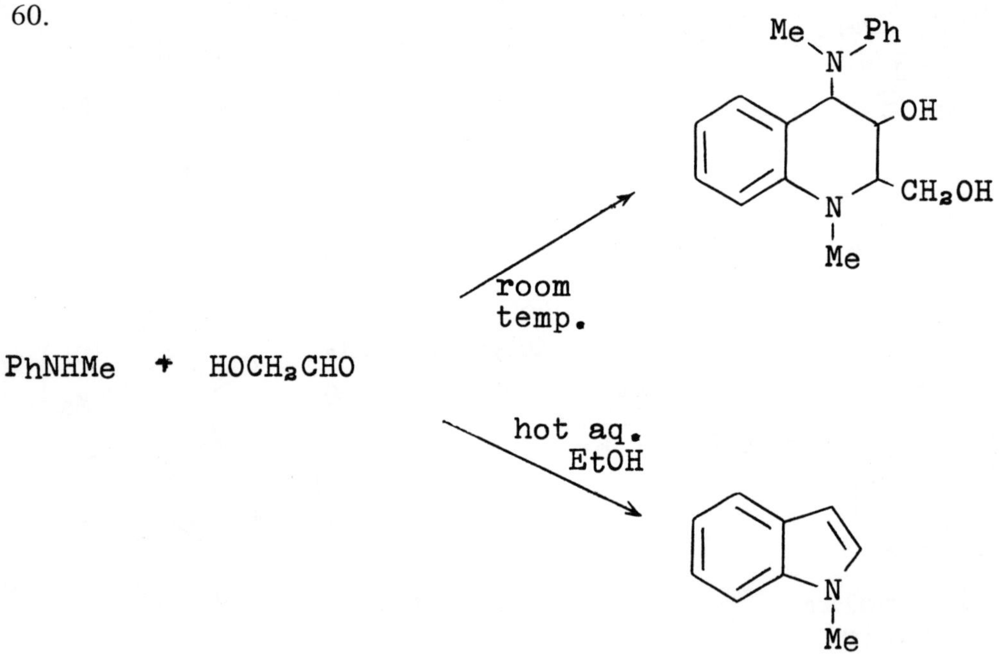

Turner, *Chem. Commun.* **1968**, 1659.

61.

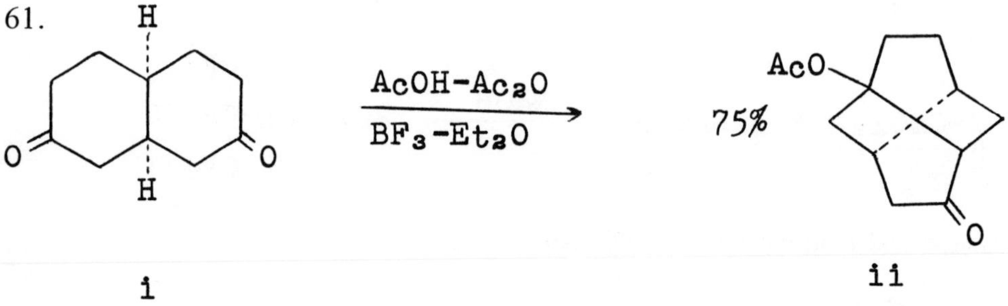

H H O O

i

AcOH-Ac_2O

BF_3-Et_2O

AcO

75%

O

ii

8-acetoxy-4-twistanone

62.

$$\text{7-MeO-indol-2-yl-}CH_2CH_2OTs + EtOOCCH_2CN \xrightarrow[\text{EtOH}]{\text{NaOEt}}$$

OMe H

N, H, C-COOEt, CN, OMe

Sakan, Matsubara, Takagi, Tokunaga, and Miwa, *Tetrahedron Lett.* **1968**, 4925.

63.

NO_2, CHO + $CH_2=C$ (N, O morpholino)$_2$ $\xrightarrow[\text{reflux 16 hr}]{\text{dry THF}}$

47% NO_2, CH=CH-C(=O)-N O + H-N O

Barton, Hewitt, and Sammes, *J. Chem. Soc., C* **1969**, 16.

64.

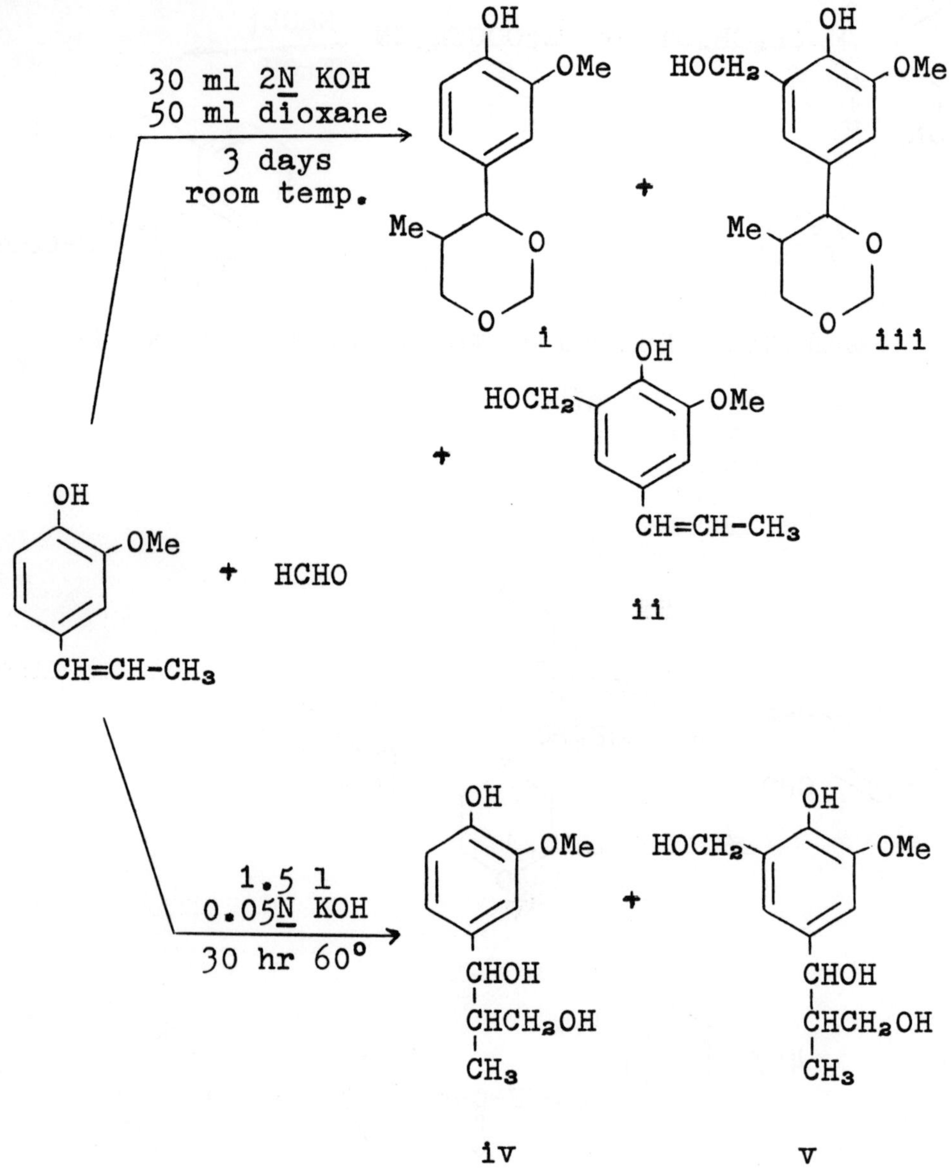

30 ml 2N KOH
50 ml dioxane
3 days
room temp.

OH, OMe, Me, O, O — i

+ HOCH2, OH, OMe, $CH=CH-CH_3$ — ii

+ OH, HOCH2, OMe, Me, O, O — iii

OH, OMe, $CH=CH-CH_3$ + HCHO

1.5 l
0.05N KOH
30 hr 60°

OH, OMe, CHOH, $CHCH_2OH$, CH_3 — iv

+ OH, HOCH2, OMe, CHOH, $CHCH_2OH$, CH_3 — v

65.

CN, NH_2 + N, N, N

anhyd. EtOH
boiled 8 hr

64%

NH_2, N, N

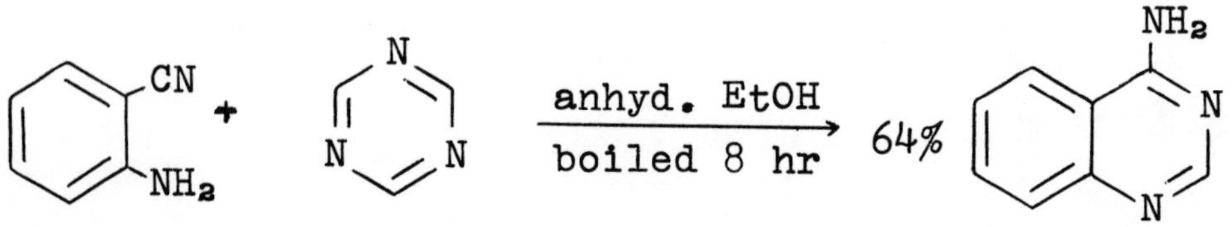

Kreutzberger and Stevens, *J. Chem. Soc., C* **1969**, 1282.

CEH

66.

$$\underset{\text{i}}{EtCOMe} + \underset{\text{ii}}{\underline{t}\text{-Bu-}\overset{\oplus}{N}\equiv\overset{\ominus}{C}} \xrightarrow[\text{2. } H_2O\text{-}H^+]{\text{1. petr. ether, } BF_3 \ 0^\circ} \underset{\text{iii}}{CH_3CH{=}C(CH_3)\text{-}C(=O)\text{-}C(=O)\text{-NH-}\underline{t}\text{-Bu}}$$

$$+ \ \underset{\text{iv}}{CH_3CH_2\text{-}C(CH_3)(OH)\text{-}C(=O)\text{-NH-}\underline{t}\text{-Bu}}$$

67.

$$\text{diphenylcyclopropenone (Ph, Ph)} + Et_2NH \longrightarrow \text{H(CONEt}_2\text{)C=C(Ph)Ph}$$

Dehmlow, *Tetrahedron Lett.* **1967**, 5177.

68.

$$PrCHO + \text{1,3,5-trioxane} \xrightarrow[\text{reflux, 4 hr}]{CHCl_3, \ H_2SO_4} \underset{20\%}{\text{5-Et-5-CHO-1,3-dioxane}} + \underset{35\%}{\text{bicyclic dioxane (Et)}}$$

(molar ratio 1:1.7)

Wesslen and Ryfors, *Acta Chem. Scand.* 22, 2071 (1968); Wesslen, *Acta Chem. Scand.* 22, 2085 (1968).

69.

MeO, OMe, OH, C=O, OMe, OMe; $\xrightarrow[\text{i-PrOH, 12 hr}]{NaBH_4}$ MeO, OMe, O, OMe, OMe; 52%

Clark-Lewis and Jemison, *Aust. J. Chem.* **21**, 2247 (1968).

70.

CH_3CO (lactone, O, =O) + $PhNHNH_2$ $\xrightarrow[\text{1 hr}]{\text{warming}}$ 39% CH_2OH, CH_3, N, N–Ph, O=

Wamhoff and Korte, *Justus Liebigs Ann. Chem.* **724**, 217 (1969).

71.

N, O, O, Et, –Ph (i) + i-PrCHO $\xrightarrow[\text{1 hr}]{70°}$ i-Pr–(O, N)–CH_2–PhCHOCOEt

i

81% ii

72.

$CH_3-\underset{\underset{O}{\|}}{C}-CH_3$ + CS_2 $\xrightarrow[\text{30 hr}]{\text{dry } NH_3}$ OH, Me, S, =S, N–H, Me Me

Takeshima, Imamoto, Yokoyama, Yamamoto, and Akano, *J. Org. Chem.* **33**, 2877 (1968).

73.

2-MeO-C_6H_4CHO + EtCOMe (excess) $\xrightarrow[\text{EtOH, 18 hr}]{\text{aq. NaOH, 95\%}}$ 64%

4-Me-C_6H_4CHO + $MeCO(CH_2)_4CH_3$ (excess) $\xrightarrow[\text{95\% EtOH, 12 days}]{\text{aq. NaOH}}$ 20%

Nielsen and Haseltine, *J. Org. Chem.* **33**, 3264 (1968); Nielsen and Dubin, *J. Org. Chem.* **28**, 2120 (1963).

74.

$EtOOC\text{-}CH\text{=}S(Me)\text{-}Me$ + Cl_3CCN $\xrightarrow[0\text{-}5^{\circ}]{Et_2O}$ 78% $EtOOC\text{-}C(Cl)\text{=}S(Me)\text{-}Me$

Payne, *J. Org. Chem.* **33**, 3517 (1968).

75.

$Me_2CH\text{-}CH\text{=}\overset{\oplus}{N}(O^{\ominus})\text{-}CH(CHMe_2)\text{-}CN$ $\xrightarrow[\text{1-2 weeks}]{\text{excess PhSH}}$ imidazole: Me_2CH, SPh, N, N-H, $CHMe_2$

Masui, Yijima, and Suda, *Chem. Commun.* **1968**, 1400.

76.

MeOOCCH$_2$COOMe + (HCHO)$_n$ → (C_6H_6, piperidine, reflux)

i

38% ii (bicyclic product with MeOOC, O, MeOOC, COOMe, O, COOMe substituents)

77.

O_2N-C$_6$H$_4$-CH=N-Ph + $\underline{t}$-Bu-$\overset{\oplus}{N}\equiv\overset{\ominus}{C}$ → (CCl_4, 110° 22 hr, sealed tube)

41% (3-NH-$\underline{t}$-Bu indole bearing 2-(C_6H_4-NO_2)) + NO_2-C_6H_4-azetidine (N-Ph, =N-$\underline{t}$-Bu, =N-$\underline{t}$-Bu)

small amount

O_2N-C$_6$H$_4$-N=CH-Ph + $\underline{t}$-Bu-$\overset{\oplus}{N}\equiv\overset{\ominus}{C}$ → (CCl_4, 100° 8 hr, sealed tube)

no indole + 51% O_2N-C_6H_4-N (azetidine with =N-$\underline{t}$-Bu, =N-$\underline{t}$-Bu, Ph)

Deyrup, Vestling, Hagan, and Yun, *Tetrahedron* **25**, 1467 (1969).

78.

p-nitrobenzaldehyde (CHO, NO_2) + $BrCH_2COOH$ + pyridine (N) ⟶

$$C_5H_5\overset{\oplus}{N}-CH_2-\underset{OH}{CH}-C_6H_4-NO_2 \quad + \quad C_5H_5\overset{\oplus}{N}-CH_3$$

i ii

Note: The total yield was 82%, of which 96% was i and 4% ii.

Phillips and Ratts, *Tetrahedron Lett.* **1969**, 1383.

79.

2-phenylazirine (Ph, N) + p-nitrobenzoyl chloride (COCl, NO_2) $\xrightarrow[\text{reflux, 2 hr}]{Et_3N,\ C_6H_6}$ 43% 5-phenyl-2-(p-nitrophenyl)oxazole (N, O, Ph, NO_2)

Sato, Kato, and Ohta, *Bull. Chem. Soc. Jap.* **40**, 2938 (1967).

80.

$$PhCH_2\overset{\oplus}{N}\equiv\overset{\ominus}{\bar{C}} \xrightarrow[2.\ PhCOPh]{1.\ BuLi} 74\% \quad \begin{matrix} Ph \\ H \end{matrix}\!\!>C=C<\!\!\begin{matrix} Ph \\ Ph \end{matrix}$$

Schöllkopf and Gerhart, *Angew. Chem. Intern. Ed. Engl.* **7**, 805 (1968) [*Angew. Chem.* **80**, 842 (1968)].

81.

$$\underset{\substack{\text{1 mole}}}{H-\underset{O}{\overset{\|}{C}}-NH_2} + \underset{\text{2 moles}}{POCl_3} \xrightarrow[\text{sealed vessel}]{120^\circ,\ 15\ hr} 43.5\%\ \text{adenine}$$

adenine (NH_2, N, N, N, N–H)

Notes: The reaction did not take place in an open vessel; other Lewis acids gave less adenine, or none at all. When $K^{14}CN$ was added, little or no ^{14}C was found in the adenine.

Ochiai, Marumoto, Kobayashi, Shimadzu, and Morita, *Tetrahedron* **24**, 5731 (1968). See also Morita, Kobayashi, Shimadzu, and Ochiai, *Tetrahedron Lett.* **1970**, 861.

82.

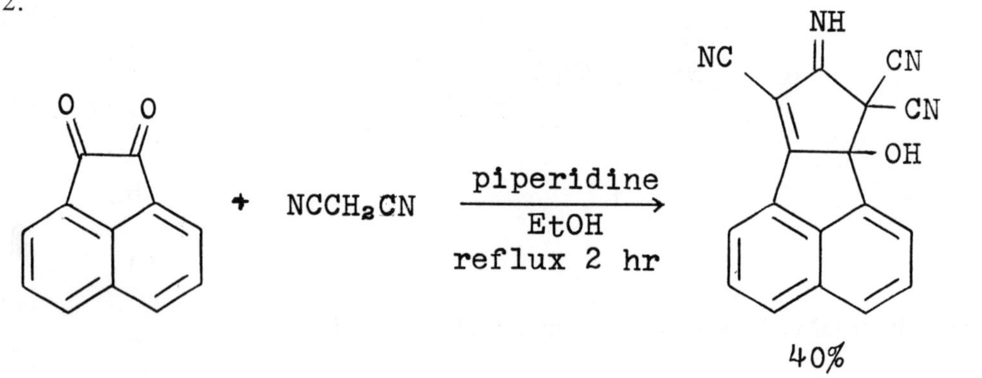

Junek, Hornischer, and Sterk, *Monatsh. Chem.* **99**, 2121 (1968).

83.

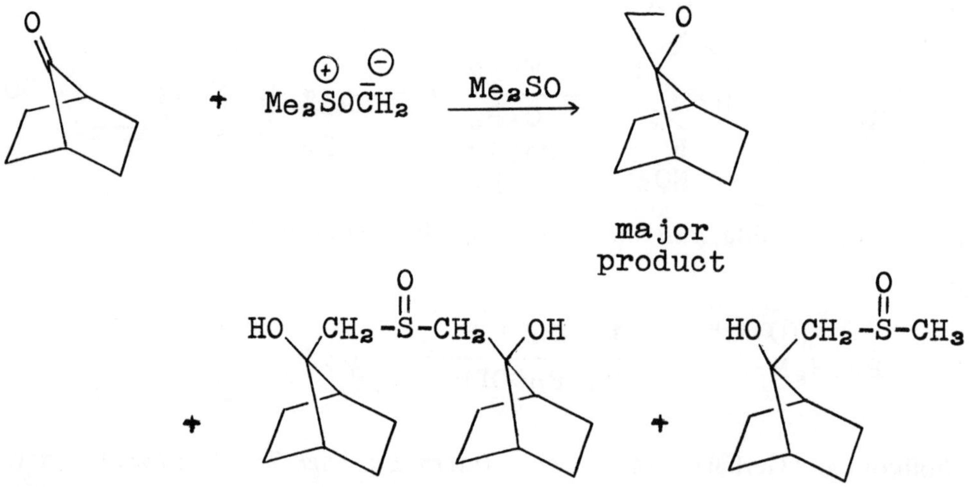

Bly and Bly, *J. Org. Chem.* **34**, 304 (1969).

In the following problems (84 to 94) give the products and suggest reasonable mechanisms.

84.

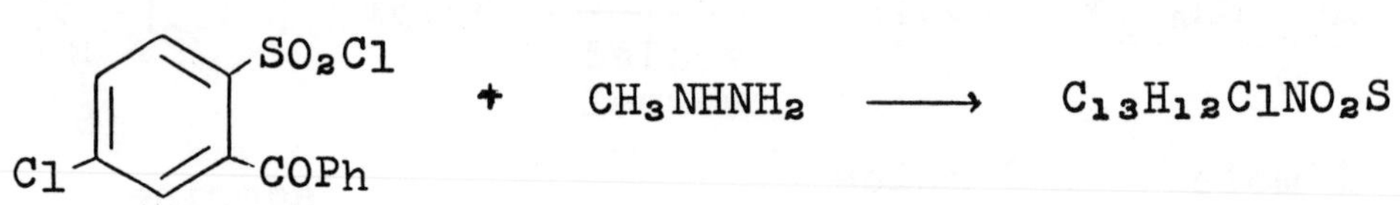

Wright, *J. Heterocycl. Chem.* **5**, 453 (1968).

85.

(1-benzylcyclohexan-1-ol: cyclohexane ring bearing CH_2Ph and OH)

i $\xrightarrow[CN^-]{H_2SO_4,\ HOAc}$ ii ($C_{14}H_{19}NO$)

86.

(benzene ring bearing CH_2CN and COOMe ortho) + Me–C(Me)(Br)–COOMe $\xrightarrow[\text{dry PhOMe},\ C_6H_6]{Zn}$ $C_{14}H_{15}NO_3$

Arsenijevic and Arsenijevic, *Bull. Soc. Chim. Fr.* **1968**, 3403.

87.

$CH_3C(=O)CH_2C(=O)CH_2C(=O)CH_3$ $\xrightarrow[THF]{\text{excess } LiN(i\text{-}Pr)_2}$ i $\xrightarrow[2.\ H^+]{1.\ CO_2}$ ii $\xrightarrow[Et_2O]{CH_2N_2}$

iii $\xrightarrow{0.5\underline{M}\ KOH}$ iv ($C_8H_8O_4$) 39% + v ($C_9H_{10}O_4$) 27%

Howarth, Murphy, and Harris, *J. Amer. Chem. Soc.* **91**, 517 (1969).

88.

Ph–N=C=O + NH_2OH $\longrightarrow$

Kreutzkamp and Messinger, *Chem. Ber.* **100**, 3463 (1967).

89.

$ClCH_2COOH$ (i) + Ph–N=CH–Ph (ii) $\xrightarrow[\text{DMF, reflux, 2 hr}]{POCl_3}$ 78% iii ($C_{15}H_{12}ClNO$)

90.

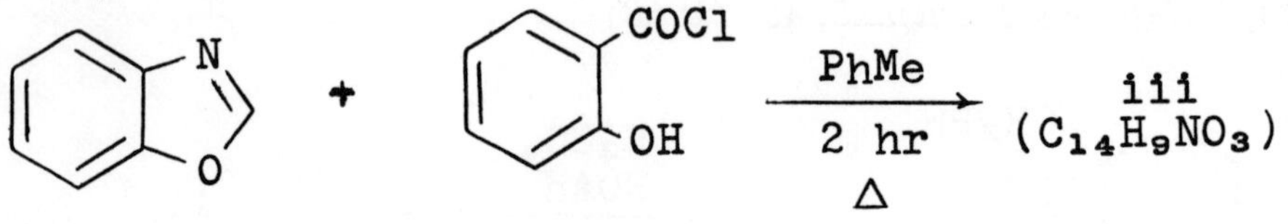

i + ii (COCl, OH) $\xrightarrow[\Delta]{\text{PhMe, 2 hr}}$ iii ($C_{14}H_9NO_3$)

i ii

91.

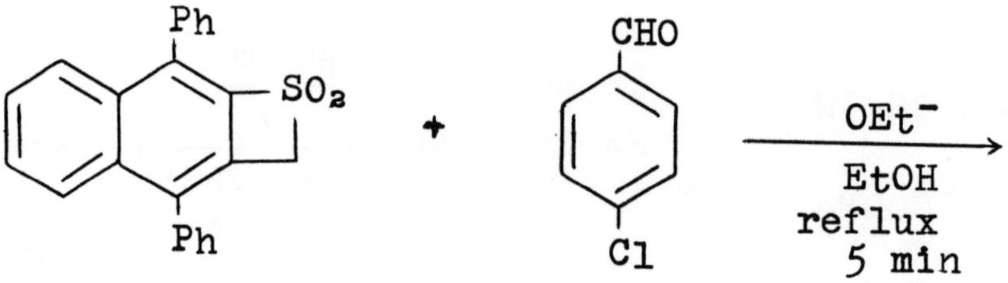

Ph, SO_2, Ph + CHO, Cl $\xrightarrow{\text{OEt}^-}$ EtOH reflux 5 min

Dittmer and Balquist, *J. Org. Chem.* **33**, 1364 (1968).

92.

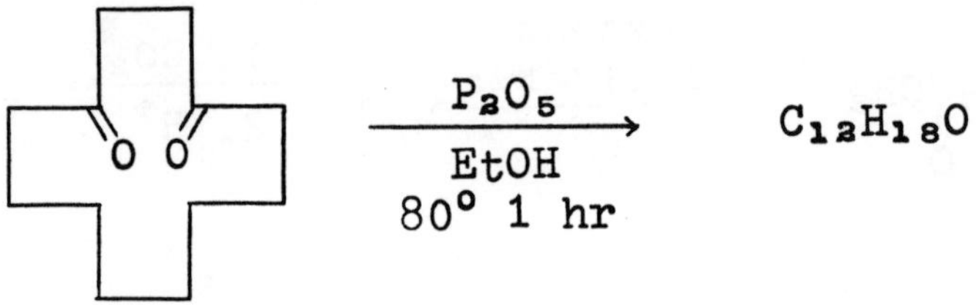

O O $\xrightarrow{P_2O_5}$ EtOH 80° 1 hr $C_{12}H_{18}O$

Nozaki, Koyama, Mori, and Noyori, *Tetrahedron Lett.* **1968**, 2181.

93.

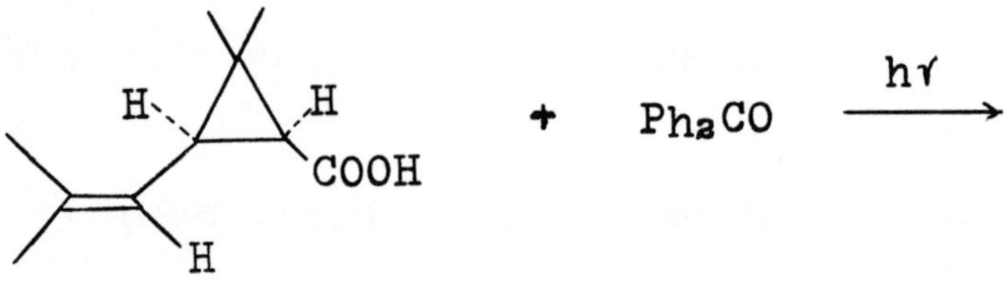

H, H, COOH, H + Ph_2CO $\xrightarrow{h\nu}$

Sasaki, Eguchi, and Ohno, *J. Org. Chem.* **33**, 676 (1968).

94.

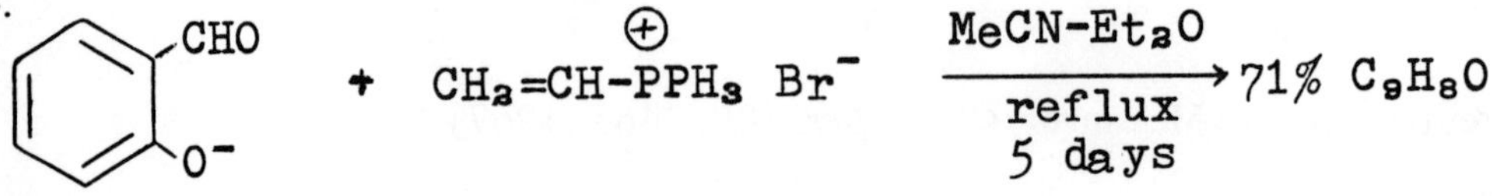

CHO, O^- + $CH_2{=}CH{-}\overset{\oplus}{P}Ph_3$ Br^- $\xrightarrow{\text{MeCN-Et}_2\text{O}}$ reflux 5 days 71% C_9H_8O

Schweizer, Liehr, and Monaco, *J. Org. Chem.* **33**, 2416 (1968).

95. When treated with B_2H_6, i gave the normal reduction product (ii), but xanthone (iii) gave xanthene (iv).

95. Cont.

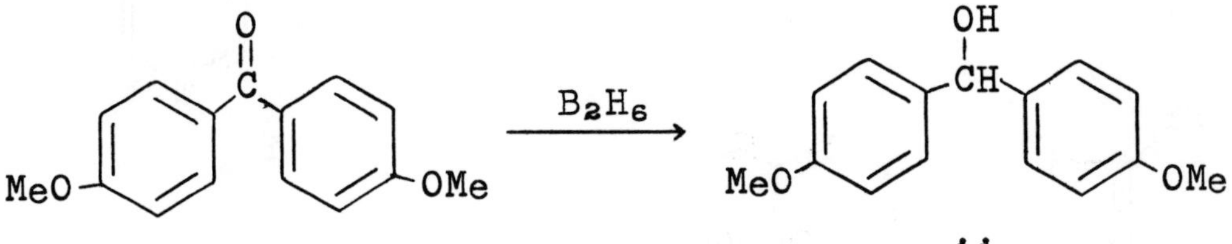

O
C
MeO OMe
B_2H_6
OH
CH
MeO OMe

i ii

O
B_2H_6
O O

iii iv

Explain.

Wechter, *J. Org. Chem.* **28**, 2935 (1963).

96.

a.
O Ph
C–N–CH_2CN
NO_2
NaOEt
EtOH
or aq.
NaOH
or piper-
idine
O
N–Ph
N O
OH
⇌
O
N–Ph
N OH
O

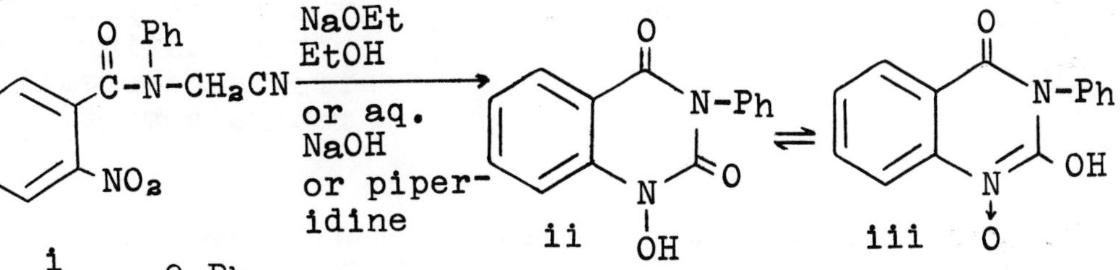

i ii iii

b.
O Ph
C–N–CH–CN
CH_3
NO_2
NaOEt
EtOH
O
N–Ph
N
H

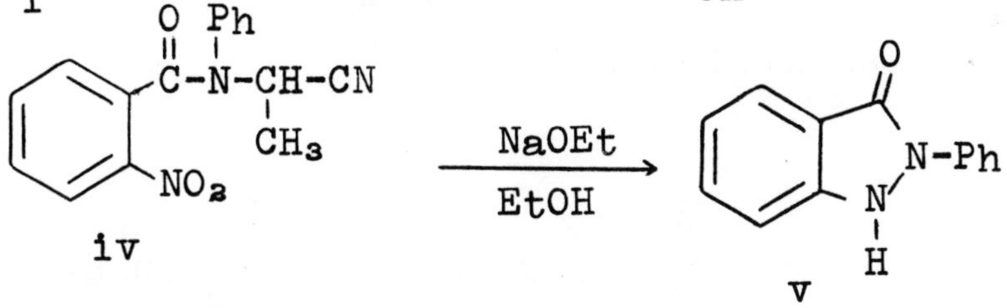

iv v

c.
O H
C–N–CH–CN
CH_3
NO_2
NaOEt
EtOH
O
N–H
N CH_3
O

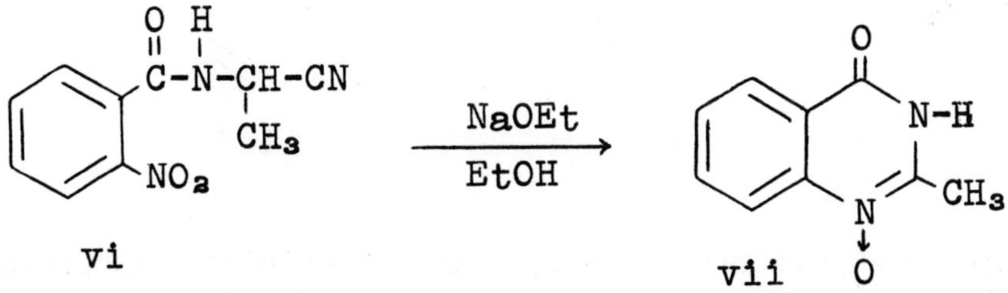

vi vii

Explain the different results.

97.

Ph-cyclobutenedione + o-phenylenediamine ⟶ (not formed) i; ⟶ ii ($PhCH_2-C(=O)-$quinoxaline)

a. Suggest a reasonable mechanism for the formation of ii.

b. Why was i, the normally expected product of di-Schiff base formation, not formed in this case?

Smutny, Caserio, and Roberts, *J. Amer. Chem. Soc.* **82**, 1793 (1960).

98.

$$PhCH(Br)CHO \xrightarrow[\text{ether}]{NaOMe} Ph\text{-}CH\text{—}CH\text{-}OMe \ (\text{epoxide, } O \text{ bridging})$$

i

(+ another product resulting from a Favorskii rearrangement)

Suggest a reasonable mechanism for the formation of i, and predict its configuration.

Riehl and Thil, *Tetrahedron Lett.* **1969**, 1913.

99.

Ph-CH=N-Ph (i)

o-HO-C_6H_4-CH=N-Ph (ii)

o-CH_3-C_6H_4-CH=N-Ph (iii)

At pH 5.6 to 6.6, i and iii hydrolyzed (to ArCHO and $PhNH_2$) at about the same rate, but ii hydrolyzed about 10 times more slowly. Explain.

Willi and Siman, *Can. J. Chem.* **46**, 1589 (1968).

100.

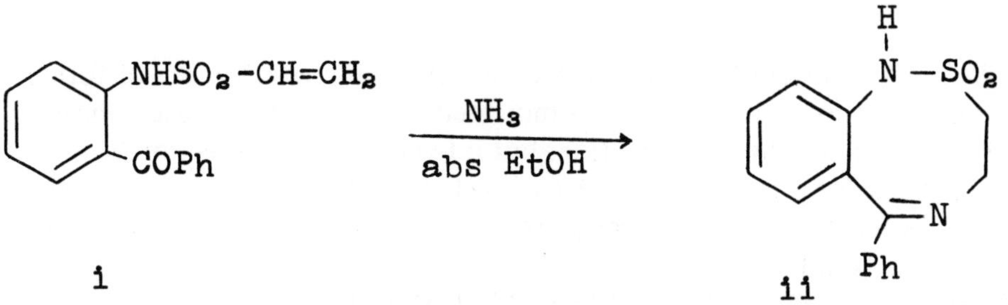

NHSO$_2$-CH=CH$_2$

COPh

$\xrightarrow[\text{abs EtOH}]{NH_3}$

H

N —SO$_2$

N

Ph

i ii

For this reaction, one may conceive two possible mechanisms:

a.

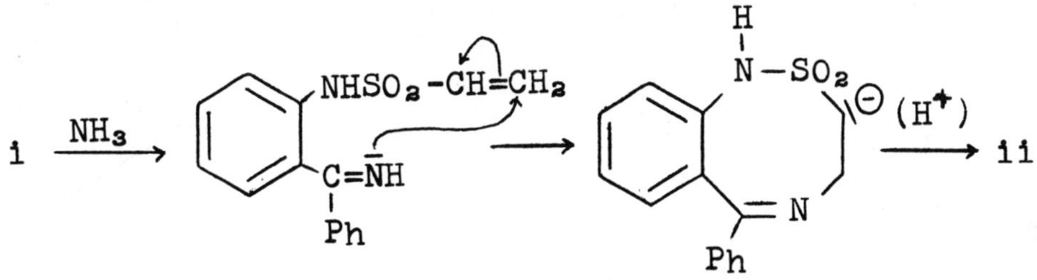

i $\xrightarrow{NH_3}$ NHSO$_2$-CH=CH$_2$ / C=NH / Ph $\longrightarrow$ H / N—SO$_2$ / ⊖ (H$^+$) / N / Ph $\longrightarrow$ ii

b.

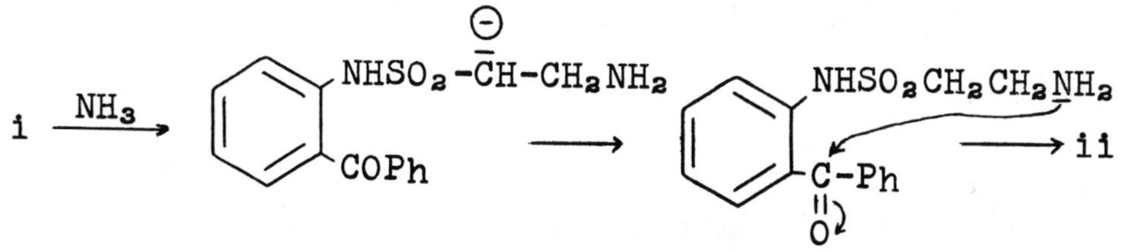

i $\xrightarrow{NH_3}$ NHSO$_2$-CH(⊖)-CH$_2$NH$_2$ / COPh $\longrightarrow$ NHSO$_2$CH$_2$CH$_2$NH$_2$ / C-Ph / O $\longrightarrow$ ii

Devise one or more experiments to distinguish these possibilities.

101.

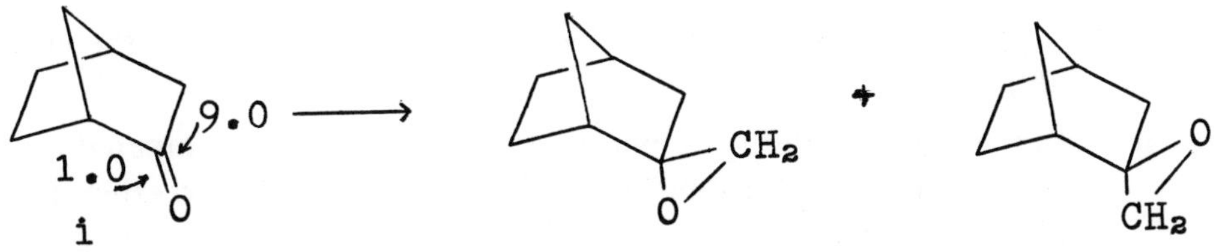

1.0, 9.0 $\longrightarrow$ CH$_2$ / O + O / CH$_2$

i

norcamphor

14, 5.8 $\longrightarrow$ CH$_2$ / O + O / CH$_2$

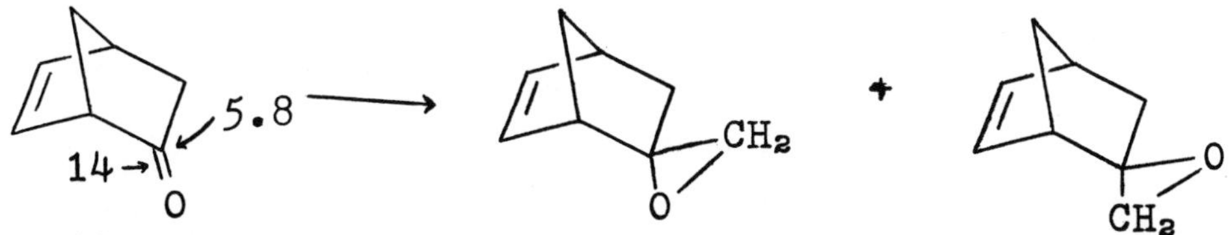

ii

dehydro-
norcamphor

101. Cont.

When i and ii were treated with $Me_2SOCH_2^-$, the relative rate of formation of the four products was as indicated above; that is, i reacts more favorably from the exo direction, but ii from the endo direction. This behavior is in sharp contrast with that of other nucleophilic reagents, such as $NaBH_4$ or Grignard reagents, which attack both i and ii from the exo side. Explain.

Bly, DuBose, and Konizer, *J. Org. Chem.* **33**, 2188 (1968).

102.

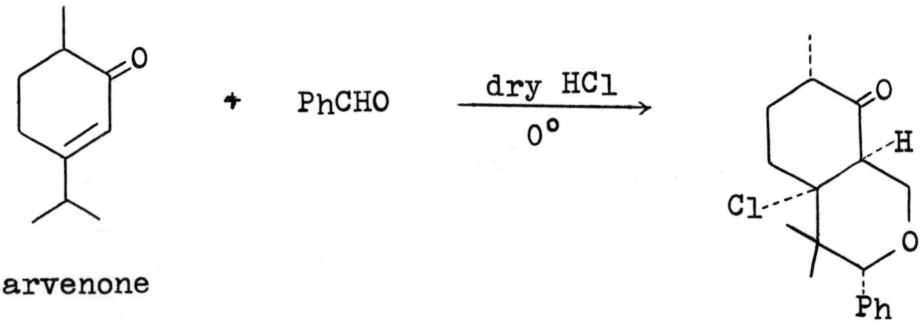

carvenone + PhCHO —(dry HCl, 0°)→ [product: Cl, H, O, Ph]

a. Suggest a reasonable mechanism.

b. Devise one or more experiments to test your mechanism.

Baxter, Forward, and Whiting, *J. Chem. Soc., C* **1968**, 1162, 1169.

103.

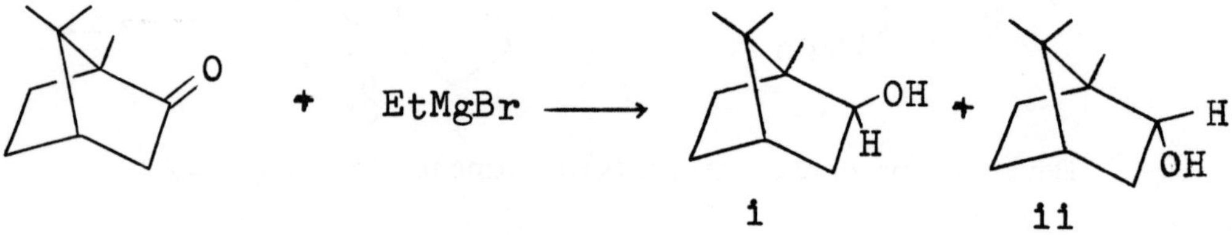

[ketone] + EtMgBr → [OH, H] i + [H, OH] ii

a. Explain why reduction takes place in this case, rather than the normal addition.

b. As the solvent polarity is increased, more i is formed compared to ii (e.g. in 100% Et_2O, i:ii ratio 35:65; in 30:70 Et_2O-petr. ether, i:ii ratio 25:75). Explain.

Suga, Watanabe, and Yamaguchi, *Aust. J. Chem.* **22**, 669 (1969).

104.

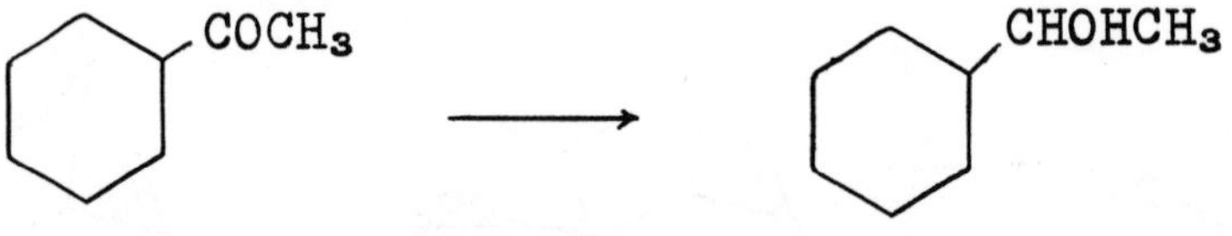

[cyclohexyl]$COCH_3$ → [cyclohexyl]$CHOHCH_3$

i

If i is reduced, in a Meerwein-Ponndorf-Verley reduction, by S-(+)-3,3-dimethyl-2-butoxide ion, which enantiomer of cyclohexylmethylcarbinol would be expected to predominate, R or S?

PROBLEM SET 11
ELIMINATIONS

Corresponds to Chapter 17 of Advanced Organic Chemistry.

The first 25 problems are synthetic. In each one you are asked to convert one compound into another. In every case the conversion has actually been carried out, and the reference is given.

1.

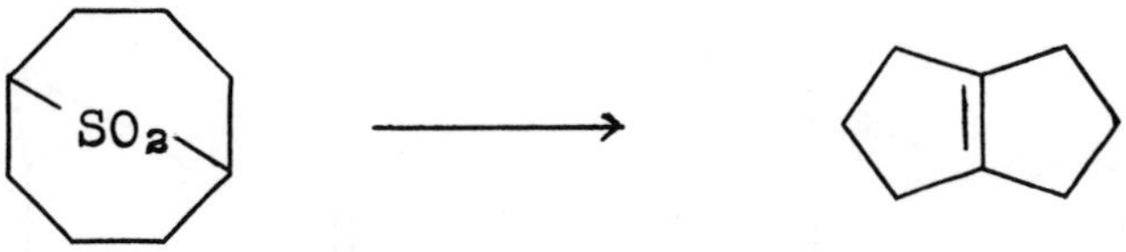

SO_2 ⟶

Corey and Block, *J. Org. Chem.* **34**, 1233 (1969).

2.

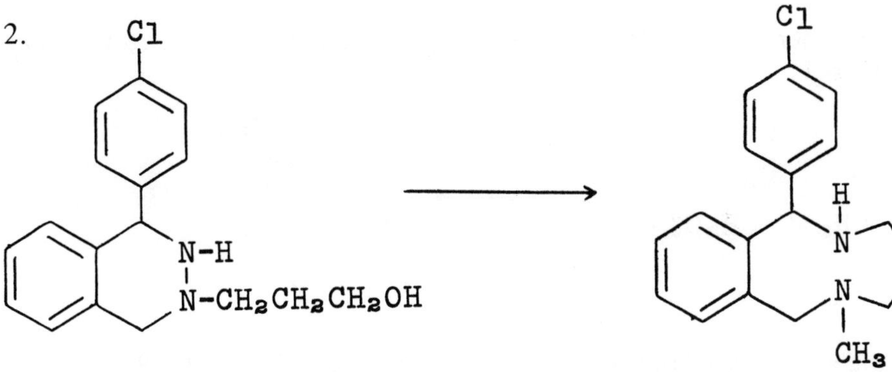

Cl Cl

N-H H

N- $CH_2CH_2CH_2OH$ ⟶ N

N

CH_3

Aeberli and Houlihan, *J. Org. Chem.* **34**, 2715 (1969).

3.

$^{13}CO_2$ ⟶ Ph-^{13}C≡CH

i

4.

CH_2=CH-CH_2Cl ⟶ O O CH_3

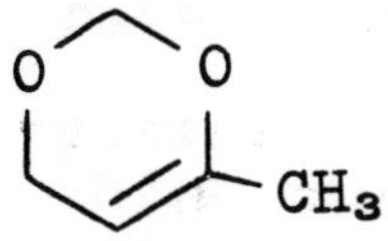

Salomaa and Hautoniemi, *Acta Chem. Scand.* 23, 709 (1969).

5.

Cl_3C-CH(OH)-NH-COOEt ⟶ Cl_3C-CH=N-COOEt

Ulrich, Tucker, and Sayigh, *J. Org. Chem.* **33**, 2887 (1968).

6.

Me COOEt / EtOOC / CH_3 / N–H → Me COOEt / EtOOC / CN / N–H

i ii

7.

[epoxide structures] →

Vogel and Klärner, *Angew. Chem. Intern. Ed. Engl.* **7**, 374 (1968) [*Angew. Chem.* **80**, 402 (1968)].

8.

$CH_2=C(CH_3)-CH_2Cl$ → CH_3 / CH_3 / N–H

Rosenmund and Grübel, *Angew. Chem. Intern. Ed. Engl.* **7**, 733 (1968) [*Angew. Chem.* **80**, 702 (1968)].

9.

OMe / SO_2H → OMe / SO_2CHN_2

m-methoxybenzene-sulfinic acid

i

ii

10.

CF_3COOAg → $CF_3CF_2CF=C=CF_2$

Banks, Braithwaite, Haszeldine, and Taylor, *J. Chem. Soc., C* **1968**, 2593.

11.

Note: this conversion cannot be accomplished by direct hydrogenation of i because the C=C bond in conjugation with the C=O bond is preferentially hydrogenated.

Unde, Hiremath, Kulkarni, and Kelkar, *Tetrahedron Lett.* **1968**, 4861.

12.

Wiberg and Hiatt, *J. Amer. Chem. Soc.* **90**, 6495 (1968).

13.

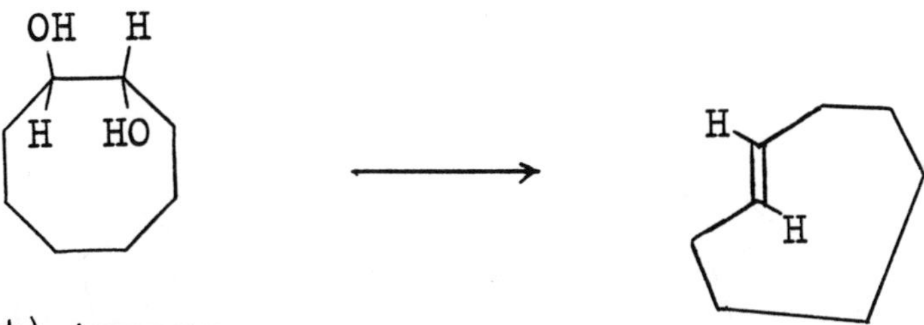

(+)-trans-1,2-cyclooctanediol

optically active trans-cyclooctene

Note: The conversion must be accomplished without resolution.

Corey and Shulman, *Tetrahedron Lett.* **1968**, 3655.

14.

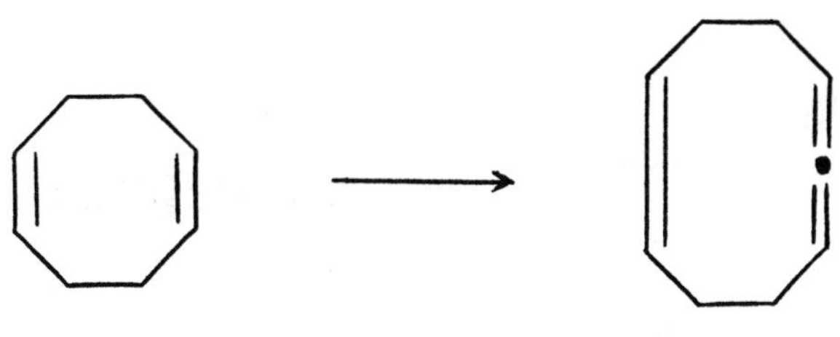

i

1,2,6-cyclononatriene

ii

15.

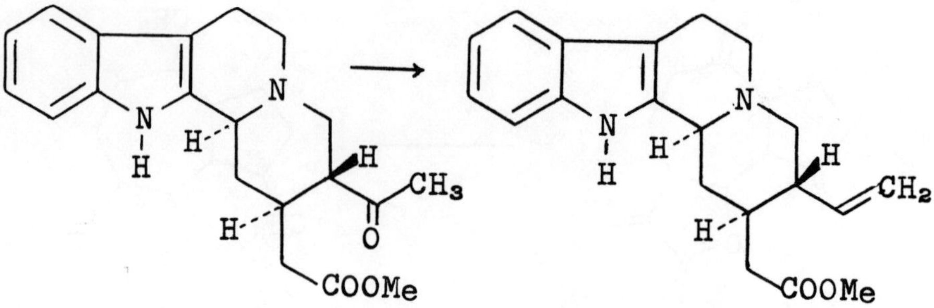

N
N-H
H
H
H
CH_3
O
H
COOMe

N
N-H
H
H
H
CH_2
H
COOMe

van Tamelen and Wright, *J. Amer. Chem. Soc.* **91**, 7349 (1969).

16.

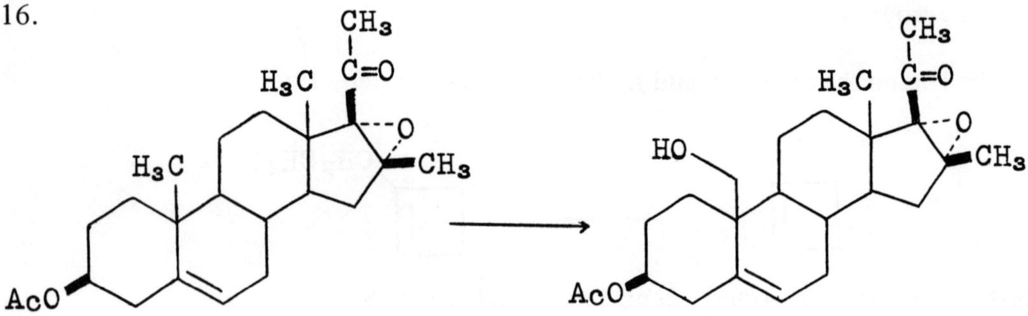

CH_3
C=O
H_3C
O
H_3C
CH_3
AcO

i

CH_3
C=O
H_3C
O
HO
CH_3
AcO

ii

17.

Cl
Cl

COOH
CN

Wilcox and Leung, *J. Org. Chem.* **33**, 877 (1968).

18.

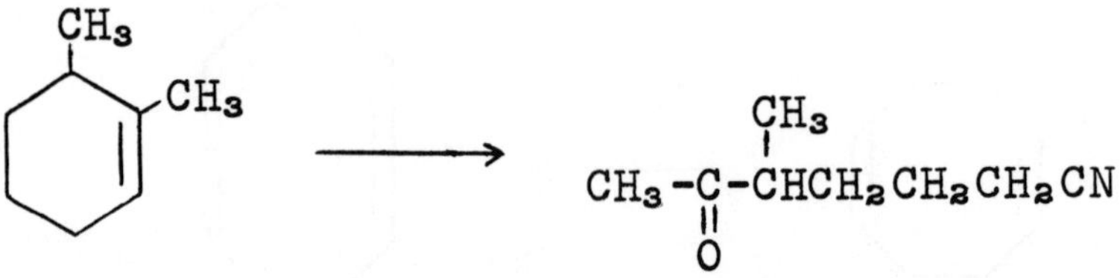

CH_3
CH_3

i

$CH_3-\underset{\underset{O}{\parallel}}{C}-\overset{\overset{CH_3}{|}}{C}HCH_2CH_2CH_2CN$

ii

19.

Paul, Johnson, Barrett, and Paquette, *Chem. Commun.* **1969**, 6.

20.

cis isomer

4,5-homotropone

Chapman and Fugiel, *J. Amer. Chem. Soc.* **91**, 215 (1969).

21.

the double bond is trans

Heathcock and Badger, *Chem. Commun.* **1968**, 1510.

22.

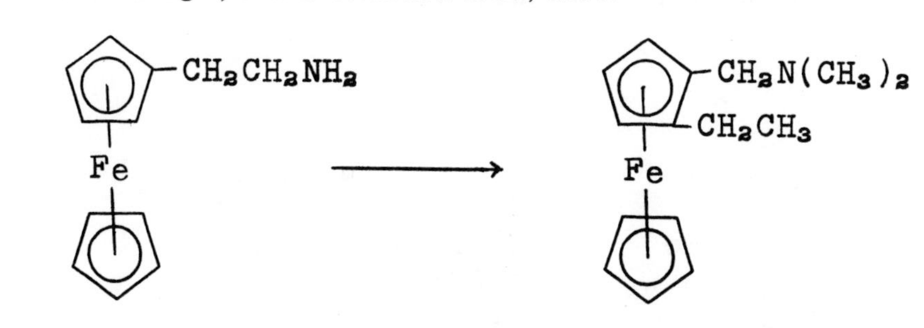

Lednicer and Hauser, *J. Org. Chem.* **24**, 43 (1959).

23.

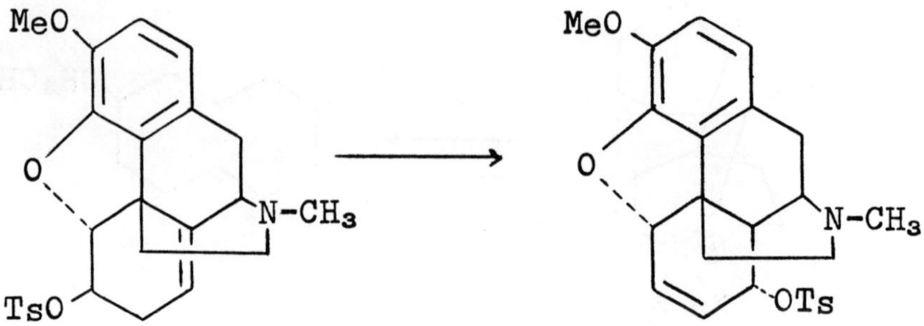

MeO, O, TsO, N–CH_3 → MeO, O, N–CH_3, OTs

isoneopine-O-tosylate

Kugita, Takeda, and Inoue, *Tetrahedron* **25**, 1851 (1969).

24.

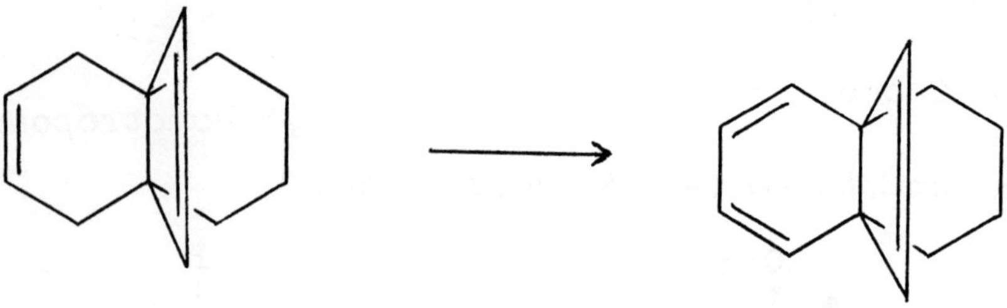

i → ii

25.

$H_2NCH_2CH_2CH_2NH_2$ → $N_2CHCH_2CHN_2$

Hart and Brewbaker, *J. Amer. Chem. Soc.* **91**, 706 (1969).

The following conversions (problems 26 to 50) have been reported in the literature. In each case suggest a reasonable mechanism.

26.

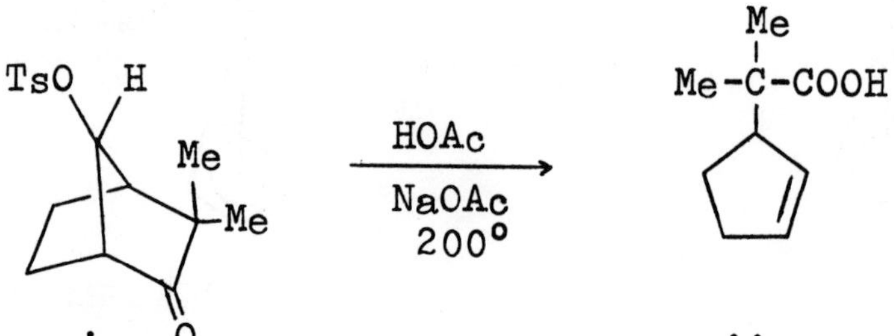

i (TsO, H, Me, Me, O) → (HOAc, NaOAc, 200°) → ii (Me–C(Me)–COOH)

27. nicotinamide + $NH_2SO_2O^- NH_4^+$ $\xrightarrow[3\ hr]{150-200°}$ 60.5% i

Note: Heating at 200° of nicotinonitrile did not give i.

Osselaere, Dejardin, and Dejardin-Duchêne, *Bull. Soc. Chim. Belges* **78**, 289 (1969).

28.

Me-N, O, N-Me, O

1. KOH-EtOH reflux 5 hr
2. HCl

83%

CONHMe, CONHMe

Collington, Hey, and Rees, *J. Chem. Soc., C* **1968**, 1026.

29.

OTs, H, CH_3, CHO, H

OH^-
$MeOH-H_2O$ reflux 5 hr

CH_3

98%

Nerdel, Barth, Frank, and Weyerstahl, *Chem. Ber.* **102**, 407 (1969).

30.

H_2N–(pyridine)–Br + NH_2^- excess $\xrightarrow[\text{-33°, 1.5 min}]{\text{liq. }NH_3}$ $NCCH{=}CHCH_2CN$

i — 5% ii

31.

(1,4-dihydronaphthalene with CH_2OTs at C-1 and C-4) $\xrightarrow[\text{5 hr}]{\text{boiling pyridine}}$ 90%

Brown and Sondheimer, *J. Amer. Chem. Soc.* **89**, 7116 (1967).

32.

HO, Ph (cyclohexenyl–cyclohexanol) $\xrightarrow{\sim 500°}$ cyclohexylidene$=CHCH_2CH_2CH_2CH_2{-}C(=O){-}Ph$

~100%

Arnold and Metzger, *J. Org. Chem.* **26**, 5185 (1961).

33.

(piperidine N–CH_2–(3,5-dimethyl-4-hydroxyphenyl); Me, Me, OH) + CS_2 $\xrightarrow[\text{reflux, 3 hr}]{\text{EtOH}}$ 90% (piperidine N–C(=S)–S–CH_2–(3,5-dimethyl-4-hydroxyphenyl); Me, Me, OH)

Notes: The reaction also took place when the CH_2 was replaced by CHMe or CMe_2, but not when the OH group was ortho instead of para, or when the OH group was replaced by either H or OMe. The reaction also failed when attempted with the methiodide of i.

Fitton, Rigby, and Hurlock, *J. Chem. Soc., C* **1969**, 230.

34.

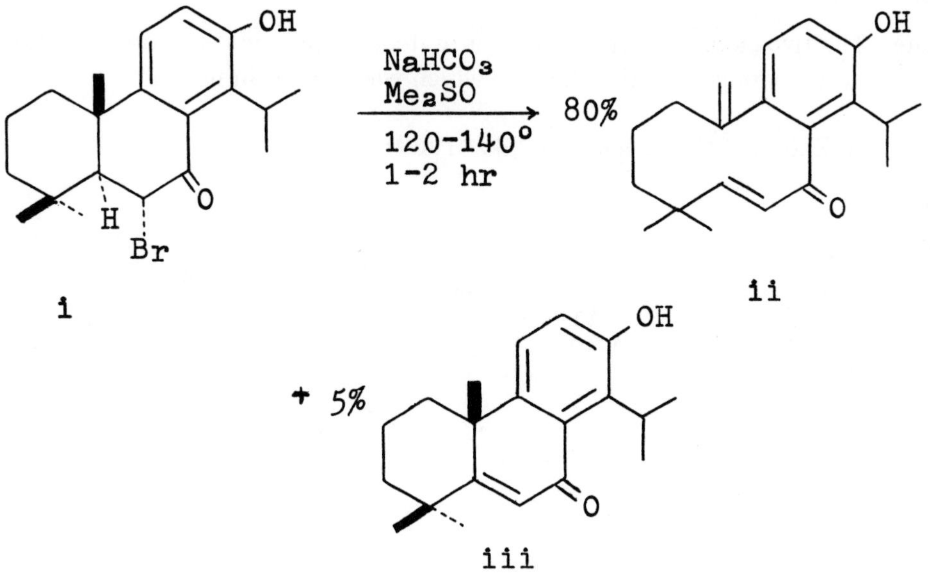

i

$NaHCO_3$, Me_2SO, 120-140°, 1-2 hr → 80% ii + 5% iii

iii is the normal elimination product. Explain how ii is formed and why the principal product is not iii, but ii.

35.

$$CH_2{=}CH{-}O{-}\underline{t}{-}Bu \xrightarrow{350^\circ} CH_3CHO + CH_3{-}C(CH_3){=}CH_2$$

Bamkole and Emovon, *J. Chem. Soc., B* **1968**, 332.

36.

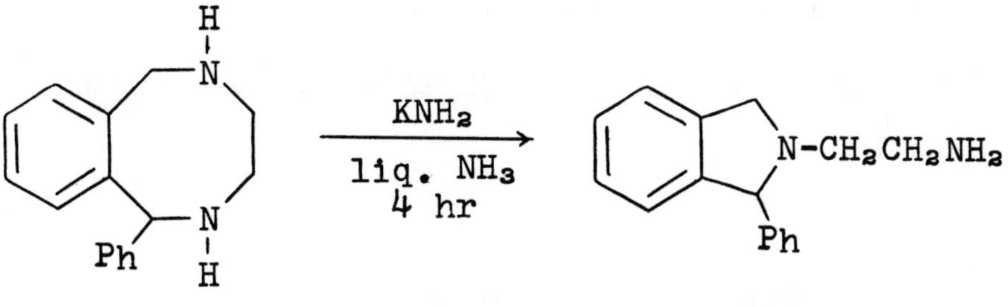

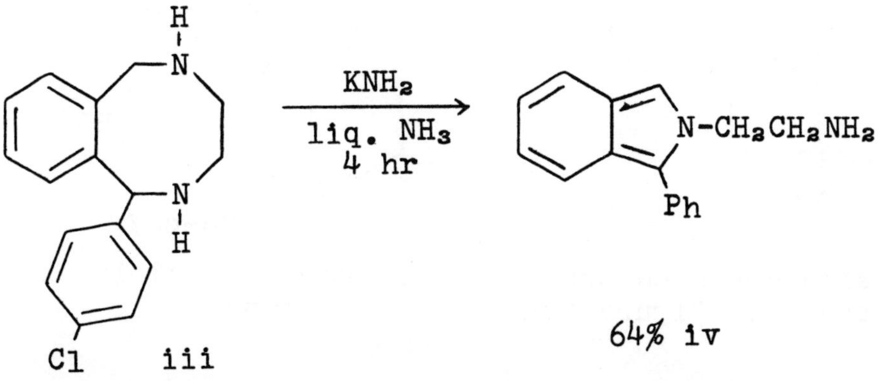

i → (KNH_2, liq. NH_3, 4 hr) → N-$CH_2CH_2NH_2$, Ph — 80% ii

iii (Cl) → (KNH_2, liq. NH_3, 4 hr) → N-$CH_2CH_2NH_2$, Ph — 64% iv

36. Cont.

Note: The five-membered ring in ii is saturated, while in iv it is aromatic. Note also that iv no longer has the chlorine which was present in iii.

Winn and Zaugg, *J. Org. Chem.* **34**, 249 (1969).

37.

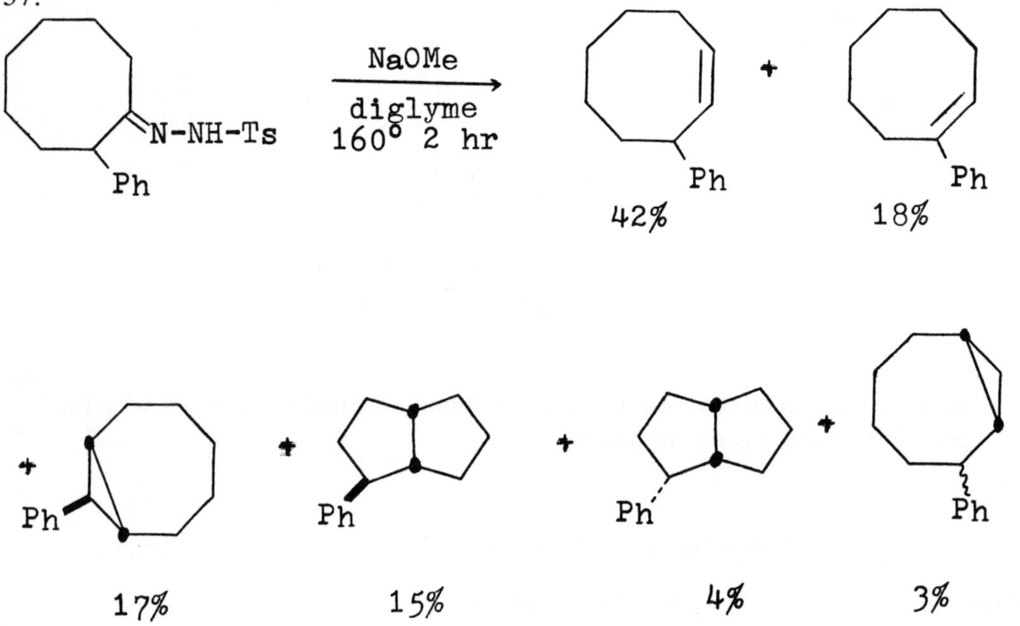

N-NH-Ts

Ph

NaOMe

diglyme

160° 2 hr

Ph 42% + Ph 18%

+ Ph 17% + Ph 15% + Ph 4% + Ph 3%

Cope and Hecht, *J. Amer. Chem. Soc.* **89**, 6920 (1967).

38. Bu-CH(N$_3$)-C(=O)-Cl (i) + Et_3N $\xrightarrow{\text{anhyd. } Et_2O}$ BuCN (ii)

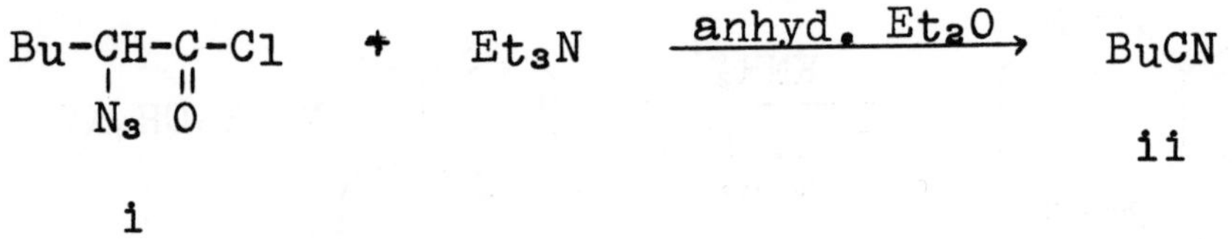

39.

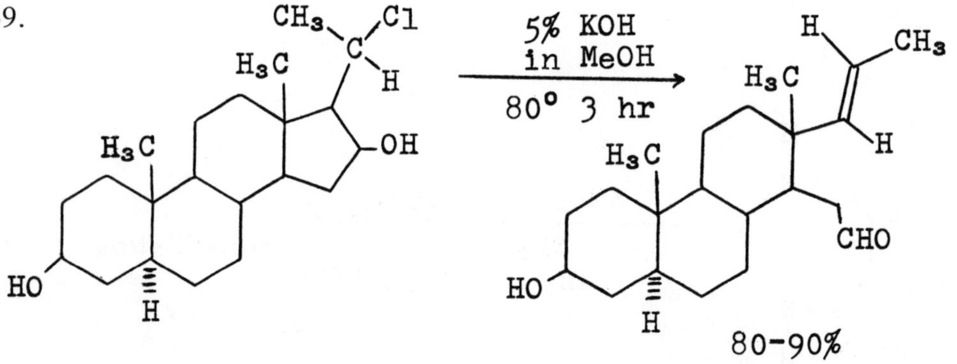

CH_3, Cl, C, H, H_3C, OH, H_3C, HO, H

5% KOH in MeOH

80° 3 hr

H, CH_3, H_3C, H, H_3C, CHO, HO, H

80-90%

The side chain carbon has the R configuration

The double bond is trans

Adam, *Angew. Chem. Intern. Ed. Engl.* **6**, 631 (1967) [*Angew. Chem.* **79**, 619 (1967)].

40.

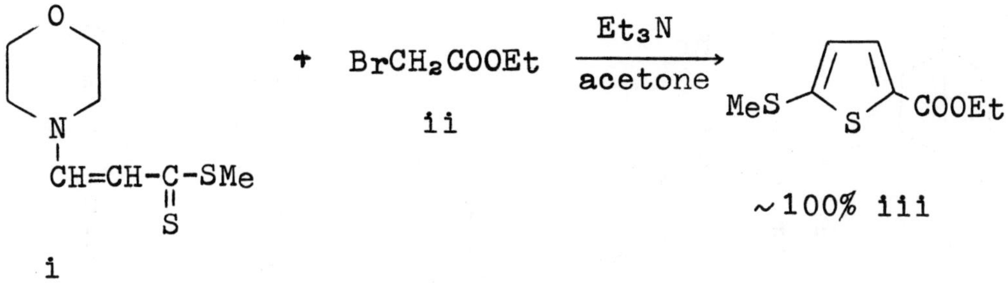

i + $BrCH_2COOEt$ (ii) $\xrightarrow[\text{acetone}]{Et_3N}$ MeS–(thiophene)–COOEt

CH=CH-C(=S)-SMe (N-morpholino)

i

ii

~100% iii

41.

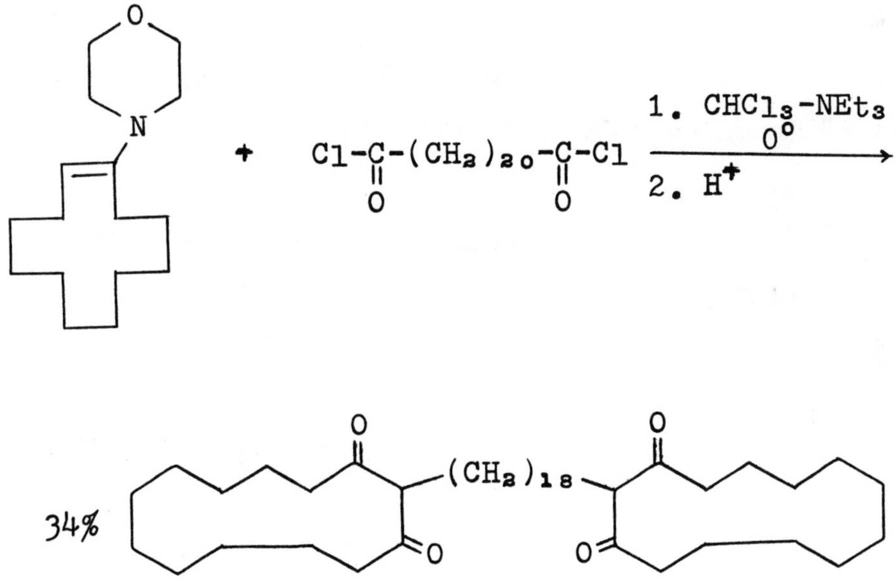

(1-morpholino-cyclododecene) + Cl-C(=O)-$(CH_2)_{20}$-C(=O)-Cl $\xrightarrow[\text{2. } H^+]{\text{1. } CHCl_3\text{-}NEt_3,\ 0^\circ}$

34% (bis(cyclododecane-1,3-dione)-linked $(CH_2)_{18}$)

Wakselman, *Bull. Soc. Chim. Fr.* **1967**, 3763; Kirrmann and Wakselman, *Bull. Soc. Chim. Fr.* **1967**, 3766. See also Hunig and Buysch, *Chem. Ber.* **100**, 4010, 4017 (1967).

42.

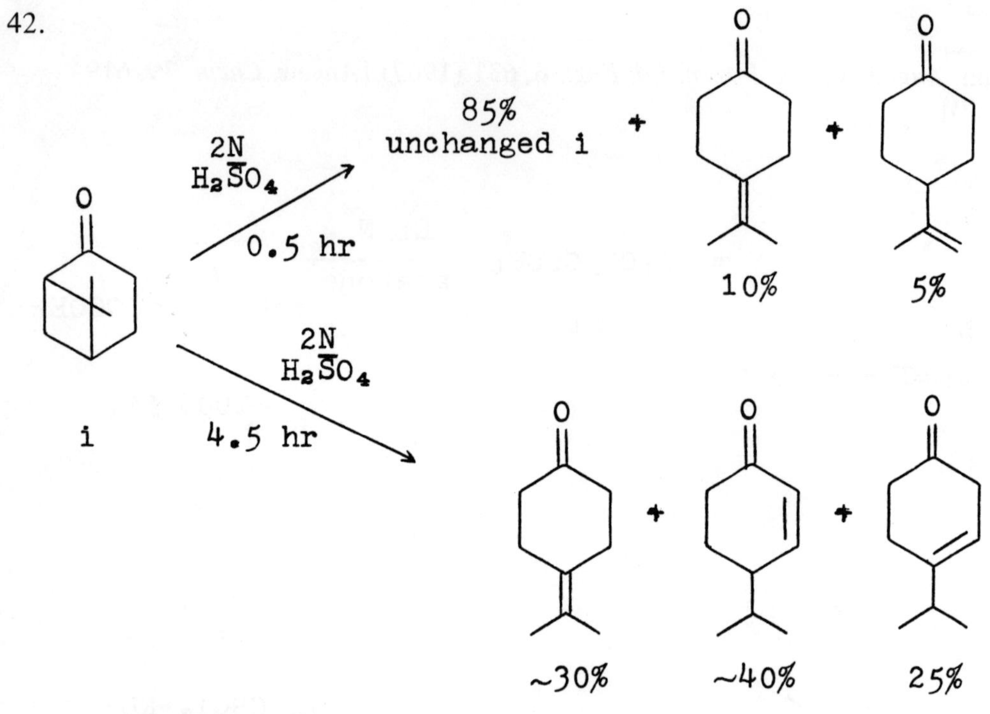

Lewis and Williams, *Aust. J. Chem.* **21**, 2467 (1968).

43.

$CF_2Cl-C(OEt)(CFCl_2)-{}^{18}O-C(=O)-CH_3 \xrightarrow{230°}$ 65% $CF_2Cl-C(={}^{18}O)-CF_2Cl$ (i) + 53% CH_3COOEt (ii)

44.

CH_2OTs

HO

$\xrightarrow{\text{NaOMe, MeOH, reflux 6 hr}}$

CH_3

CHO

85%

Marshall, Scanio, and Iburg, *J. Org. Chem.* **32**, 3750 (1967).

45.

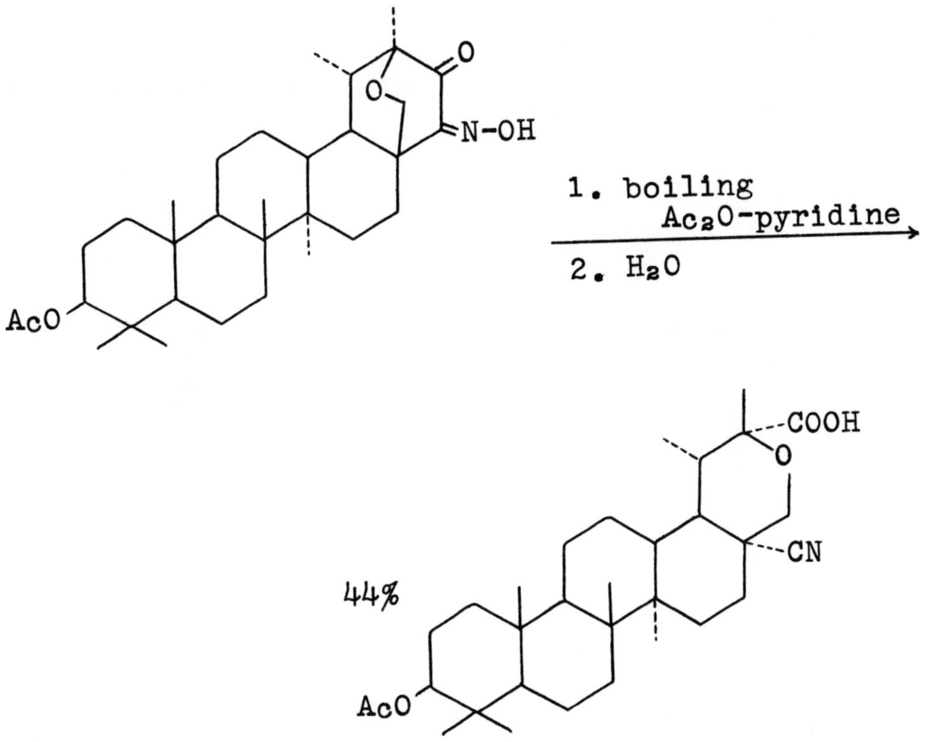

AcO

1. boiling Ac_2O-pyridine

2. H_2O

44%

Říhová and Vystrčil, *Cellect. Czech. Chem. Commun.* **34**, 240 (1969).

46.

$$MeSO_2OCH_2CH{=}CHCH_2OSO_2Me \xrightarrow[\text{dioxane-}H_2O \text{ reflux 3 hr}]{NaOH} CH_3CH{=}CHCHO$$

Hudson and Withey, *J. Chem. Soc., B* **1966**, 237.

47.

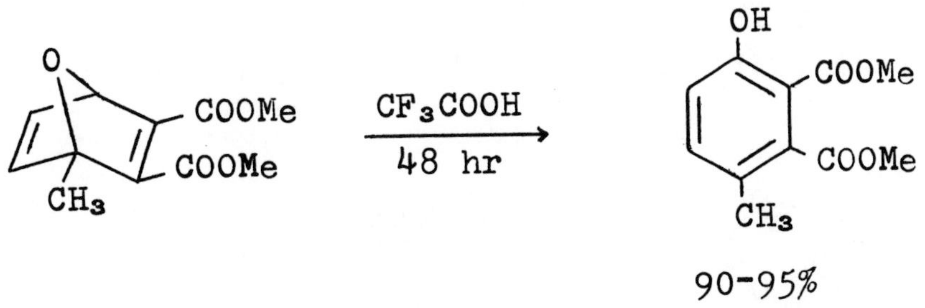

$$\xrightarrow[48 \text{ hr}]{CF_3COOH}$$

90–95%

Vogel, Willhalm, and Prinzbach, *Helv. Chim. Acta* **52**, 584 (1969).

48.

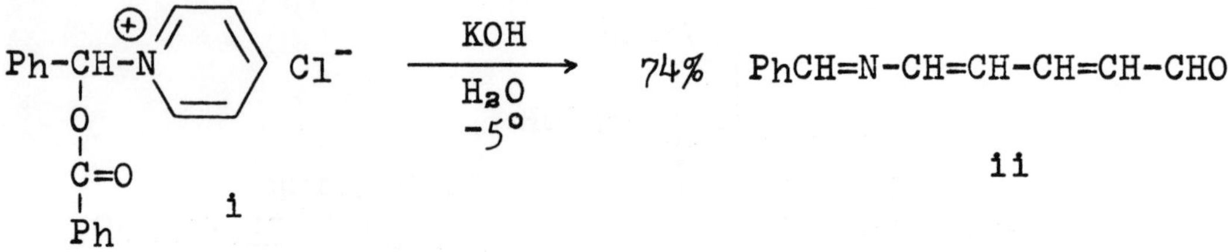

Ph-CH-N⊕(pyridinium) Cl⁻ with O-C(=O)-Ph on CH — **i**

$\xrightarrow[\text{H}_2\text{O},\ -5^\circ]{\text{KOH}}$ 74% PhCH=N-CH=CH-CH=CH-CHO

ii

49.

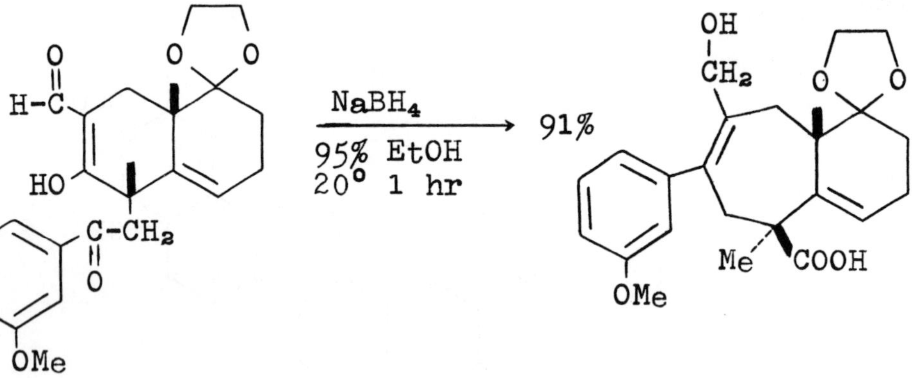

$\xrightarrow[\text{95\% EtOH, } 20^\circ \text{ 1 hr}]{\text{NaBH}_4}$ 91%

Hand and Los, *Chem. Commun.* **1969**, 673.

50.

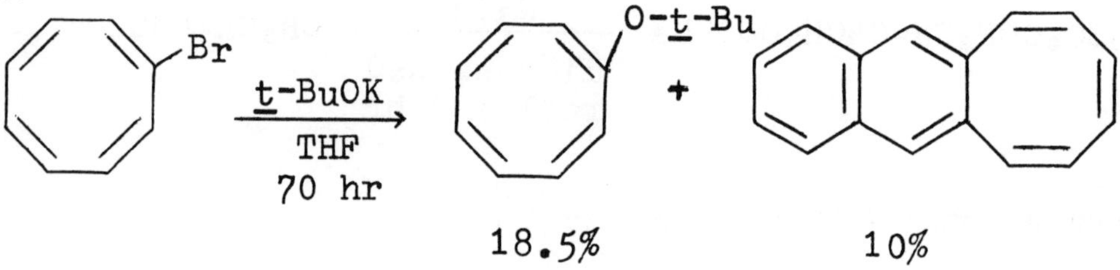

Br (bromocyclooctatetraene) $\xrightarrow[\text{THF, 70 hr}]{t\text{-BuOK}}$ O-t-Bu (18.5%) + (10%)

Krebs and Byrd, *Justus Liebigs Ann. Chem.* **707**, 66 (1968).

In the following problems (51 to 53) give the products and suggest reasonable mechanisms.

51.

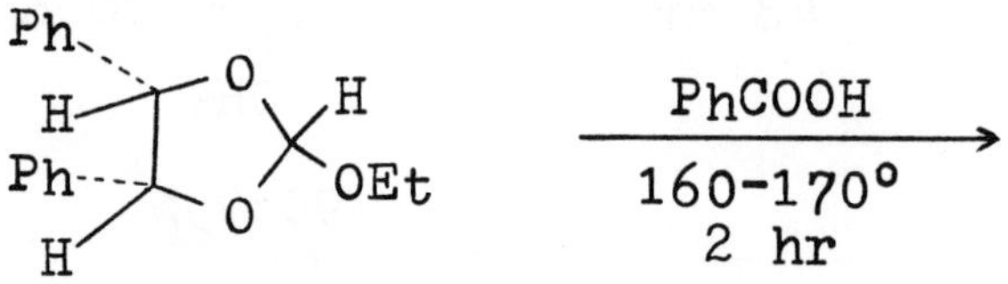

Ph, H, Ph, H, O, O, H, OEt $\xrightarrow[\text{160-170}^\circ\text{, 2 hr}]{\text{PhCOOH}}$

Josan and Eastwood, *Aust. J. Chem.* **21**, 2013 (1968).

52.

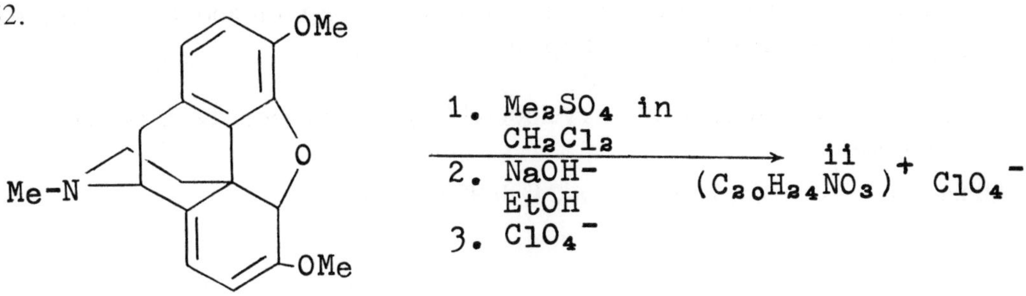

OMe

Me-N

O

OMe

i

thebaine

1. Me_2SO_4 in CH_2Cl_2
2. NaOH-EtOH
3. ClO_4^-

→ ii $(C_{20}H_{24}NO_3)^+$ ClO_4^-

53.

Ph, N, NH_2 + $COCH_3$, O

AcOH, 1.5-3 hr, 0° → 77% i → 140-50°, 60 torr, 1-1.5 hr →

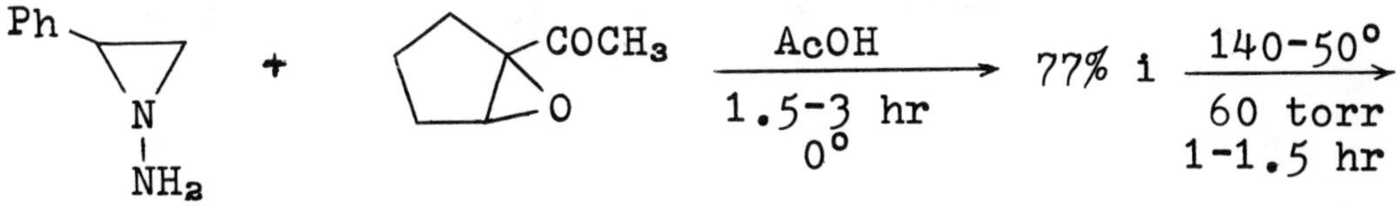

94% $CH_3C{\equiv}CCH_2CH_2CH_2CHO$ + $PhCH{=}CH_2$ + N_2

Felix, Schreiber, Piers, Horn, and Eschenmoser, *Helv. Chim. Acta* **51**, 1461 (1968).

54.

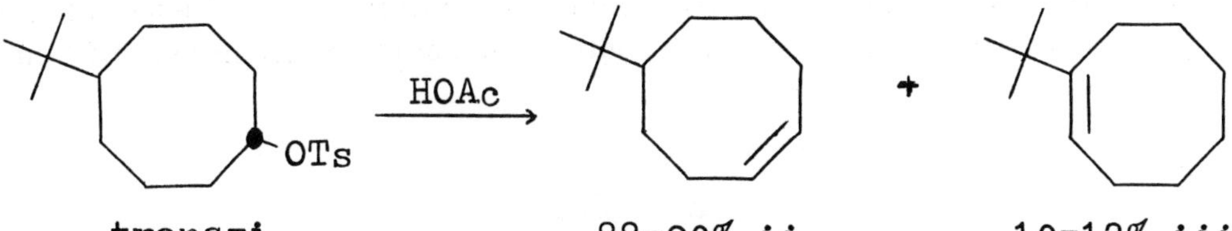

OTs

HOAc →

+

trans-i

88-90% ii

10-12% iii

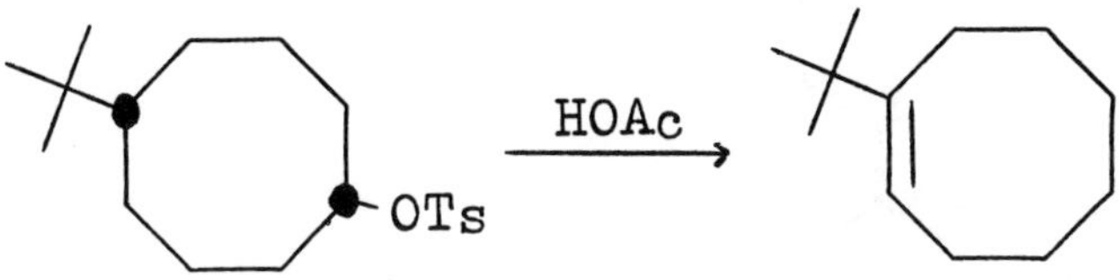

OTs

HOAc →

cis-i

99% iii

Acetolysis of *trans*-i gave mostly ii. Acetolysis of *cis*-i was about 50 times faster and gave more than 99% iii. Explain.

Note: Acetolysis of unsubstituted cyclooctyl tosylate was about twice as fast as that of *trans*-i.

Allinger and Szkrybalo, *Tetrahedron* **24**, 4699 (1968).

55.

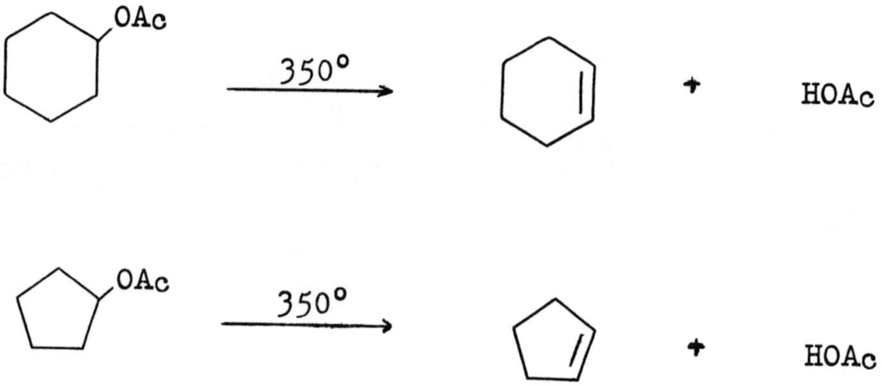

OAc → 350° → + HOAc

OAc → 350° → + HOAc

Cyclopentyl acetate underwent the above reaction 4.8 times as fast as cyclohexyl acetate. Explain why.

Bamkole and Emovon, *J. Chem. Soc., B* **1969**, 187.

56. The 4 thujols, called (-)-thujol, (-)-neothujol, (+)-isothujol, and (+)-neoisothujol all have structure i, but differ in configuration at the positions marked 1 and

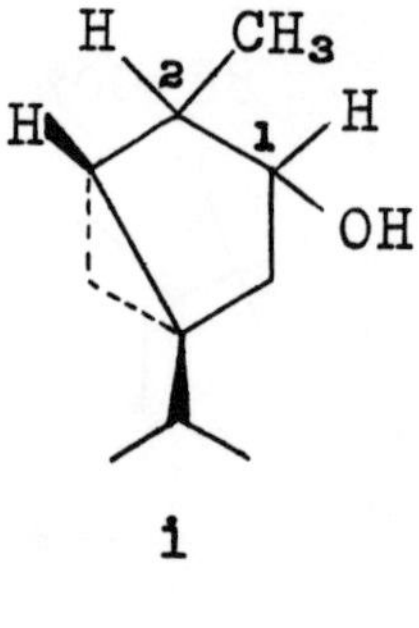

H, CH_3, H, OH, H, 2, 1

i

56. Cont.

2. Each of the 4 was dehydrated (E2 mechanism) with $POCl_3$-pyridine at 0°, under conditions where double bond isomerization of the products does not occur. The following results were obtained:

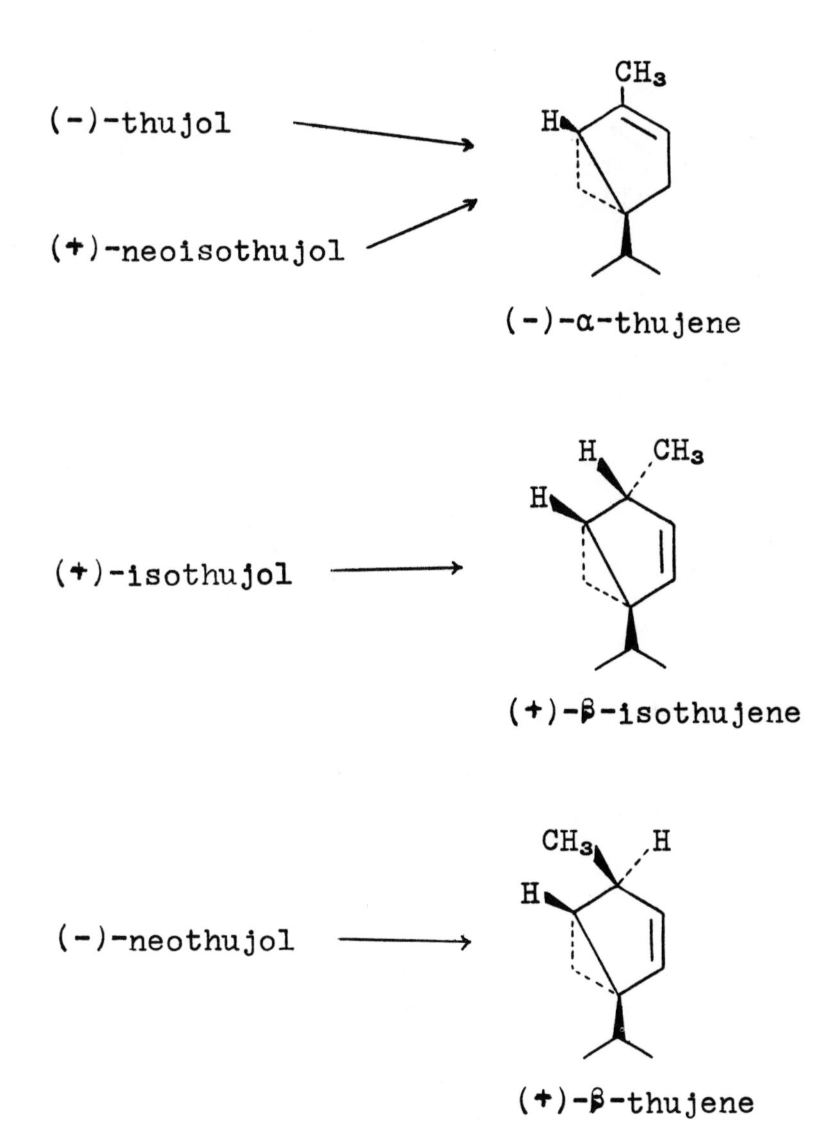

Write structures for those thujols for which it is possible to do so on the basis of the data given. Is it possible for all four?

57. In the conversion of i to ii, with cold H_2SO_4, one can envision two mechanisms:

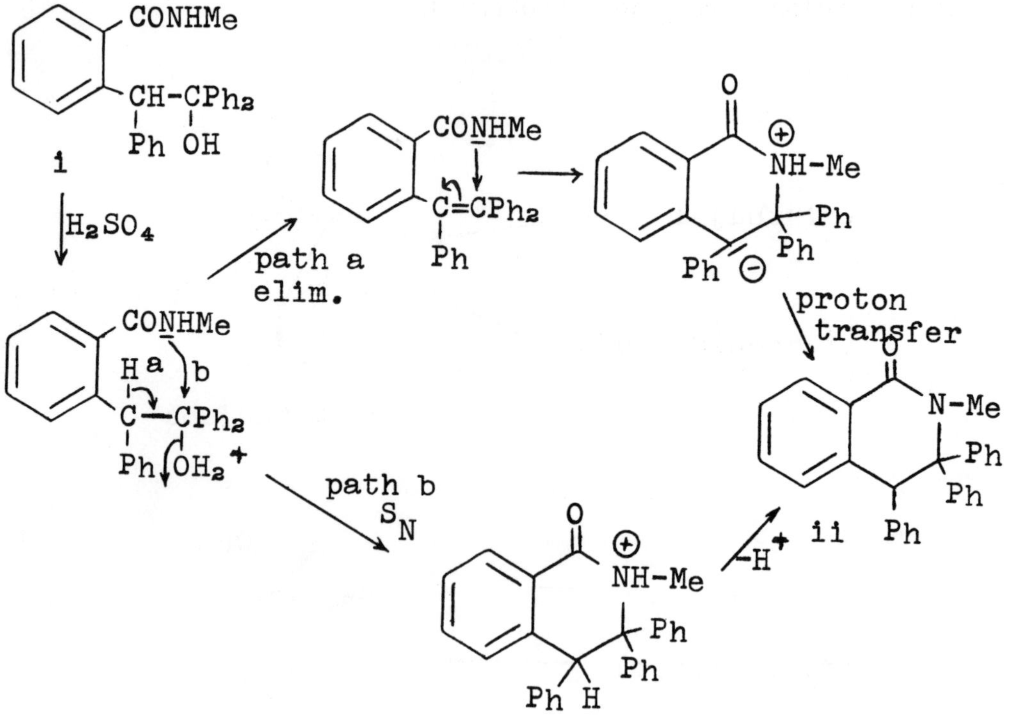

Devise one or more experiments to distinguish these possibilities.
Mao, Henoch, and Hauser, *Chem. Commun.* **1968**, 1595.

58. If you had samples of the 2 isomers of 6β-methylsulfinyl-5α-cholestane (i) how could you determine in which of the two the configuration of the SOMe group

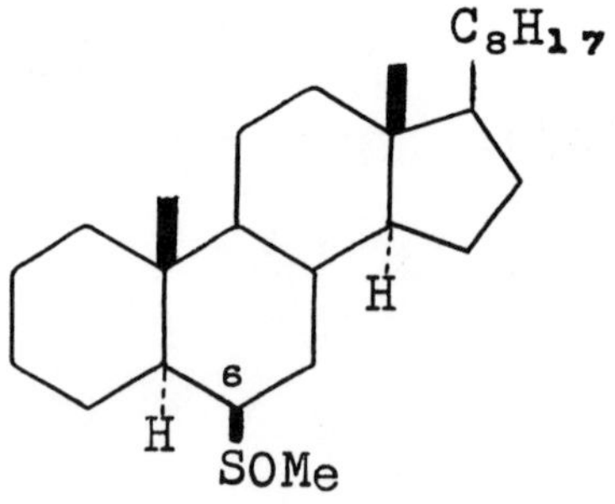

at position 6 if R? (That is, the configuration of the 6-carbon is known: it is β. The question is: what is the configuration at the sulfur?).
Jones, Green, and Whitehouse, *Chem. Commun.* **1968**, 1634; Jones and Higgins, *Chem. Commun.* **1968**, 1685.

59.

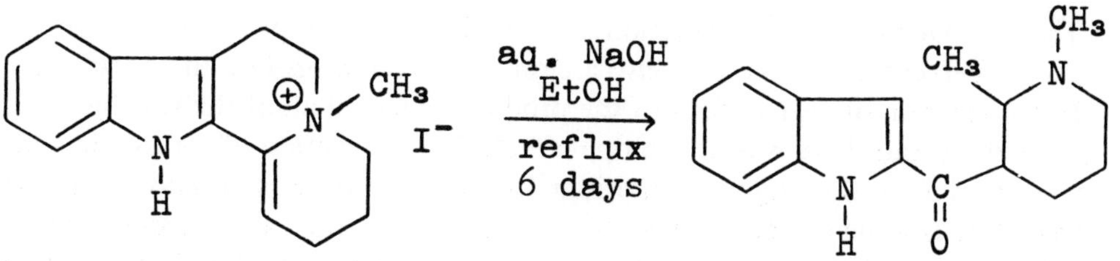

i → (aq. NaOH, EtOH, reflux 6 days) → 41% ii

Suggest a reasonable mechanism for the above reaction, taking into account the following results:

a. Under the same reaction conditions, iii gave iv:

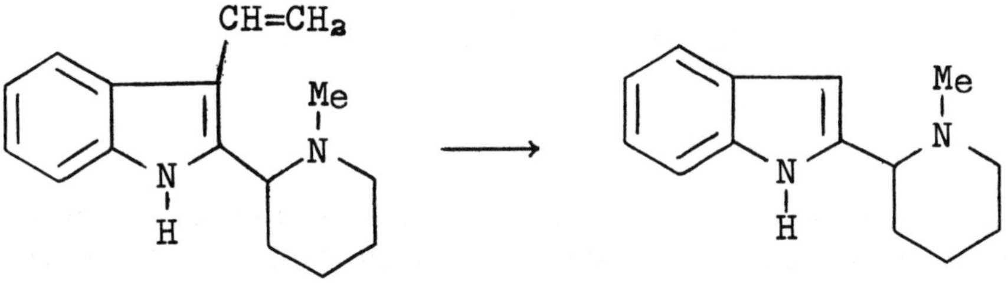

iii → iv

b. When labeled ethanol ($CH_3{}^{14}CH_2OH$) was used, the label appeared in ii. Dolby and Gribble, *Tetrahedron* 24, 6377 (1968).

60.

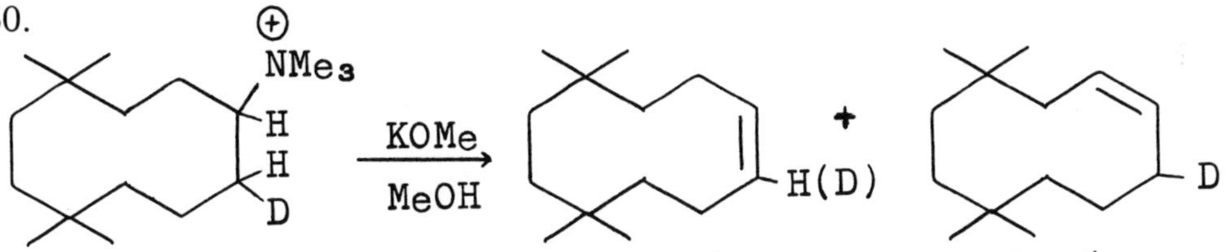

cis and trans-i → (KOMe, MeOH) → ii + iii + iv + v

	ii	iii	iv	v
	6.1%	9.7%	45.8%	38.4%
	2.2%	13.2%	28.7%	55.9%

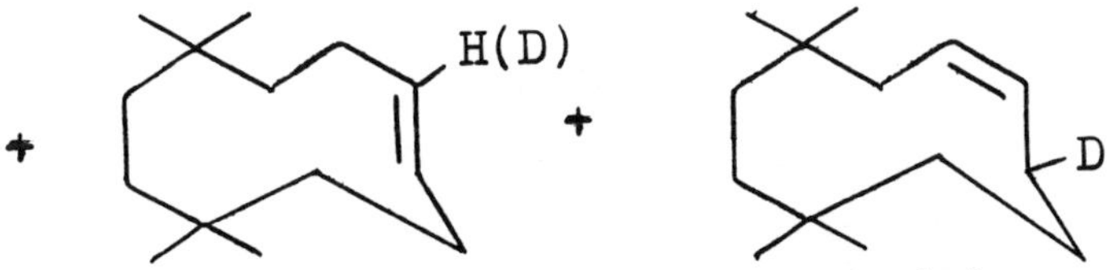

60. Cont.

The Hofmann elimination reaction of *cis* and *trans*-i gave products as shown above (percentage yields from *cis*-i are on top; from *trans*-i on the bottom). In the reaction of *cis*-i all four of the products retained all of the starting deuterium, and there was no isotope effect. In the reaction of *trans*-i, iii, and v retained all of the original deuterium, but ii and iv were essentially deuterium-free. The isotope effect for this case was 2.7-3.6. In the case of *trans*-i the other product, NMe_3, was investigated for deuterium content and found to be deuterium-free. What conclusions can be drawn from these results? Consider the possibility of an α', β (ylide) mechanism.

61.

CH_3
$\oplus$ N–CH_2–CMe_3
O $\ominus$
CH_3

i

Because of the asymmetric nitrogen, i has four stereoisomers, a trans *dl* pair (not shown) and a cis *dl* pair. If you had samples of the 2 cis enantiomers, how could you determine which sample had the R and which the S configuration at the nitrogen atom?

Goldberg and Lam, *J. Amer. Chem. Soc.* **91**, 5113 (1969).

PROBLEM SET 12
REARRANGEMENTS

Corresponds to Chapter 18 of Advanced Organic Chemistry.

The first 24 problems are synthetic. In each one you are asked to convert one compound into another. In every case the conversion has actually been carried out and the reference is given.

1.

Barton, Kumari, Welzel, Danks, and McGhie, *J. Chem. Soc., C* **1969**, 332.

2.

Orr and Johnson, *Can. J. Chem.* **47**, 47 (1969).

3.

Stoll and Troxler, *Helv. Chim. Acta* **51**, 1864 (1968); Troxler, Stoll, and Niklaus, *Helv. Chim. Acta* **51**, 1870 (1968).

4.

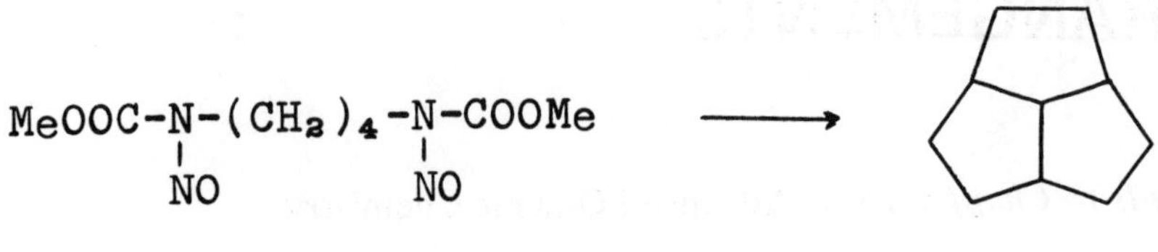

$MeOOC-N(NO)-(CH_2)_4-N(NO)-COOMe$ ⟶

i ii

5.

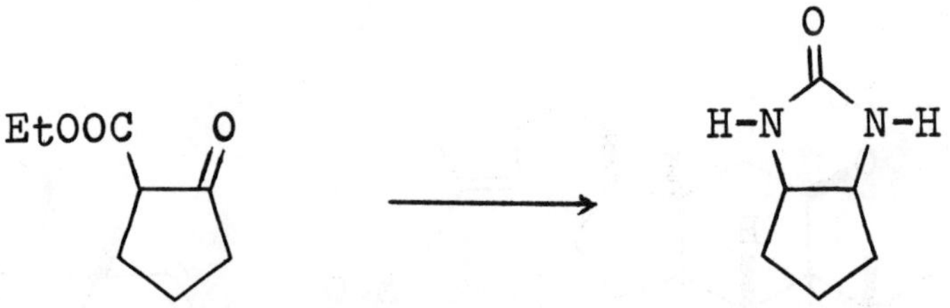

Takaya, Yoshimoto, and Imoto, *Bull. Chem. Soc. Jap.* **40**, 2844 (1967).

6.

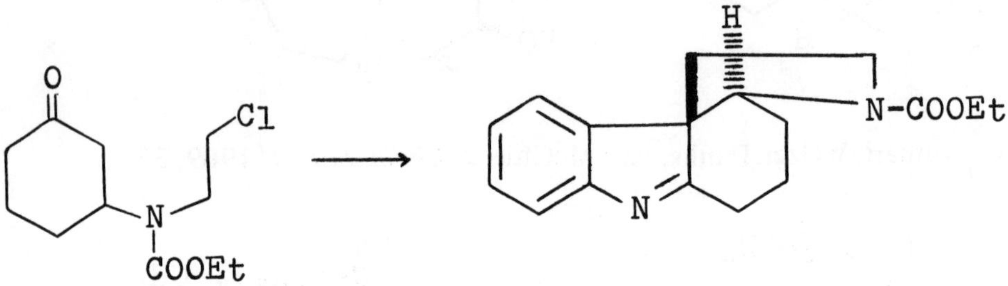

Fritz and Rubach, *Justus Liebigs Ann. Chem.* **715**, 135 (1968).

7.

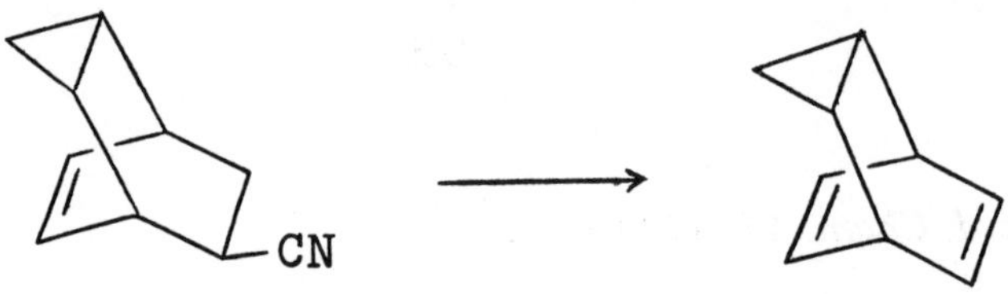

Daub and Schleyer, *Angew. Chem. Intern. Ed. Engl.* **7**, 468 (1968); [*Angew. Chem.* **80**, 446(1968)]

8.

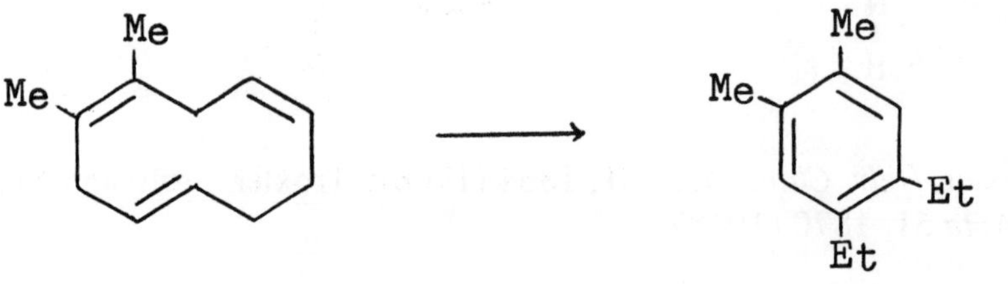

i ii

9.

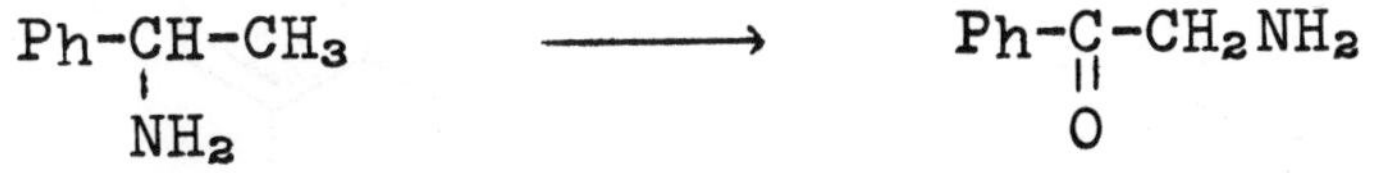

$$Ph\text{-}CH(NH_2)\text{-}CH_3 \longrightarrow Ph\text{-}C(=O)\text{-}CH_2NH_2$$

Baumgarten and Petersen, *Org. Syn.* **41**, 82.

10.

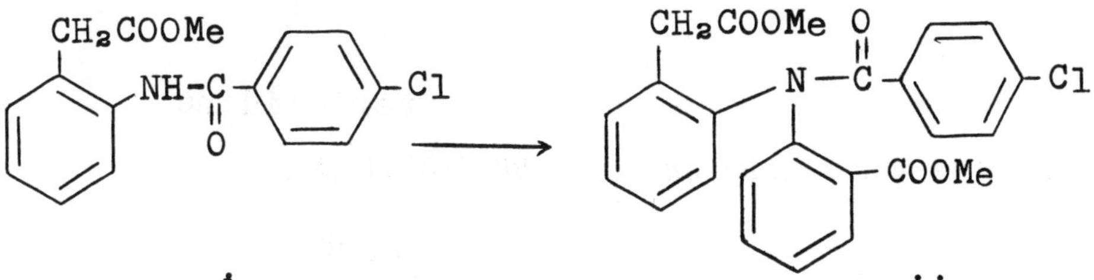

CH₂COOMe NH-C(=O) Cl → CH₂COOMe N-C(=O) Cl COOMe

i ii

11.

H CH_2OH

geraniol → HO bakuchiol

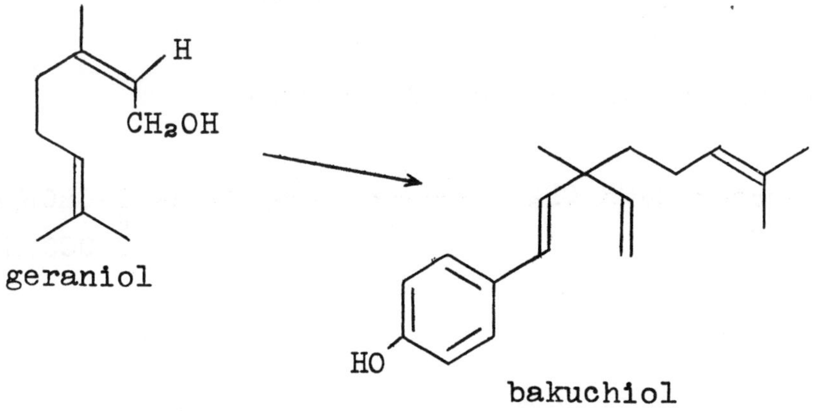

Cardnuff and Miller, *J. Chem. Soc., C* **1968**, 2671.

12.

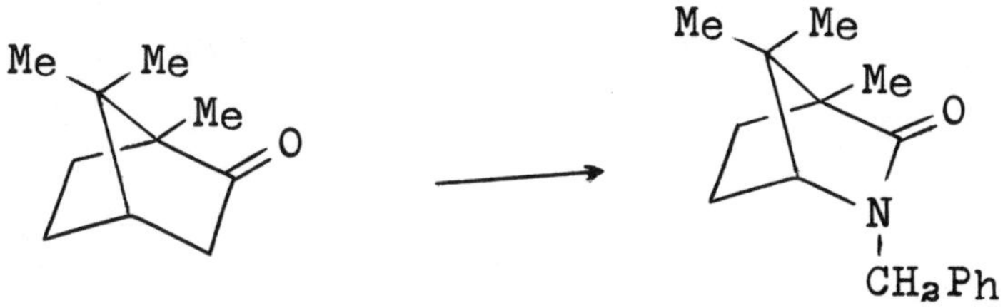

Me Me Me O → Me Me Me O N CH_2Ph

Gassman and Cryberg, *J. Amer. Chem. Soc.* **91**, 2047 (1969).

13.

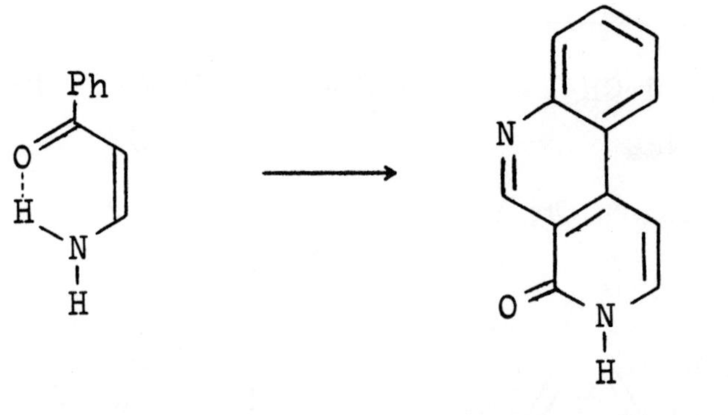

perlolidine

Powers and Ponticello, *J. Amer. Chem. Soc.* **90**, 7102 (1968).

14.

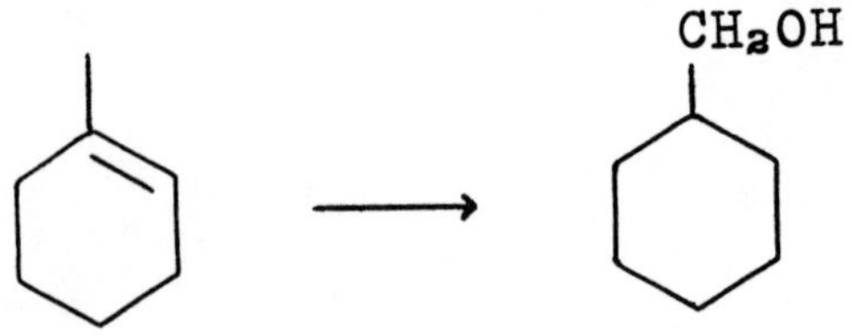

Brown and Zweifel, *J. Amer. Chem. Soc.* **82**, 1504 (1960).

15.

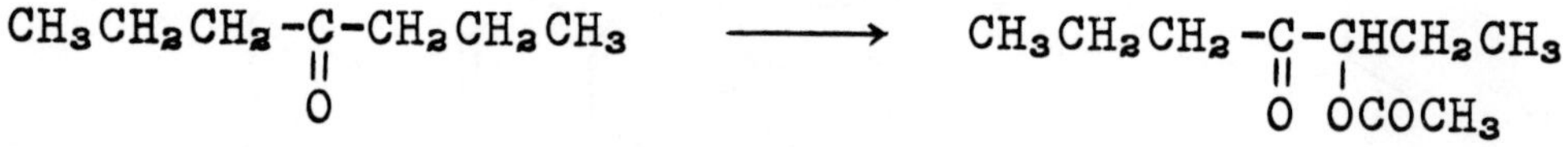

$CH_3CH_2CH_2-C(=O)-CH_2CH_2CH_3 \longrightarrow CH_3CH_2CH_2-C(=O)-CH(OCOCH_3)CH_2CH_3$

i ii

16.

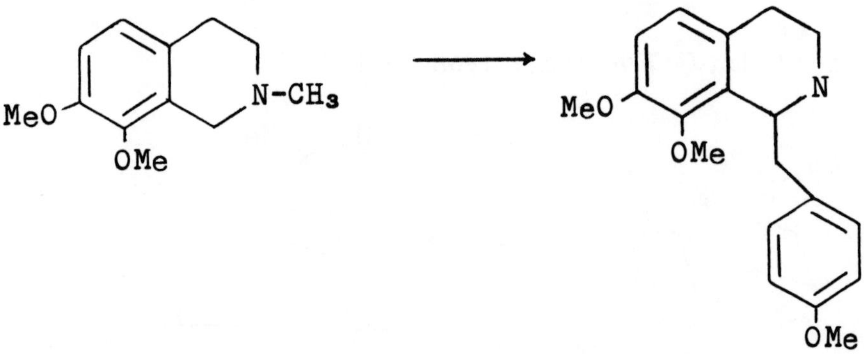

Grethe, Lee, and Uskoković, *Tetrahedron Lett.* **1969**, 1937.

17.

Hammons, Probasco, Sanders, and Whalen, *J. Org. Chem.* **33**, 4493 (1968).

18.

cyclododecanone → cycloundecanone

Garbisch and Wohllebe, *J. Org. Chem.* **33**, 2157 (1968).

19.

Malzieu, *Bull. Soc. Chim. Fr.* **1968**, 4145.

20.

i → ii

21.

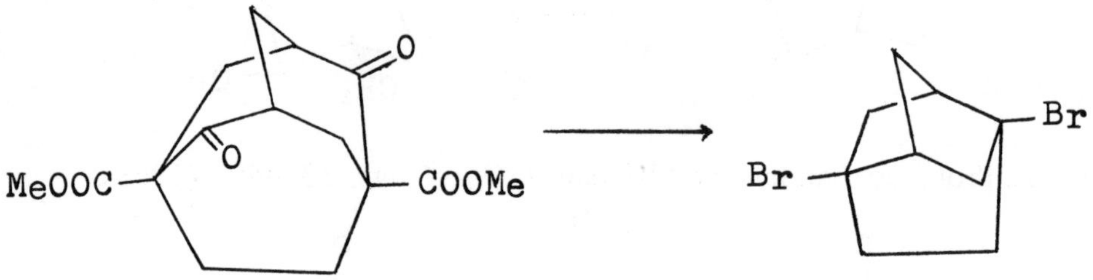

MeOOC, COOMe, O, O → Br, Br

i → ii

22.

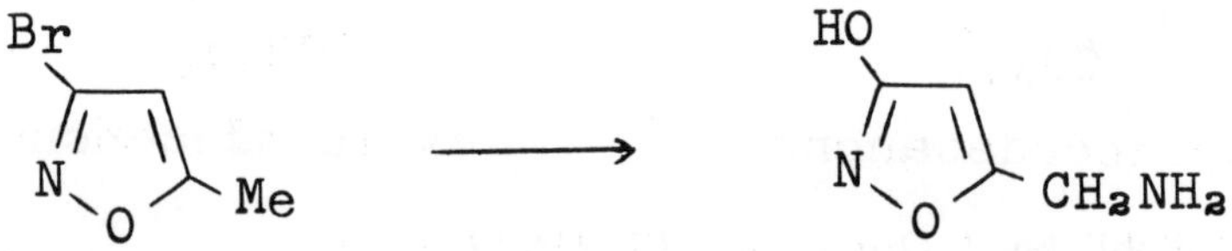

Br, N, O, Me → HO, N, O, CH_2NH_2

pantherine

Bowden, Crank, and Ross, *J. Chem. Soc., C* **1968**, 172.

23.

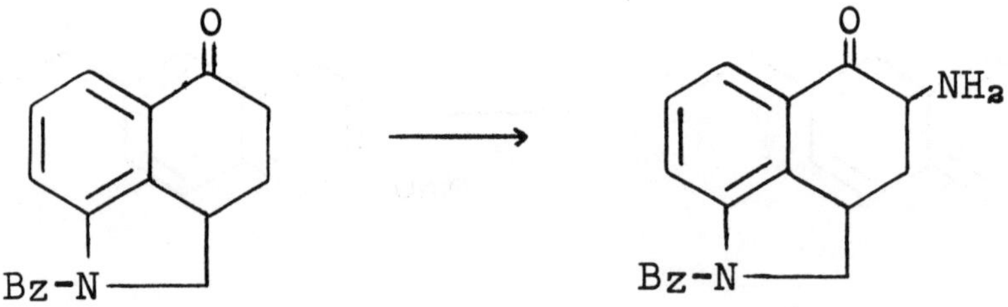

O, Bz-N → O, NH_2, Bz-N

i → ii

24.

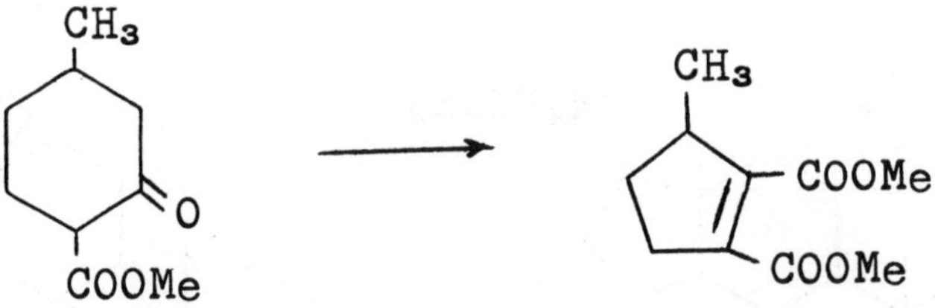

CH_3, O, COOMe → CH_3, COOMe, COOMe

Schorno, Adolphen, and Eisenbraun, *J. Org. Chem.* **34**, 2801 (1969).

The following conversions (problems 25 to 86) have been reported in the literature. In each case suggest a reasonable mechanism.

25.

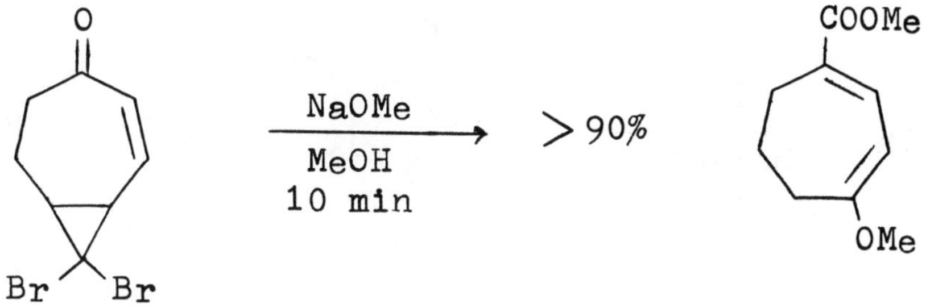

NaOMe / MeOH / 10 min → >90%

Iskander and Stansfield, *J. Chem. Soc., C* **1969**, 669.

26.

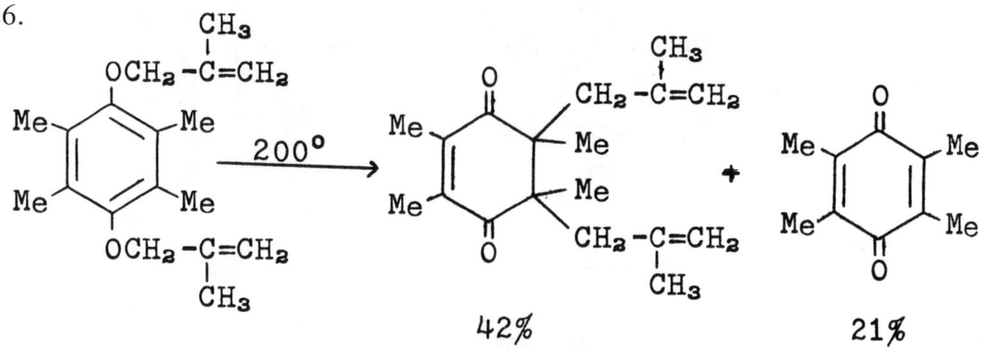

200° → 42% + 21%

Newman and Hetzel, *J. Org. Chem.* **34**, 1216 (1969).

27.

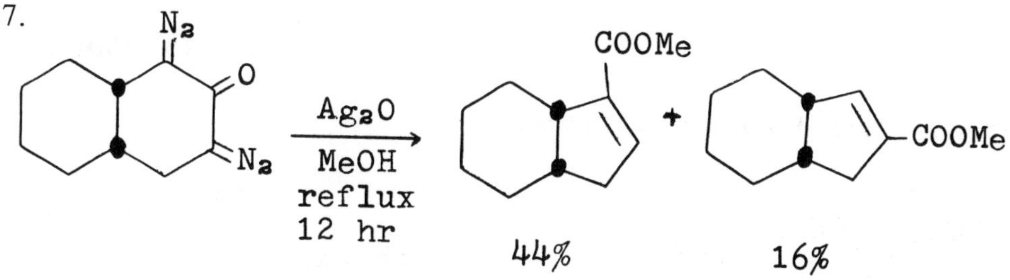

Ag_2O / MeOH / reflux / 12 hr → 44% + 16%

Borch and Fields, *J. Org. Chem.* **34**, 1480 (1969).

28.

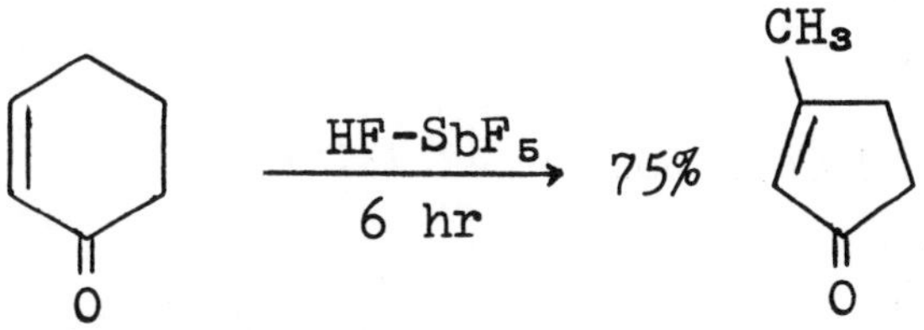

HF-SbF_5 / 6 hr → 75%

Note: The pseudo first-order rate "constant" increased with increasing concentration of SbF_5.

Hogeveen, *Rec. Trav. Chim.* **87**, 1295 (1968).

29.

PhC≡CPh $\xrightarrow[\text{hexane, 1 week}]{h\nu}$ i + ii + iii + iv

i: 1,2,3-triphenylazulene (Ph, Ph, Ph)

ii: Ph, Ph, Ph

iii: Ph, Ph, Ph, Ph, Ph, Ph

iv: Ph, Ph, Ph, Ph, Ph, Ph, Ph, Ph

30.

$CH_3CH_2CH_2{}^{13}CH_3 \underset{}{\overset{SbF_5\text{-}HF}{\rightleftharpoons}} CH_3CH_2{}^{13}CH_2CH_3$

+ no isobutane

Brouwer, *Rec. Trav. Chim.* **87**, 1435 (1968).

31.

PhCH2, CH2Ph, COOEt, O, CH3, N–H $\xrightarrow[\text{Ac}_2\text{O}]{\text{H}_2\text{SO}_4}$ OH, COOEt, PhCH2, PhCH2, CH3, N–H

36%

Suehiro and Eimura, *Bull. Chem. Soc. Jap.* **42**, 737 (1969).

32.

i (longifolene) $\xrightarrow[170^\circ\ 6\ \text{hr}]{\text{ZnCl}_2}$ 28% ii

i

longifolene

ii

33.

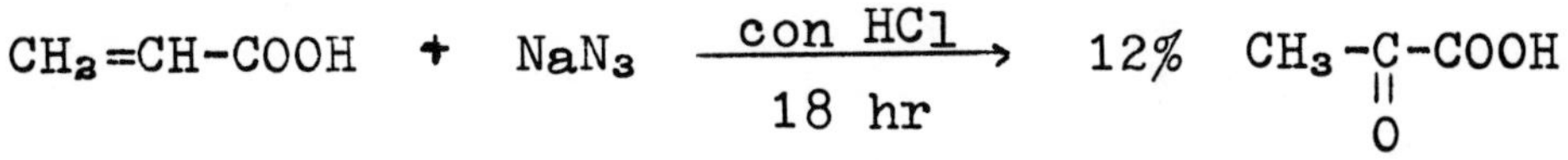

$$CH_2{=}CH\text{-}COOH + NaN_3 \xrightarrow[18\ hr]{con\ HCl} 12\%\ CH_3\text{-}\underset{\underset{O}{\|}}{C}\text{-}COOH$$

Davies and Marks, *J. Chem. Soc., C* **1968**, 2703.

34.

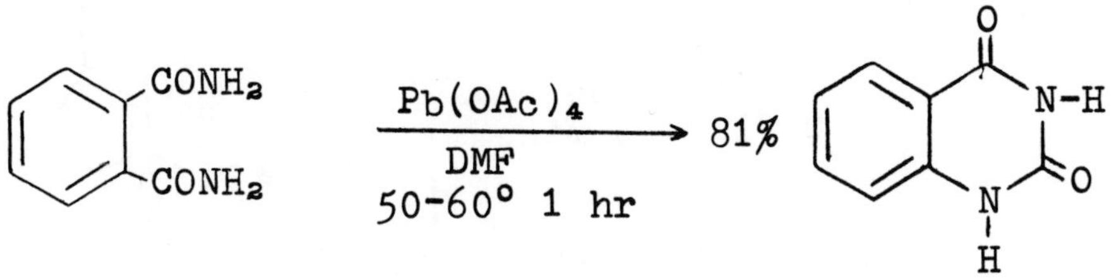

$$C_6H_4(CONH_2)_2 \xrightarrow[DMF,\ 50\text{-}60°\ 1\ hr]{Pb(OAc)_4} 81\%$$

CONH₂ / CONH₂ → N-H, N-H, O, O

Beckwith and Hickman, *J. Chem. Soc., C* **1968**, 2756.

35.

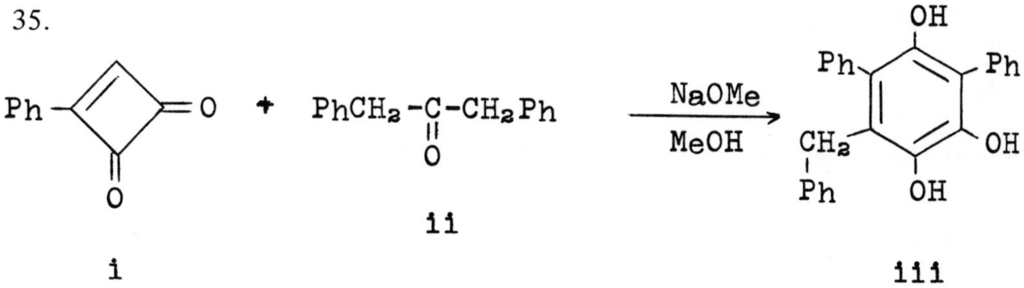

Ph, O, O (i) + $PhCH_2\text{-}\underset{\underset{O}{\|}}{C}\text{-}CH_2Ph$ (ii) $\xrightarrow[MeOH]{NaOMe}$ OH, Ph, Ph, CH_2, Ph, OH, OH (iii)

i ii iii

Note: What was actually isolated, in 50-60% yield, was the 1,4-quinone which is formed by air oxidation of iii.

36.

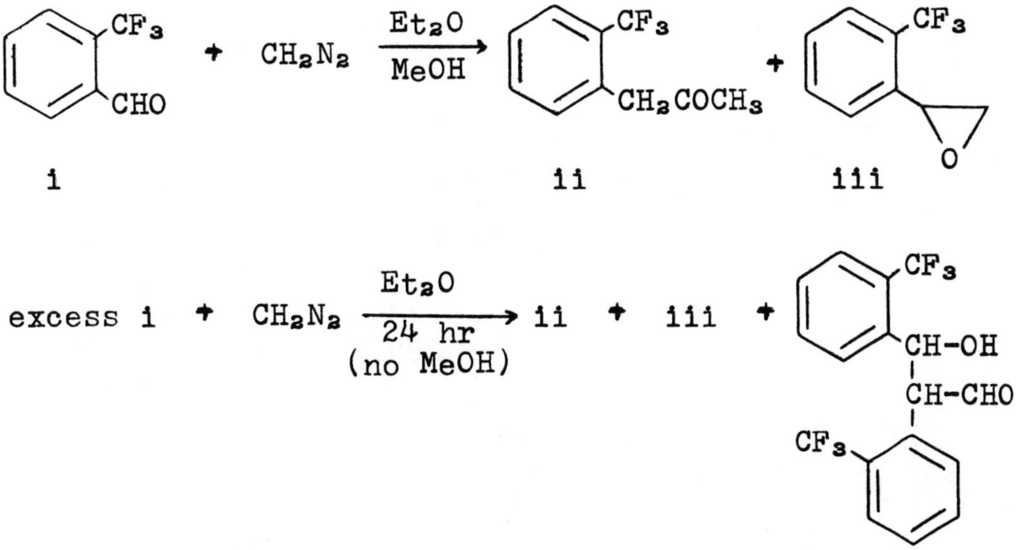

CF_3, CHO (i) + CH_2N_2 $\xrightarrow[MeOH]{Et_2O}$ CF_3, CH_2COCH_3 (ii) + CF_3, O (iii)

i ii iii

excess i + CH_2N_2 $\xrightarrow[24\ hr\ (no\ MeOH)]{Et_2O}$ ii + iii + CF_3, CH-OH, CH-CHO, CF_3

Eistert, Schade, and Mecke, *Justus Liebigs Ann. Chem.* **717**, 80 (1968).

37.

Me C_8H_{17} Me H H H

TsOH-AcOH
cyclohexane
reflux 15 min

C_8H_{17} H H H Me Me

Me C_8H_{17} Me H H H

Blunt, Hartshorn, and Kirk, *Tetrahedron* **25**, 149 (1969).

38.

OMe =O Br + CH_3MgI $\xrightarrow{Et_2O}$ OH CH_3-C-CH_3 OMe + CH_3 CH-CH_3 OH OMe

Tsuruta and Mukai, *Bull. Chem. Soc. Jap.* **41**, 2489 (1968).

39.

Et COPh

i

excess $AlCl_3$
xylene
reflux 48 hr

35%

90%

O Et Ph

iii

Ph COEt

ii

$AlCl_3$
xylene
reflux 4 hr

40.

CH₂OH + CH-OEt, CH₃ → $Hg(OAc)_2$ → CHO

Thomas, *Chem. Commun.* **1968**, 1657.

41.

Ph, Ph, O → excess HN_3, $AcOH-H_2SO_4$, 75° → Ph, Ph, O, N, H + Ph, Ph, N-H, O + Ph, N, H-N + Ph, N-H, NH_2, O

Marsili, *Tetrahedron* **24**, 4981 (1968).

42.

O, N, $CH_3CH_2-C=CHCH_3$ + $CH_2=CH-C-Cl$ (O) → 1. dry C_6H_6 reflux 6-16 hr; 2. H_2O-ice → O, Me, Me, O 41%

Hargreaves, Hickmott, and Hopkins, *J. Chem. Soc., C* **1968**, 2599.

43.

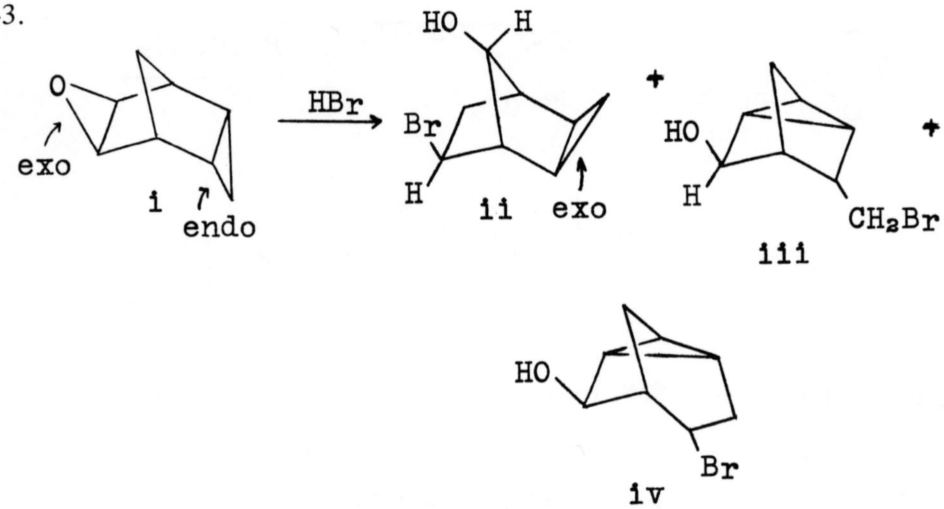

HO H

O

exo

i endo

HBr

Br

H

ii exo

+

HO

H

iii

CH₂Br

+

HO

Br

iv

44.

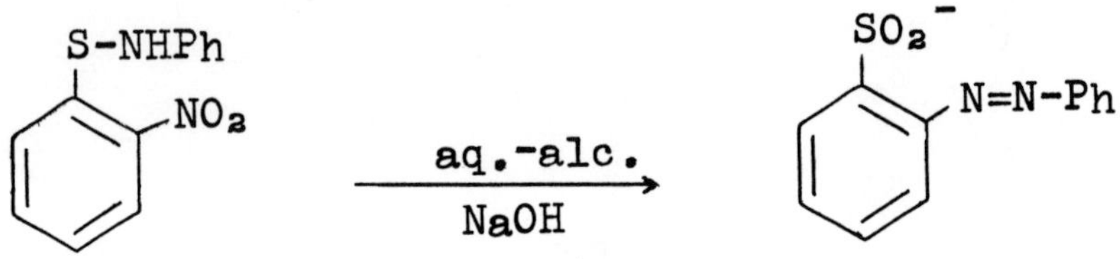

S-NHPh

NO_2

aq.-alc.

NaOH

SO_2^-

N=N-Ph

Notes: When $^{18}OH^-$ in $H_2{}^{18}O$ was used, no label appeared in the SO_2^- group of the product. The conversion could also be smoothly accomplished in dry alcoholic NaOEt (that is, in the absence of H_2O).

Brown, *Chem. Commun.* **1969**, 100.

45.

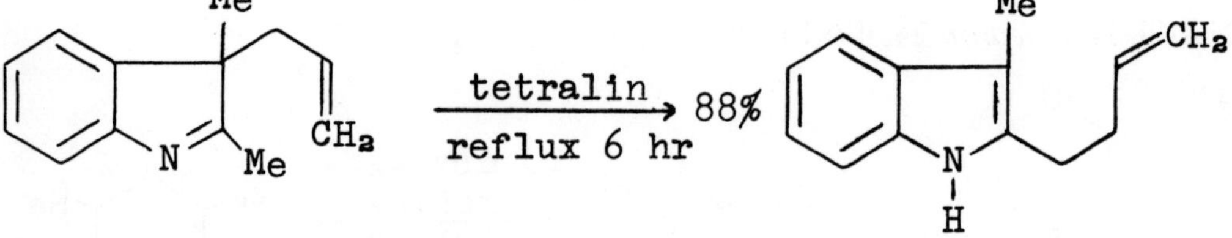

Me

N

Me

CH_2

tetralin

reflux 6 hr

88%

Me

CH_2

N

H

Bramley and Grigg, *Chem. Commun.* **1969**, 99.

46.

Br

Ph

O

H

Br

$\underline{t}$-BuOK

$\underline{t}$-BuOH

42° 6 min

76%

Br

Ph

O

i

ii

The reaction is stereospecific. That is, the *endo*-phenyl-*exo*-*p*-bromophenyl isomer of i gave the *endo*-phenyl-*exo*-*p*-bromophenyl isomer of ii.

Zimmerman, Crumrine, Döpp, and Huyffer, *J. Amer. Chem. Soc.* **91**, 434 (1969).

47.

O

O

COOH

HOAc

O

C–O

O

i

santonic acid

ii

santonide

48.

Me

CH_2OBu

excess PrLi

$Me_2NCH_2CH_2NMe_2$

pentane

Bu

CHO

Me

30%

\+

Me

CH–OH

Bu

Felkin and Tambuté, *Tetrahedron Lett.* **1969**, 821.

49.

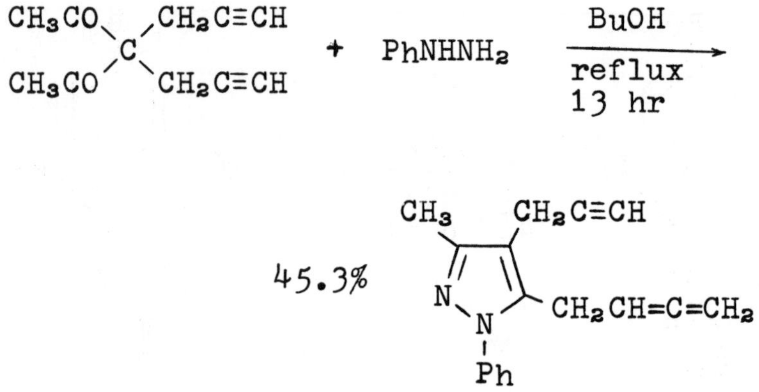

$(CH_3CO)_2C(CH_2C\equiv CH)_2$ + $PhNHNH_2$ $\xrightarrow[\text{reflux, 13 hr}]{\text{BuOH}}$

45.3% — pyrazole: CH_3, $CH_2C\equiv CH$, N, N, Ph, $CH_2CH{=}C{=}CH_2$

Manning, Coleman, and Langdale-Smith, *J. Org. Chem.* **33**, 4413 (1968).

50.

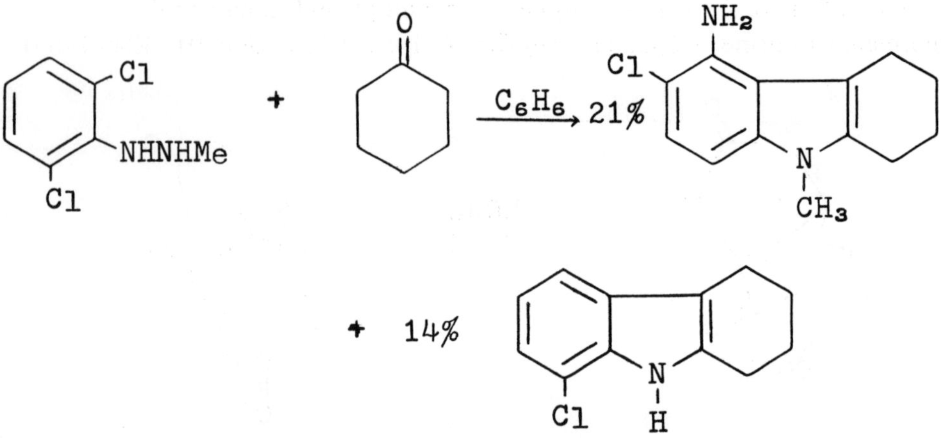

Cl, NHNHMe, Cl + O $\xrightarrow{C_6H_6}$ 21% NH_2, Cl, N, CH_3

+ 14% N, H, Cl

Bajwa and Brown, *Can. J. Chem.* **46**, 1927 (1968); Robinson and Brown, *Can. J. Chem.* **42**, 1940 (1964).

51.

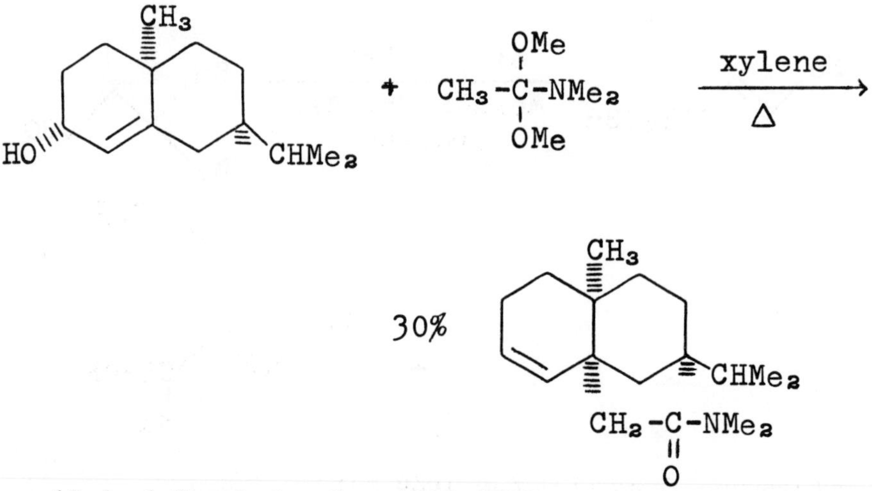

CH_3, HO, $CHMe_2$ + $CH_3-C(OMe)_2-NMe_2$ $\xrightarrow[\Delta]{\text{xylene}}$

30% CH_3, $CHMe_2$, $CH_2-C(=O)-NMe_2$

Dawson and Ireland, *Tetrahedron Lett.* **1968**, 1899.

52.

C_8H_{17} $\xrightarrow[\text{THF-reflux, 5 hr}]{CH_3C^{18}O_2K}$ C_8H_{17}

Br, O, Br — i; CH_3-C(=^{18}O)-O, ^{18}O — 34% ii

Only the two oxygens shown carried the label. The ether oxygen of the acetoxy group was not labeled.

53.

OMe, N $\xrightarrow[\text{irreversible}]{125\text{–}130°}$ N, MeO

The Reference is the same as that for Problem 54.

54.

OMe, N $\xrightarrow[\text{acetone}]{h\nu}$ OMe, N + MeO, N

> 95% < 5%

O, H–N $\xrightarrow[\text{acetone}]{h\nu}$ O, NH + O, H–N

17% 50%

Paquette and Malpass, *J. Amer. Chem. Soc.* **90**, 7151 (1968).

55.

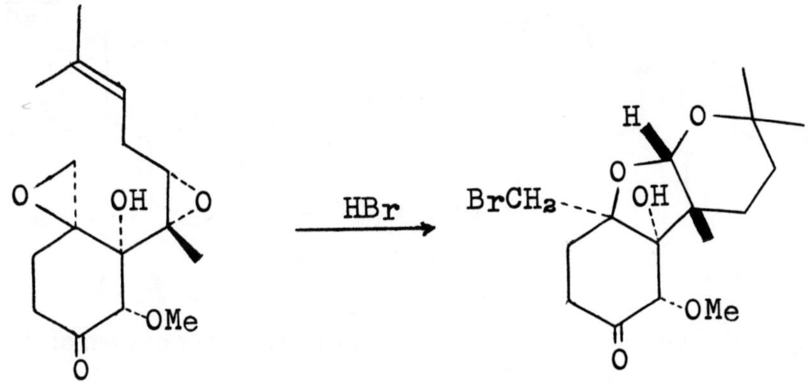

O OH O

HBr

$BrCH_2$ H O O OH

OMe

O

Sigg and Weber, *Helv. Chim. Acta* **51**, 1395 (1968).

56.

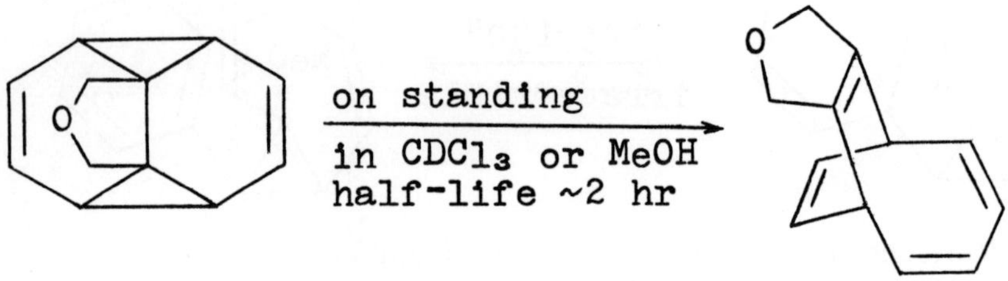

O

on standing

in $CDCl_3$ or MeOH
half-life ~2 hr

O

i

100% ii

57.

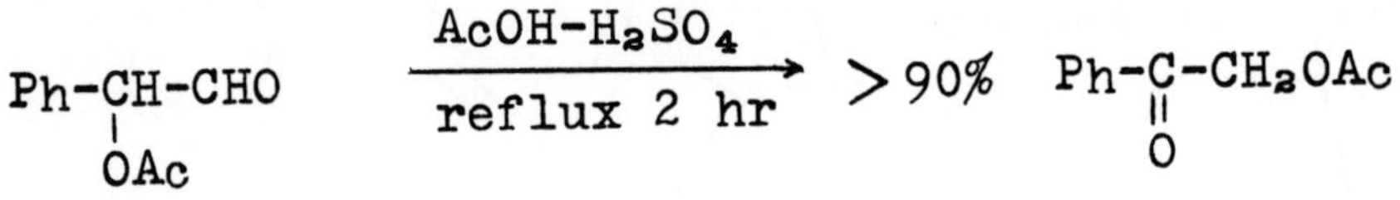

Ph-CH-CHO
OAc

$AcOH-H_2SO_4$

reflux 2 hr

>90% Ph-C(=O)-CH_2OAc

i

ii

58.

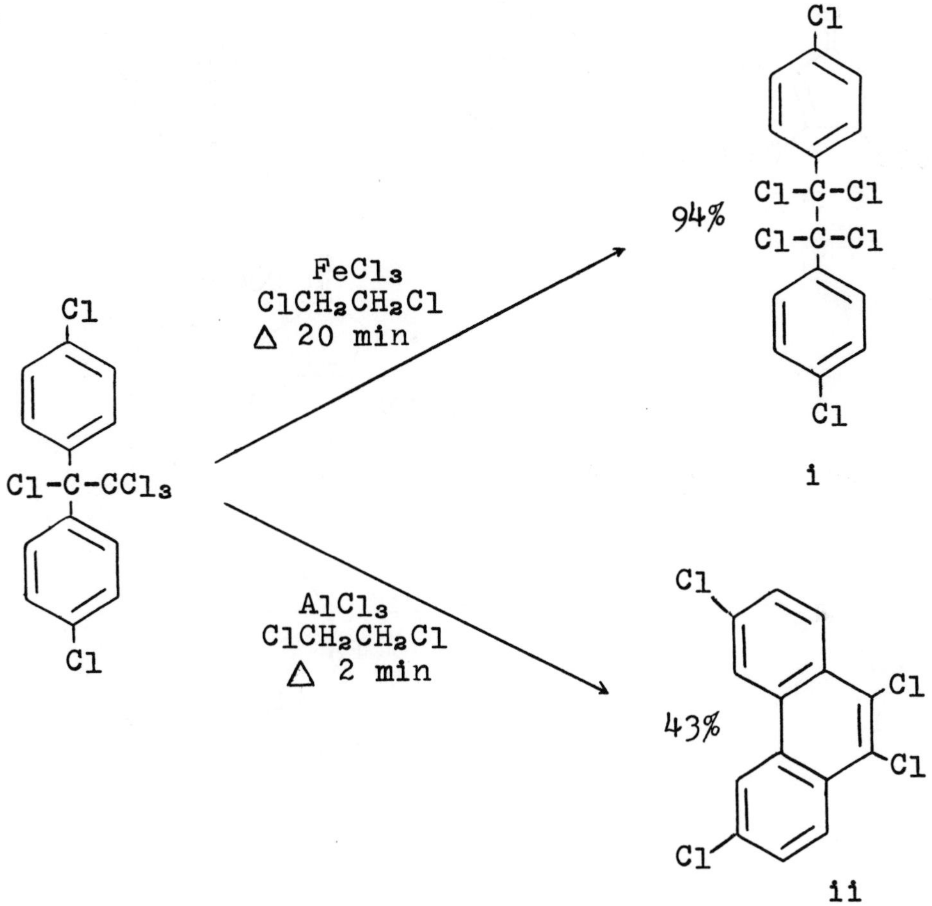
Cl
Cl-C-CCl3
Cl
FeCl3
ClCH2CH2Cl
Δ 20 min
AlCl3
ClCH2CH2Cl
Δ 2 min
94%
Cl
Cl-C-Cl
Cl-C-Cl
Cl
i
43%
Cl
Cl
Cl
Cl
ii

Note: i could be converted to ii by heating with $AlCl_3$ in ethylene chloride.

Weis, *Helv. Chim. Acta* **51**, 1572 (1968).

59.

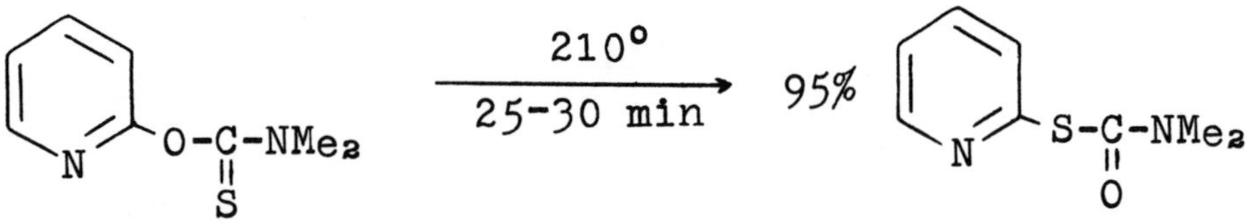
N
O-C-NMe2
S
210°
25-30 min
95%
N
S-C-NMe2
O

Newman and Karnes, *J. Org. Chem.* **31**, 3980 (1966).

60.

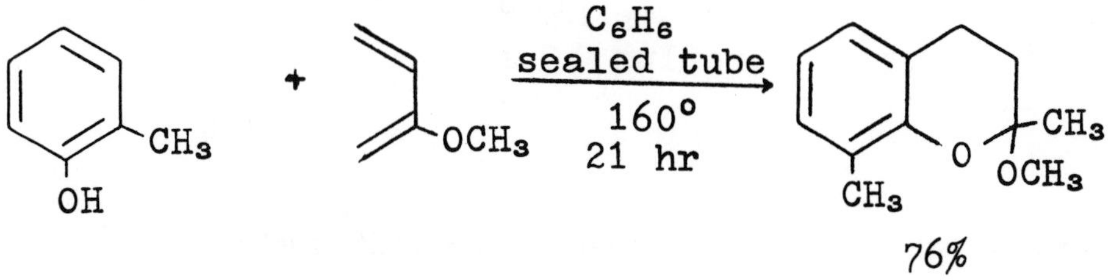
CH3
OH
+
OCH3
C6H6
sealed tube
160°
21 hr
O
CH3
OCH3
CH3
76%

The reference is the same as that for Problem 61.

61.

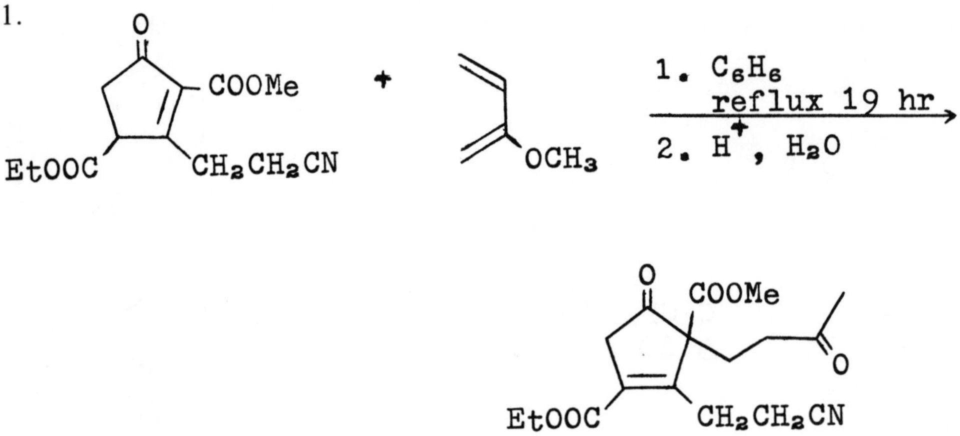

O, COOMe, EtOOC, CH_2CH_2CN + OCH_3

1. C_6H_6 reflux 19 hr
2. H^+, H_2O

O, COOMe, O, EtOOC, CH_2CH_2CN

Dolby, Elliger, Esfandiari, and Marshall, *J. Org. Chem.* **33**, 4508 (1968).

62.

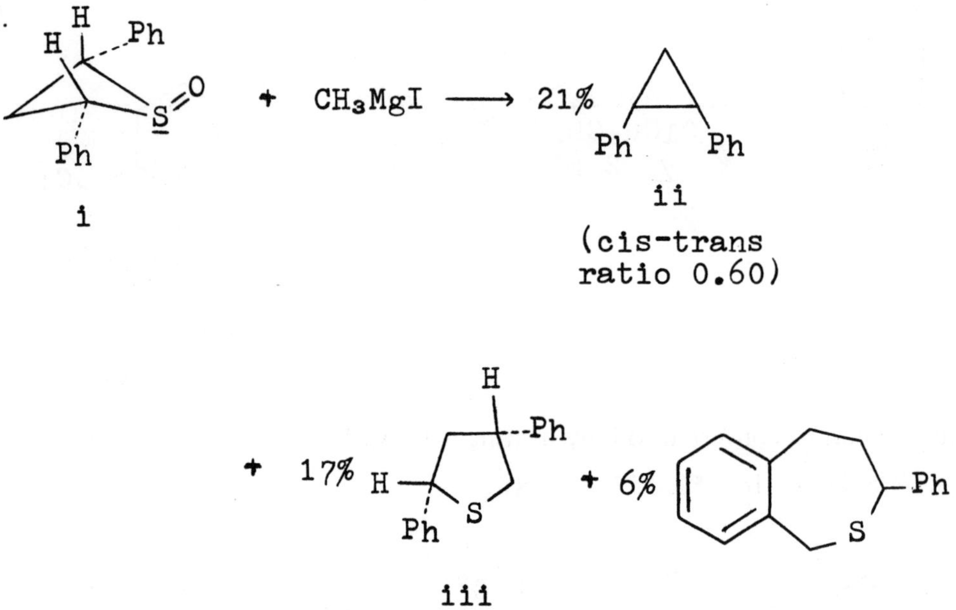

H, H, Ph, S=O, Ph

i

+ CH_3MgI ⟶ 21%

Ph Ph

ii

(cis-trans ratio 0.60)

+ 17% H, H, Ph, S, Ph + 6% S, Ph

iii

63.

O, Ph, N, CH_3

1. $PhNHNH_2$ EtOH reflux 30 min
2. HCl in EtOH reflux 15 min

N, N-CH_3, Ph

71%

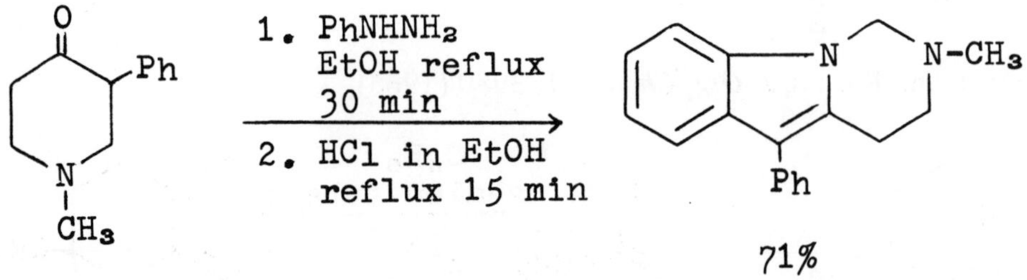

Ebnöther, Niklaus, and Süess, *Helv. Chim. Acta* **52**, 629 (1969).

64.

Et$_2$NH
reflux
20 hr

i

36% ii

65.

C_8H_{17}

+ NCN_3
cyanogen
azide

EtOAc

Scribner, *Tetrahedron Lett.* **1967**, 4737.

66.

quinoline
350° 1 hr

38% ~35%

16.5% 10% 8%

Kwart and Cohen, *Chem. Commun.* **1968**, 1296.

67.

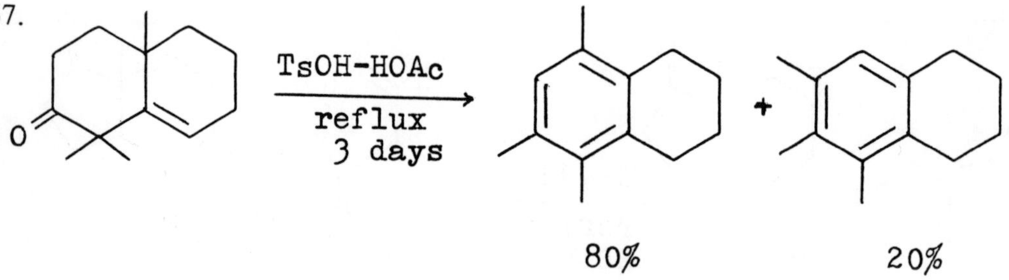

TsOH-HOAc, reflux, 3 days

80%   20%

Note: The two products are not interconvertible under the reaction conditions.

Bey and Ourisson, *Bull. Soc. Chim. Fr.* **1968**, 1411.

68. HO CH=CH_2 + perbenzoic acid $\xrightarrow{Et_2O}$ 32% O CH_2OH

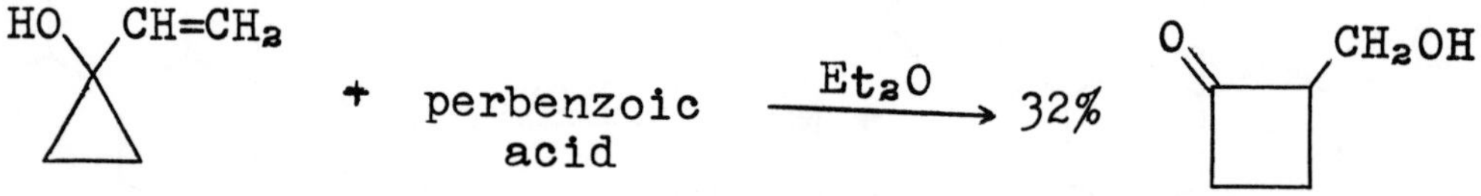

Wasserman, Cochoy, and Baird, *J. Amer. Chem. Soc.* **91**, 2375 (1969).

69.

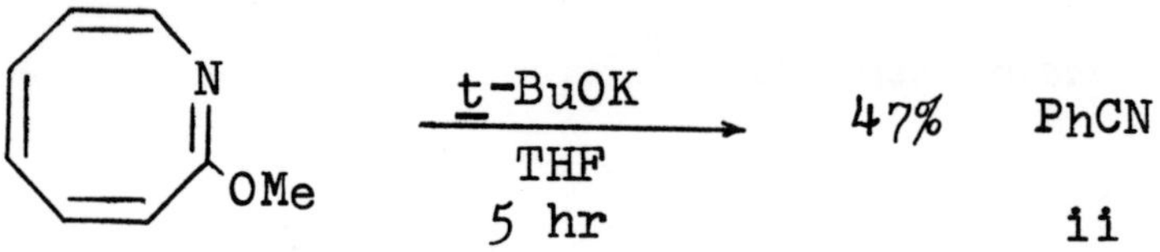

N OMe

i

t-BuOK, THF, 5 hr → 47% PhCN

ii

70.

NH_2 H MeO

HONO, HOAc-H_2O, 0° → 50%

CH_3 OAc H H MeO

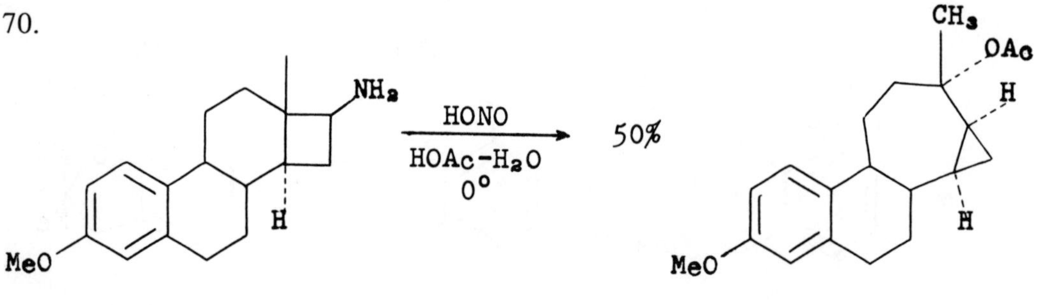

Meinwald and Ripoll, *J. Amer. Chem. Soc.* **89**, 7075 (1967).

71. HC≡CCH_2-C(CH_3)(OH)-CH=CH_2 $\xrightarrow{375°}$ 16% CH_2=C=$CHCH_2CH_2C$(=O)CH_3 + 58% CH_2=C=CH_2

+ 58% CH_2=CH-C(=O)-CH_3 + 22% (cyclopropane with CH=CH_2 and $COCH_3$)

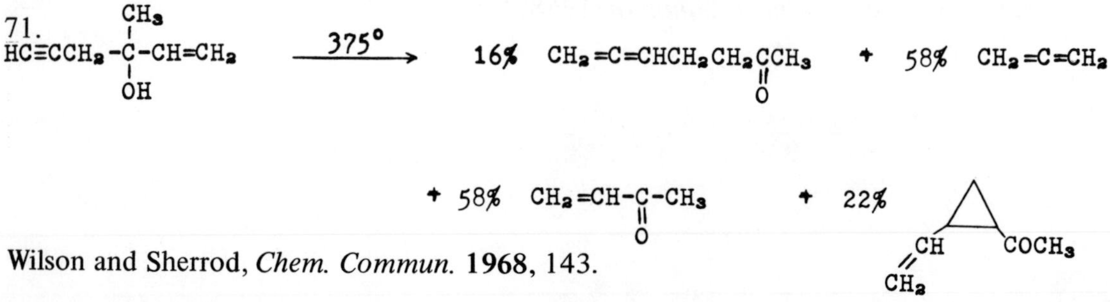

Wilson and Sherrod, *Chem. Commun.* **1968**, 143.

72.

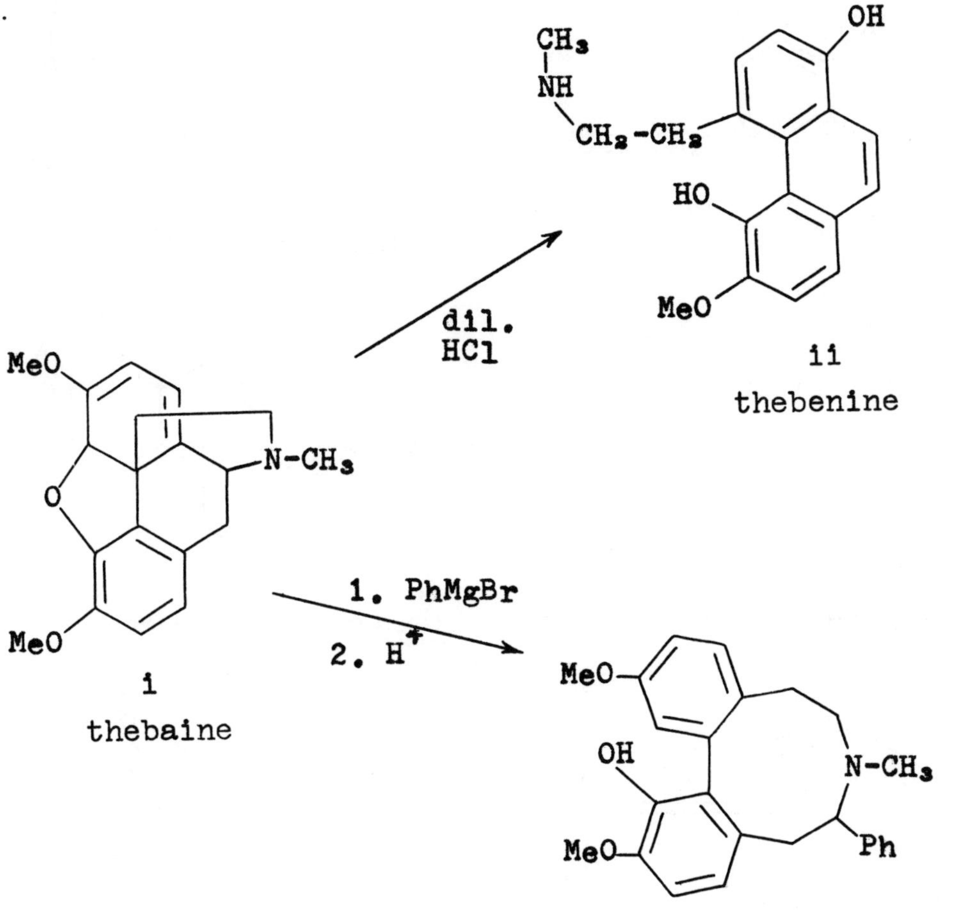

MeO

O

N-CH_3

MeO

i

thebaine

dil. HCl

CH_3

NH

CH_2-CH_2

OH

HO

MeO

ii

thebenine

1. PhMgBr

2. H^+

MeO

OH

N-CH_3

MeO

Ph

iii

phenyldihydrothebaine

73.

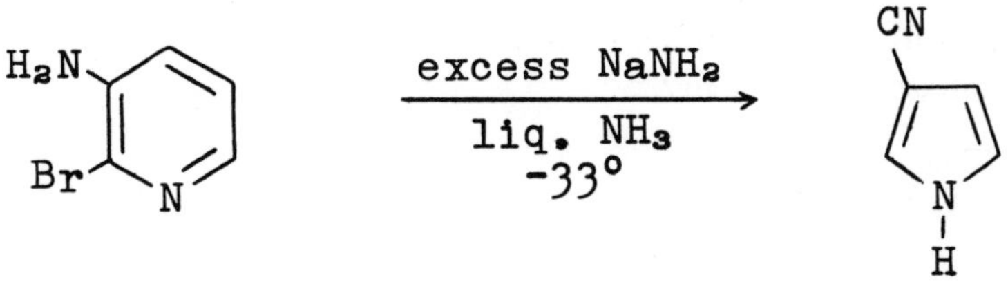

H_2N

Br

N

excess $NaNH_2$

liq. NH_3

-33°

CN

N

H

den Hertog and Buurman, *Tetrahedron Lett.* **1967**, 3657.

74.

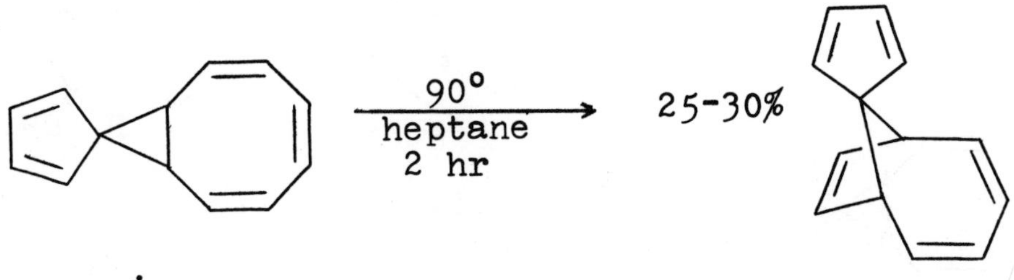

i

90°

heptane

2 hr

25-30%

ii

75.

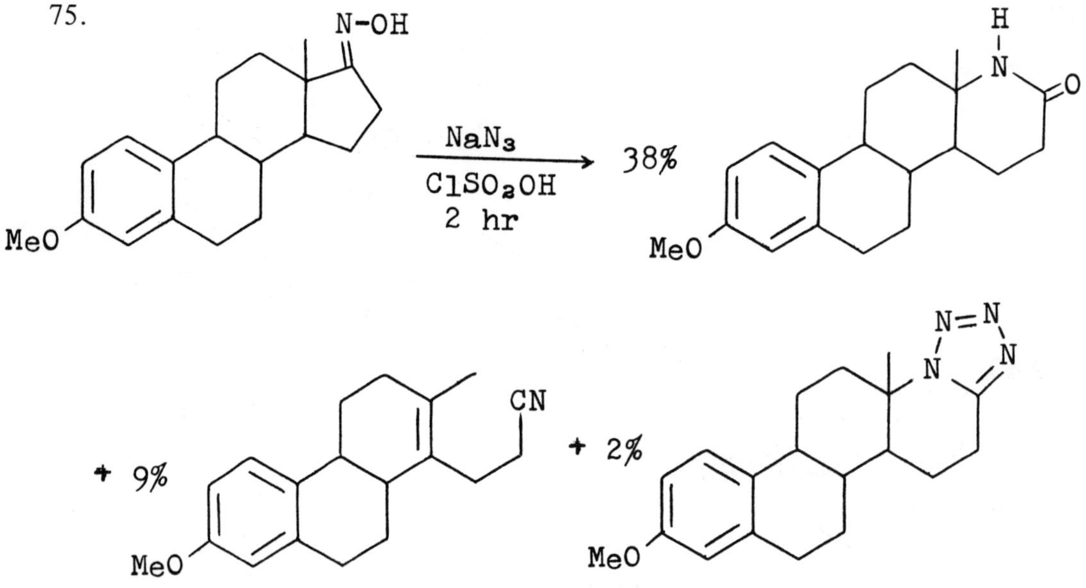

N-OH

$\xrightarrow[\text{ClSO}_2\text{OH, 2 hr}]{\text{NaN}_3}$ 38%

H N O

MeO

+ 9% CN + 2%

N=N N N

MeO MeO

Cervantes, Crabbé, Iriarte, and Rosenkranz, *J. Org. Chem.* **33**, 4294 (1968).

76.

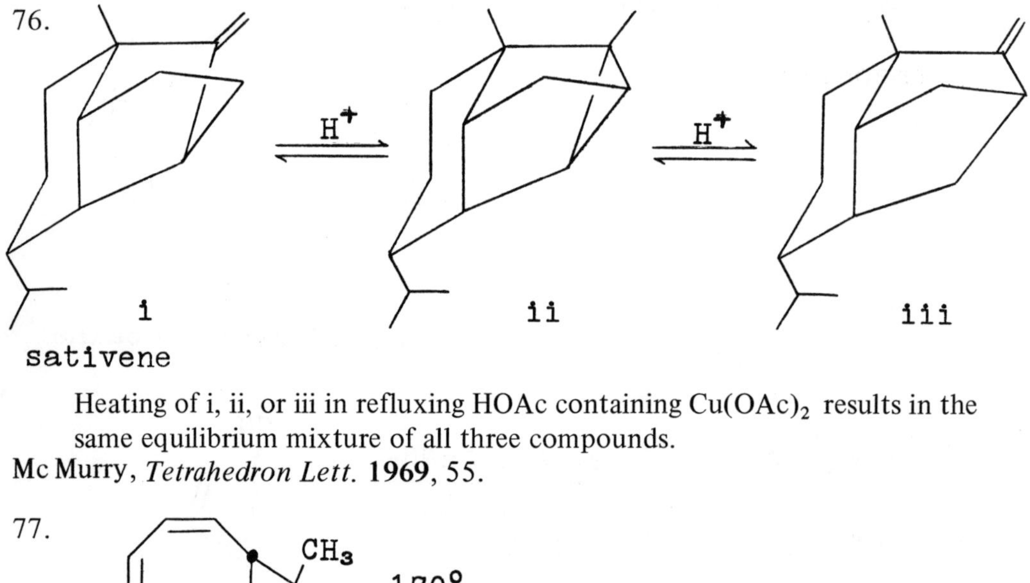

$\underset{}{\overset{H^+}{\rightleftharpoons}}$ $\overset{H^+}{\rightleftharpoons}$

i ii iii

sativene

Heating of i, ii, or iii in refluxing HOAc containing $Cu(OAc)_2$ results in the same equilibrium mixture of all three compounds.

Mc Murry, *Tetrahedron Lett.* **1969**, 55.

77.

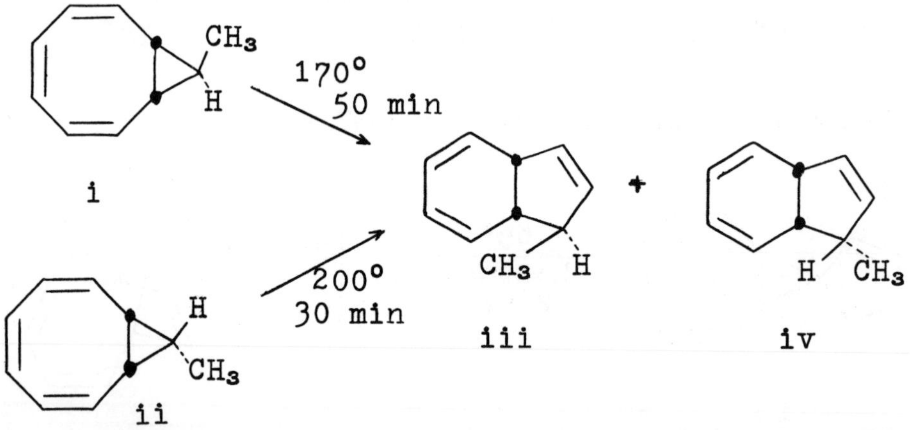

CH_3 H

170° 50 min

i

+

200° 30 min

H CH_3

CH_3 H H CH_3

iii iv

ii

77. Cont.

Either i or ii gave the same equilibrium mixture (iii:iv ratio about 4:1) of iii and iv. Total yield 95%.

Radlick and Fenical, *J. Amer. Chem. Soc.* **91**, 1560 (1969).

78.

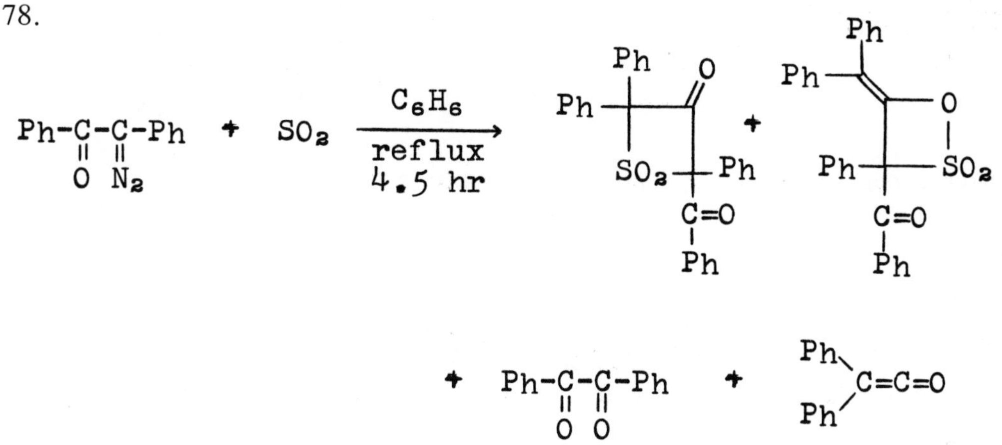

Nagai, Tanaka, and Tokura, *Tetrahedron Lett.* **1968**, 6293.

79.

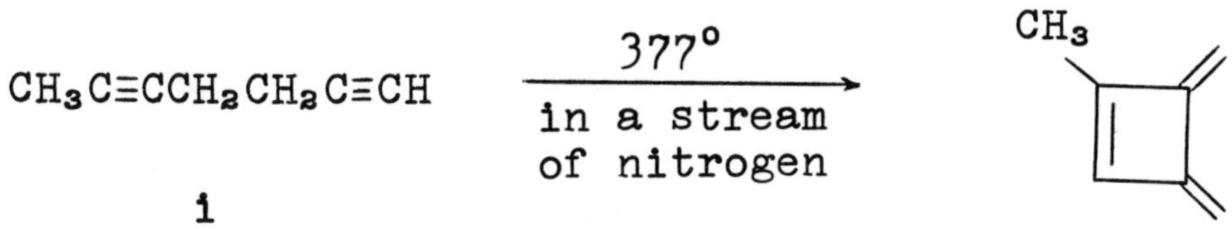

i ii

80.

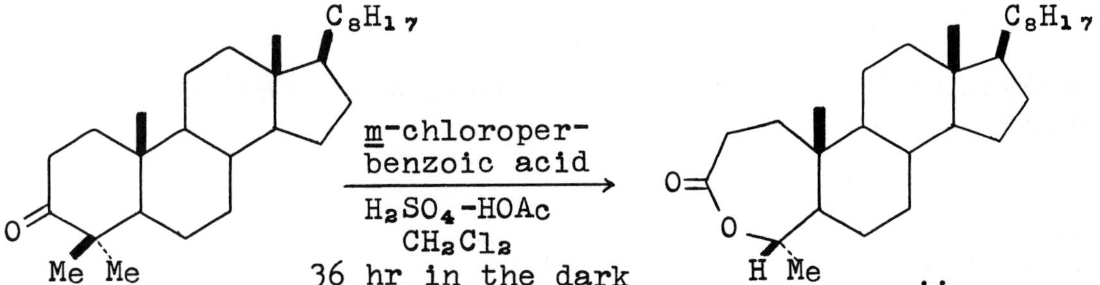

ii

i

Notes: 1. One of the methyl groups originally in position 4 has been lost. 2. i, treated with *m*-chloroperbenzoic acid, but without mineral acid, gave iii. 3. iii, treated with 10% H_2SO_4 in HOAc gave iv. 4. iii or iv, subjected to the original conditions, gave ii.

80. Cont.

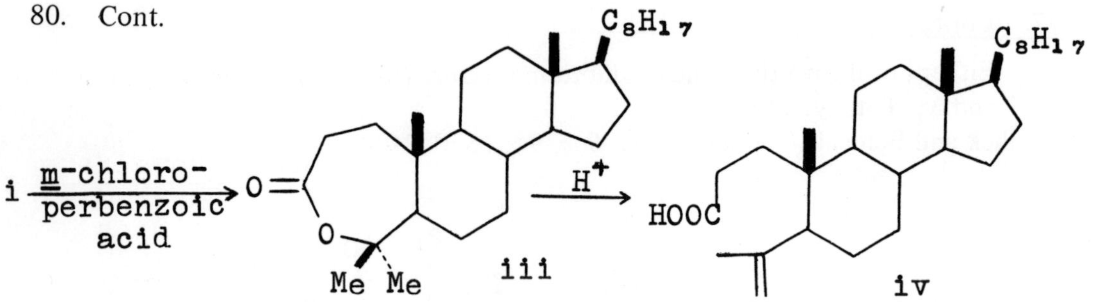

i —m-chloroperbenzoic acid→ iii —H^+→ iv

Holker, Jones, and Ramm, *J. Chem. Soc., C* **1969**, 357.

81.

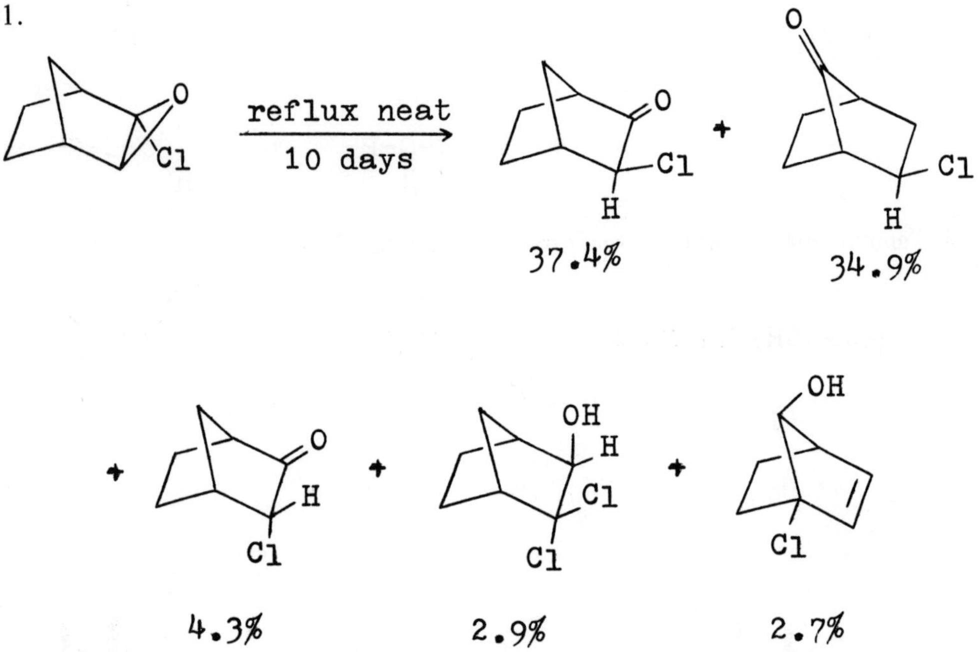

reflux neat, 10 days

37.4% 34.9%

4.3% 2.9% 2.7%

McDonald and Tabor, *J. Org. Chem.* **33**, 2934 (1968); McDonald and Steppel, *J. Amer. Chem. Soc.* **91**, 782 (1969).

82.

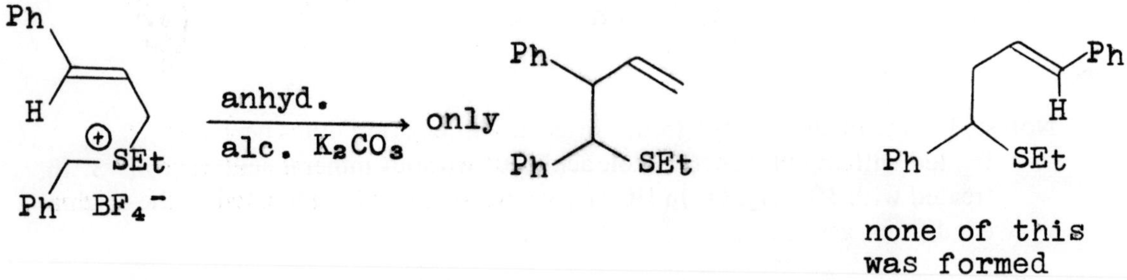

anhyd. alc. K_2CO_3 → only

none of this was formed

82. Cont.

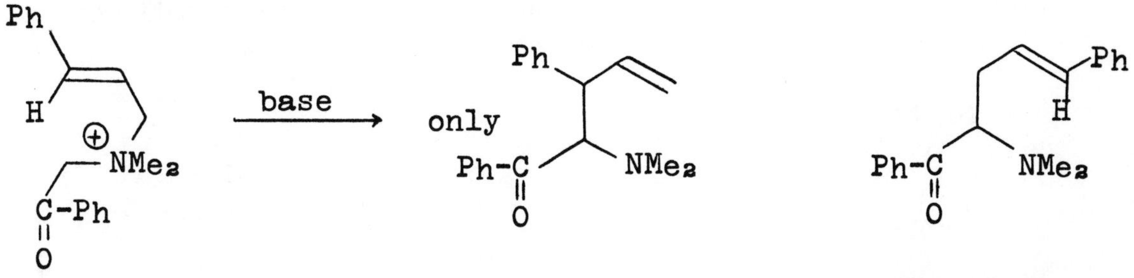

Ph H ⊕ NMe$_2$ C-Ph O base → only Ph Ph-C NMe$_2$ O

Ph H Ph-C NMe$_2$ O

none of this was formed

Cose, Davies, Ollis, Smith, and Sutherland, *Chem. Commun.* **1969**, 293; Jemison and Ollis, *Chem. Commun.* **1969**, 294.

83. CH_3 O → Ac_2O / H_2SO_4 → OAc CH_3

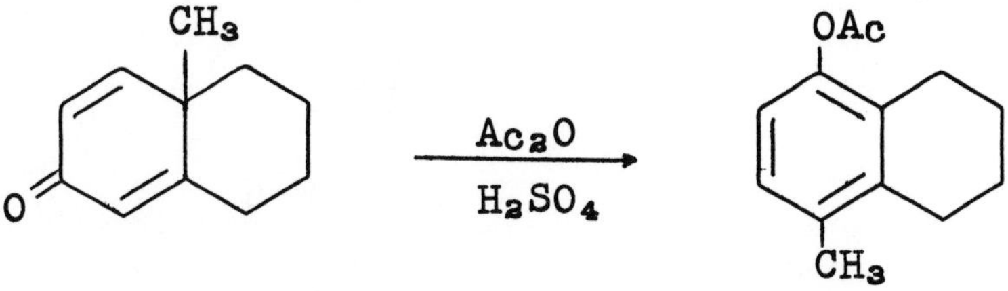

Woodward and Singh, *J. Amer. Chem. Soc.* 72, 494 (1950).

84.

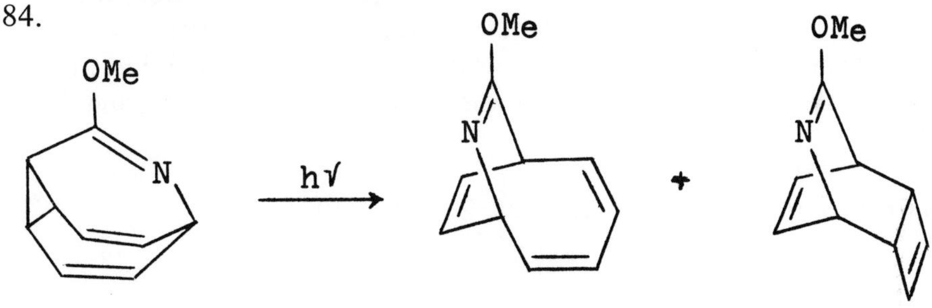

OMe N → hν → OMe N + OMe N

i an azabullvalene ii iii

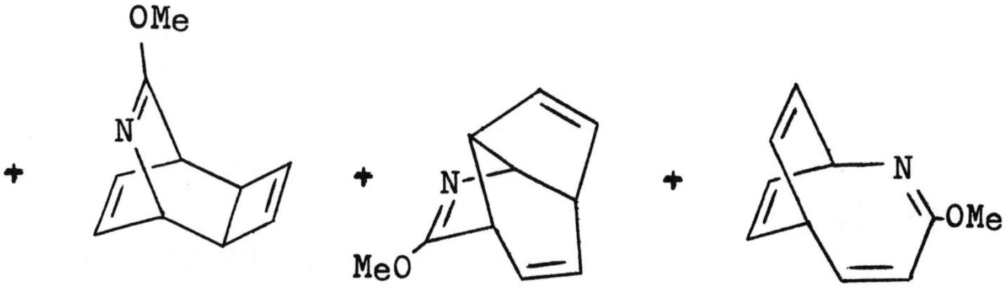

+ OMe N + N MeO + N OMe

iv v vi

84. Cont.

Note: When ii was treated with uv light, only i, iii, iv, and vi were formed for the first 3 hr, after which v was formed, with the yield of i decreasing.

85. Me Me Me N-Cl i → MeOH reflux 36 hr → Me Me Cl H N Me 59% ii + Me Me MeO H N Me 20% iii

+ Me Me Me N-H 7% iv

Note: When i was solvolyzed in the presence of $AgClO_4$, the rate was accelerated by a factor of 2×10^3 (the reaction took place at room temperature and was complete in less than 10 min) and there was a small precipitate of AgCl, but the products were the same; in fact *more* ii was formed (78% ii, 8.5% iii, and 4% iv).

Gassman and Cryberg, *J. Amer. Chem. Soc.* **91**, 2047 (1969).

86. O Cl Cl i → 1. OH^- THF 2. H^+ → COOH Cl 92.6% ii

In the following problems (87 to 90) give the products and suggest reasonable mechanisms.

87. 2,2,3,3-Tetramethylcyclopropanone (O; Me, Me, Me, Me) $\xrightarrow[\text{CH}_2\text{Cl}_2,\ -78^\circ]{\text{CH}_3\text{CHN}_2}$ i ($C_8H_{14}O$) + ii ($C_8H_{14}O$)

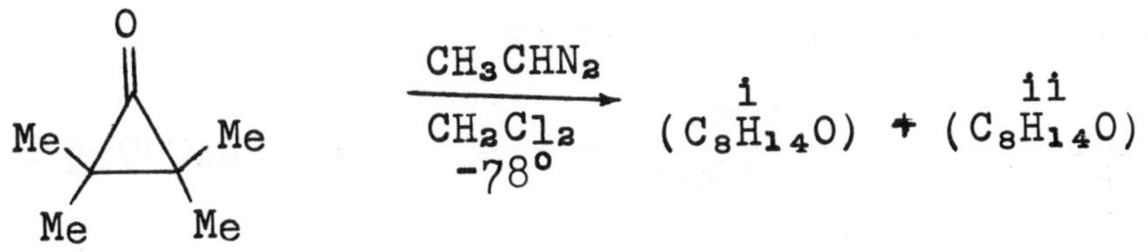

Give also the configurations of the products (models will help). Turro and Gagosian, *Chem. Commun.* **1969**, 949.

88. β-ionone $\xrightarrow{\text{perbenzoic acid}}$ i ($C_{13}H_{20}O_3$) $\xrightarrow[\text{H}_2\text{O}]{\text{OH}^-}$ (bicyclic hemiacetal product, OH)

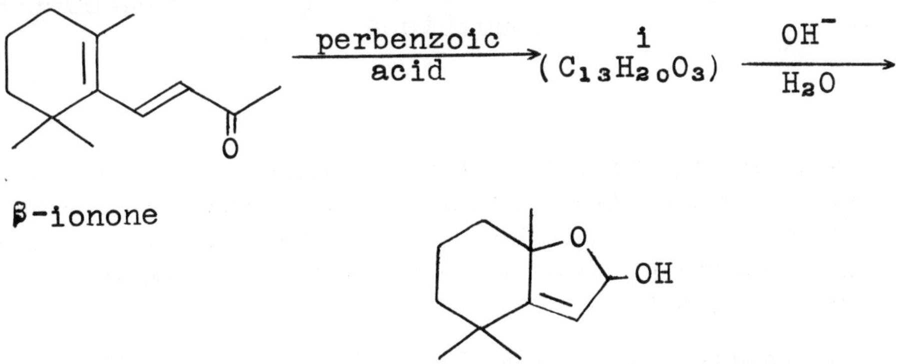

Isoe, Hyeon, Ichikawa, Katsumura, and Sakan, *Tetrahedron Lett.* **1968**, 5561.

89. i (Me, Me; CCl_3; OH; O; Me, Me) + OH^- $\longrightarrow$ ii ($C_{14}H_{18}O_3$)

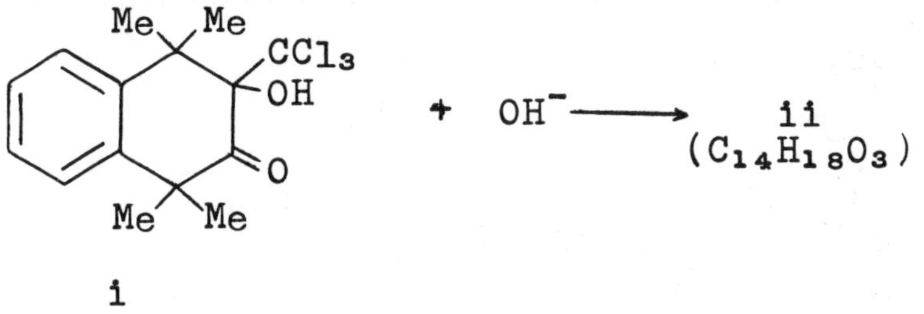

90. (Br; O) $\xrightarrow[\text{reflux 4 hr}]{\text{30\% aq. KOH}}$

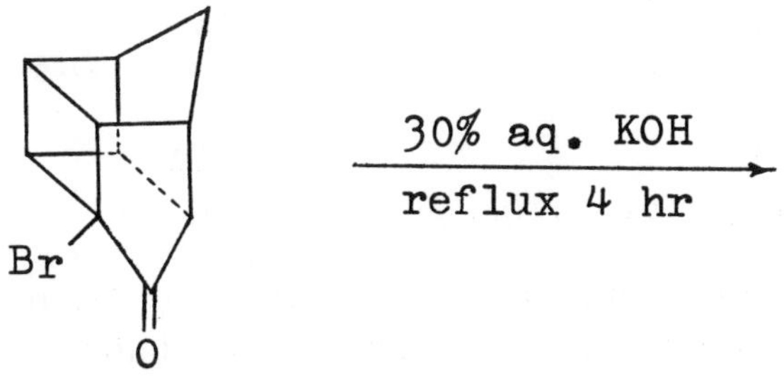

Dunn, DiPasquo, and Hoover, *J. Org. Chem.* **33**, 1454 (1968).

91. $\xrightarrow[\text{2 hr}]{220^\circ}$ ~75%

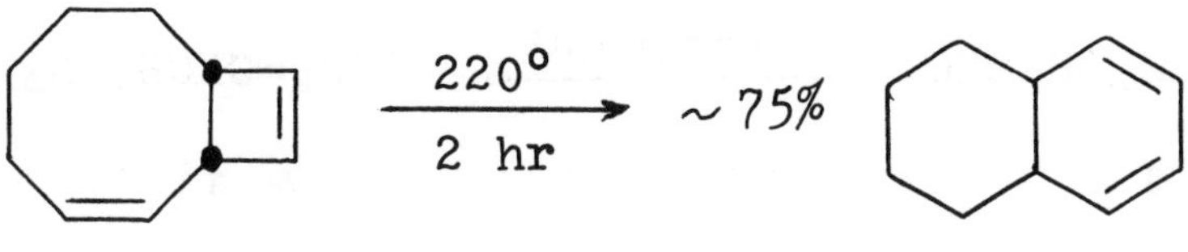

91. Cont.

a. Suggest a reasonable mechanism. b. What is the configuration at the ring junction of the product?

Radlick and Fenical, *Tetrahedron Lett.* **1967**, 4901.

92.

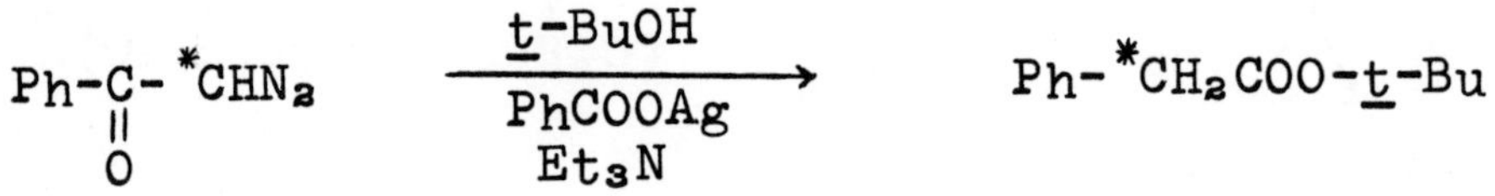

$$\underset{\mathbf{i}}{\text{Ph-C(=O)-}^{*}\text{CHN}_2} \xrightarrow[\text{PhCOOAg, Et}_3\text{N}]{t\text{-BuOH}} \text{Ph-}^{*}\text{CH}_2\text{COO-}t\text{-Bu}$$

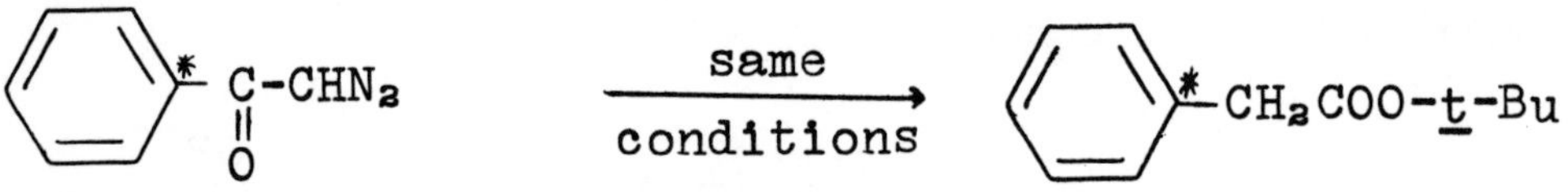

$$\underset{\mathbf{ii}}{\text{C}_6\text{H}_5{}^{*}\text{-C(=O)-CHN}_2} \xrightarrow[\text{conditions}]{\text{same}} \text{C}_6\text{H}_5{}^{*}\text{-CH}_2\text{COO-}t\text{-Bu}$$

i and ii were subjected to the Wolff rearrangement as shown above and the rates compared with that of unlabeled α-diazo-acetophenone. In neither case was there an isotope effect. It was previously known that N_2 is lost in the rate-determining step. What implications can be drawn about the mechanism from these results?

93. Hexamethyl Dewar benzene dissolves in HF or HF-BF_3 at $-80°$ to give a solution containing an equilibrium mixture of the epimers i and ii, which rapidly interconvert. Suggest a reasonable mechanism for this interconversion.

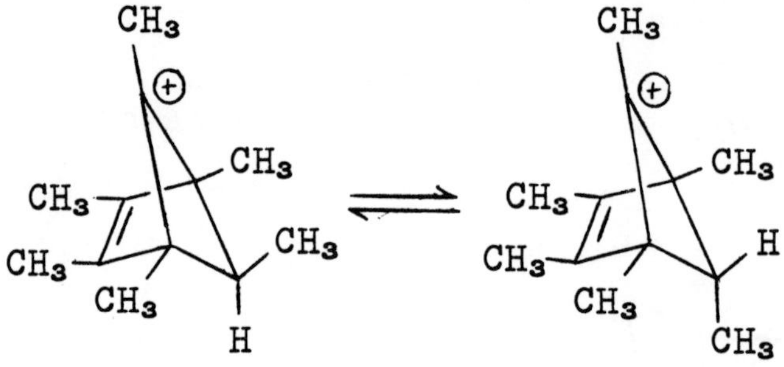

CH3, ⊕, CH3, CH3, CH3, CH3, CH3, H ⇌ CH3, ⊕, CH3, CH3, CH3, CH3, H, CH3

i ii

Hogeveen and Volger, *Rec. Trav. Chim.* **88**, 353 (1969).

94. Decarbonylation of 4-pentenal (i) gave 1-butene (ii). A very small amount of methylcyclopropane was also formed, but no cyclobutane could be be detected. When 4-pentenal labeled in the 2 position (iii) was used, the label ap-

$$\underset{\mathbf{i}}{CH_2{=}CHCH_2CH_2CHO} \xrightarrow{(t\text{-Bu})_2O} \underset{\mathbf{ii}}{CH_2{=}CHCH_2CH_3}$$

94. Cont.

$$CH_2{=}CHCH_2CD_2CHO \xrightarrow{(\underline{t}\text{-Bu})_2O} CH_2{=}CHCH_2CD_2H + CH_2{=}CHCD_2CH_3$$

iii

peared in both the 3 and 4 positions. When the initial concentration of iii was varied from 0.50 to 6.0 *M*, the ratio of deuterons in the 4 position to those in the 3 position increased monotonically from 0.96 to 1.49. Decarbonylation of iv gave an approximately 50:50 mixture of the cis and trans isomers v and vi. Give a mechanism consistent with these results.

$$(H)(D)C{=}C(H)CH_2CH_2CHO \xrightarrow{(\underline{t}\text{-Bu})_2O} (H)(D)C{=}C(H)CH_2CH_3 + (D)(H)C{=}C(H)CH_2CH_3$$

iv → v + vi

Montgomery and Matt, *J. Amer. Chem. Soc.* **89**, 6556 (1967).

95.

i (trans, cis, cis, trans) → iii

ii (trans, cis, trans) → iv

95. Cont.

In the thermal rearrangement above: a. Predict the configuration of the products (that is, are the methyl groups cis or trans?); b. Which compound, i or ii, should cyclize more easily (milder conditions)?

96. OMe

OMe

N

$\xrightarrow[\text{diglyme}]{\text{NaH}}$ 7-11%

OMe

N

OMe

12a

12

(-)-R isomer

a. Suggest a reasonable mechanism; b. What is the configuration (R or S) of the product at the 12*a* position? (Models will help.)

Brewster and Jones, *J. Org Chem.* **34**, 354 (1969).

97. Treatment of i with TsOH gave ii.

$\text{Ph-C(CH}_3)_2\text{-CH=CH}_2 \xrightarrow[\text{C}_6\text{H}_6 \text{ reflux}]{\text{TsOH}} (\text{Ph})(\text{CH}_3)\text{C=C}(\text{CH}_3)_2$

i ii

a. ii could arise either by migration of Ph or of Me. Devise one or more experiments to determine which.

b. It was determined that both Me and Ph migrated, to about the same extent. However, when iii was dehydrated with TsOH or $HClO_4$; or when the tosyl derivative of iii was treated with TsOH, ii was formed by

$\text{Ph-C(CH}_3)_2\text{-CH(OH)-CH}_3 \xrightarrow[\text{or HClO}_4]{\text{TsOH}} \text{ii}$

iii

exclusive migration of phenyl. What mechanistic conclusion can be drawn from these results?

Grimaud and Laurent, *Bull. Soc. Chim. Fr.* **1967**, 3599.

98. Tropones which contain good leaving groups in the α position are converted by many bases into benzoic acid derivatives. When two leaving groups are present mixtures are often obtained. Suggest one or more mechanisms to account for

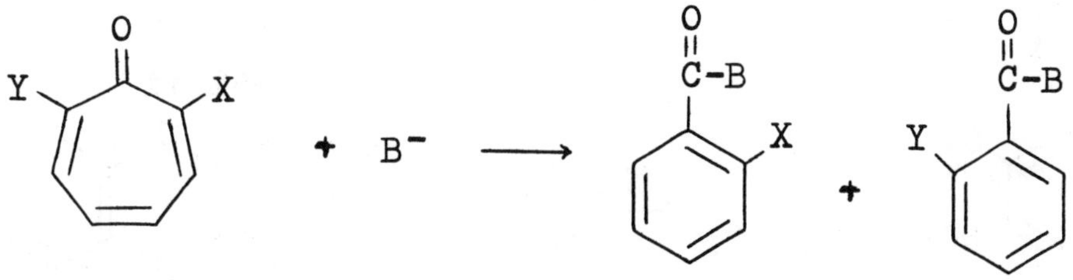
O
Y
X
+ B⁻
O
C-B
X
+
O
C-B
Y

this reaction, and devise one or more experiments to test your mechanisms.
Magid, Grayson, and Cowsar, *Tetrahedron Lett.* **1968**, 4819.

99.

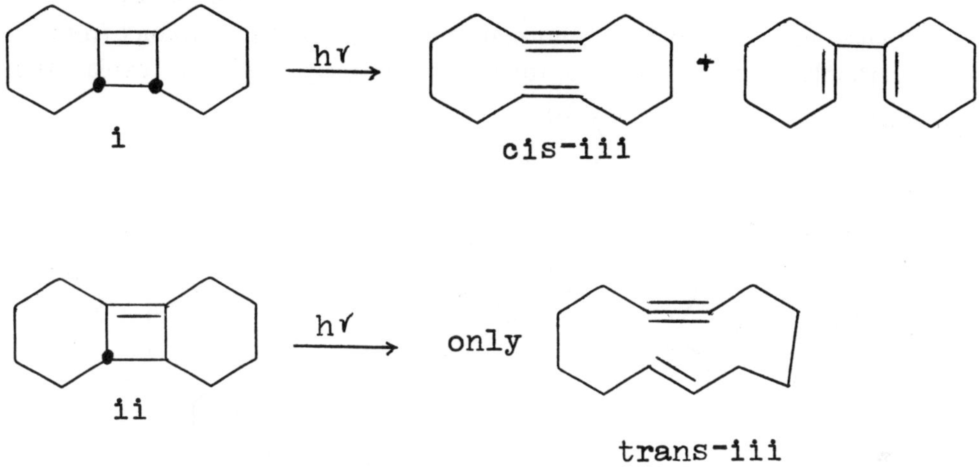
hν
+
i
cis-iii
hν
only
ii
trans-iii

Why was no bicyclohexenyl formed in the case of ii?

100.

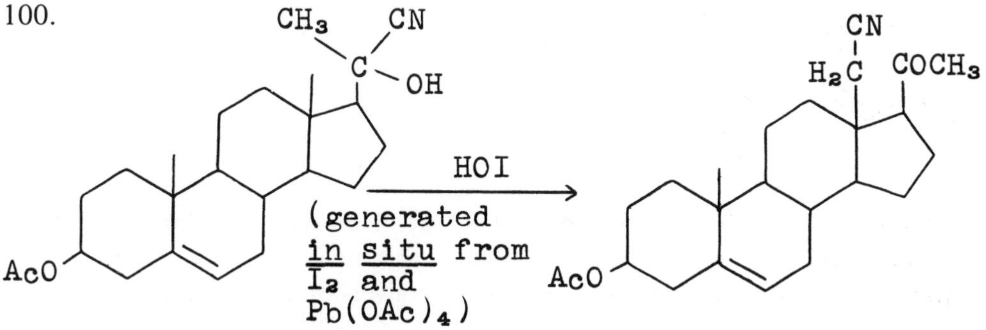
CH3
CN
C
OH
AcO
HOI
(generated in situ from I2 and Pb(OAc)4)
CN
H2C
COCH3
AcO

a. Devise one or more experiments to determine whether the CN group migrates inter or intramolecularly.

b. Assume that the migration is intramolecular (it has been so found). Suggest a reasonable mechanism.

Kalvoda, *Helv. Chim. Acta* **51**, 267 (1968).

101.

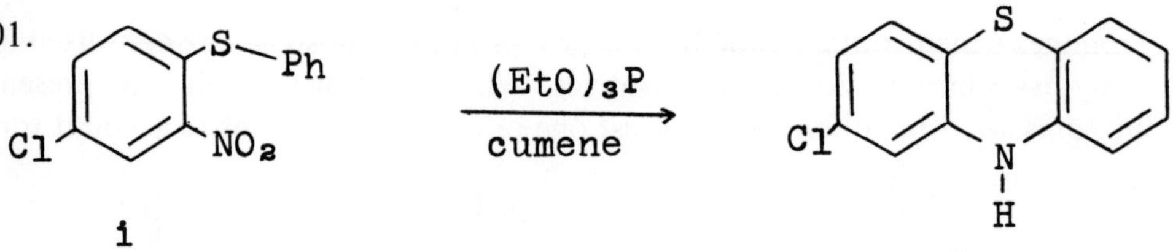

i → $(EtO)_3P$, cumene → 55% ii

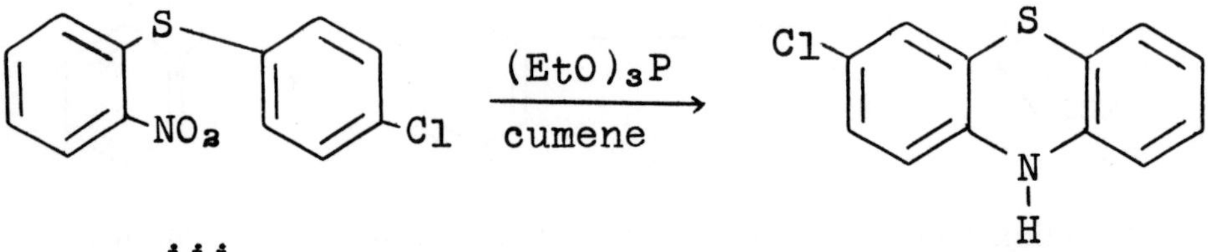

iii → $(EtO)_3P$, cumene → 63% iv

i gives what appears to be a normal product (ii); but the product from iii is certainly rearranged (ii is not formed in this case). Suggest an explanation.

102.

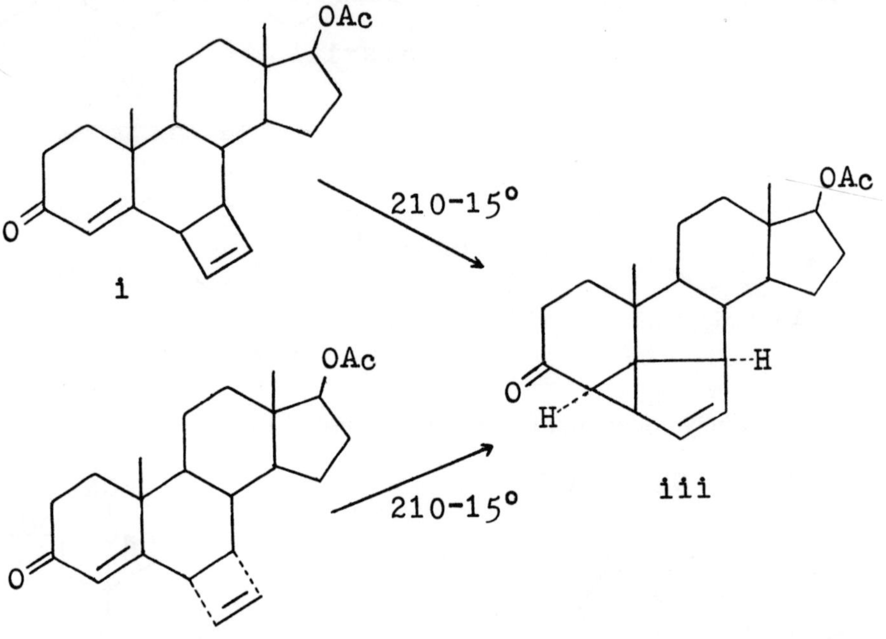

i → 210-15° → iii

ii → 210-15° → iii

Both i and ii give iii when heated at 210-215°. However, iv is stable at this temperature, and does not react until 250°, at which point it is converted to v. Suggest a suitable mechanism for the conversion of i and ii to iii.

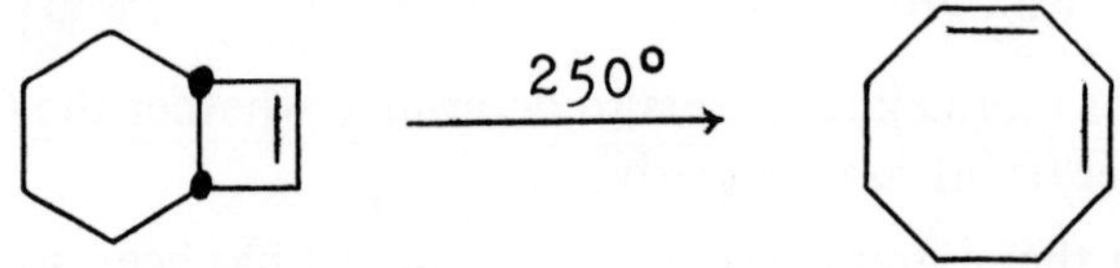

Nelson, Murphy, Edwards, and Fried, *J. Amer. Chem. Soc.* **90**, 5572 (1968).

PROBLEM SET 13

OXIDATIONS AND REDUCTIONS

Corresponds to Chapter 19 of Advanced Organic Chemistry.

The first 51 problems are synthetic. In each one you are asked to convert one compound to another. In every case the conversion has actually been carried out and the reference is given.

1.

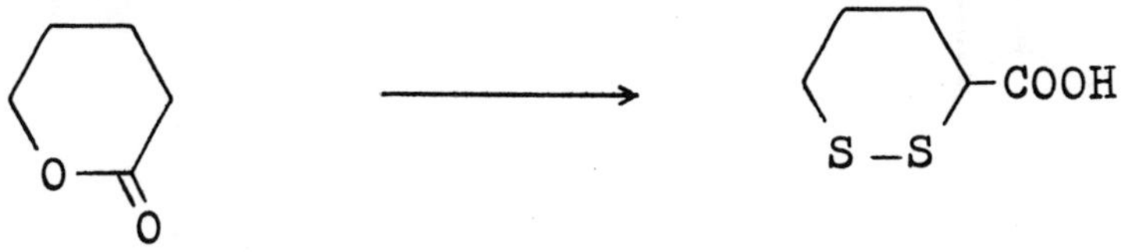

i → ii (COOH; S –S)

2.

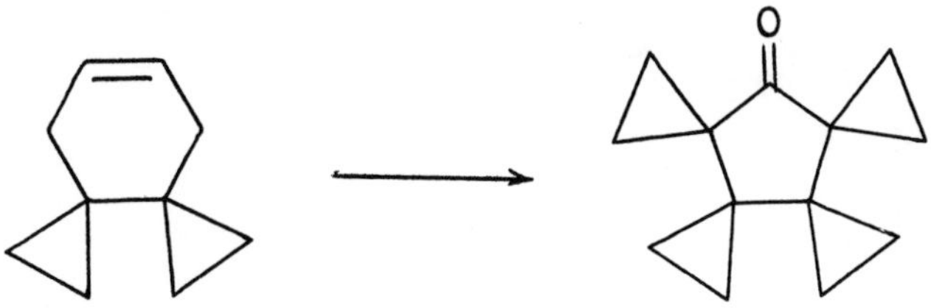

Ripoll and Conia, *Tetrahedron Lett.* **1969**, 979.

3.

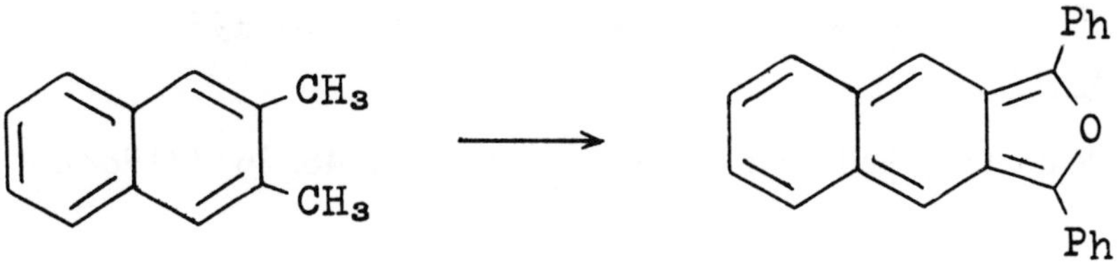

CH_3, CH_3 → Ph, O, Ph

Cava and VanMeter, *J. Org. Chem.* **34**, 538 (1969).

4.

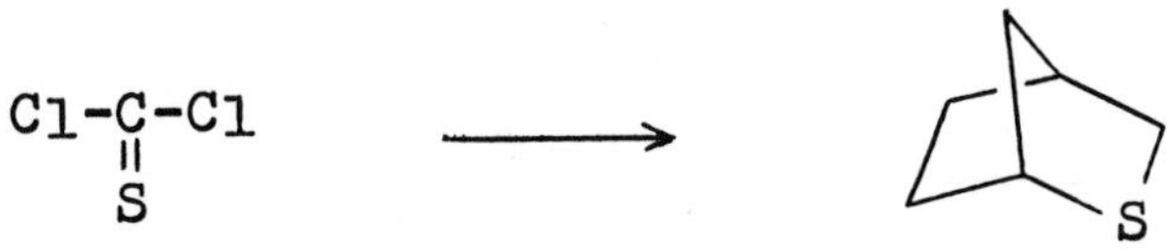

Cl-C-Cl, S → S

i thiophosgene → ii

5.

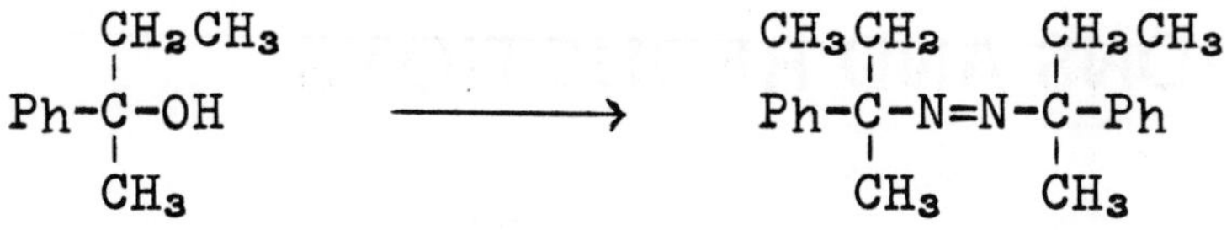

$$Ph{-}C(CH_2CH_3)(CH_3){-}OH \longrightarrow Ph{-}C(CH_3CH_2)(CH_3){-}N{=}N{-}C(CH_2CH_3)(CH_3){-}Ph$$

Severn and Kosower, *J. Amer. Chem. Soc.* **91**, 1710 (1969).

6.

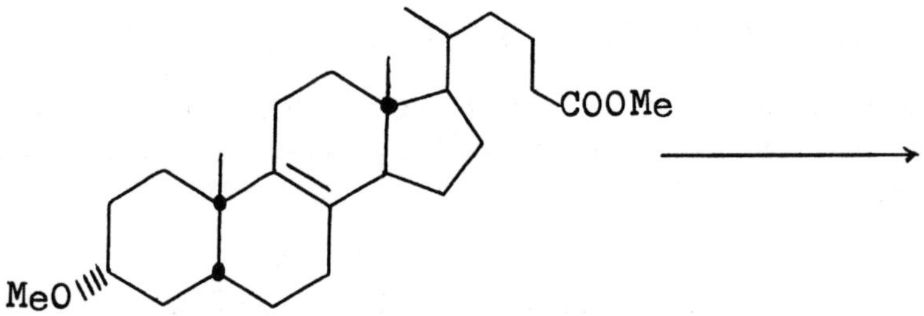

COOMe

MeO

i

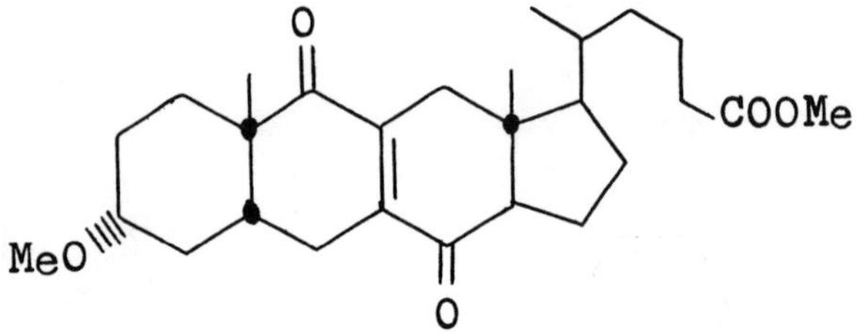

→

O

COOMe

MeO

O

ii

7.

$$\text{cyclopentanone} \longrightarrow CH_3(CH_2)_6{-}CH(Br){-}(CH_2)_3CH_2Br$$

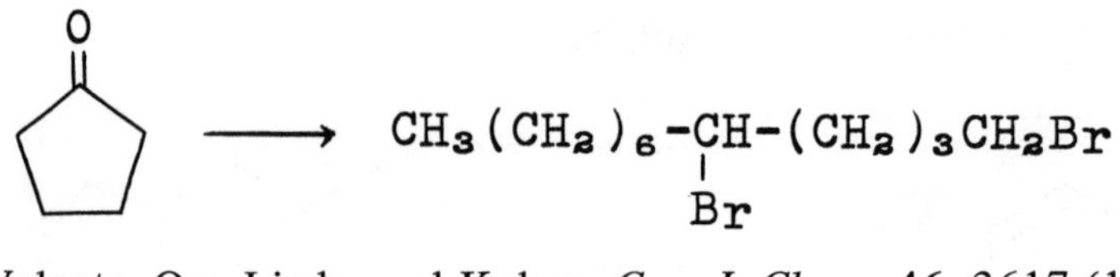

Wiesner, Valenta, Orr, Liede, and Kohan, *Can. J. Chem.* **46**, 3617 (1968).

8.

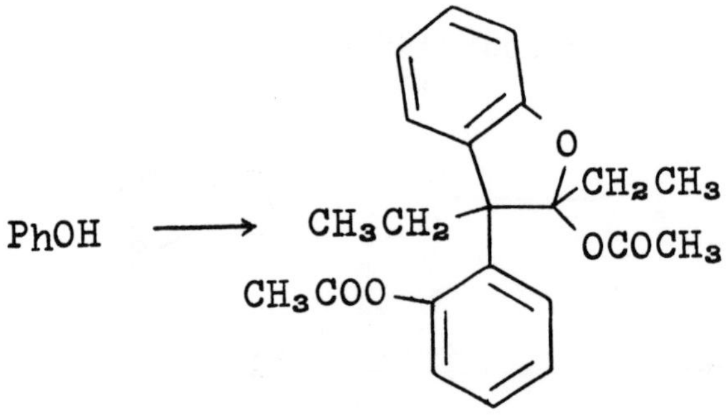

PhOH →

O

CH_2CH_3

CH_3CH_2

$OCOCH_3$

CH_3COO

Devis and Depovere, *Bull. Soc. Chim. Fr.* **1967**, 3185.

9.

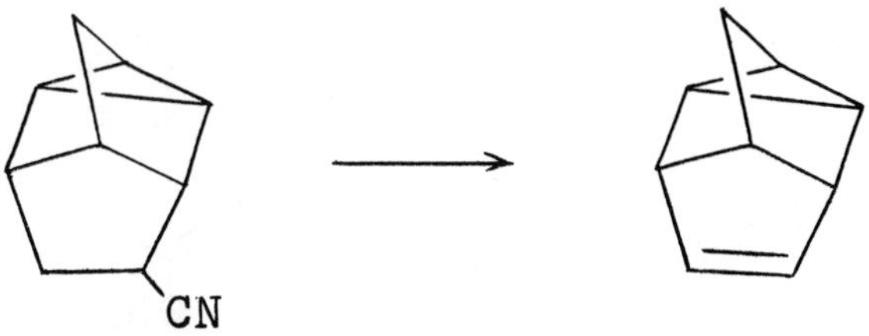

Wiscott and Schleyer, *Angew. Chem. Intern. Ed. Engl.* **6**, 694 (1967) [*Angew. Chem.* **79**, 680 (1967)].

10.

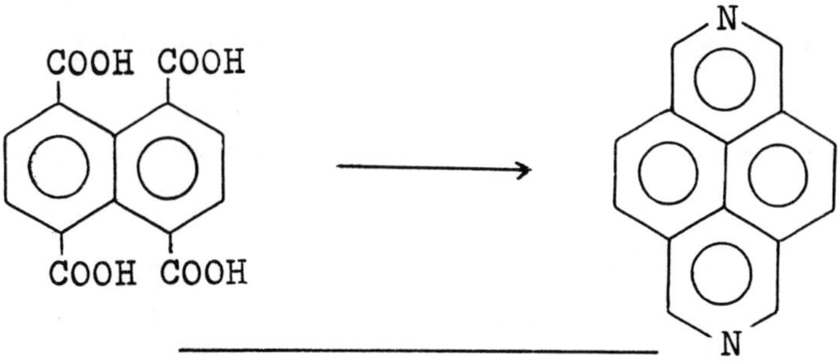

Lier, Hünig, and Quast, *Angew. Chem. Intern. Ed. Engl.* **7**, 814 (1968) [*Angew. Chem.* **80**, 799 (1968)].

11.

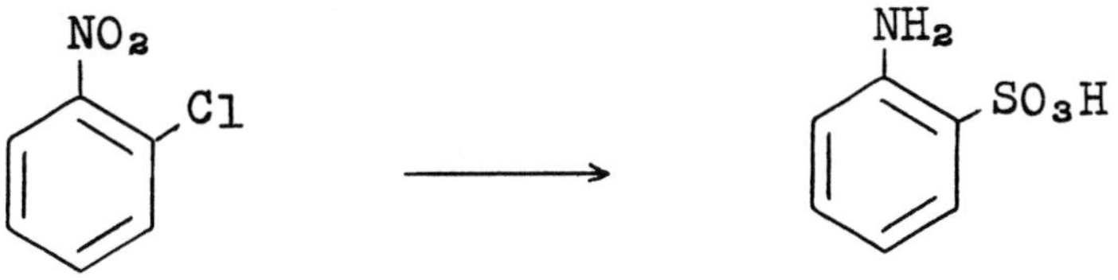

Bogert and Stull, *Org. Syn.* **I**, 220; Wertheim, *Org. Syn.* **II**, 471.

12.

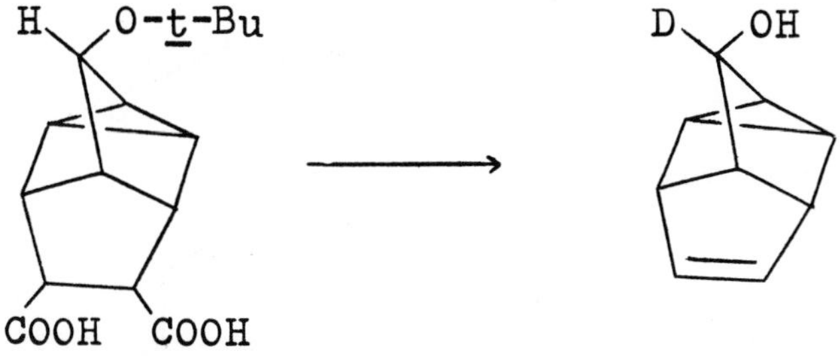

13.

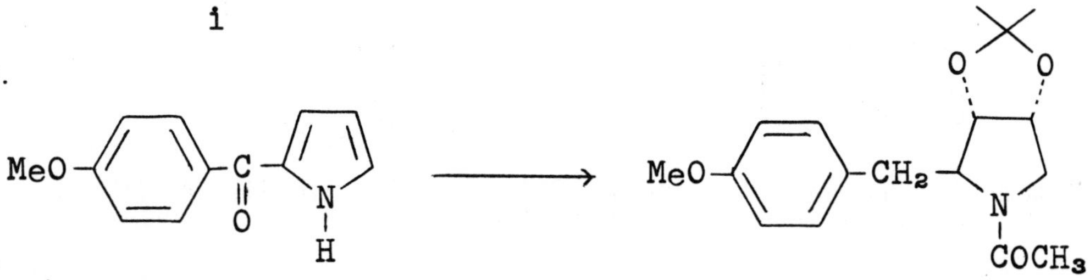

Salmón and Walls, *Chem. Commun.* **1969**, 63.

14.

OCH_2Ph

$PhCH_2O$

$COCH_3$

OMe

OH

O

Ph

OMe

obtusaquinone

Gregson, Ollis, Redman, Sutherland, and Dietrichs, *Chem. Commun.* **1968**, 1395.

15. $EtOOCCH_2COOEt \longrightarrow EtOOC\text{-}CH(NH_2)\text{-}COOEt$

Zambito and Howe, *Org. Syn.* **40**, 21; Hartung, Beaujon, and Cocolas, *Org. Syn.* **40,** 24.

16.

OMe

COOEt

MeOOC

i

OMe

O

MeOOC

ii

17.

Cl

O

O

Hata and Tanida, *J. Amer. Chem. Soc.* **91,** 1170 (1969).

18.

N

Kesavan, Devanathan, and Arumugam, *Bull. Chem. Soc. Jap.* **41,** 3008 (1968).

19.

$CH_3-{}^{14}C(=O)-OEt$ ⟶ [lactone with CH_3, OH, O, 14]

Pichat, Blagoev, and Hardovin, *Bull. Soc. Chim. Fr.* **1968**, 4489.

20.

[Cl, N, CH_2Cl, N→O, Ph] ⟶ [O, N, CH_2Cl, N, Ph]

Sternbach, Kaiser, and Reeder, *J. Amer. Chem. Soc.* **82**, 476 (1960).

21. [HOOC, Me, COOH, Me, Me, COOH] ⟶ [Me, Me, Me, Me, Me, Me]

Canonne and Regnault, *Tetrahedron Lett.* **1969**, 243.

22.

i ⟶ ii

23.

Paquette and Philips, *Chem. Commun.* **1969**, 680.

24.

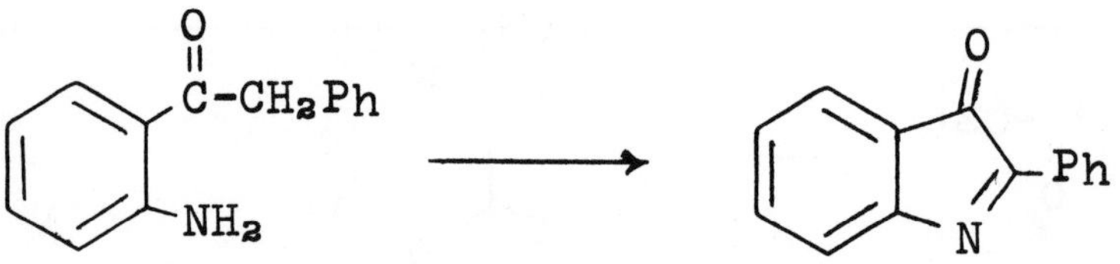

Schmitt, Perrin, Langlois, and Suquet, *Bull. Soc. Chim. Fr.* **1969**, 1227.

25.

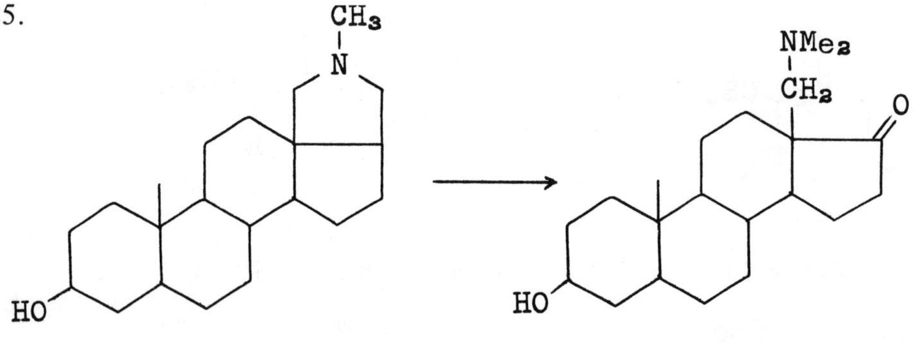

26.

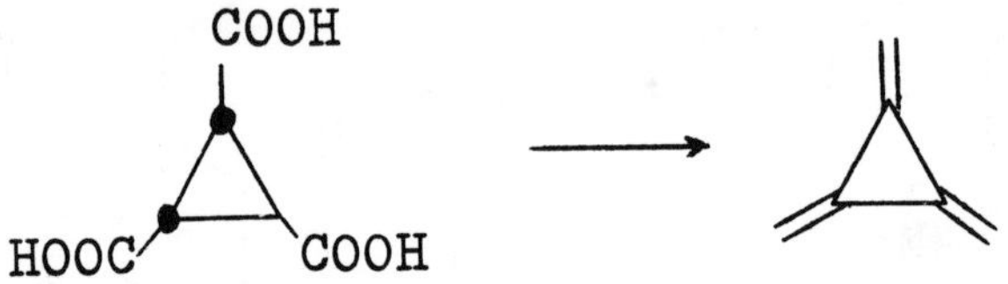

Waitkus, Sanders, Peterson, and Griffin, *J. Amer. Chem. Soc.* **89**, 6318 (1967).

27.

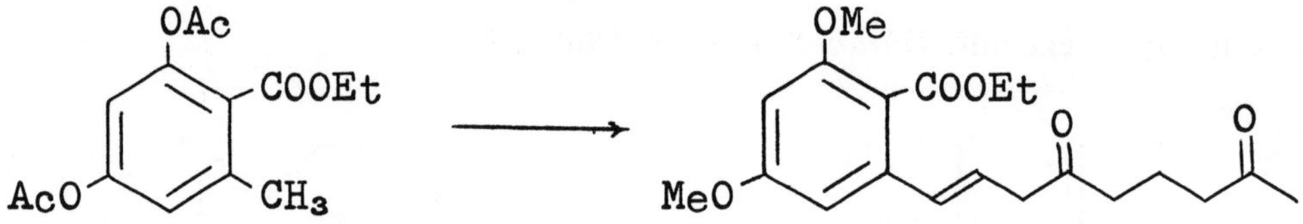

Vlattas, Harrison, Tökés, Fried, and Cross, *J. Org. Chem.* **33**, 4176 (1968).

28.

29.

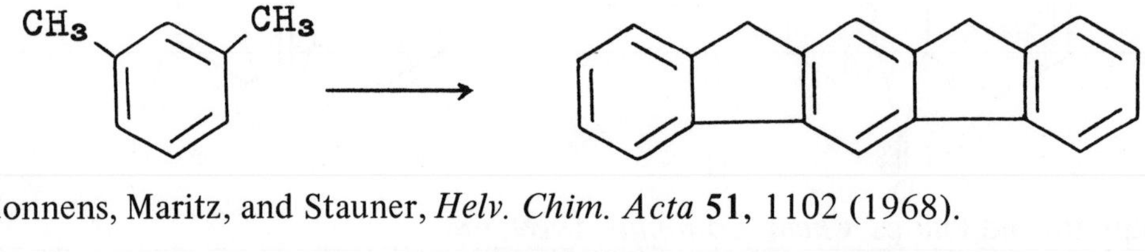

Chardonnens, Maritz, and Stauner, *Helv. Chim. Acta* **51**, 1102 (1968).

30.

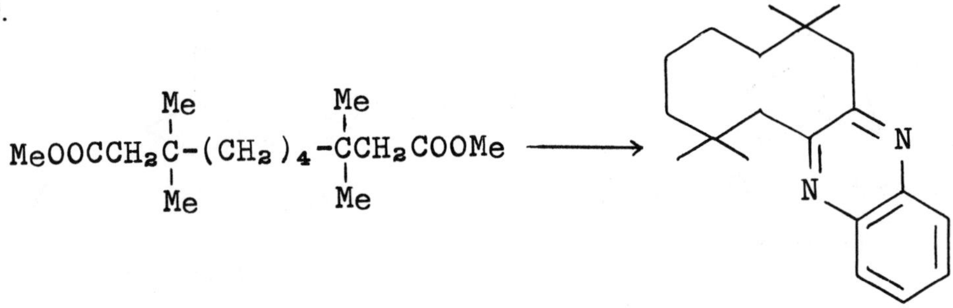

$MeOOCCH_2C(Me)_2-(CH_2)_4-C(Me)_2CH_2COOMe$ ⟶

Cordes, Prelog, Troxler, and Westen, *Helv. Chim. Acta* **51**, 1663 (1968).

31.

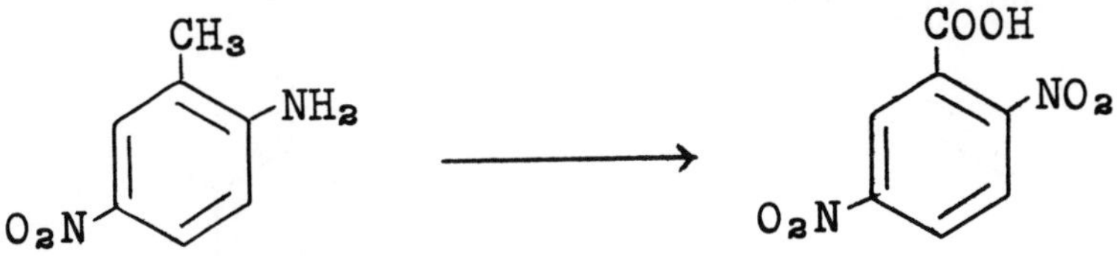

Langley, *Org. Syn.* **III**, 334.

32.

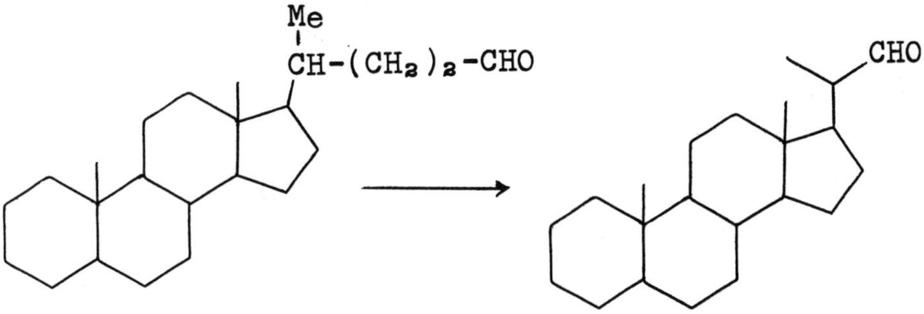

$CH(Me)-(CH_2)_2-CHO$ ⟶ CHO

Yanuka, Katz, and Sarel, *Chem. Commun.* **1968**, 851.

33.

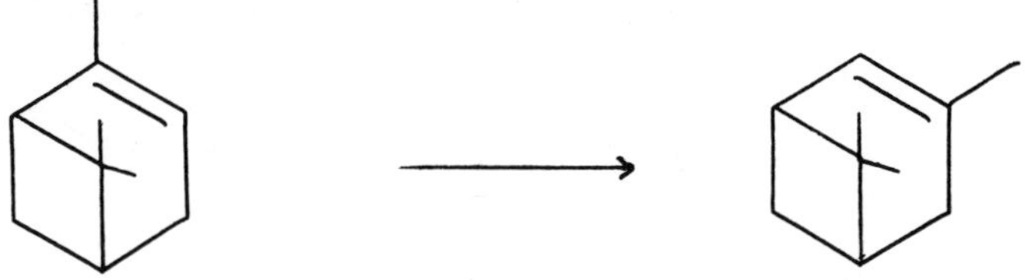

i ⟶ ii

α-pinene ⟶ orthodene

34.

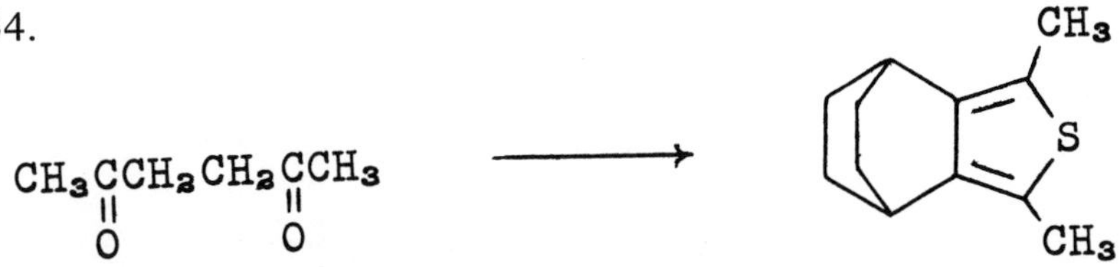

$CH_3CCH_2CH_2CCH_3$ (O, O) ⟶ CH_3, S, CH_3

Wynberg and Klunder, *Rec. Trav. Chim.* **88**, 328 (1969).

35.

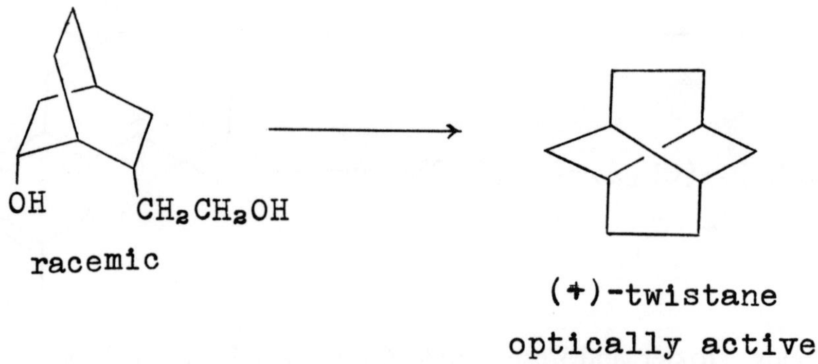

OH CH$_2$CH$_2$OH

racemic

(+)-twistane

optically active

Adachi, Naemura, and Nakazaki, *Tetrahedron Lett.* **1968**, 5467.

36.

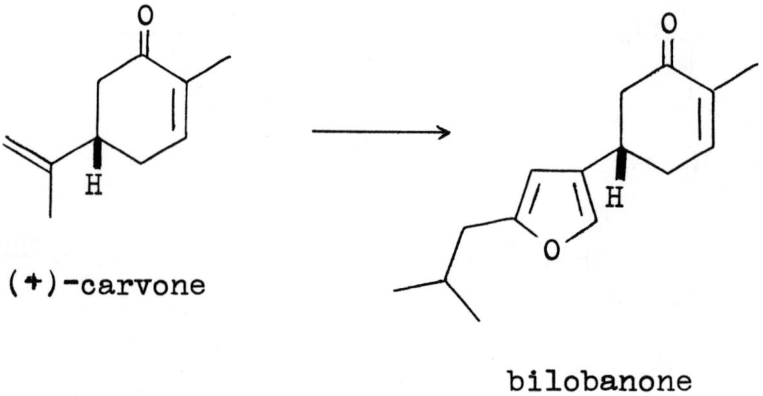

O

H

(+)-carvone

O

H

O

bilobanone

Büchi, and Wüest, *J. Org. Chem.* **34**, 857 (1969).

37.

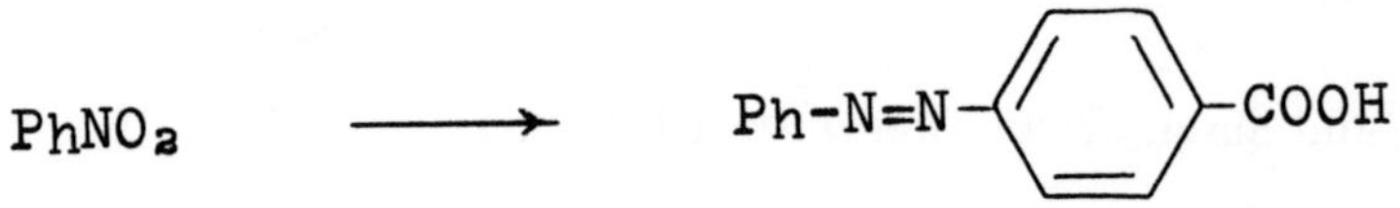

PhNO$_2$ ⟶ Ph-N=N-C$_6$H$_4$-COOH

Anspon, *Org. Syn.* **III**, 711; Coleman, McCloskey, and Stuart, *Org. Syn.* **III**, 668.

38.

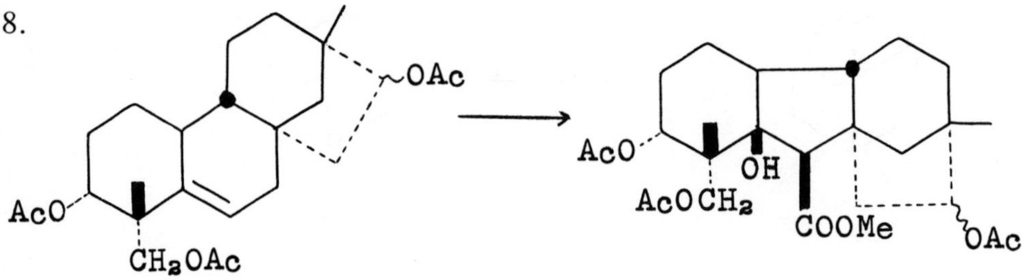

~OAc

AcO

CH$_2$OAc

AcO

OH

AcOCH$_2$

COOMe

OAc

Ogawa, Mori, Matsui, and Sumiki, *Tetrahedron Lett.* **1968**, 2551.

39.

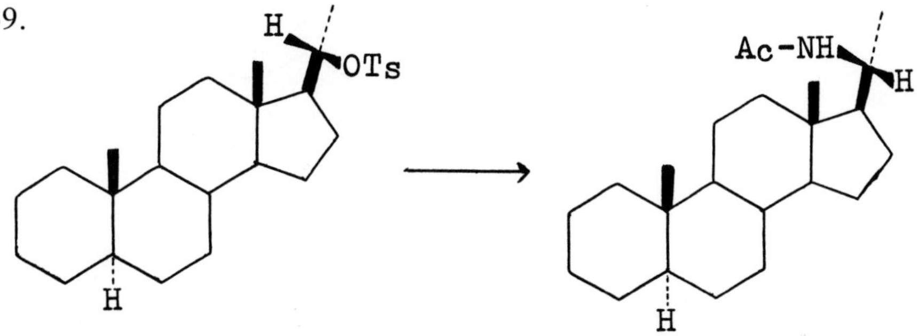

H, OTs → Ac−NH, H

Leboeuf, Cavé, and Goutarel, *Bull. Soc. Chim. Fr.* **1969**, 1619.

40.

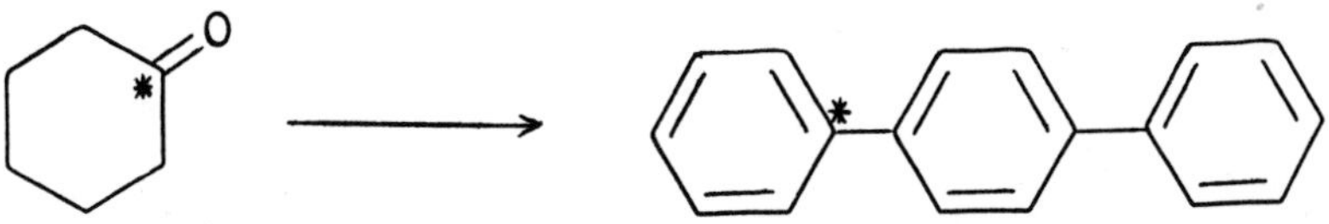

cyclohexanone-1-^{14}C → p-terphenyl-1-^{14}C

Juppe, Hafferl, and Bloss, *Chem. Ber.* **101**, 1917 (1968).

41.

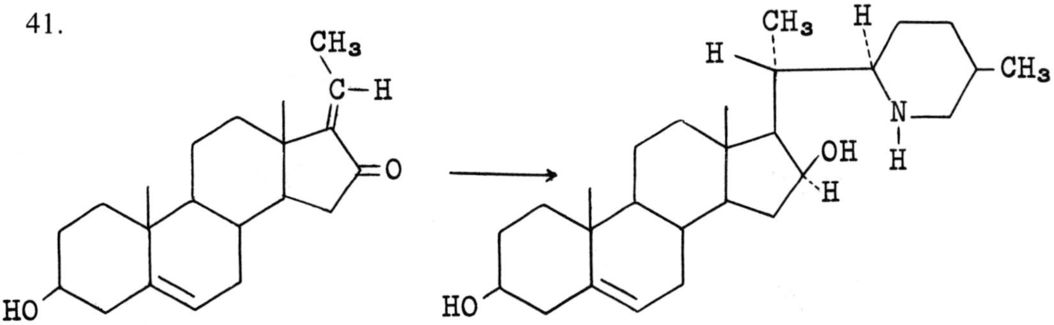

Kessar, Bhatti, and Mahajan, *Tetrahedron Lett.* **1969**, 603.

42.

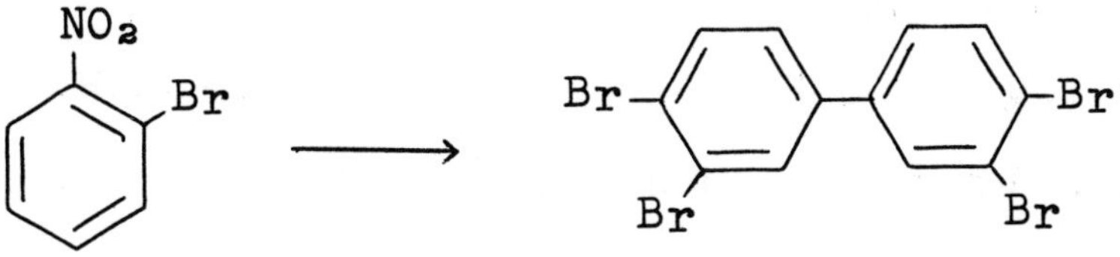

i → ii

43.

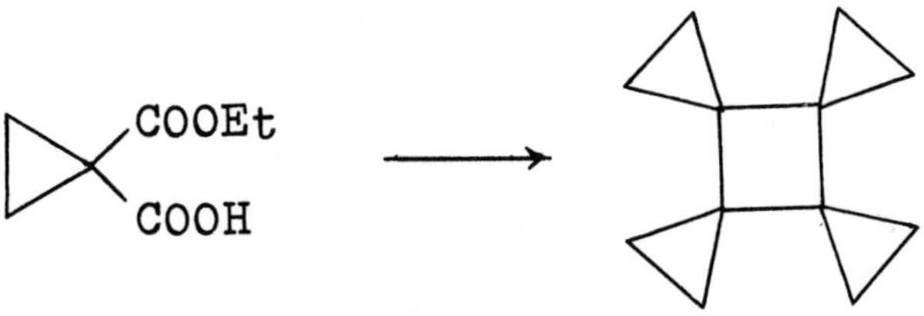

Conia and Denis, *Tetrahedron Lett.* **1969**, 3545.

44.

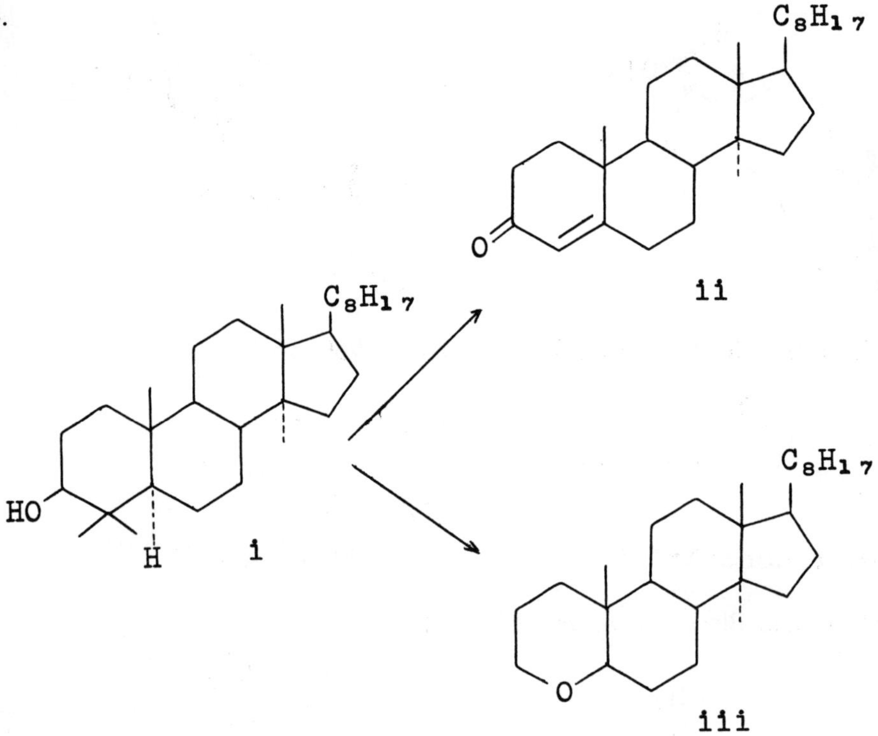

C_8H_{17}

HO

H

i

ii

iii

45.

$COCH_3$ (p-bromo) → OH, CH-COOH (p-bromo)

Br

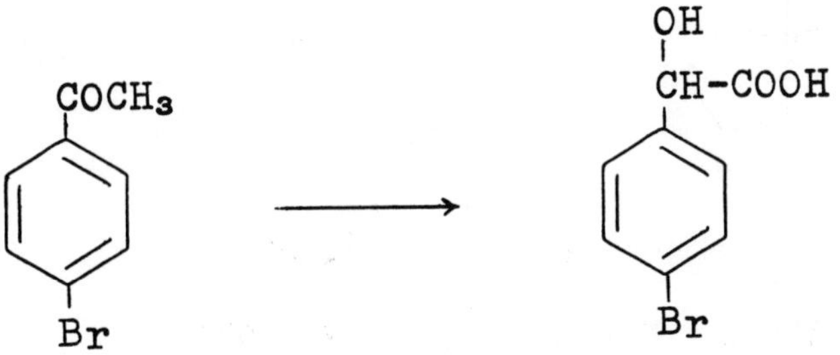

Klingenberg, *Org. Syn.* **IV**, 110.

46. $CH_2{=}CH{-}(CH_2)_8{-}COOH \longrightarrow HO_3S{-}(CH_2)_{10}{-}COOH$

Showell, Russell, and Swern, *J. Org. Chem.* **27**, 2853 (1962).

47.

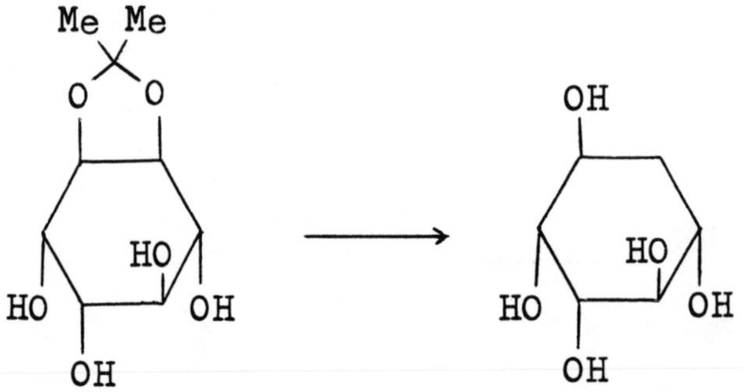

Me Me

O O

HO HO OH OH → OH HO HO OH OH

McCasland, Naumann, and Durham, *J. Org. Chem.* **33**, 4220 (1968).

48.

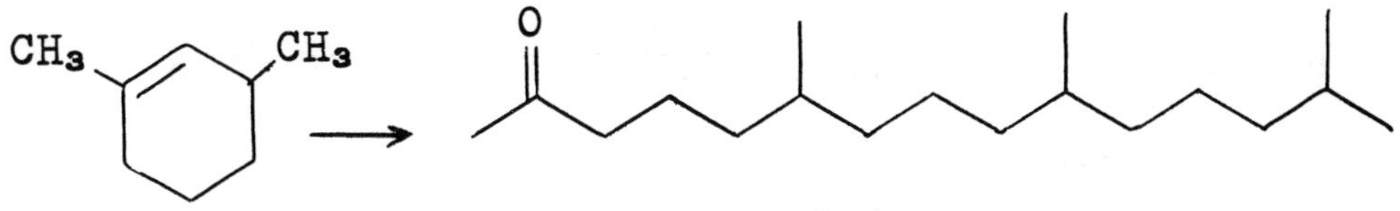

CH_3 CH_3 → O

phytone

Ichikawa and Kato, *Bull. Chem. Soc. Jap.* **41**, 1232 (1968).

49.

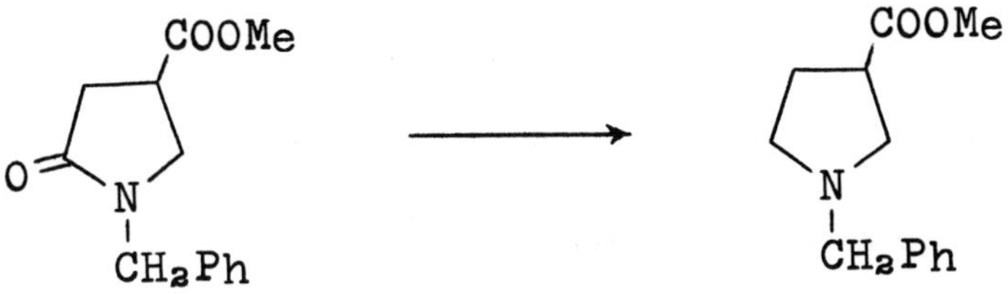

COOMe → COOMe

O N N

CH_2Ph CH_2Ph

i ii

50.

OH → NH_2 OH

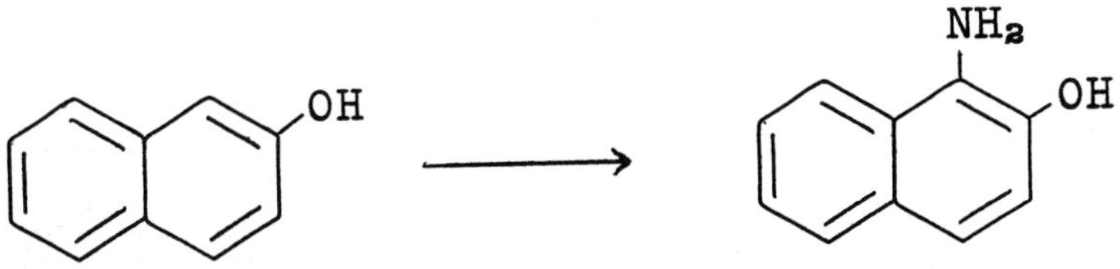

Fieser, *Org. Syn.* **II**, 35; or Marvel and Porter, *Org. Syn.* **I**, 411; Conant and Corson, *Org. Syn.* **II**, 33.

51.

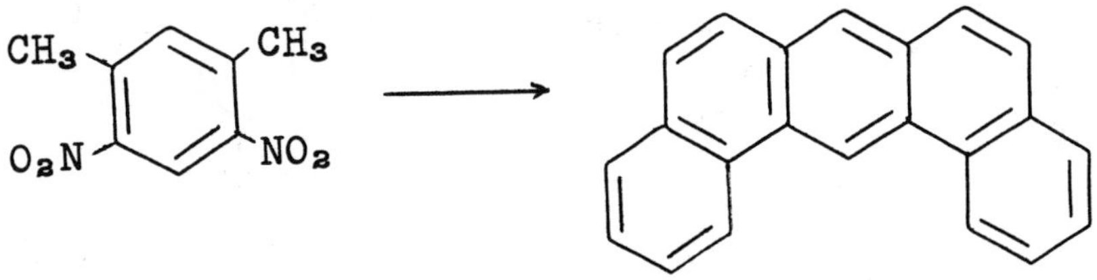

CH_3 CH_3

O_2N NO_2 →

Vögtle and Staab, *Chem. Ber.* **101**, 2709 (1968).

The following conversions (problems 52 to 68) have been reported in the literature. In each case suggest a reasonable mechanism.

52.

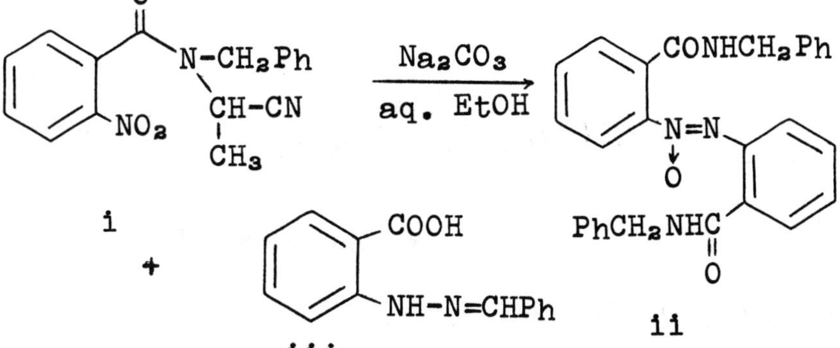

O

$N-CH_2Ph$

NO_2 $CH-CN$

CH_3

i

$\xrightarrow[\text{aq. EtOH}]{Na_2CO_3}$

$CONHCH_2Ph$

N=N

↓

O

$PhCH_2NHC$

O

ii

+

COOH

NH-N=CHPh

iii

53.

$$Ph\text{-}cyclopropane + PdCl_2 \xrightarrow[75^\circ\ 2\ hr]{H_2O} PhCOCH_2CH_3$$

1:1 ratio — 60% i

+ $PhCH_2COCH_3$ (35% ii) + $PhCH{=}CHCH_3$ (iii)

Note: iii, treated with $PdCl_2$ under the same conditions gives only ii.

Oullette and Levin, *J. Amer. Chem. Soc.* **90**, 6889 (1968).

54.

2,4-di-tert-butyl-6-ethylphenol (OH, CH_2CH_3) $\xrightarrow[C_6H_6]{\text{excess alkaline } K_3Fe(CN)_6}$ spiro products: 21% + 15%

Note: The same products were formed when PbO_2 was used instead of $K_3Fe(CN)_6$

Cook and Butler, *J. Org. Chem.* **34**, 227 (1969).

55.

$$PhCH_2OH + Hg(OAc)_2 \xrightarrow[130^\circ\ 1\text{-}2\ hr]{Me_2SO} 86\%\ PhCHO$$

$$+ Me_2S + HgO$$

Tien, Tien, and Ting, *Tetrahedron Lett.* **1969**, 1483.

56.

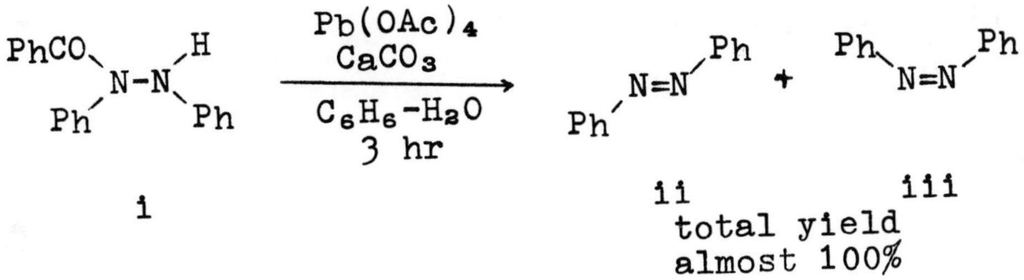

PhCO(Ph)N–NH(Ph) → [$Pb(OAc)_4$, $CaCO_3$; C_6H_6-H_2O, 3 hr] → PhN=NPh (ii) + PhN=NPh (iii)

i

ii iii

total yield almost 100%

57.

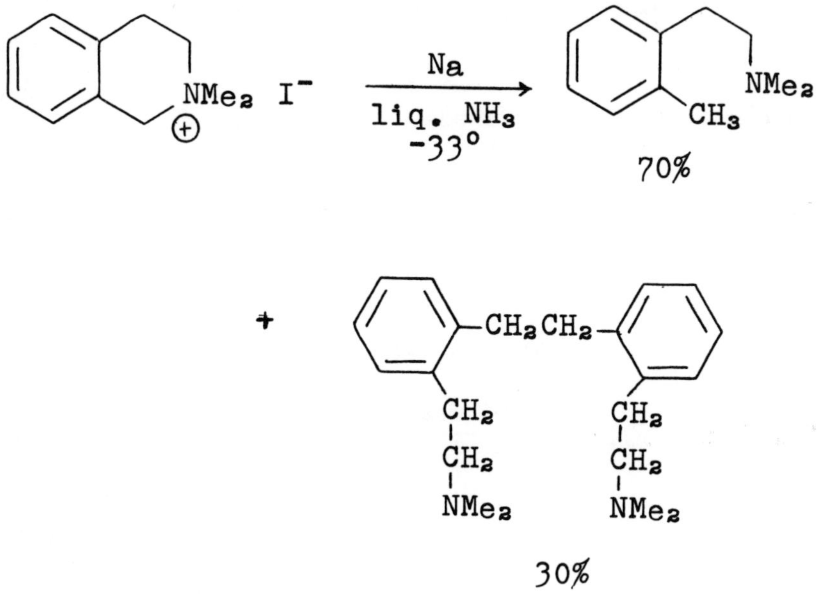

$\xrightarrow[\text{liq. } NH_3,\ -33^\circ]{Na}$

NMe₂ I⁻ → NMe₂, CH_3 70%

+ CH_2CH_2; CH_2, CH_2, NMe_2; CH_2, CH_2, NMe_2

30%

Wrŏbel, Bieñ, and Pazdro, *Chem. Ind.* (London) **1968**, 1569.

58.

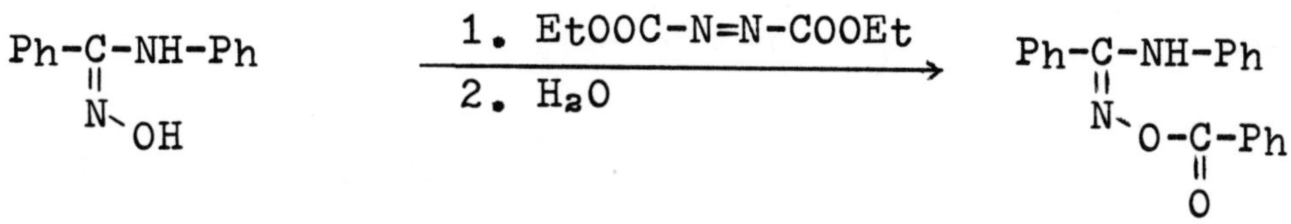

Ph–C(=N–OH)–NH–Ph → [1. EtOOC-N=N-COOEt; 2. H_2O] → Ph–C(=N–O–C(=O)–Ph)–NH–Ph

The same product was obtained, though with lower yield, when lead tetraacetate or N-bromosuccinimide was used as the reagent.

Boyer and Frints, *J. Org. Chem.* **33**, 4554 (1968).

59.

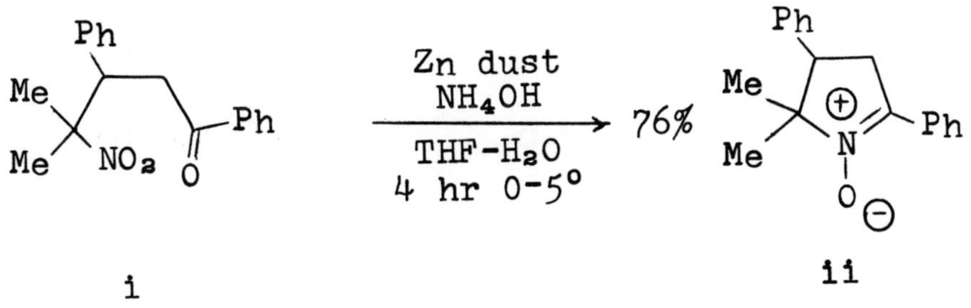

i → [Zn dust, NH_4OH; THF-H_2O, 4 hr 0–5°] → 76% ii

Ph, Me, Me, NO_2, O, Ph → Ph, Me, Me, Ph, N⁺, O⁻

i ii

60.

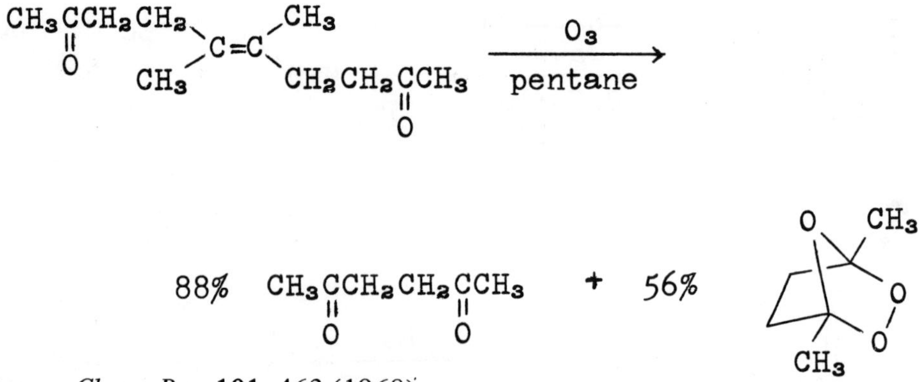

$$CH_3\underset{\underset{O}{\|}}{C}CH_2CH_2(CH_3)C{=}C(CH_3)CH_2CH_2\underset{\underset{O}{\|}}{C}CH_3 \xrightarrow[\text{pentane}]{O_3}$$

$$88\%\ CH_3\underset{\underset{O}{\|}}{C}CH_2CH_2\underset{\underset{O}{\|}}{C}CH_3 \quad + \quad 56\%$$

Griesbaum, *Chem. Ber.* **101**, 463 (1968).

61.

$$(PhCH_2)_2NNH_2 + Pb(OAc)_4 \xrightarrow{C_6H_6} PhCHO + PhCH_2N_3$$

$$+ (PhCH_2)_2NH$$

Koga and Anselme, *J. Amer. Chem. Soc.* **91**, 4323 (1969).

62.

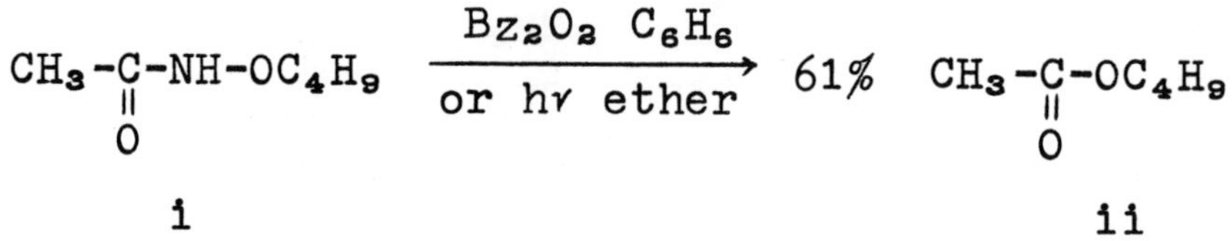

$$\underset{\textbf{i}}{CH_3-\underset{\underset{O}{\|}}{C}-NH-OC_4H_9} \xrightarrow[\text{or } h\nu \text{ ether}]{Bz_2O_2\ C_6H_6} 61\%\ \underset{\textbf{ii}}{CH_3-\underset{\underset{O}{\|}}{C}-OC_4H_9}$$

63.

MeO, MeO, OMe, OMe, MeO, OMe

$$\xrightarrow{Na_2Cr_2O_7}$$

O, O, OMe, OMe, MeO, MeO, OMe, OMe, O

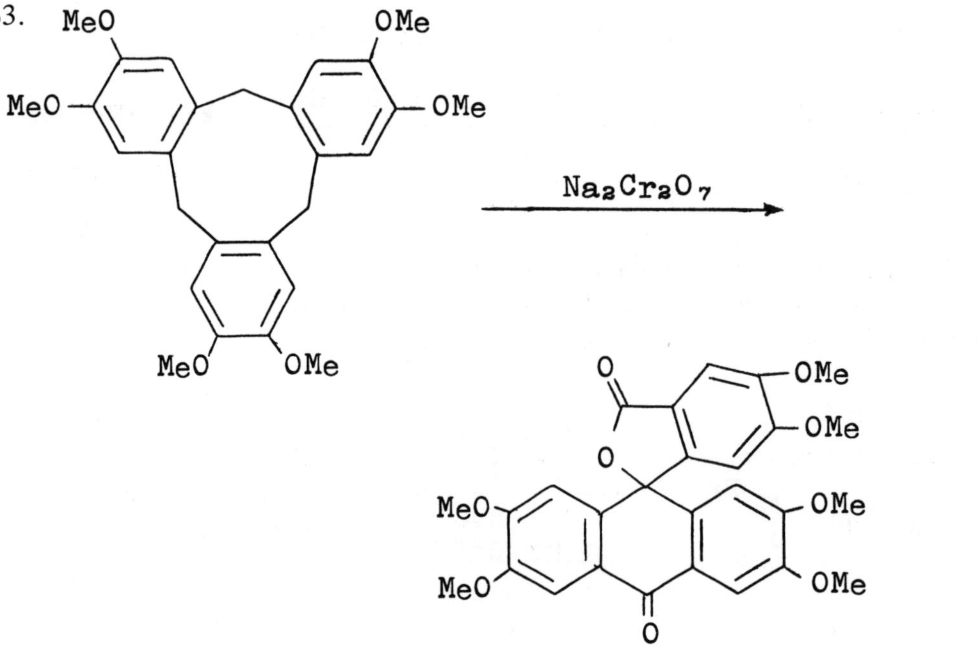

Baldwin and Kelly, *Chem. Commun.* **1968**, 1664.

64.

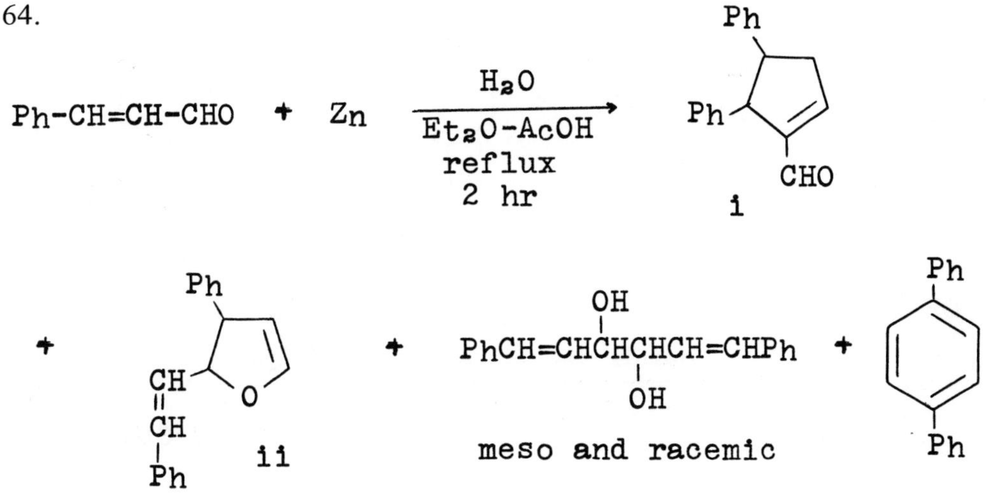

Ph-CH=CH-CHO + Zn $\xrightarrow[\text{Et}_2\text{O-AcOH, reflux, 2 hr}]{H_2O}$ i (Ph, Ph, CHO) + ii (Ph, CH=CH-Ph, O) + PhCH=CHCH(OH)CH(OH)CH=CHPh (meso and racemic) + Ph–C₆H₄–Ph

i

ii

meso and racemic

Note: ii, heated 30 min at 270°, gives i.

Chuche and Wiemann, *Bull. Soc. Chim. Fr.* **1968**, 1497.

65.

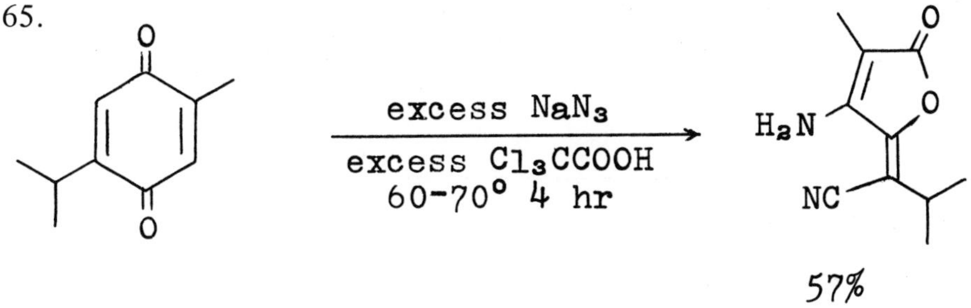

excess NaN_3 / excess Cl_3CCOOH, 60-70° 4 hr

H_2N, NC, O, O

57%

Moore and Shelden, *J. Org. Chem.* **33**, 4019 (1968).

66.

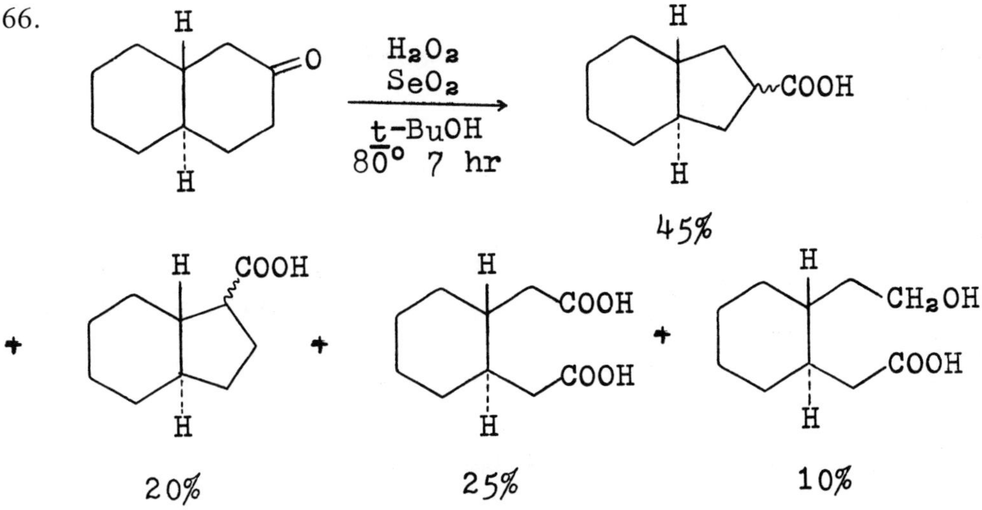

H_2O_2, SeO_2 / t-BuOH, 80° 7 hr

H, O, H → H, COOH, H

45%

+ (H, COOH, H) 20% + (H, COOH, COOH, H) 25% + (H, CH_2OH, COOH, H) 10%

Note: H_2SeO_4 is a satisfactory representation of the reaction mixture.

Granger, Boussinesq, Girard, and Rossi, *Bull. Soc. Chim. Fr.* **1968**, 1445.

67. OH

$NaBH_4$-NaOH, reflux 3 hr

COCH$_3$ i → CH$_2$CH$_3$ 94% ii + CH–CH$_3$ / OH 6% iii

OH, COCH$_3$ → $NaBH_4$-NaOH, reflux 3 hr → OH, CH–CH$_3$ / OH 100%

68. CHO i → 30% H_2O_2, 5% NaOH, $HOCH_2CH_2OH$ → 55-62% O ii

i ______________ ii

69. Predict the principal product.

NO_2 → 2 moles O_3, 162 min →

Pappas, Keaveney, Berger, and Rush, *J. Org. Chem.* 33, 787 (1968).

70. Predict the principal products.

→ O_3, 90% aq. HOAc → i + ii (both $C_{20}H_{12}O_6$)

Anastassiou and Griffin, *J. Org. Chem.* 33, 3441 (1968).

71.

$$\text{C}_6\text{H}_4(\text{S-CH}_3)(\text{COOH}) \xrightarrow[\text{H}_2\text{O}]{\text{I}_2} \text{C}_6\text{H}_4(\text{S(=O)-CH}_3)(\text{COOH})$$

The rate of the above reaction is dependent on the pH: it is very slow at low pH and sharply rises as the pH increases from 3.5 to about 6, above which little or no further rate increase is seen. The para isomer, however, oxidized very slowly and the rate did not increase with increasing pH. When the reaction was conducted in $H_2{}^{18}O$, the ^{18}O was incorporated in the sulfoxy group of the product. When the reaction was conducted in D_2O, the isotope effect was 2.3; that is, the rate was 2.3 times as slow as the rate with H_2O under the same conditions. Suggest one or more mechanisms compatible with these results.

Tagaki, Ochiai, and Oae, *Tetrahedron Lett.* **1968**, 6131.

72. Pregeijerene ($C_{12}H_{18}$) (i) is a compound obtained from the leaves of *Geijera parviflora.* Upon ozonolysis of i, followed by treatment with H_2O_2, there was isolated 1.36 moles of levulinic acid (ii), some levulinaldehyde and a low yield

$CH_3\text{-}C(=O)\text{-}CH_2CH_2COOH$

ii

$CH_3\text{-}C(=O)\text{-}(CH_2)_5\text{-}CH(CH_3)CH_2CH_2COOH$

iii

$CH_3\text{-}C(=O)\text{-}(CH_2)_3\text{-}CH(CH_3)\text{-}(CH_2)_4\text{-}COOH$

iv

of oxalic acid. Hydrogenation of i gave a dihydropregeijerene (v) and a mixture (vi) of tetrahydropregeijerenes. Ozonolysis of vi followed by oxidation with H_2O_2 gave a mixture of iii and iv. What is the structure of i?

73. Assume that you have a sample of cyclohexene which could be labeled with radioactive ^{14}C at positions a, b, or c. Devise a method for determining the ^{14}C content of each of these positions.

[Cyclohexene ring with positions labeled a, b, c]

Tjan, Steinberg, and de Boer, *Rec. Trav. Chim.* **88**, 673 (1969).

74. Ozonides can be quantitatively reduced by Ph_3P. Devise an experiment to distinguish between these two possible mechanisms:

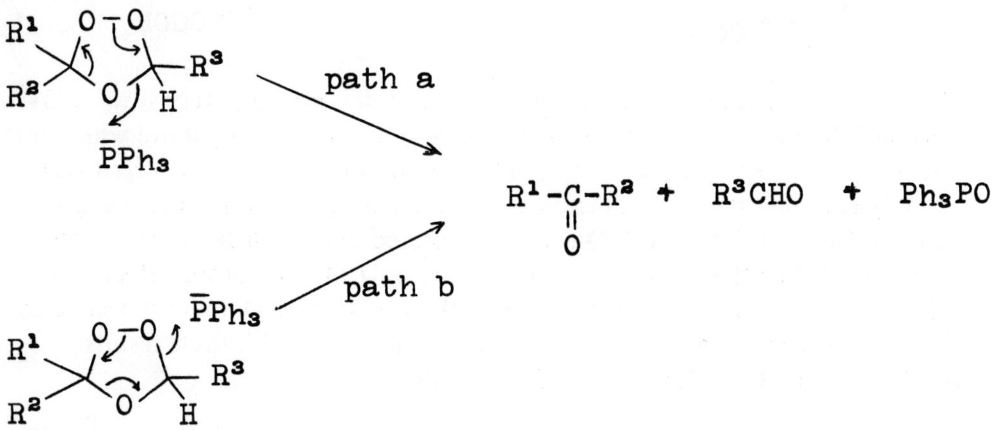

path a

path b

$R^1-\overset{\displaystyle O}{\overset{\|}{C}}-R^2 + R^3CHO + Ph_3PO$

PROBLEM SET 14

ADDITIONAL PROBLEMS IN SYNTHESIS AND MECHANISMS

The first 17 problems are synthetic. In each one you are asked to convert one compound to another. In every case the conversion has been carried out and the reference is given.

1.

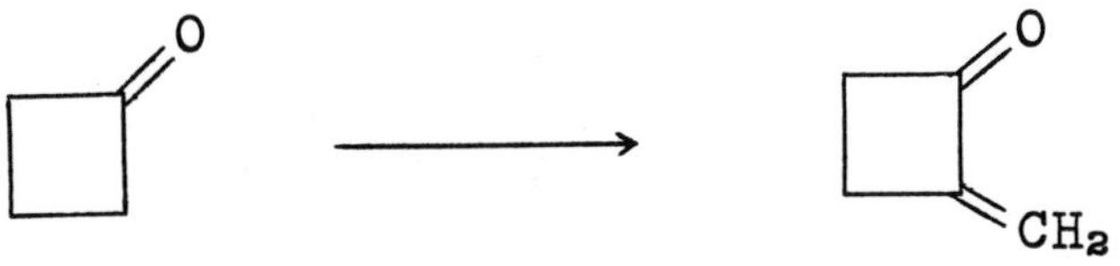

i → ii

2.

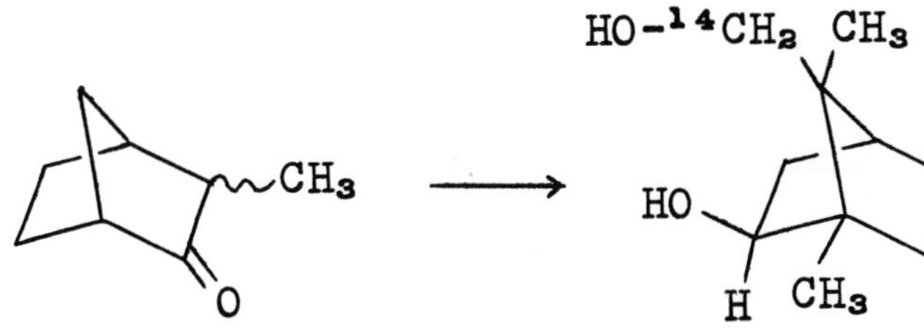

The only source of ^{14}C is $^{14}CO_2$.
Finch and Vaughan, *J. Amer. Chem. Soc.* **91**, 1416 (1969).

3.

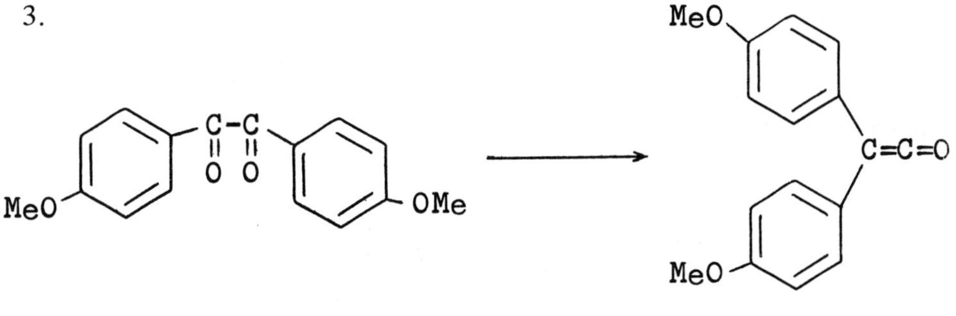

Droescher and Jenny, *Helv. Chim. Acta* **51**, 643 (1968).

4.

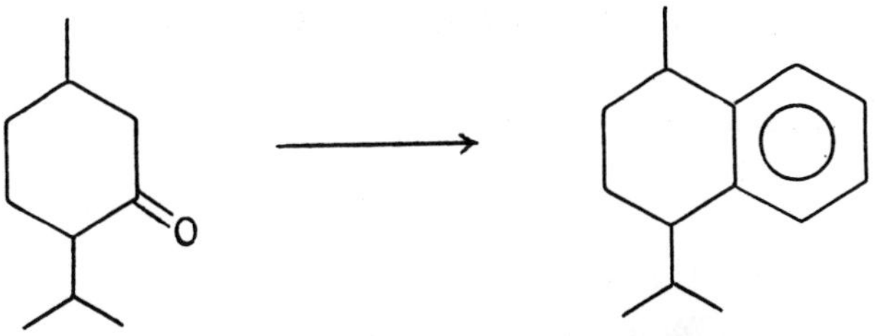

menthone → calamenene

Ladwa, Joshi, and Kulkarni, *Chem. Ind.* (London) **1968**, 1601.

5.

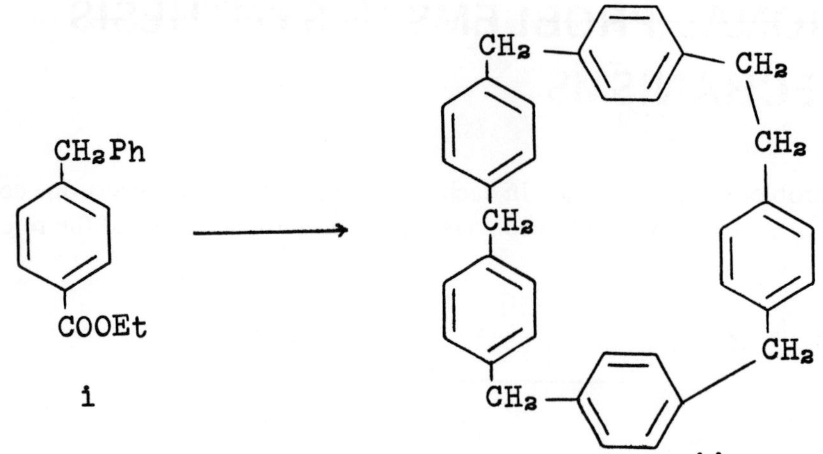
CH₂Ph
COOEt
CH₂
CH₂
CH₂
CH₂
CH₂
CH₂

i

ii

[2.1.1.1.1] paracyclophane

6.

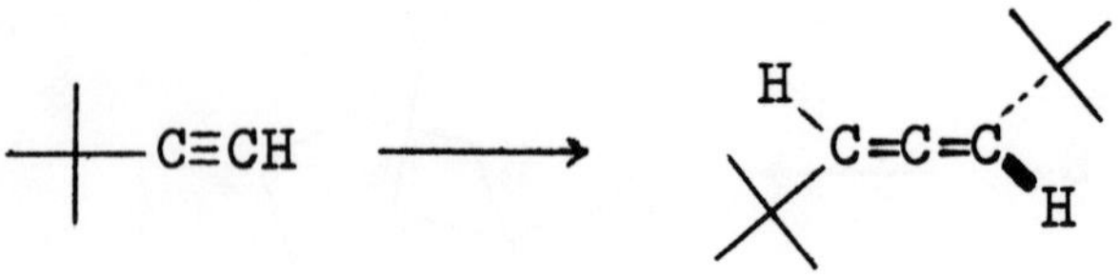
C≡CH
H
C=C=C
H

(+) isomer

Borden and Corey, *Tetrahedron Lett.* **1969**, 313.

7.

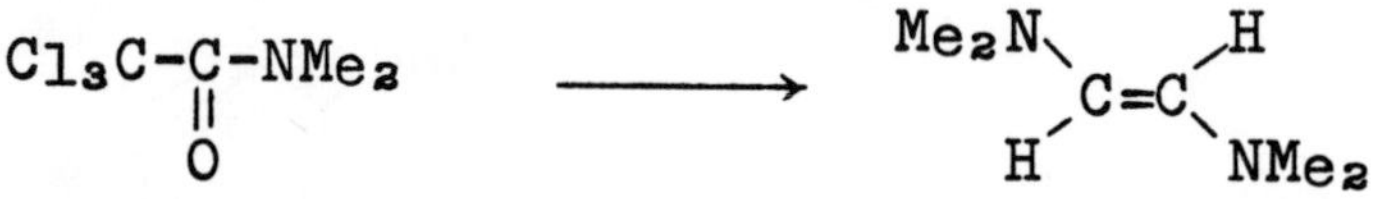
Cl₃C-C-NMe₂
O
Me₂N
H
C=C
H
NMe₂

Halleux and Viehe, *J. Chem. Soc., C* **1968**, 1726.

8.

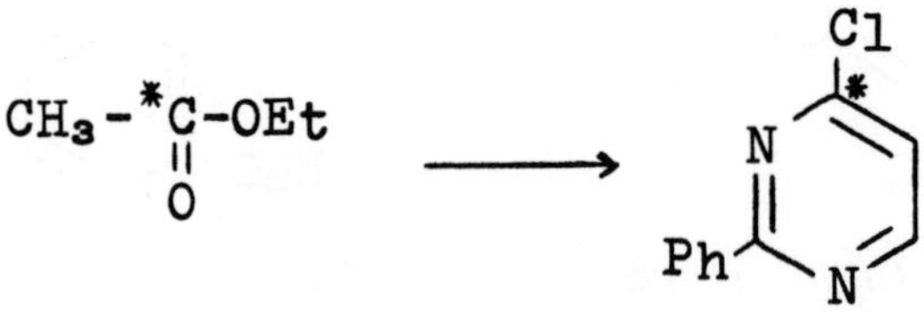
CH₃-*C-OEt
O
Cl
N
Ph
N

van Meeteren and van der Plas, *Rec. Trav. Chim.* **86**, 15 (1967).

9.

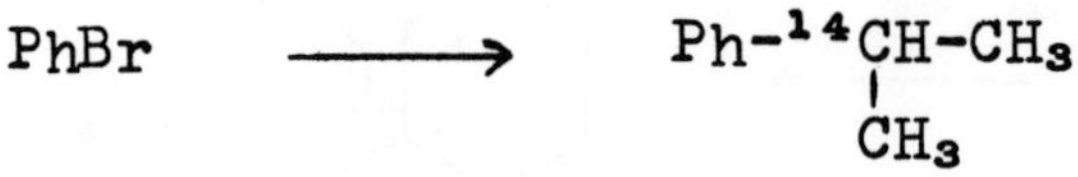
PhBr
Ph-¹⁴CH-CH₃
CH₃

The only source of ^{14}C is $^{14}CO_2$.

Pines and Abramovici, *J. Org. Chem.* **34**, 70 (1969).

10.

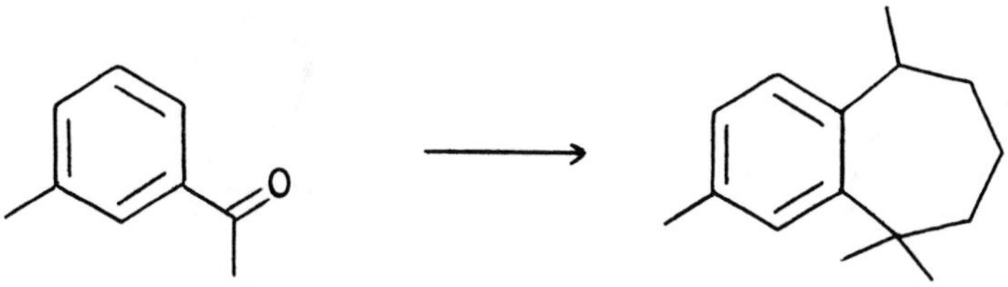

ar-himachalene

Pandey and Dev, *Tetrahedron* **24**, 3829 (1968).

11.

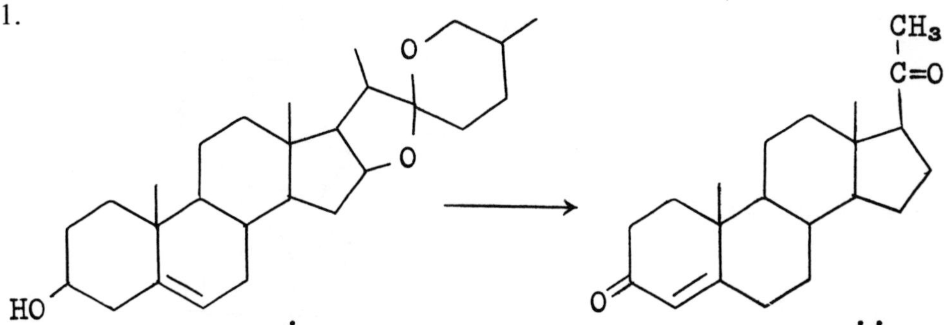

HO

CH_3
C=O

i
diosgenin

ii
progesterone

12.

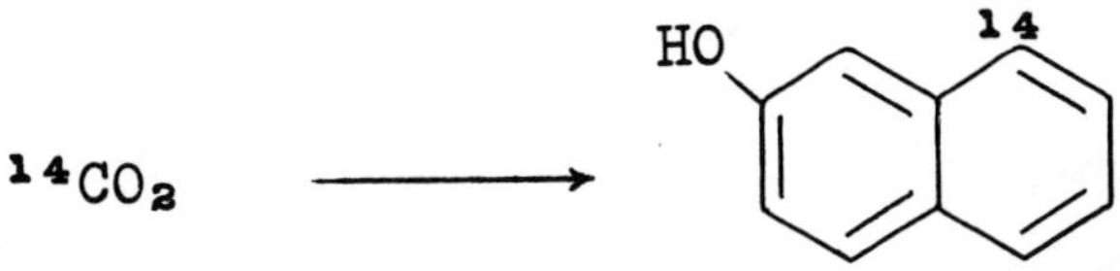

$^{14}CO_2$

HO
14

Hoang-Nam, Dat-Xuong, Herbert, and Pichat, *Bull. Soc. Chim. Fr.* **1967**, 4632.

13.

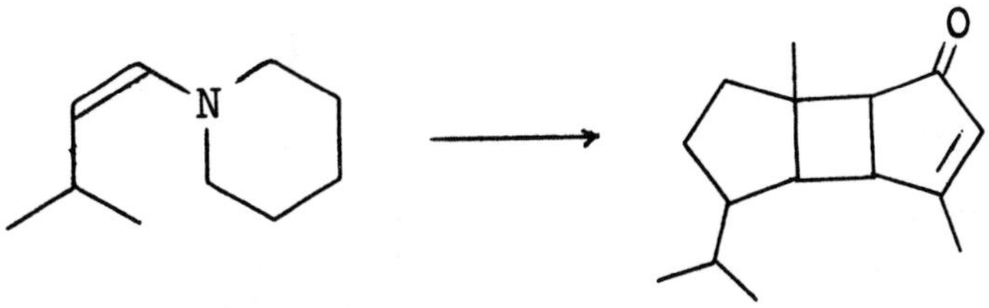

N

O

Brown, *J. Org. Chem.* **33**, 162 (1968).

14.

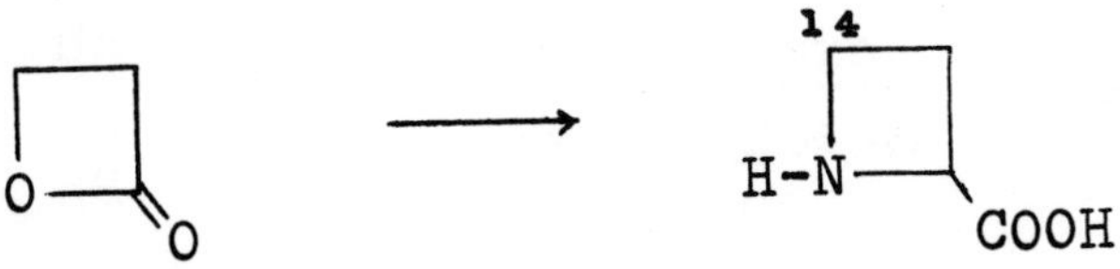

O
O

14
H-N
COOH

unlabeled

Pichat, Liem, and Guermont, *Bull. Soc. Chim. Fr.* **1968**, 4079.

15.

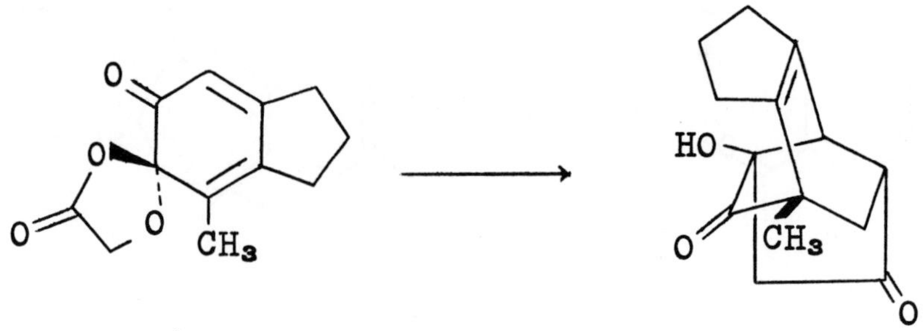

O O O O CH$_3$

i

HO O CH$_3$ O

ii

16.

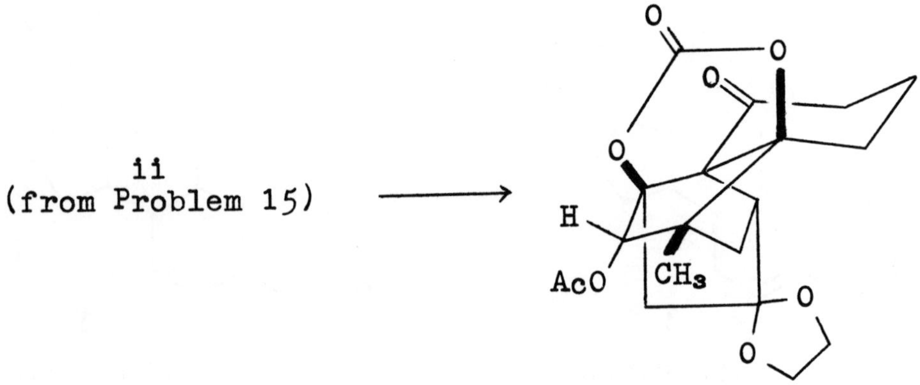

ii
(from Problem 15)

O O O O H AcO CH$_3$ O O

iii

17.

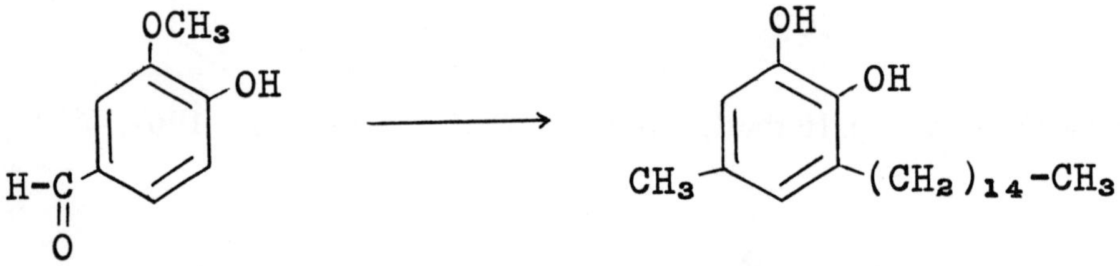

OCH$_3$ OH H-C=O

vanillin

OH OH CH$_3$ (CH$_2$)$_{14}$-CH$_3$

Byck and Dawson, *J. Org. Chem.* **33**, 2451 (1968).

The following conversions (problems 18 to 51) have been reported in the literature. In each case suggest a reasonable mechanism.

18.

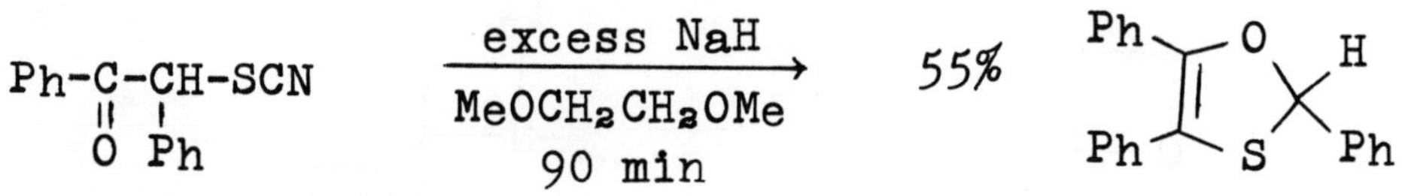

Ph-C(=O)-CH(Ph)-SCN $\xrightarrow[\text{MeOCH}_2\text{CH}_2\text{OMe, 90 min}]{\text{excess NaH}}$ 55% Ph, Ph, O, S, H, Ph

Kuhlmann and Dittmer, *J. Org. Chem.* **34**, 2006 (1969).

19.

Cl

+ Et_2CO $\xrightarrow{NaNH_2}$ 30%

O

Caubere and Brunet, *Tetrahedron Lett.* **1969**, 3323.

20.

$O-CH_2-C{\equiv}CH$

CH_3 CH_3

i

$\xrightarrow[\Delta]{}$

O

CH_3 CH_3

ii

21.

PhCO H
C=C
$BrCH_2$ Ph

+ NaSH ⟶

Ph
PhCO
S
S Ph

45%

+

PhCO COPh
S Ph

30%

+

PhCO Ph
S
S

18%

Padwa and Gruber, *Chem. Commun.* **1969**, 5.

22.

CHO

citral

+

HO OH
OH

phloroglucinol

$\xrightarrow{pyridine}$

35%

O OH
O

Crombie and Ponsford, *Tetrahedron Lett.* **1968**, 4557; Kane and Razdan, *Tetrahedron Lett.* **1969**, 591.

23.

NC, Me, CN, Me, N, Me, Me — boiling PhMe → ~50% NC, Me, Me, CN, C, Me, NHMe

Mahendran and Johnson, *Chem. Commun.* **1970**, 10.

24.

+ $ClSO_2NCO$ $\xrightarrow[-65°]{Et_2O}$

100% O, $=N-SO_2Cl$

Doyle and Conway, *Tetrahedron Lett.* **1969**, 1889.

25.

S, CH_3, N, CH_3-N, S, $PhCH_2$-N, S, CH_3, O — liq. NH_3, sealed tube, 20 hr, room temp. → Ph, N, S, CH_3, N, S, S, N, CH_3

83% ii

i

Takamizawa and Matsumoto, *Tetrahedron Lett.* **1969**, 2875.

26.

Ph, Ph, N, Ph, CH_3 — 1. hν O_2, sensitized by methylene blue, 2. hydrolysis → O, H, Ph, O=C, C=O, Ph, Ph + PhCOOH

65% 12%

Wasserman and Miller, *Chem. Commun.* **1969**, 199.

27.

2-methyl-1,4-naphthoquinone + o-phenylenediamine $\xrightarrow[\text{reflux, 15 hr}]{\text{AcOH}}$

Immer, Kunesch, Polonsky, and Wenkert, *Bull. Soc. Chim. Fr.* **1968**, 2420.

28.

$$EtO\text{-}\underset{\underset{O}{\|}}{C}\text{-}CHN_2 + Me_2CHOH \xrightarrow[0.5\ hr]{h\nu} \underset{29\%}{CH_3CH_2OCH_2COOCHMe_2}$$

$$+ \underset{25\%}{Me_2CHOCH_2COOCH_2CH_3} + \underset{12\%}{Me_2CHOCH_2COOCHMe_2}$$

$$+ \underset{9\%}{Me_2\underset{\underset{OH}{|}}{C}CH_2COOCH_2CH_3}$$

Strausz, DoMinh, and Gunning, *J. Amer. Chem. Soc.* **90**, 1660 (1968).

29.

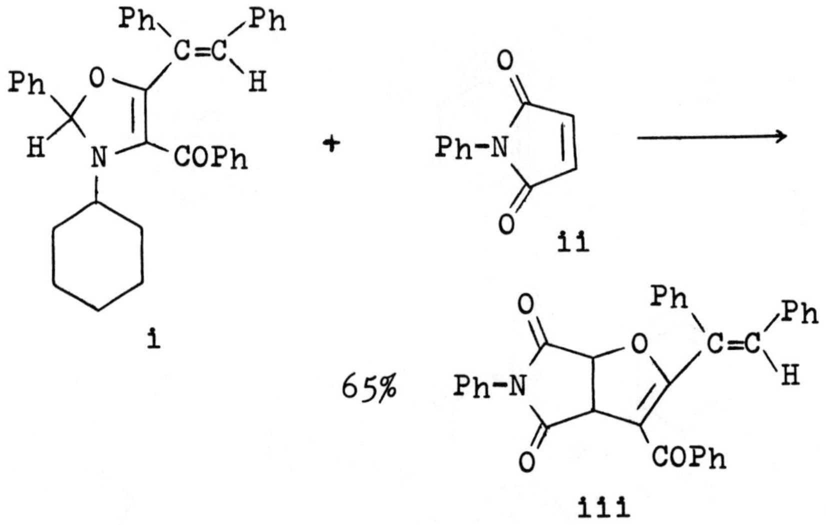

Ph, C=C, Ph, H, Ph, O, H, N, COPh, i + Ph-N (O, O) ii ⟶

65% Ph-N (O, O) O, Ph, C=C, Ph, H, COPh iii

30.

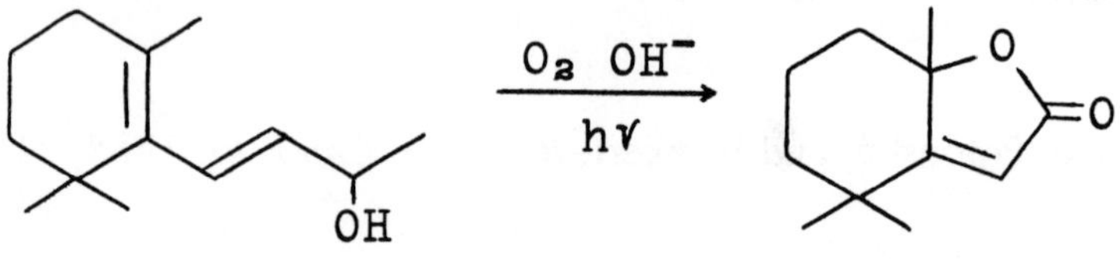

OH

β-ionol

O_2 OH^- / hν ⟶ O, =O

The reference is the same as that for problem 31.

31.

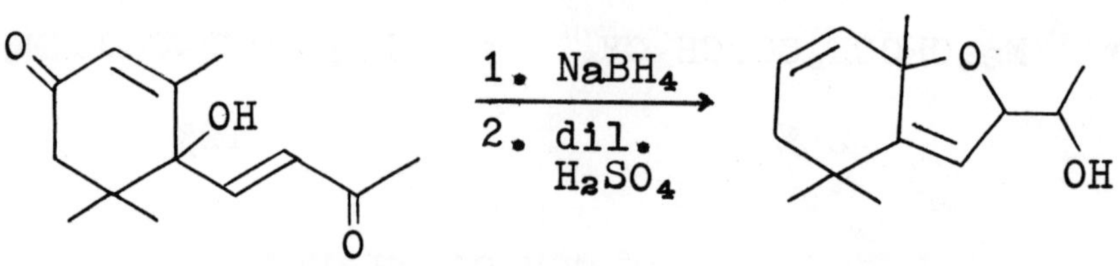

O, OH, O

1. $NaBH_4$ / 2. dil. H_2SO_4 ⟶ O, OH

Isoe, Hyeon, Ichikawa, Katsumura, and Sakan, *Tetrahedron Lett.* **1968**, 5561.

32.

Ph, $COCH_3$, COOEt, N=N (i) $\xrightarrow{Et_2NH}$ ether, 4-13 days in refrigerator → 95% Ph, COOEt, N, N, $COCH_3$

+ a small amount of Ph, COOEt, N, N–H

Note: When i was treated with Et_2NH in the presence of excess $EtCONEt_2$ the product was unchanged.

Danion-Bougot and Carrié, *Bull. Soc. Chim. Fr.* **1968**, 4241.

33.

i (−)-tabersonine: N, N–H, COOMe → xylene 205° sealed tube

ii (±)-pseudocatharanthine: N, N–H, COOMe → xylene 140° sealed tube

iii (+)-catharanthine: N, N–H, COOMe → xylene 175° sealed tube

→ iv (N–H, OH, CH_3) + v (N, Et)

34

0.1M NaOH
95% aq. EtOH
steel bomb
200-250°
several days

C=O
O
CN
CN
+
COO^-
O^-

Russell and Topping, *J. Chem. Soc., C* **1969**, 1134.

35.

CH_3 N N CH_3 + $NH_2NH_2 \cdot H_2O$ $\xrightarrow[\text{190° 5 hr, sealed tube}]{H_2O}$ Me Me N N

van der Plas and Jongejan, *Rec. Trav. Chim.* **87**, 1065 (1968).

36.

$HC{\equiv}C{-}(CH_2)_4{-}C{\equiv}CH$ $\xrightarrow[\text{diglyme, reflux 8 hr}]{t\text{-BuOK}}$ + + + +

+ equilibrium mixture of methylcycloheptatrienes

Eglinton, Raphael, and Zabkiewicz, *J. Chem. Soc., C* **1969**, 469.

37.

$H-N$ + $CH_2=CH-COOH$ $\xrightarrow[2\ hr]{135°}$

63%

Wiesner, Poon, Jirkovský, and Fishman, *Can. J. Chem.* **47**, 433 (1969).

38.

Br, NO_2

H_2SO_4 0-5°

Br, O, N

H_2SO_4 room temp.

Br

Ranganathan and Raman, *Tetrahedron Lett.* **1969**, 3747.

39. CH_3 CH_3 S CH C=O Br

$\xrightarrow[\text{reflux, 2 hr}]{H_2O}$

64%

$CH_3-S-CH_2-O-C=CH_2$

Br

Ratts and Yao, *J. Org. Chem.* **33**, 70 (1969).

40

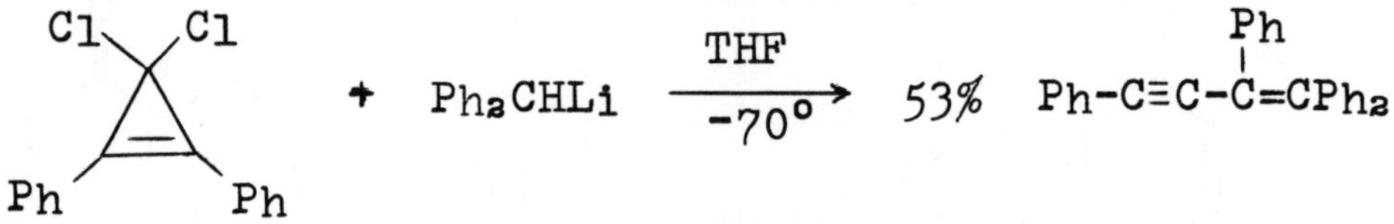

Cl Cl Ph Ph + Ph_2CHLi $\xrightarrow[-70°]{THF}$ 53% Ph-C≡C-C(Ph)=CPh_2

Melloni and Ciabattoni, *Chem. Commun.* **1968**, 1505.

41.

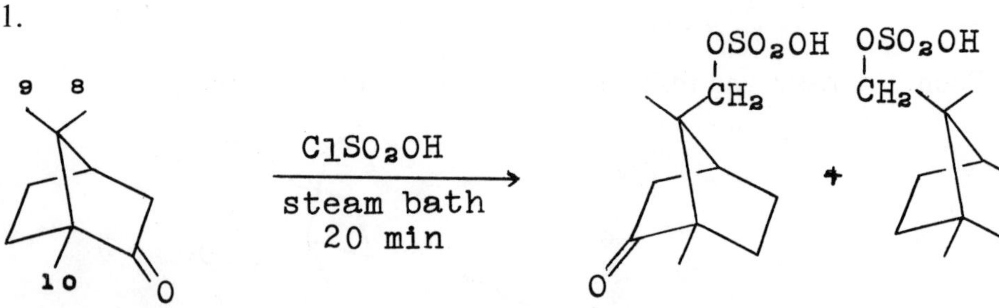

9 8 10 O

$ClSO_2OH$, steam bath, 20 min

OSO_2OH CH_2 O + OSO_2OH CH_2 O

(+)-camphor

racemic mixture

Note: When (+)–camphor labeled with ^{14}C in the 8 position was subjected to the reaction, the racemized product contained the label about equally distributed between the 8 and the 10 positions.

Finch and Vaughan, *J. Amer. Chem. Soc.* **91**, 1416 (1969).

42.

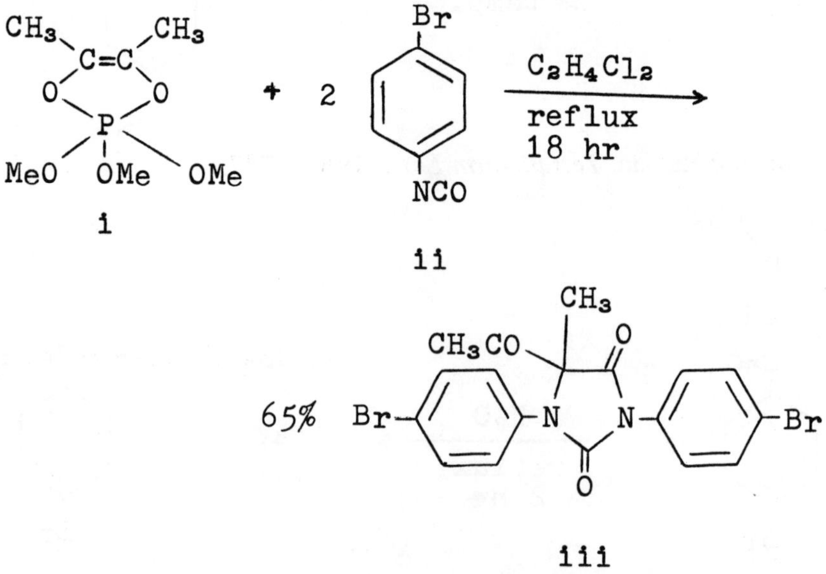

CH_3 CH_3 C=C O O P MeO OMe OMe

i

+ 2 Br NCO

ii

$C_2H_4Cl_2$, reflux, 18 hr

65% Br CH_3CO CH_3 O N N O Br

iii

43.

NO_2

N

N Ph

100° 5 mm / 12 hr → 51%

N

N Ph

+ O_2N CH=CH NO_2

Heine and Henzel, *J. Org. Chem.* **34**, 171 (1969).

44.

$Ph_2C(Cl)-C(=O)-NH_2$ $\xrightarrow[\text{MeOH}]{\text{KOH}}$ 20% Ph_2CO + 53% $Ph_2CHNHCOOCH_3$

+ 15% $Ph_2C(OMe)-CONH_2$

Breuer, Berger, and Sarel, *Chem. Commun.* **1968**, 1596.

45. $Et_2N(H)C{=}C(H)NEt_2$ (i) + $MeOOC-C{\equiv}C-COOMe$ (ii) $\xrightarrow[\text{reflux 2 hr}]{\text{cyclohexane}}$

80% $Et_2NCH{=}C(COOMe)-C(COOMe){=}CHNEt_2$

iii

46.

Oxirane + PhN=CPh$_2$ + CHCl$_3$ $\xrightarrow[150^\circ \; 5 \text{ hr}]{Et_4N^+ \; Br^-}$

44% 3,3-diphenyl-1-(2-chloroethyl)oxindole (N–CH$_2$–CH$_2$Cl)

Klamann, Wache, Ulm, and Nerdel, *Chem. Ber.* **100**, 1870 (1967); Nerdel, Weyerstahl, and Dahl, *Justus Liebigs Ann. Chem.* **716**, 127 (1968).

47.

PhCH=NPh $\xrightarrow[NaCN]{\text{dry } Me_2SO \text{ or DMF}}$ 77% Ph-C(=N-Ph)-C(=N-Ph)-Ph + 3-5% Ph-C(=O)-NHPh

Notes: NaCN was present in catalytic amounts. No reaction was observed in the absence of CN$^-$ or in the presence of NaOH. The compound Ph–N=CPh–CHPh –NH–Ph is unchanged in basic Me$_2$SO solution.

Walia, Singh, Chattha, and Satyanarayana, *Tetrahedron Lett.* **1969**, 195.

48.

i (COOEt, H, H) + ii (Cl, COOOH) $\xrightarrow{CH_2Cl_2}$ iii (O, O, H, COOEt)

+ iv (O, O, OH, COOEt) + v (COOEt, HOCH$_2$, O)

49. CH_3 O_2N NO_2 NO_2 + N-CH_2-N dioxane →

NO_2 N-CH_2CH_2- NO_2 NO_2 + N H

Note: 3,4,5-Trinitrotoluene did not give this reaction.

Fernandez, Mones, Schwartz, and Wulff, *J. Chem. Soc., B* **1969**, 506.

50.

F F N N F F

820° silica wool 40 sec →

F N F N F F 60%

hν →

F N F F N F almost 100%

Allison, Chambers, Cheburkov, MacBride, and Musgrave, *Chem. Commun.* **1969**, 1200.

51.

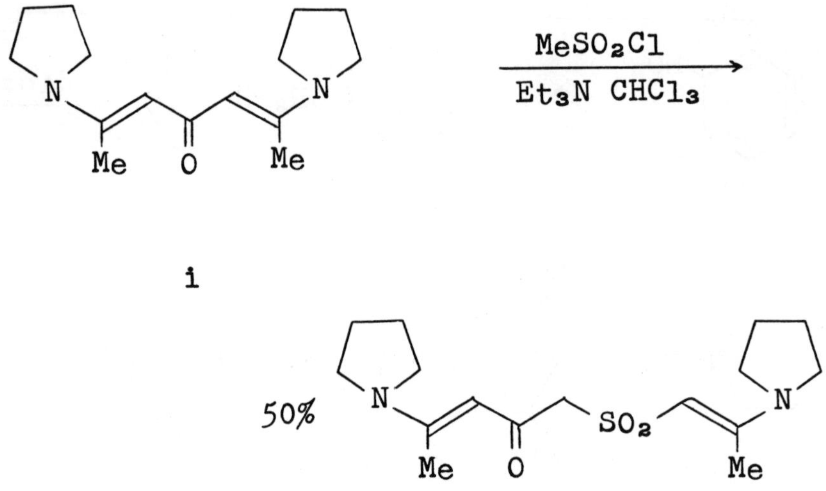

$\xrightarrow[Et_3N\ CHCl_3]{MeSO_2Cl}$

i

50%

SO_2

Me O Me

ii

52.

MeO + MeOOC–C≡C–COOMe $\xrightarrow{Et_2O}$ i

i $\xrightarrow[90°\ 5\ min]{HOAc-H_2O}$ MeO– –OH

MeO–C(=O) C(=O)–OMe

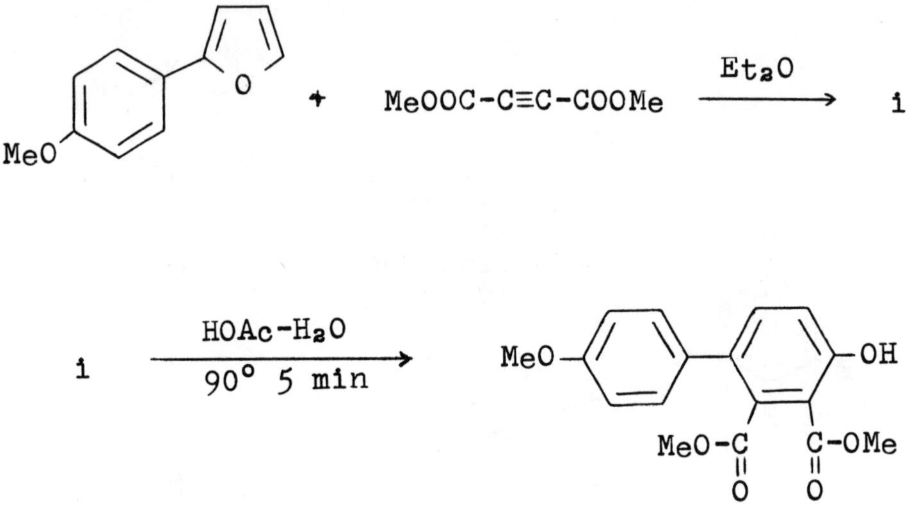

Give the structure of i, and suggest reasonable mechanisms.
Ayres and Smith. *J. Chem. Soc., C* **1968**, 2737.

53.

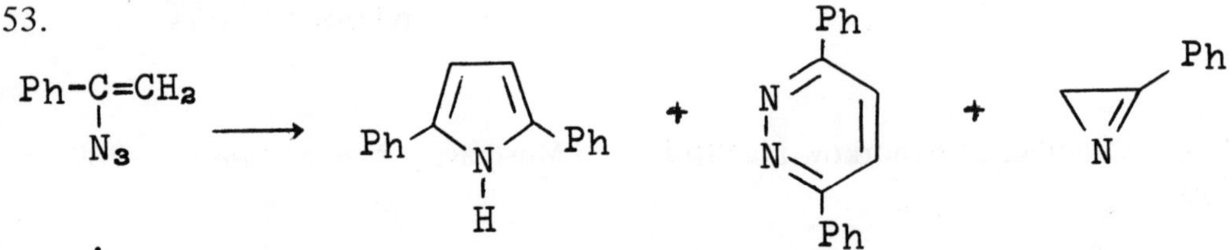

Ph–C(N$_3$)=CH$_2$ ⟶ Ph Ph N H + Ph N N Ph + Ph N

i

32% ii 7% iii ~10% iv

53. Cont.

On standing for one month in tightly closed bottle (which was opened daily to allow escape of the accumulated gas), i gave the products shown. Explain the formation of ii and iii. iv does not dimerize to v.

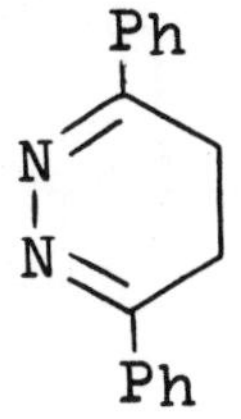

Ph

N

N

Ph

v

Boyer, Krueger, and Modler, *Tetrahedron Lett.* **1968**, 5979.

The following problems (54 to 63) are devoted to the total synthesis of natural products. In each case the naturally-occurring compound has been synthesized by the route outlined. Each arrow stands for one, or more usually, several steps, in some cases as many as a half dozen or more. Where a capital letter appears over an arrow, show how to proceed from one compound to the next.

54. Caryophyllene

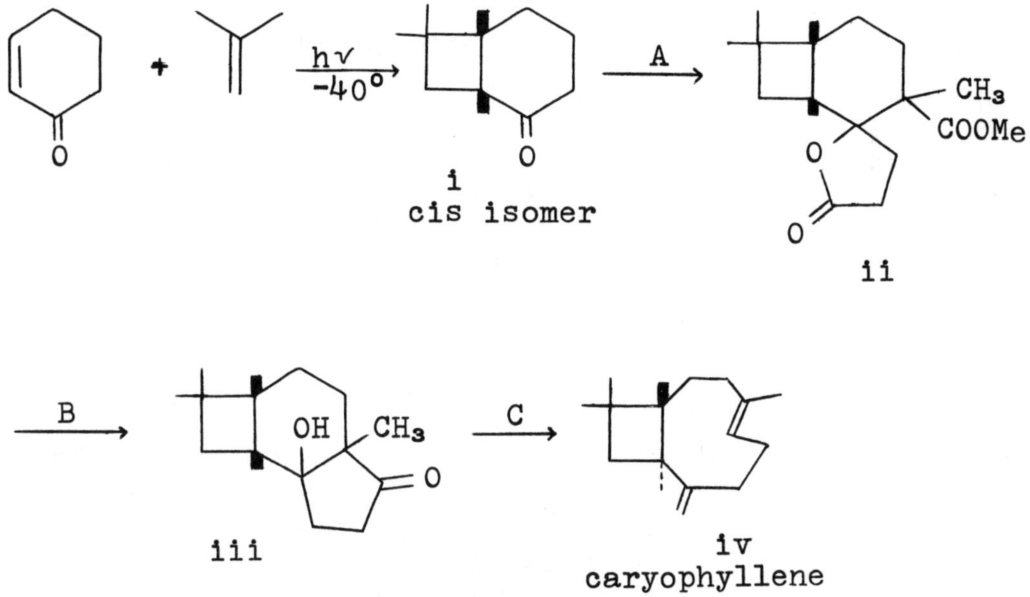

hν -40°

A

i cis isomer

CH_3 COOMe

O O

ii

B

OH CH_3 O

iii

C

iv caryophyllene

In the photochemical 2 + 2 cycloaddition shown above, the less stable trans isomer is chiefly formed, but it was epimerized to the cis compound (i), which was used in the next step.

55. Aromadendrene

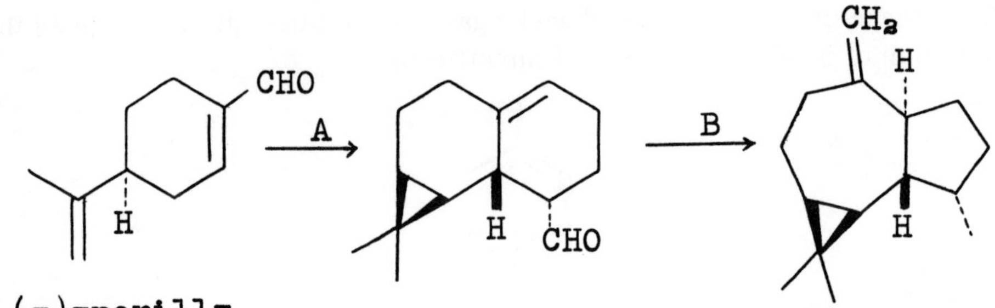

(-)-perill-
aldehyde

A

B

aromadendrene

Büchi, Hofheinz, and Paukstelis, *J. Amer. Chem. Soc.* **91**, 6473 (1969).

56. Lysergic acid

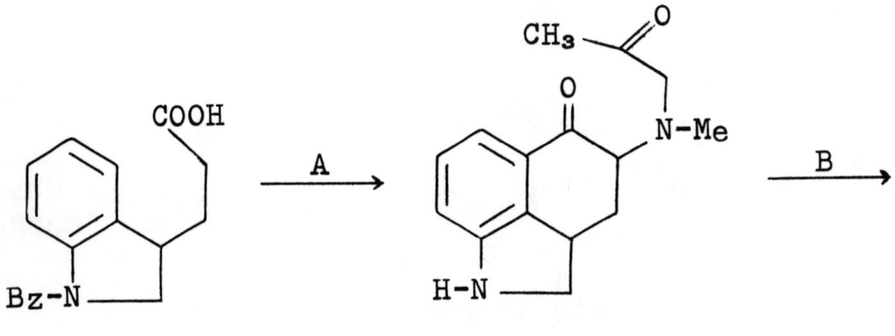

i

A

ii

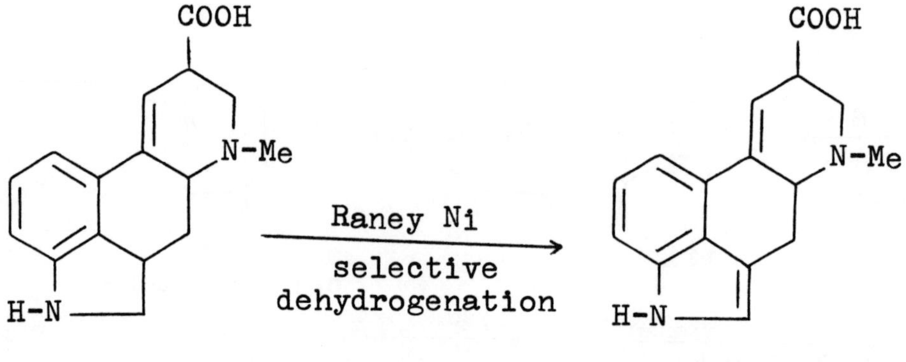

B

iii

Raney Ni
selective
dehydrogenation

lysergic acid

57. Cholestanol

O
CH3
CH3
O
A
O
OAc
Zn
hot Ac_2O
Boord type reaction

O
B
O
C

O
O
CN
O
CH-N-Ph
Me
D
O
O
O

E
COOMe
O
F
OH
HO
H

G
HO
H
cholestanol

57. Cont.

Woodward, Sondheimer, Taub, Heusler, and McLamore, *J. Amer. Chem. Soc.* **74**, 4223 (1952).

58. Sirenin

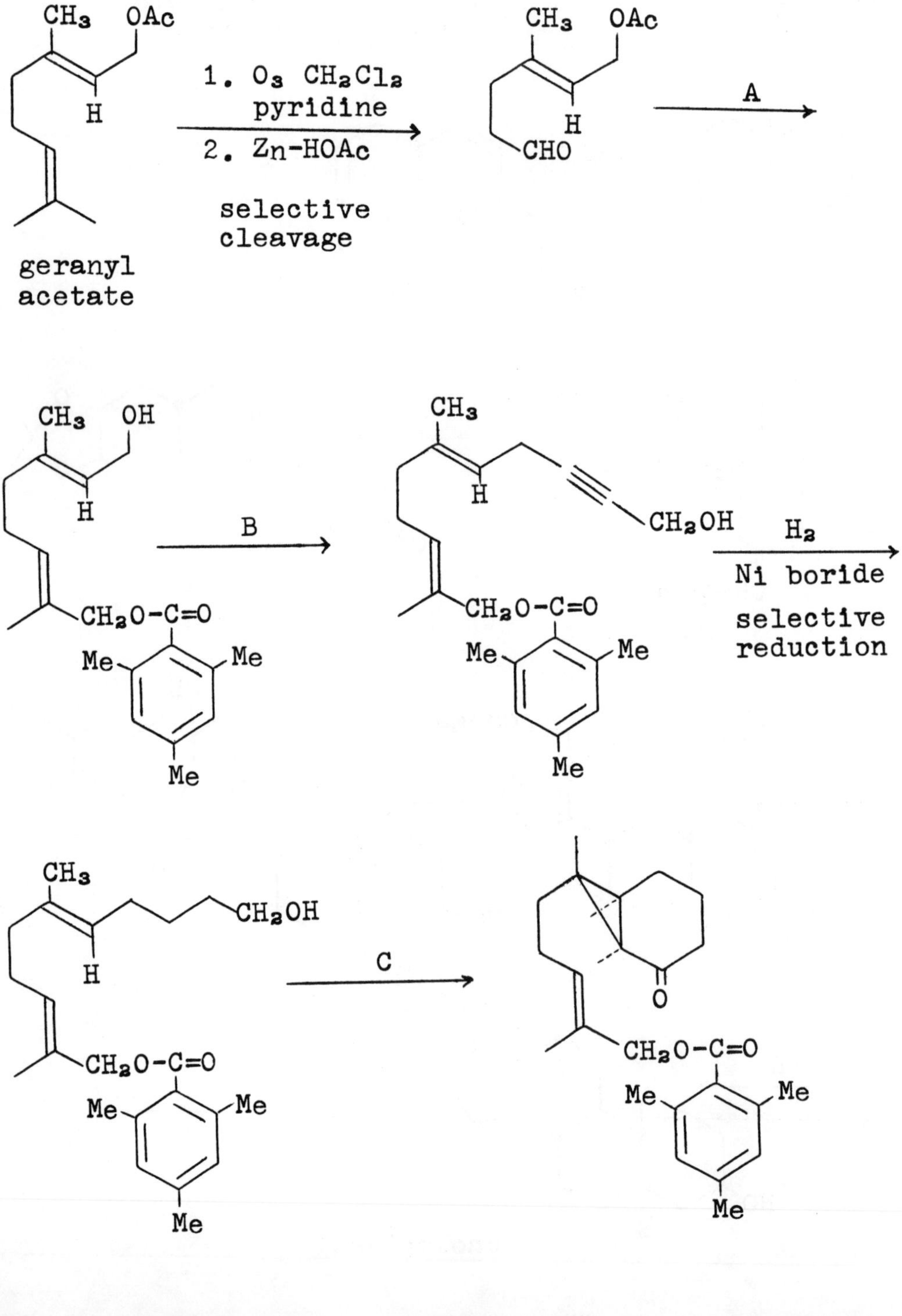

CH_3 OAc — H — geranyl acetate

1. O_3 CH_2Cl_2 pyridine
2. Zn-HOAc

selective cleavage

CH_3 OAc — H — CHO

A

CH_3 OH — H — $CH_2O-C=O$ — Me Me Me

B

CH_3 — H — CH_2OH — $CH_2O-C=O$ — Me Me Me

H_2

Ni boride

selective reduction

CH_3 — CH_2OH — H — $CH_2O-C=O$ — Me Me Me

C

O — $CH_2O-C=O$ — Me Me Me

58. Cont.

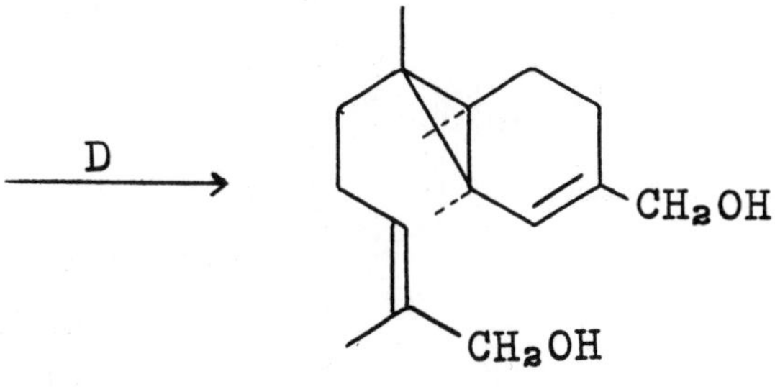

D →

CH_2OH

CH_2OH

sirenin

Corey, Achiwa, and Katzenellenbogen, *J. Amer. Chem. Soc.* **91**, 4318 (1969).

59. Longifolene

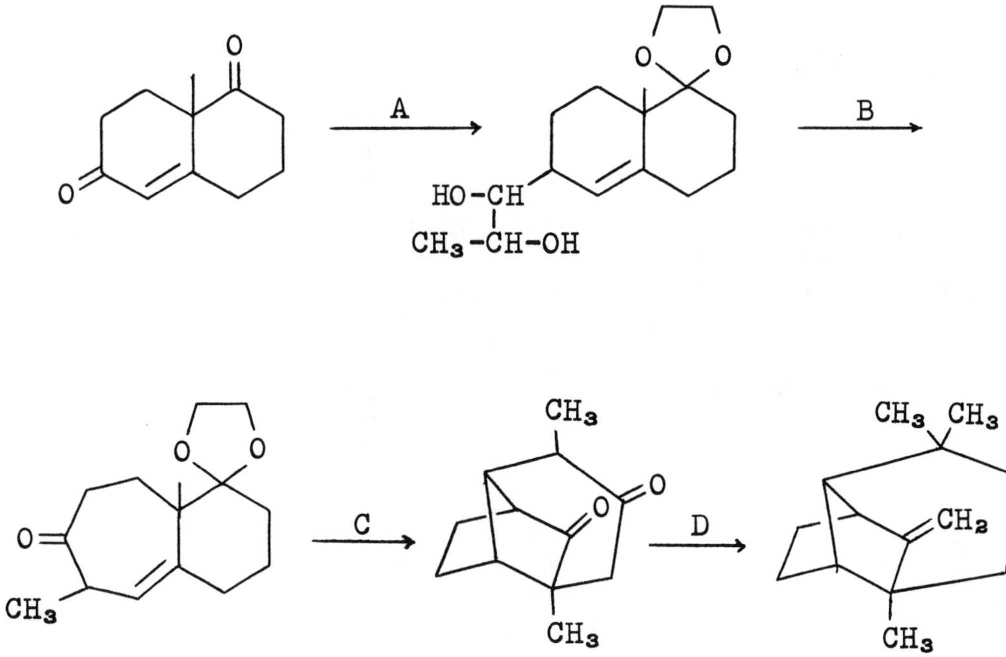

O O O O

A → B →

HO-CH
CH_3-CH-OH

O O

O=

CH_3

C → D →

CH_3 O O

CH_3

CH_3 CH_3

CH_2

CH_3

longifolene

Corey, Ohno, Mitra, and Vatakencherry, *J. Amer. Chem. Soc.* **86**, 478 (1964).

60. Lycoramine

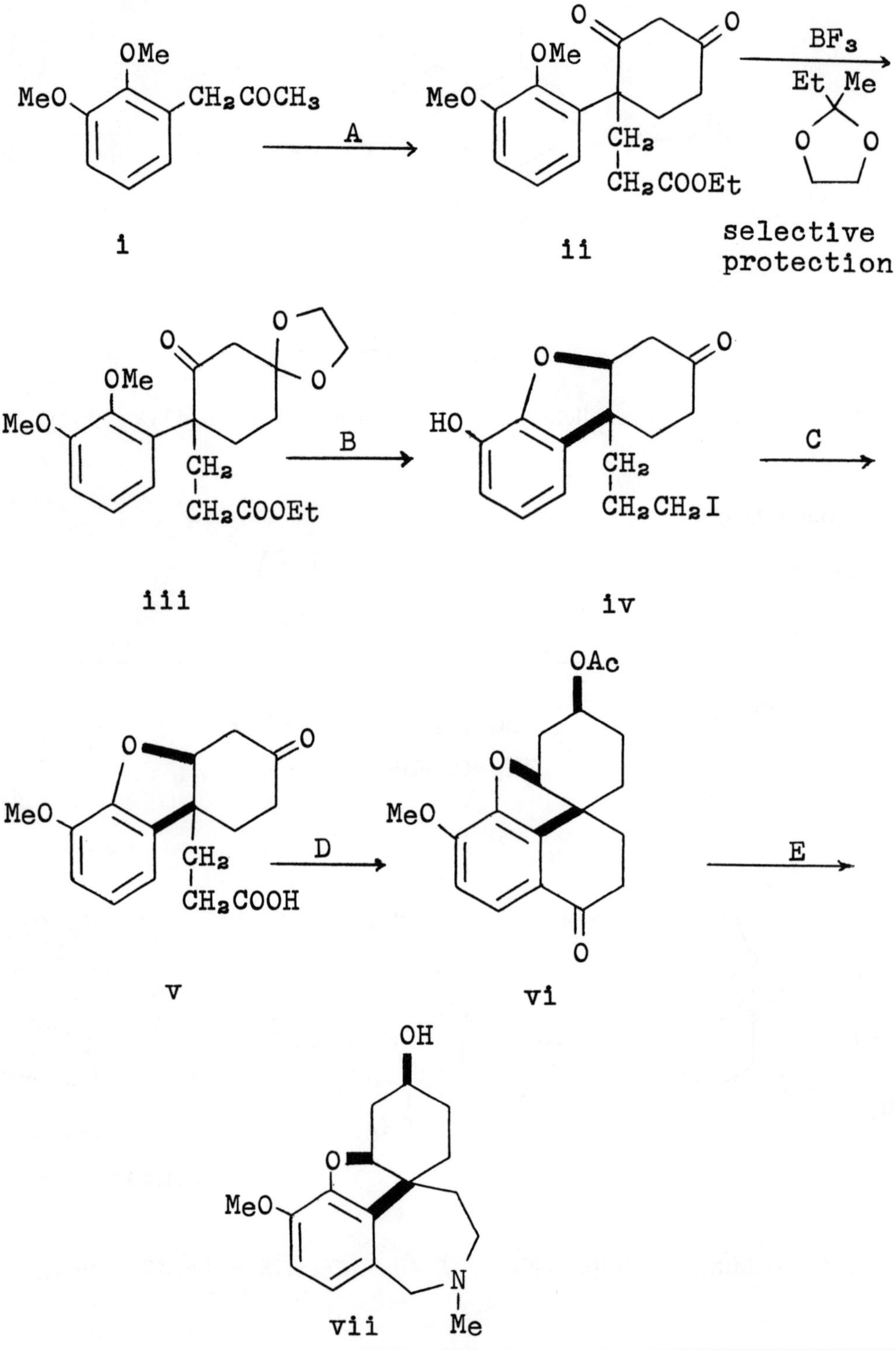

$$\text{i} \xrightarrow{\text{A}} \text{ii} \xrightarrow[\text{selective protection}]{BF_3,\ \text{Et(Me)C(OCH}_2\text{CH}_2\text{O)}} \text{iii} \xrightarrow{\text{B}} \text{iv} \xrightarrow{\text{C}} \text{v} \xrightarrow{\text{D}} \text{vi} \xrightarrow{\text{E}} \text{vii}$$

OMe

MeO

CH_2COCH_3

i

O O

CH_2

CH_2COOEt

ii

BF_3

Et Me

O O

selective
protection

iii

HO

CH_2CH_2I

iv

MeO

CH_2COOH

v

OAc

vi

OH

N

Me

vii

lycoramine

61. *dl*-6-Demethyl-6-deoxytetracycline

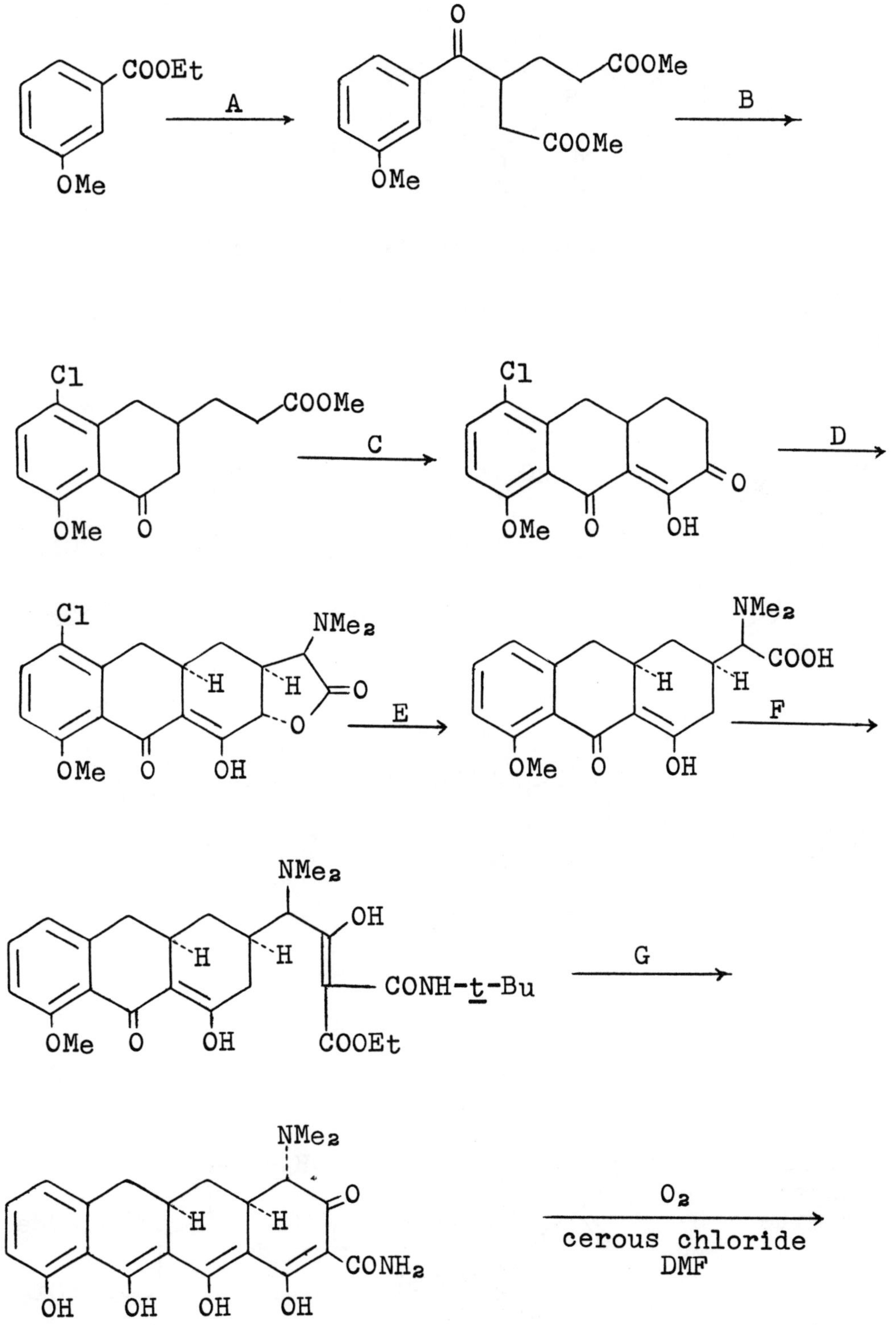

COOEt
OMe
A
O
COOMe
COOMe
OMe
B

Cl
COOMe
OMe O
C
Cl
O
OMe O OH
D

Cl
NMe_2
H H
O
OMe O OH
E
NMe_2
COOH
H H
OMe O OH
F

NMe_2
OH
H H
CONH-t-Bu
OMe O OH COOEt
G

NMe_2
O
H H
$CONH_2$
OH OH OH OH
O_2
cerous chloride
DMF

61. Cont.

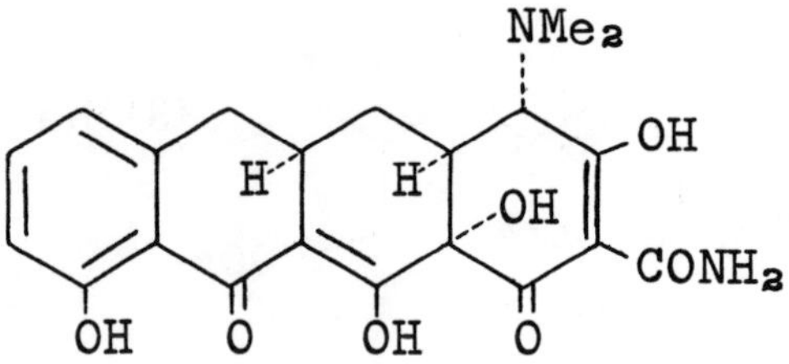

NMe_2 OH H H OH $CONH_2$ OH O OH O

<u>dl</u>-6-demethyl-6-deoxytetracycline

Korst, Johnston, Butler, Bianco, Conover, and Woodward, *J. Amer. Chem. Soc.* **90**, 439 (1968).

62. Morphine

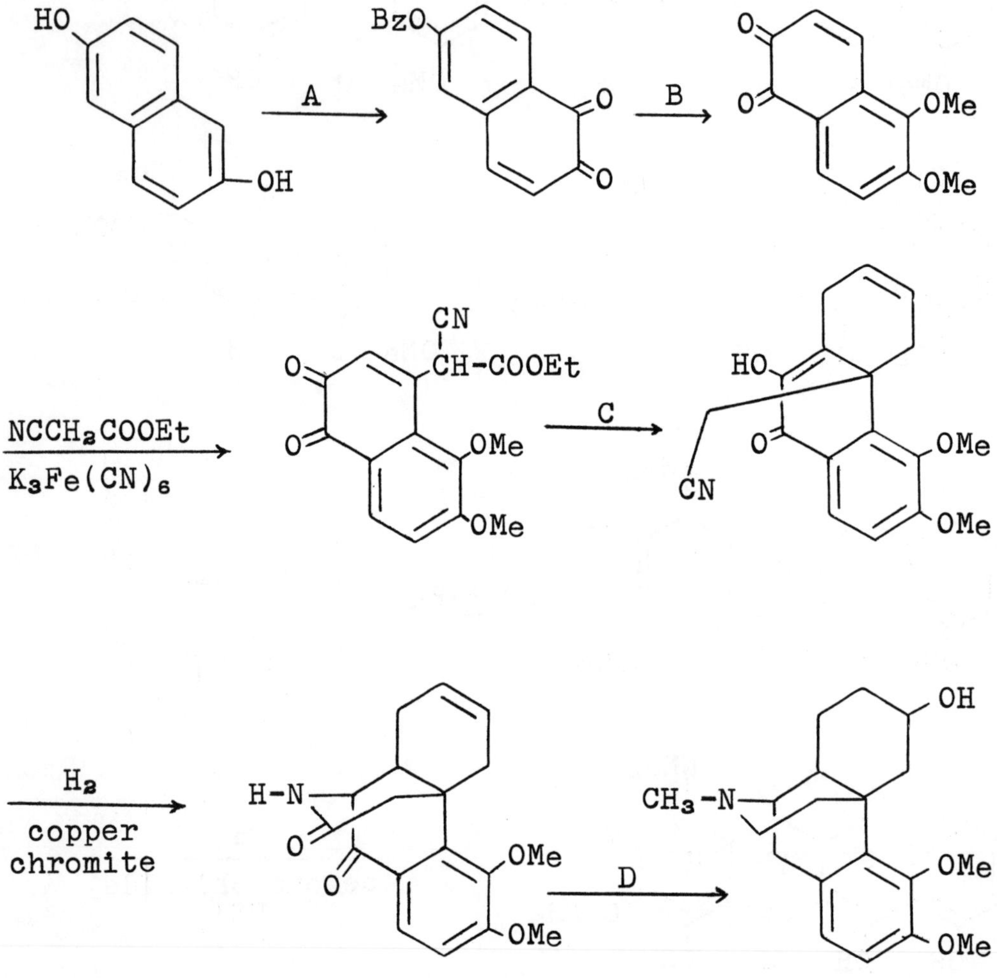

HO OH →(A) BzO O O →(B) O O OMe OMe

→ $NCCH_2COOEt$ / $K_3Fe(CN)_6$ → O O CN CH-COOEt OMe OMe →(C) HO O CN OMe OMe

→ H_2 / copper chromite → H-N O O OMe OMe →(D) OH CH_3-N OMe OMe

62. Cont.

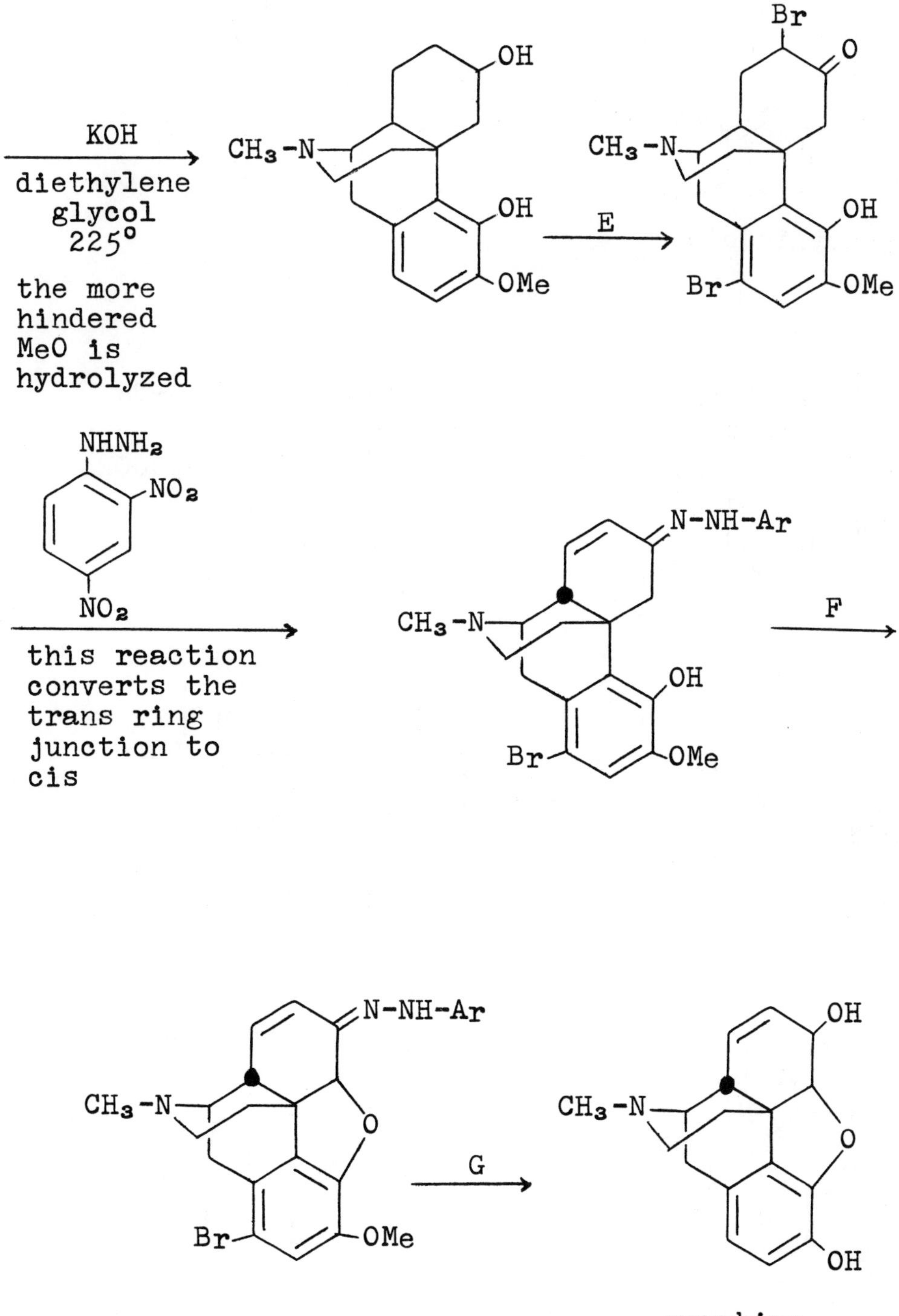

KOH

diethylene glycol 225°

the more hindered MeO is hydrolyzed

CH_3-N

OH

OH

OMe

E

Br

O

CH_3-N

OH

Br

OMe

$NHNH_2$

NO_2

NO_2

this reaction converts the trans ring junction to cis

N-NH-Ar

CH_3-N

OH

Br

OMe

F

N-NH-Ar

CH_3-N

O

Br

OMe

G

OH

CH_3-N

O

OH

morphine

Gates and Tschudi, *J. Amer. Chem. Soc.* **78**, 1380 (1956); Gates, *J. Amer. Chem. Soc.* **72**, 228 (1950).

63. Strychnine

2-veratrylindole

strychnine

Woodward, Cava, Ollis, Hunger, Daeniker, and Schenker, *Tetrahedron* **19**, 247 (1963).

PROBLEM SET 15
NOMENCLATURE

Corresponds to Appendix A of Advanced Organic Chemistry

1. Give suitable names for the following compounds.

a.

CH_3 CH_3

b.

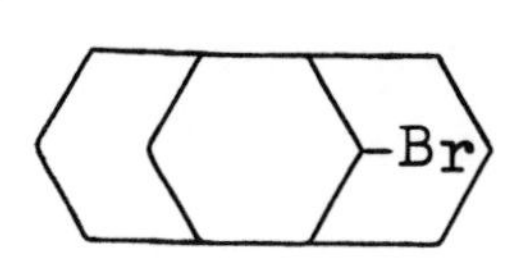

c.

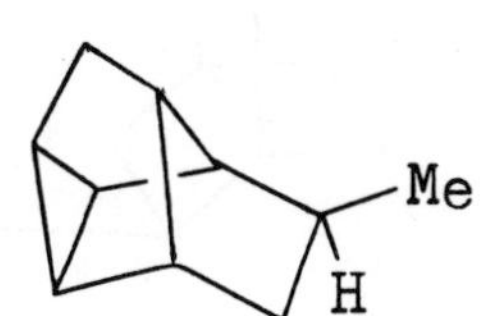

d.

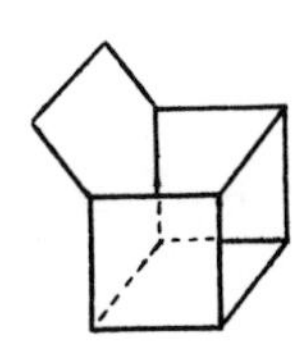

e.

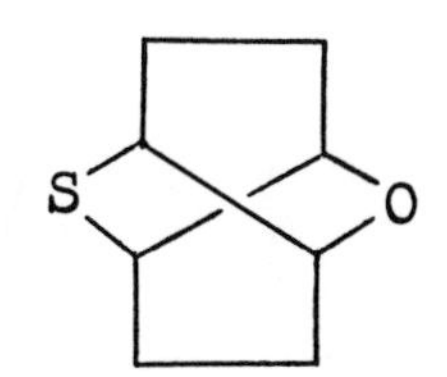

f.

HC — CH — CH
O — CH_2 - CH
O — CH_2 - CH — CH_2
O
HC — CH — CH

g.

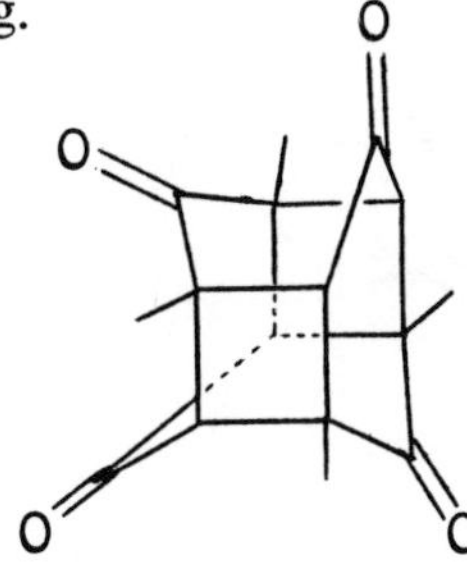

h.

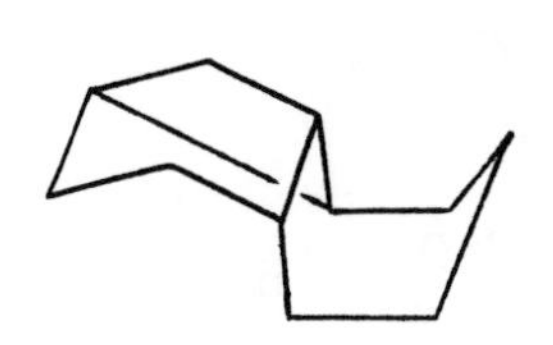

i.

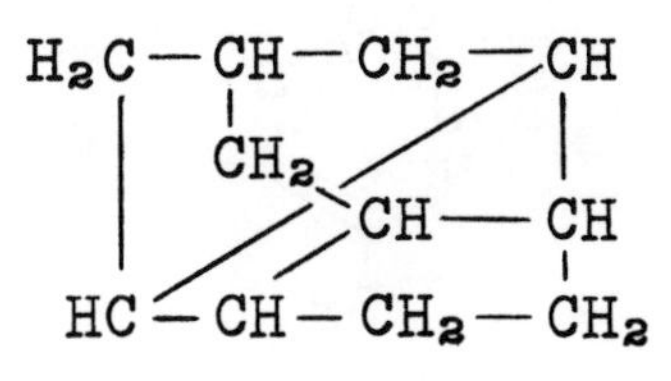

j.

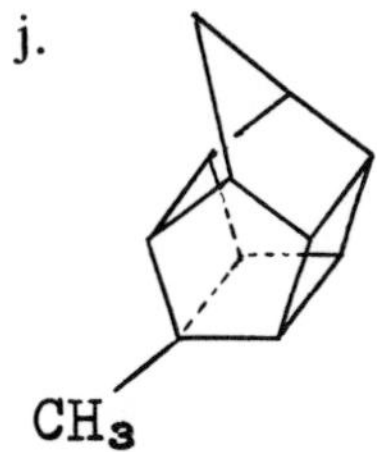

k.

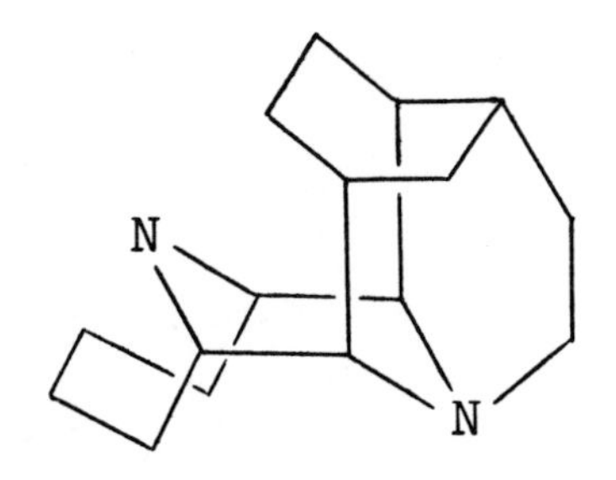

1. Cont.

l.

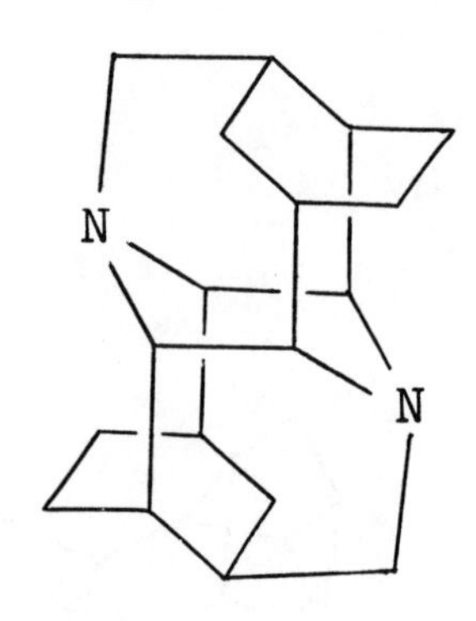

m.

n.

Se
Me

o.

CH_3
O—N
N

p.

S
P-H
N
H

q.

O
N
N⊕
H
Br^-

r.

O
S
N
Br

s.

O
O
P
H

t.

S
S
S
H-N
N
H

u.

O
O
S
S

v.

1. Cont.

w.

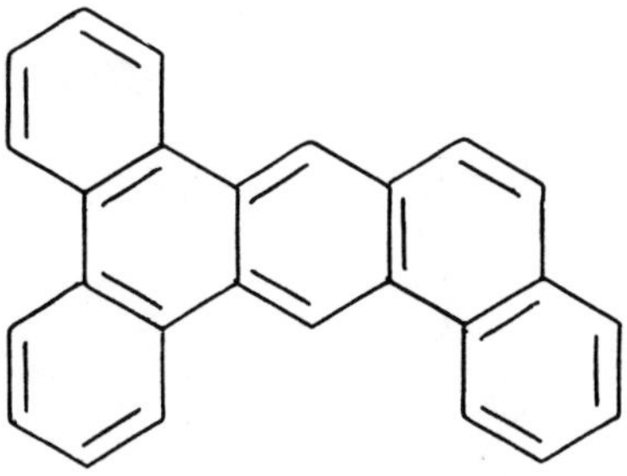

x.

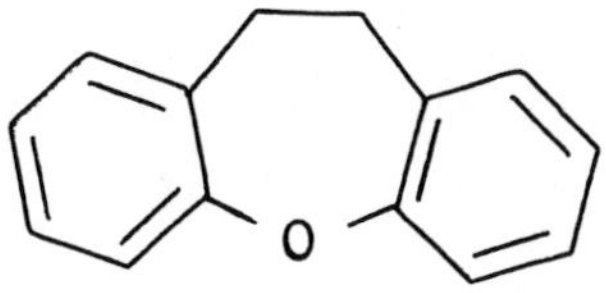

y.

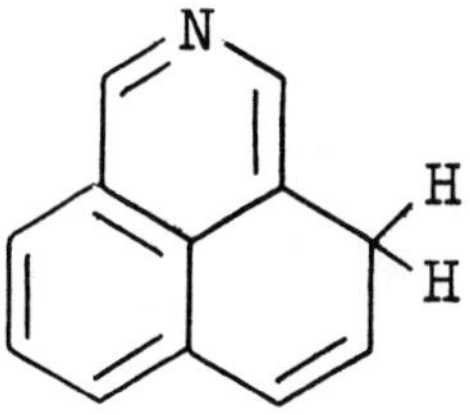

z.

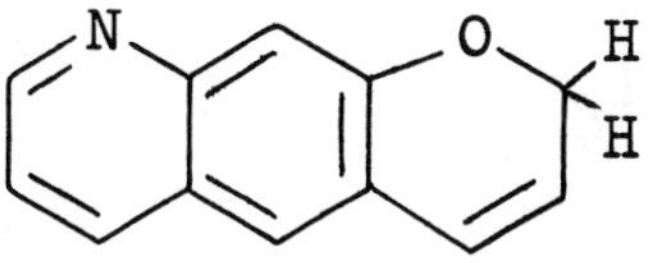

aa.

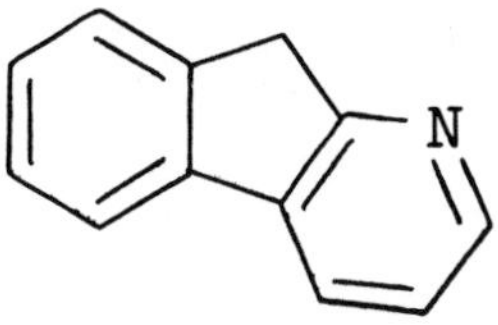

bb.

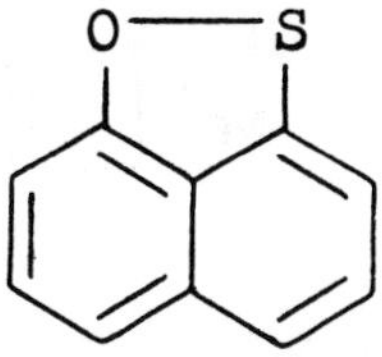

cc.

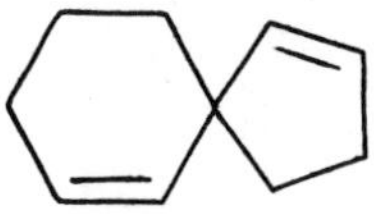

dd.

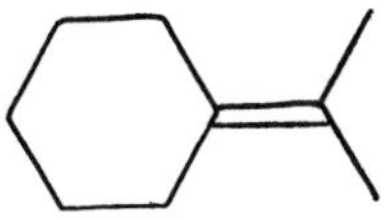

ee.

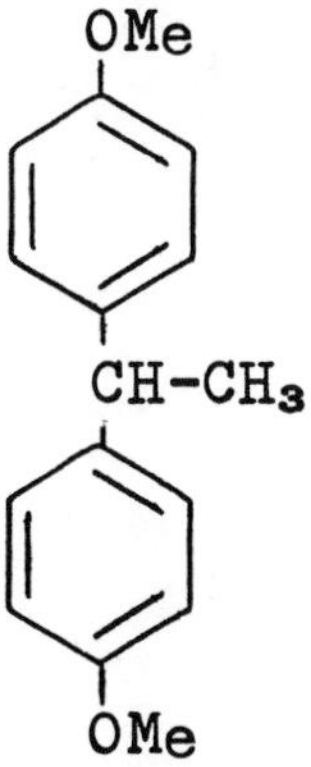

OMe

$CH-CH_3$

OMe

ff.

SO_2Cl

SO_2Cl

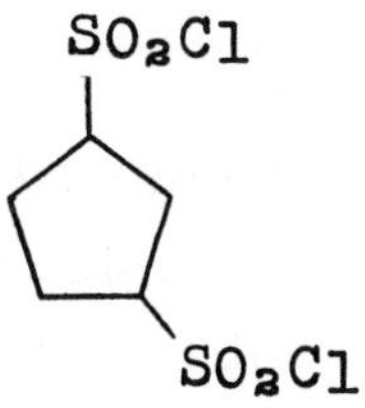

1. Cont.

gg.

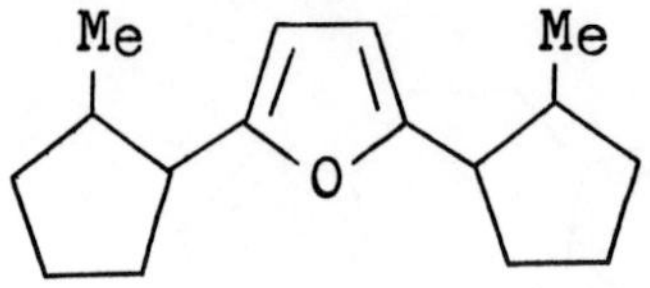

hh.

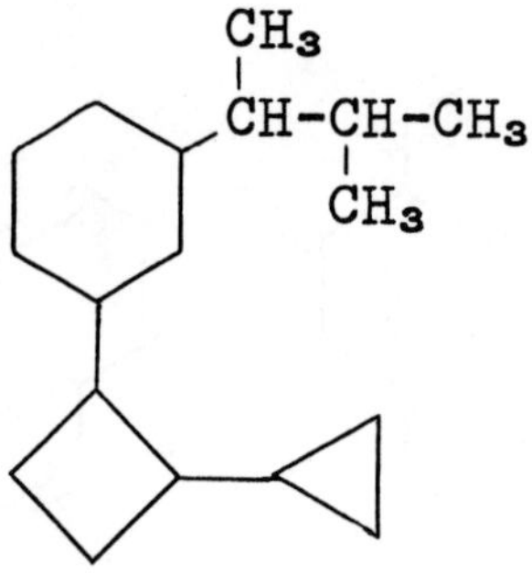

ii.

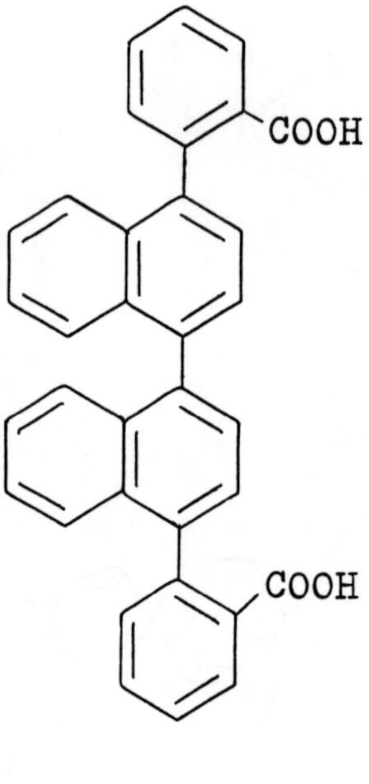

jj.

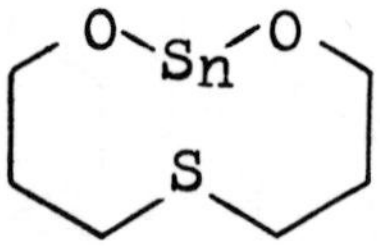

kk.

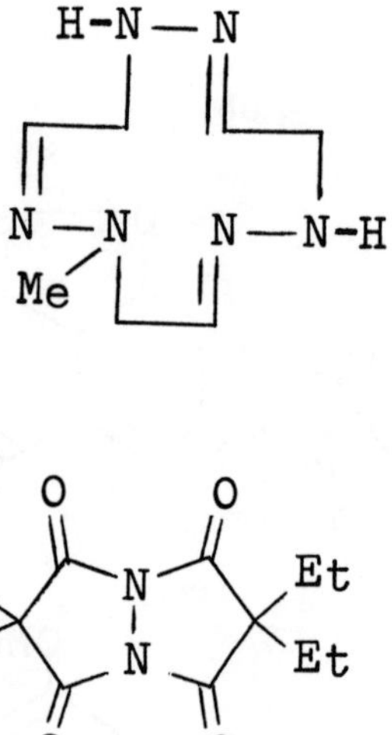

ll.

2. Draw structural formulas

a. 1,2,3-dioxazetidine

b. thiirene

c. 2*H*-tetrazole

d. 1,3,2-dioxaborolane

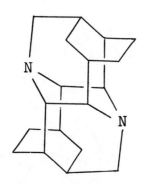

e. 1,3,2,4-dioxadithiane

2. Cont.

   f. hexahydro-1,3,5-triazine
   g. 2,8-dimethyl-1,3,5-triazocine
   h. tricyclo[4.2.1.$0^{2,5}$]nonane
   i. tricyclo[3.3.0.$0^{2,6}$]octan-3-one
   j. 1,5-dimethyl-6-methylenetricyclo[3.2.1.$0^{2,7}$]oct-3-en-8-one
   k. hexacyclo[6.2.1.$1^{3,6}$.$0^{2,7}$.$0^{4,10}$.$0^{5,9}$]dodecane
   l. naphtho[2,3-*g*]chrysene
   m. 8*H*-dibenz[*a,de*]anthracene
   n. 11*H*-ideno[2,1-*a*]azulene
   o. naphtho[1,2-*c*]thiophene
   p. 1*H*-benzo[*a*]cyclopent[*j*]anthracene
   q. propylene glycol

PROBLEM SET 16

LITERATURE

Corresponds to Appendix B of Advanced Organic Chemistry

1. Locate the earliest reference in the original literature (paper or patent) to each of the following compounds which has been reported. For each compound which is listed in Beilstein, also give the Beilstein locations (Hauptwerk and all supplements). If the compound is not in Beilstein, give a. the volume of Beilstein it would be in and b. the earliest *Chemical Abstract* reference (if in *CA*).

a.

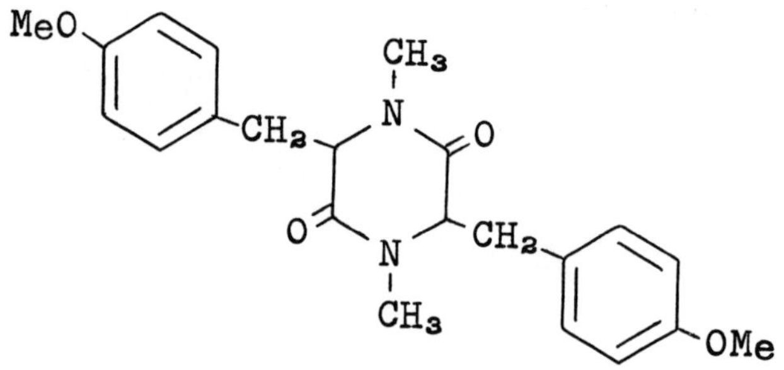

MeO, CH_3, N, O, CH_2, O, N, CH_3, CH_2, OMe

b.

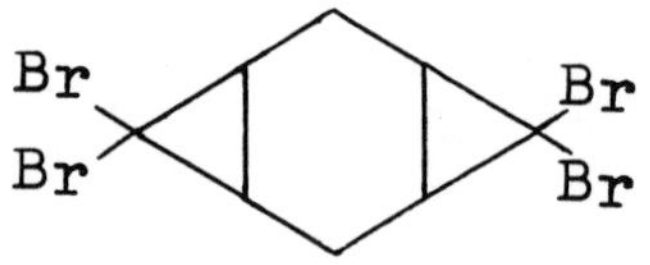

Br, Br, Br, Br

c.

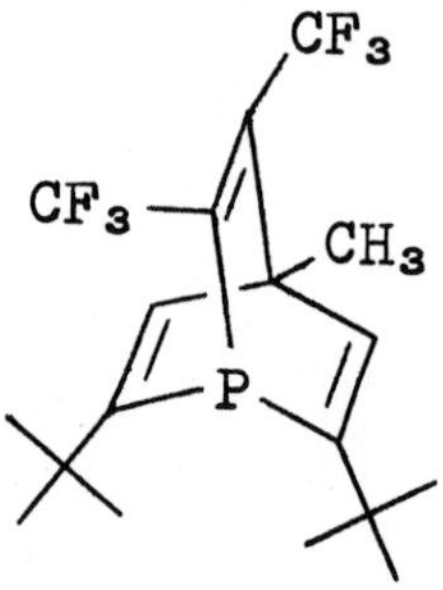

CF_3, CF_3, CH_3, P

d.

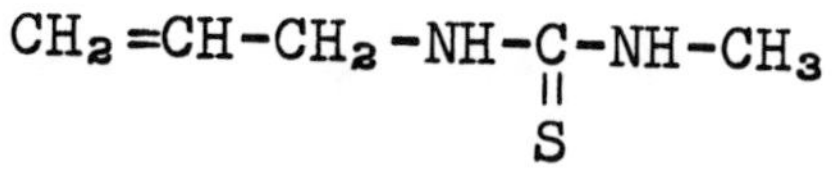

$CH_2=CH-CH_2-NH-C(=S)-NH-CH_3$

e.

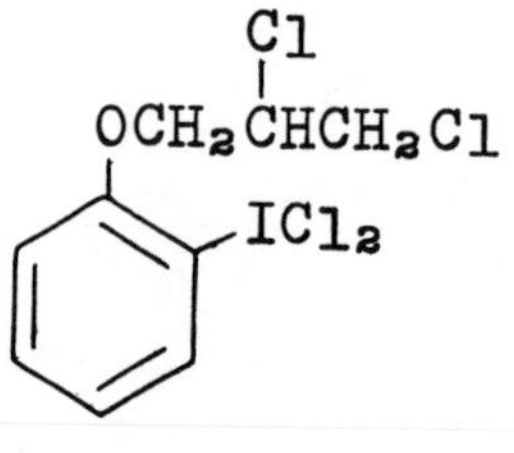

Cl, OCH_2CHCH_2Cl, ICl_2

f.

Ph, NO_2, C, OH, O_2N

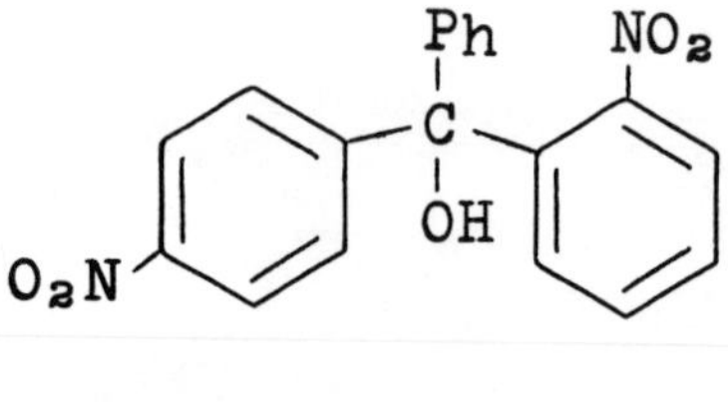

1. Cont.

g.

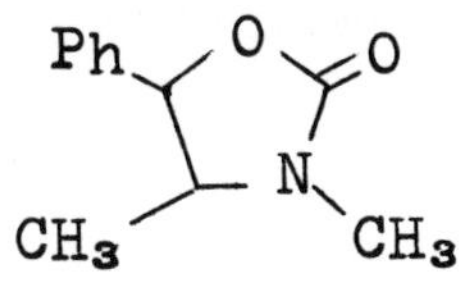

Ph O O N CH_3 CH_3

h.

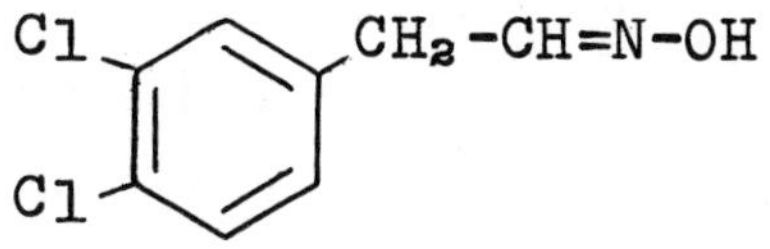

Cl Cl CH_2-CH=N-OH

i.

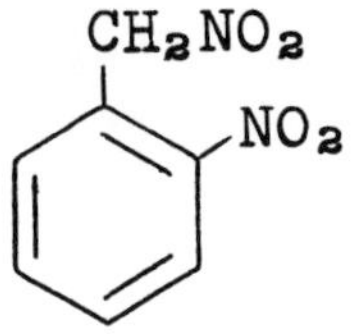

CH_2NO_2 NO_2

j.

$CH_3-C{\equiv}C-CH_2-O-(CH_2)_6-CH_3$

k.

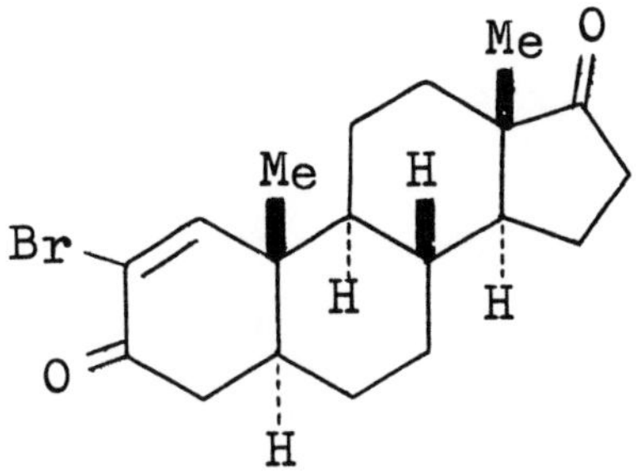

Me O Me H Br H H O H

l.

O CH_3 COOH

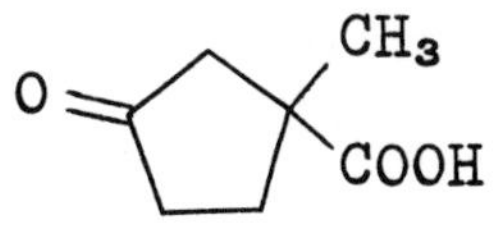

2. Given the following *CA* abbreviations, what is the full name and English translation (if not in English) of each of these journals, abstracts from which have appeared in the organic sections of *CA* in early 1970.* (Before doing this problem, see Problem 3).

   a. *Liet. TSR Mokslu Akad. Darb., Ser. B*

   b. *Trans. Ky. Acad. Sci.*

   c. *Zesz. Nauk. Uniw. Jagiellon, Pr. Chemi.*

   d. *Tohoku Yakka Daigaku Kenkyu Nempo*

   e. *Sb. Pr. Vyzk. Chem. Vyuziti Uhli, Dehtu Ropy*

*The 1969 edition of the CA list of periodicals abstracted is called "Access."

3. In which of the following libraries would you expect to find 1969 issues of the journals listed in Problem 2?

   a. Library of Congress, Washington, D.C. (DLC)

   b. New York Public Library, Science and Technology Division (NN)

   c. Los Angeles Public Library (CL)

   d. Center for Research Libraries, Chicago, Ill. (ICRL)

4. Each of the following compounds appears in only one of the four Beilstein series (*Hauptwerk* and three supplements). Given that location, on what page would it be found in each of the other three series?

a. NH_2; O_2N; NH-NH-Ph; NO_2

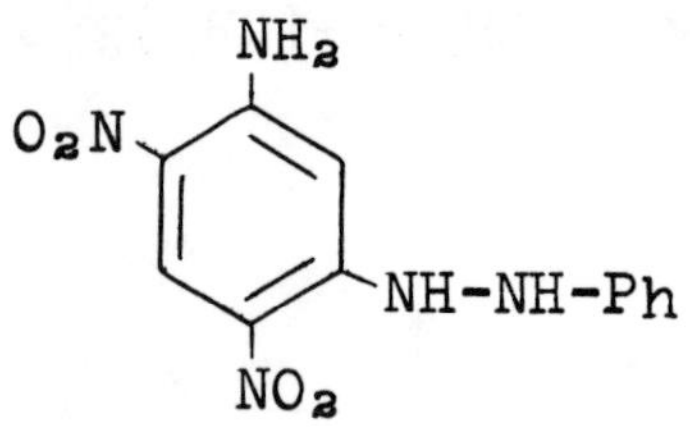

b. O; $CHMe_2$; Br; O

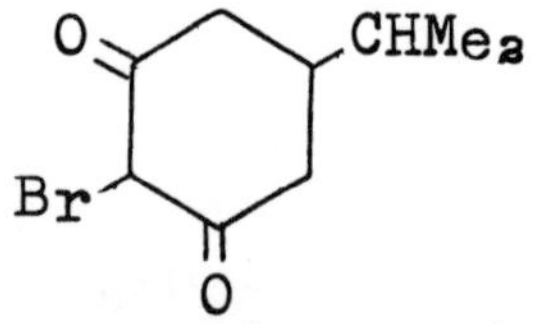

c. Et; OAc; AcO

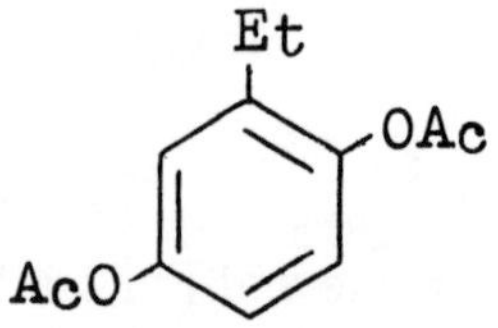

d. $HSe-CH_2CH_2COOH$

a. **15 E I**, 214

b. **7 H**, 564

c. **6 E II**, 885

d. **3 E III**, 559

5. Compounds containing which of the following ring systems (saturated or unsaturated) will be found in Volume **68** (Jan.-June, 1968) of *CA*?

Note: Two of these are in Vol. **68** and the other two are not.

a.

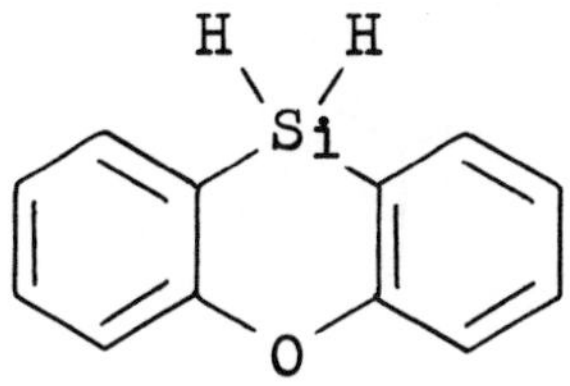

b.

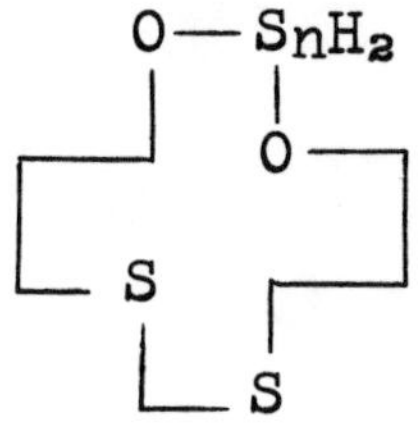

c.

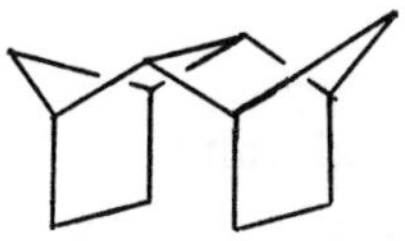

d.

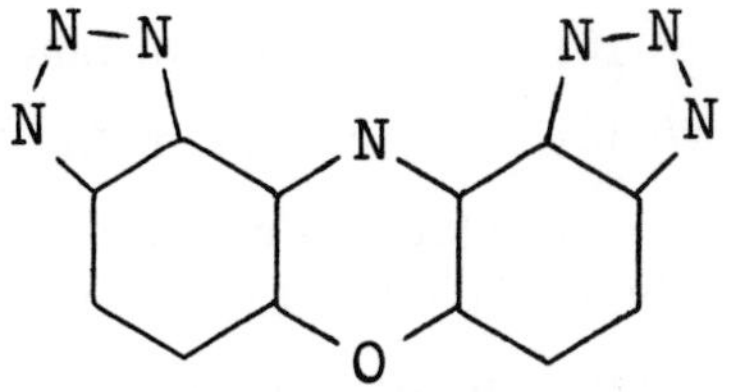

6. The answers to the following questions may be obtained from one (or more than one) of the reference books (not treatises or textbooks) discussed in Appendix B of *Advanced Organic Chemistry* under Compendia and Tables of Information, and Other Books. Answer each question (do not use Beilstein).
   a. What is the boiling point of benzoyl bromide?
   b. What would be a suitable derivative for 2,4-dinitro-1-naphthol, m.p. 138°?
   c. What would be a suitable derivative for *p*-methylsulfonyl benzaldehyde? (Note: this compound is not listed in "Handbook of Tables for Organic Compound Identification," 3d edition.)

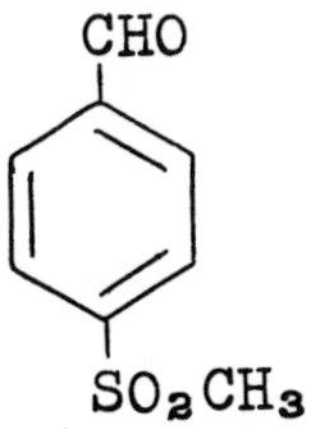

d. What is the name and ring numbering for this system:

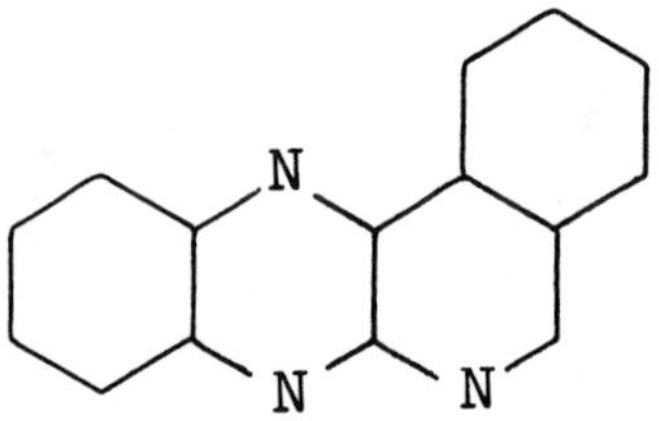

e. What is the chemical name and structure of the drug whose trade name is Pelazid?

f. What is the medical use of the drug whose generic name is piperidolate?

g. What is the C-F bond distance in CH_3F?

h. What is the dipole moment of 1,10-dichlorodecane?

i. If an aromatic aldehyde boils at 186° at 760 mm., at what temperature would you expect it to boil at 10 mm.?

j. What is the melting point of $(PhCH_2)_3SnOAc$?

k. Locate a literature reference for the transformation:

$$ArNO_2 \longrightarrow ArNHCH_2Ph$$

l. How would you prepare the reagent: Raney cobalt?

m. What is the reagent dicyclohexylamine used for?

n. How could the compound α-bromo-*n*-valeraldehyde be prepared?

o. Give the boiling point and percentage composition of the azeotrope, if any, formed by the following pairs:

   i. 2-pentanol and benzene.

   ii. ethanol and cyclopentane.

7. *Organic Syntheses* is a source of tested procedures for the synthesis of specific compounds. Therefore when one wishes to prepare a given compound one often consults *Org. Syn.* to see if a procedure is given there. Determine if each of the following compounds is found in *Org. Syn.* and if so, give the volume and page number.

a. 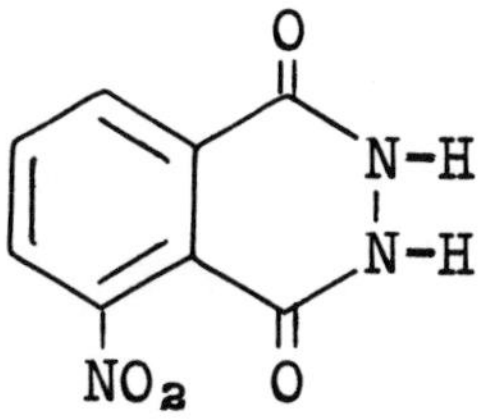 O, N-H, N-H, NO_2, O

b. 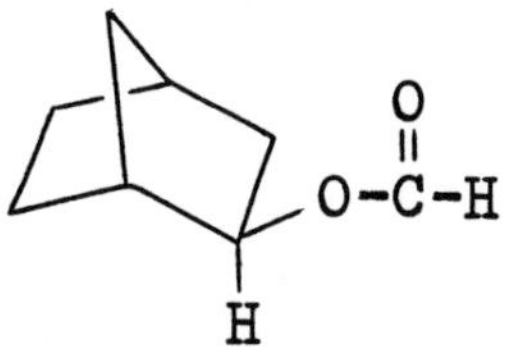 O-C(=O)-H, H

c. $ClCH_2-C(=O)-CH_2Cl$

d. 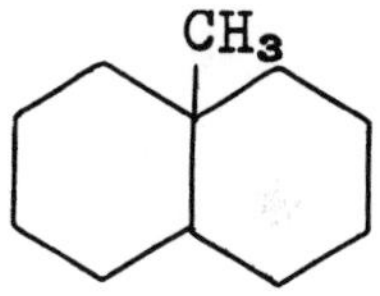 CH_3

e. 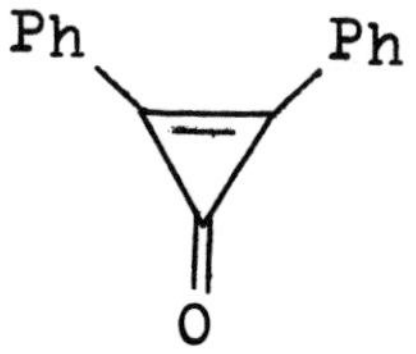 Ph, Ph, O

ANSWERS TO PROBLEM SET 1

BONDING

1. a. Linear
   b. Planar, with 120° angles
   c. Linear
   d. The three carbons linear. The hydrogens in perpendicular planes (see p. 79, *Advanced Organic Chemistry*)
   e. Pyramidal, angles of about 105-110°
   f. Tetrahedral
   g. C–O–C angle about 111°
   h. Pyramidal, angles of about 105-110°

2.

a. 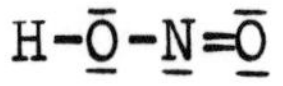 H-O-N=O

b.  $H_2C=N^{\oplus}=N^{\ominus}$ ⟷ $H_2C^{\ominus}-N^{\oplus}\equiv N$

c. 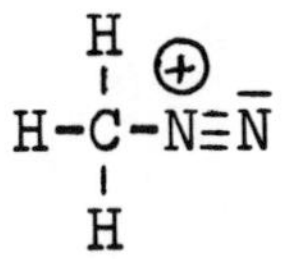 $H_3C-N^{\oplus}\equiv N$

d. 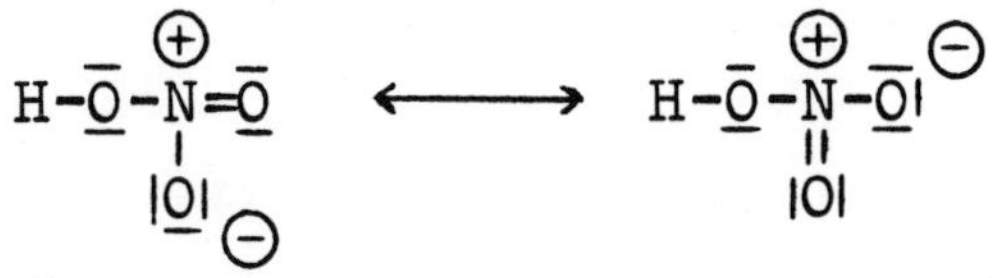 $H-O-N^{\oplus}(=O)-O^{\ominus}$ ⟷ $H-O-N^{\oplus}(=O)-O^{\ominus}$

e. 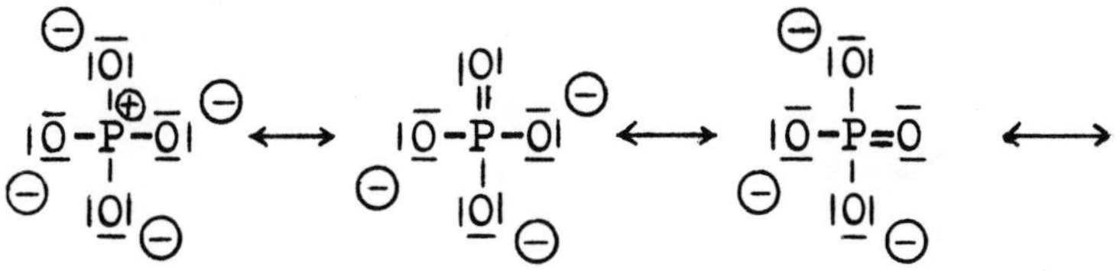 $^{\ominus}O-P^{\oplus}(O^{\ominus})(O^{\ominus})-O^{\ominus}$ ⟷ $^{\ominus}O-P(=O)(O^{\ominus})-O^{\ominus}$ ⟷ $^{\ominus}O-P(=O)(O^{\ominus})-O^{\ominus}$ ⟷

two other forms

f.  $^{\ominus}N=N^{\oplus}=O$ ⟷ $N\equiv N^{\oplus}-O^{\ominus}$

g. 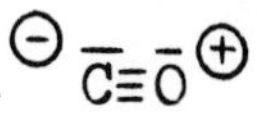 $^{\ominus}C\equiv O^{\oplus}$

2. Cont.

h. O=C=O

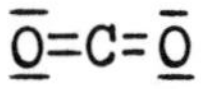

i. |O=N−O• ⟷ •O−N=O| ⟷ ⊕ ⊖ |O=N−O| ⟷ ⊖ ⊕ |O−N=O|

j.

```
        C-O-H             C-O-H        ⊕     C-O-H
 (ring) ||        ⟷ (ring) ||    ⟷  (ring)= |
        O|                O|                |O| ⊖

⟷  (ring)=C-O-H      ⟷  (ring)=C-O-H
   ⊕        |               ⊕    |
           |O| ⊖                |O| ⊖
```

k. ⊖ N≡C−CH−C≡N ⟷ ⊖ N=C=CH−C≡N ⟷ N≡C−CH=C=N ⊖

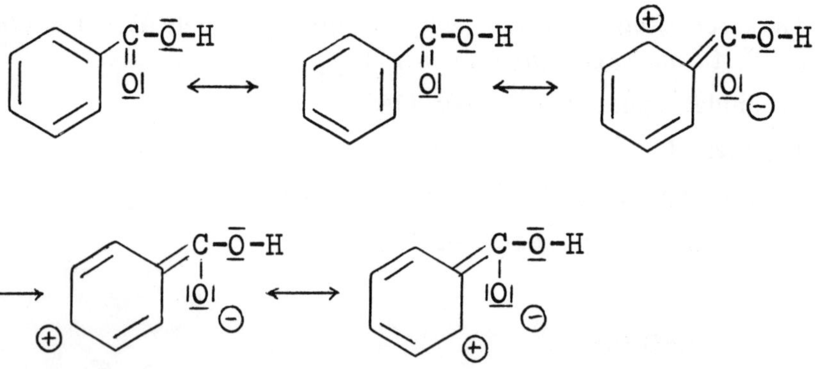

l. ⊖ |O−Cl−O| ⊖ (⊕ on Cl)

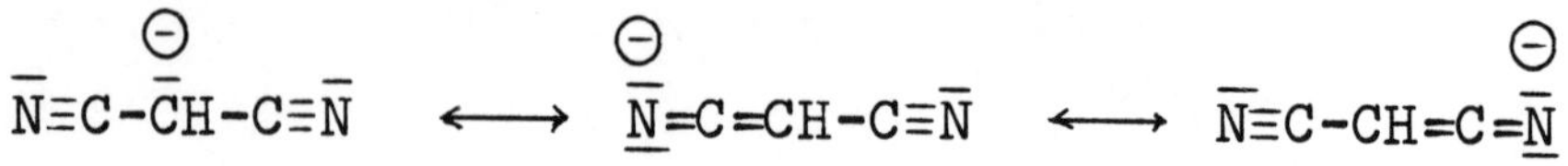

m. CH_2=CH−Cl| ⟷ ⊖ CH_2−CH=Cl| ⊕

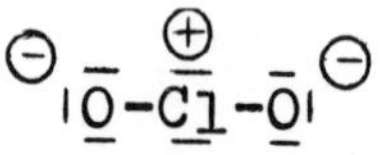

n.
```
   ⊕
CH3-S-CH3
    |
   CH3
```

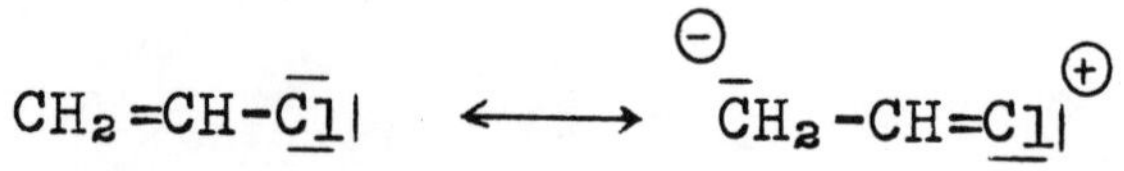

o.
```
  H  H
⊖ |  | ⊕
H-B-N-H
  |  |
  H  H
```

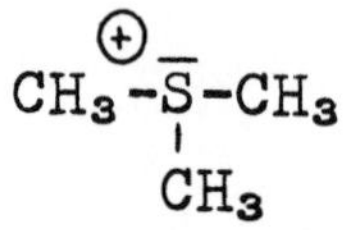

p.
```
   Cl
   |
Cl-P-Cl
  / \
 Cl  Cl
```

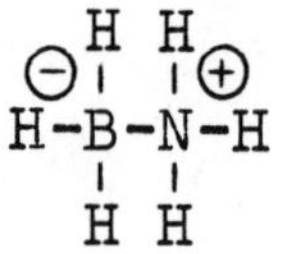

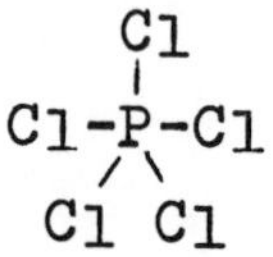

2. Cont.

q.

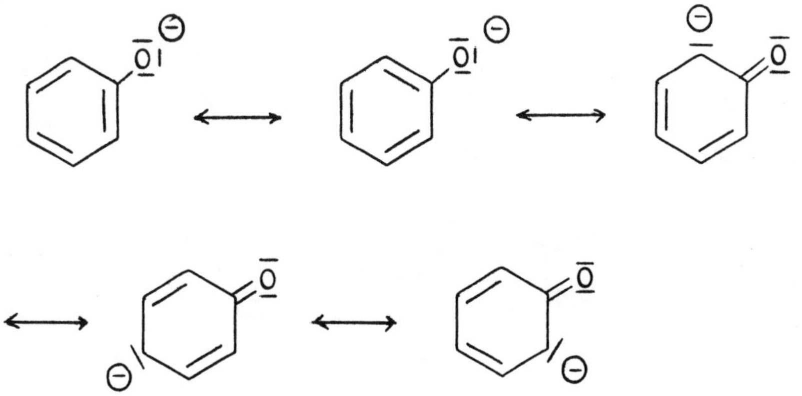

r.

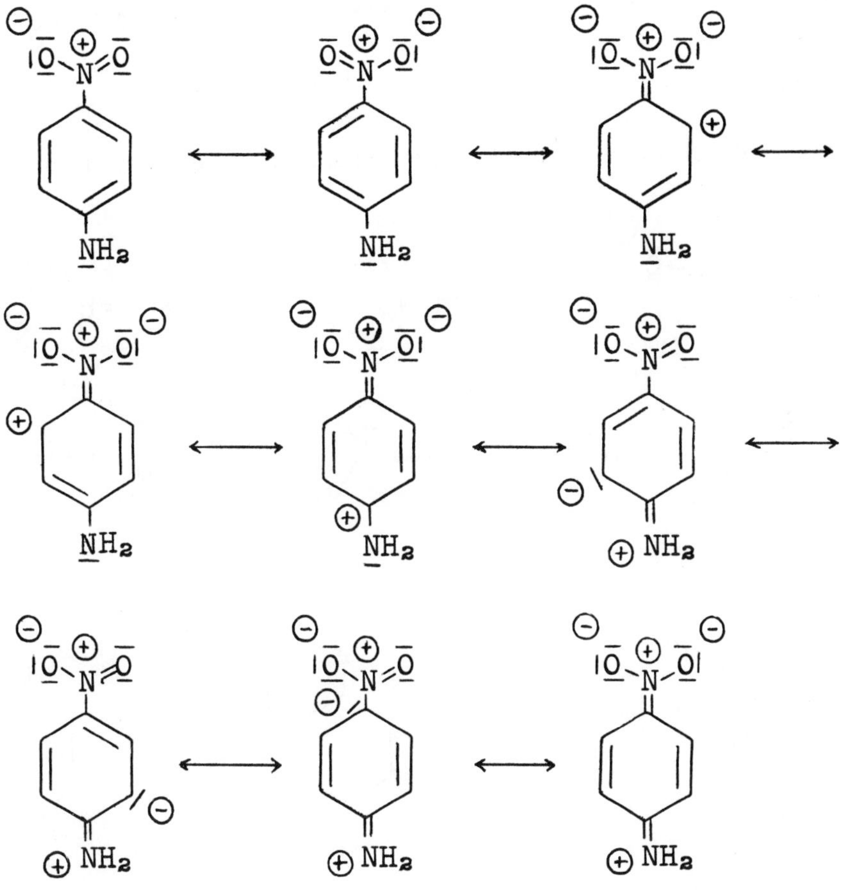

s.

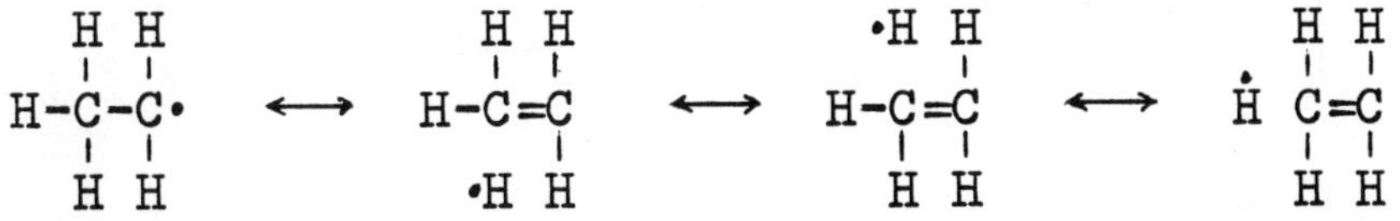

2. Cont.

t.

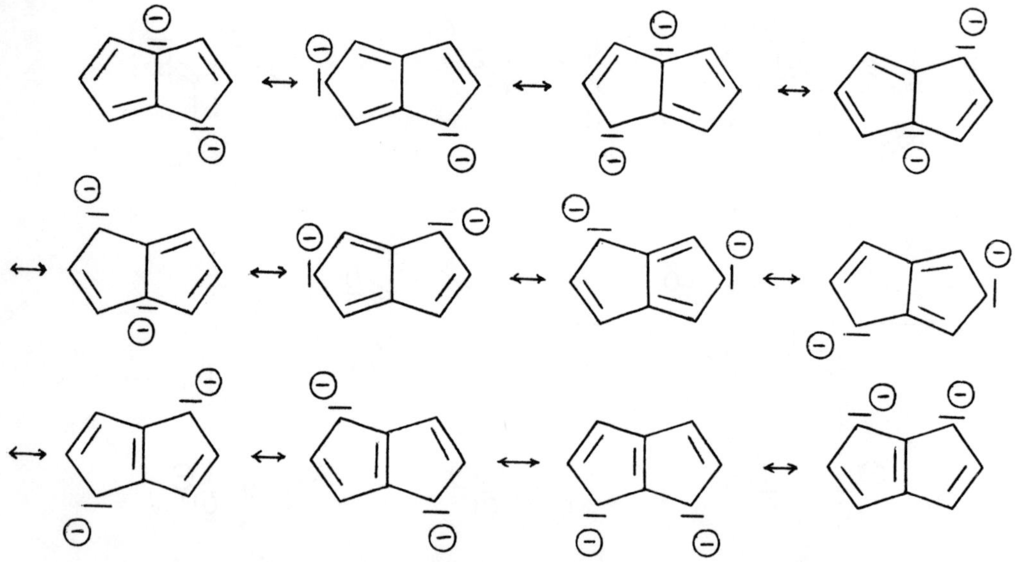

↔ ↔ ↔

↔ ↔ ↔ ↔

↔ ↔ ↔ ↔

Canonical forms with negative charges on adjacent atoms have been omitted.

u.

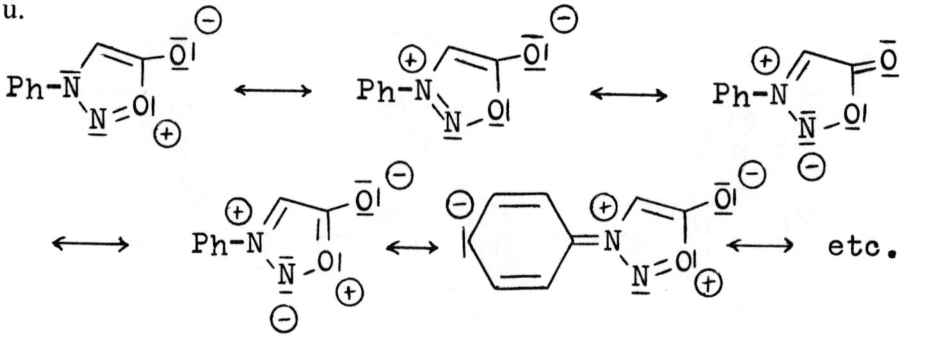

Ph–N ⟷ Ph–N ⟷ Ph–N

⟷ Ph–N ⟷ ⟷ etc.

3. No dipole moments: a, c, d, f.

b. H, Cl, Cl, Cl

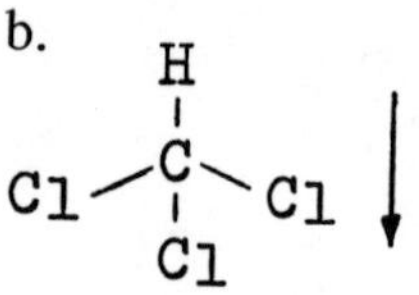

e. F

g. CH_3, CH_3

h. H, C, O + H, C, O ⟶

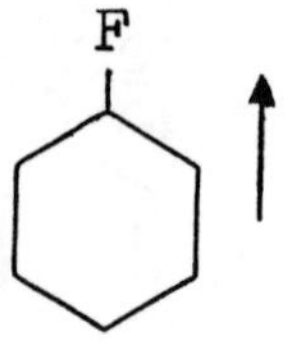

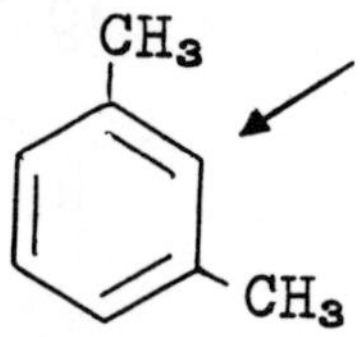

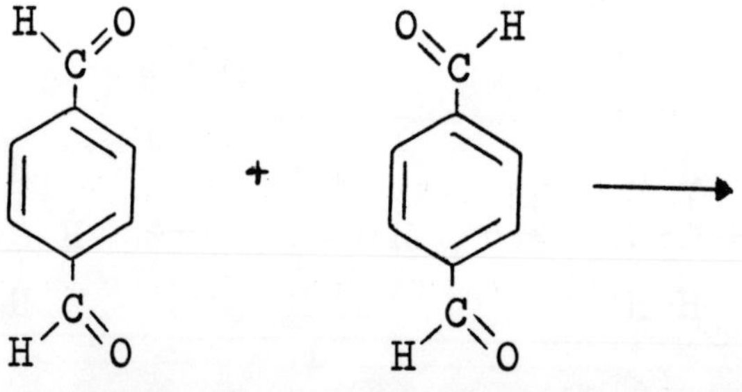

3. Cont.

The arrow points to the negative end. In the case of h, there are two extreme conformations, one with a dipole moment and one without. The actual compound is a mixture of these and thus has a dipole moment in the direction shown. The value is 2.37 D.

4. a. Compound i: 2.5 D; Compound ii: 3.4 D
   b. Compound i: 6.3 D; Compound ii: 2.6 D
   c. Compound i: 3.48 D; Compound ii: 2.68 D
   d. Compound i: no moment; Compound ii: 1.9 D
   e. Compound i has a higher moment, since both negative groups are pointing in the same direction.
   f. Compound i: 2.8 D; Compound ii: 1.5 D

5. The methyl group is an electron donor. Since it is increasing the dipole moment in both cases, the original moments must have been in the same direction.

i ↑ ii ↑ iii ↓ iv ↓

6.

$C_4H_{8\,(gas)} + 6O_{2\,(gas)}$	=	$4CO_{2\,(gas)}$	+	$4H_2O_{(liq)}$	+ 649.5 kcal
$4CO_{2\,(gas)}$	=	$4C_{(graphite)}$	+	$4O_{2\,(gas)}$	– 376.4
$4H_2O_{(liq)}$	=	$4H_{2\,(gas)}$	+	$2O_{2\,(gas)}$	– 273.2
$4H_{2\,(gas)}$	=	$8H_{(gas)}$			– 416.8
$4C_{(graphite)}$	=	$4C_{(gas)}$			– 686.8
$C_4H_{8\,(gas)}$	=	$8H_{(gas)}$	+	$4C_{(gas)}$	–1103.7 kcal

The heat of atomization of 1-butene is thus 1103.7 kcal/mole;

1-Butene has 8 CH bonds:	8 X 98	=	784
and 2 C–C bonds:	2 X 84	=	168
			952

$$\begin{array}{r} 1104 \\ -952 \\ \hline \end{array}$$

152 kcal/mole is the energy of the C=C bond.

7. a. Form iii does not contribute at all, because the atoms are not in the same place as in the other forms. i, a hyperconjugation form, contributes much less than ii or iv, because it has fewer covalent bonds. ii and iv are equivalent, and contribute equally.

7. Cont.

b. Forms iii and iv are completely incorrect and contribute nothing. iii has 9 electron pairs, whereas 3 N atoms and 1 H atom have a total of 8 valence pairs. iv has 10 electrons around the middle N and is thus not a bona fide Lewis structure. i and v are bona fide Lewis structures, but contribute little and can be ignored. v has two adjacent atoms with positive charges as well as an atom with a charge of −2. i has only 6 electrons around the middle N as well as a charge of +2 on that atom. ii and vi are the important canonical forms.

c. v is an inadmissable form because the atoms are not in the same place, the H having moved. iii is also incorrect because it does not have the same number of unpaired electrons as the others (it is a separate species in its own right, but is not a contributing form to the resonance of i). ii is an exceedingly high energy form because of the long bond between C-1 and C-4. The only important canonical forms among the 5 shown, are i and iv. iv contributes more because it has the negative charge on the more electronegative atom.

8. Most acidic iii pK = 17.5
i pk = 20.5
least acidic ii pk = 21.4

Kuhn and Rewicki, *Justus Liebigs Ann. Chem.* **704**, 9 (1967).

11. If ii underwent a similar reaction, it would cause loss of aromaticity in one benzene ring.

Grohmann and Sondheimer, *J. Amer. Chem. Soc.* **89**, 7119 (1967).

12. a. Not aromatic

Bindra, Elix, Garratt, and Mitchell, *J. Amer. Chem. Soc.* **90**, 7372 (1968).

b. Aromatic

Cava and Husbands, *J. Amer. Chem. Soc.* **91**, 3952 (1969).

c. Aromatic

Hafner and Tappe, *Angew. Chem. Intern. Ed. Engl.* **8**, 593 (1969) [*Angew. Chem.* **81**, 564 (1969)].

d. Should be aromatic, but not yet prepared.

e. Not aromatic

Brown and Sondheimer, *J. Amer. Chem. Soc.* **91**, 760 (1969).

f. Aromatic. The parent has not been made, but derivatives are known.

Jones and Pyron, *J. Amer. Chem. Soc.* **87**, 1608 (1965); Kende and Izzo, *J. Amer. Chem. Soc.* **87**, 1609 (1965); Prinzbach, Seip, and Fischer, *Angew. Chem. Intern. Ed. Engl.* **4**, 242 (1965) [*Angew. Chem.* **76**, 412 (1964)].

12. Cont.

g. Not aromatic

Doering, in "Theoretical Organic Chemistry, The Kekule Symposium," pp. 35-48, Butterworth Scientific Publications, London, 1959.

h. Not aromatic

Hafner, *Angew. Chem. Intern. Ed. Engl.* **3,** 165 (1964), p. 170 (*Angew. Chem.* **75,** 1041 (1963), p. 1047).

i. Aromatic

Lawson and Miles, *J. Chem. Soc.* **1959,** 2865.

16. Aromatic properties are enhanced in ii because of additional resonance not present in i:

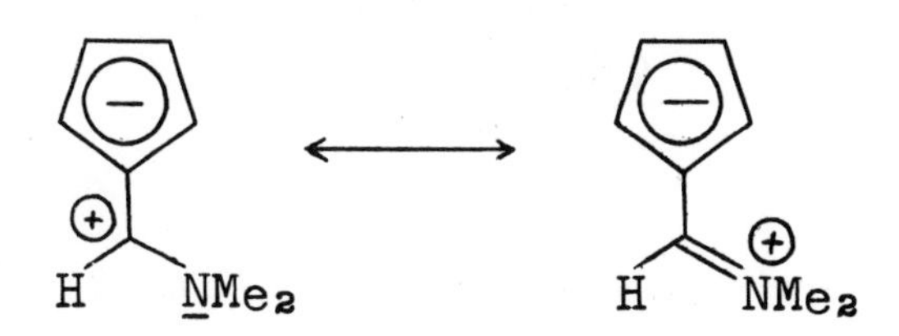

18. Alternant with a nonbonding energy level: b, c

Alternant without a nonbonding energy level: e, h

Nonalternant: a, d, f, g

19. No tautomerism: b (see Nonhebel, *J. Chem. Soc., C* **1967,** 1716), k

a. $Ph{-}CH{=}C(OH){-}C({=}O){-}CH_3$

c. $HOCH_2{-}CH(OH){-}CH(OH){-}CH(OH){-}CH(OH){-}CHO$

d.

Ph
O
Ph
H–N
N
Me

19. Cont.

e.

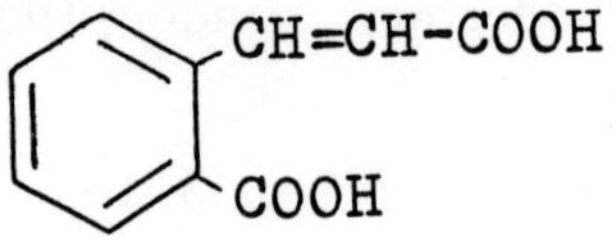

$CH{=}CH{-}COOH$

$COOH$

f.

H, Ph

N - N

CH_3, O

N

CH_3

H

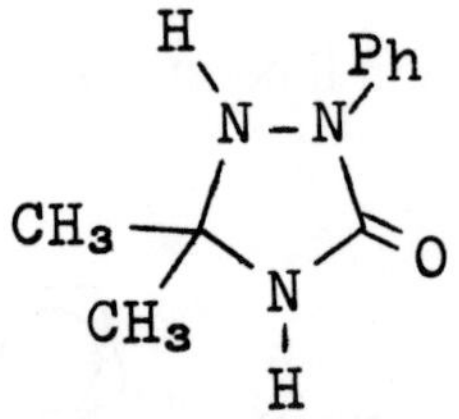

Schildknecht and Hatzmann, *Justus Liebigs Ann. Chem.* **724**, 226 (1969).

g. $Ph{-}C{=}N{-}CH_3$

NH_2

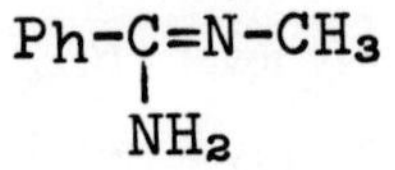

h. $Ph{-}C{=}CH{-}C{=}CH{-}Ph$

OH----F

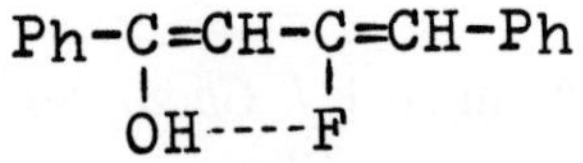

i.

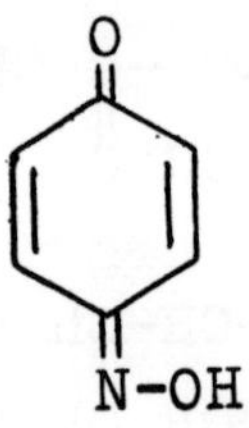

O

N-OH

j.

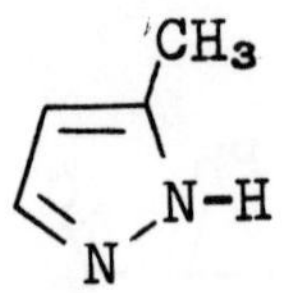

CH_3

N-H

N

l.

$Ph{-}NH{-}N{=}C{-}N{=}N{-}$ $-NO_2$

CH_3

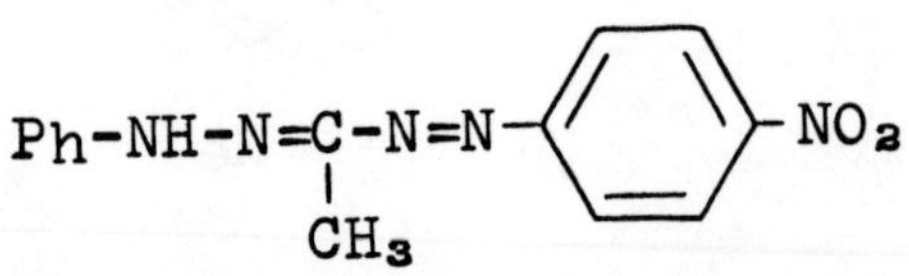

Neugebauer and Jenne, *Tetrahedron Lett.* **1969**, 791.

20. Intramolecular hydrogen bonds will form: a, c, e, g

    They will not form: b, d, f

21. No bonds will form: b, h

    a. HCN- - - -HCN

    c. H_2CO- - -HOH

    d. Cl^-- - - -HOH

    e. $(CH_3)_2CO$- - -$HOOCC_6H_4OH$

    f. $O(CH_2CH_2)_2O$- - - -$HCFCl_2$

    g. $H_2NCH_2CH_2O$-H- - - - - -$NH_2CH_2CH_2OH$

22. The alcoholic OH group hydrogen bonds with the more basic C=O rather than the less basic OH oxygen, so that i is found, but not iii. The OH of the COOH is a stronger acid than is the OH of the alcohol moiety so that structure ii competes even though there is no C=O for this group to bond with.

    Mori, Asano, Irie, and Tsuzuki, *Bull. Chem. Soc. Jap.* **42**, 482 (1969).

ANSWERS TO PROBLEM SET 2

STEREOCHEMISTRY

1. Resolvable: a, b, d, e, f, i, j, k, n, p (see Tochtermann, Küppers, and Franke, *Chem. Ber.* **101**, 3808 (1968)), q (see Haas and Prelog, *Helv. Chim. Acta* **52**, 1202 (1969)).

   Not resolvable: c, g, h, l, m (two cis-trans isomers, but neither is resolvable), o.

2. a. 2 isomers (a *dl* pair)

   b. 4 isomers (a syn *dl* pair; an anti *dl* pair)

   c. 4 isomers (2 *dl* pairs)

   d. 6 isomers (3 *dl* pairs: RRR–SSS; RSR–SRS; RRS–SSR)

   e. No isomerism. Only one compound.

   f. 4 isomers (a cis *dl* pair; a trans *dl* pair)

   g. No isomerism. Only one compound.

   h. 3 isomers:

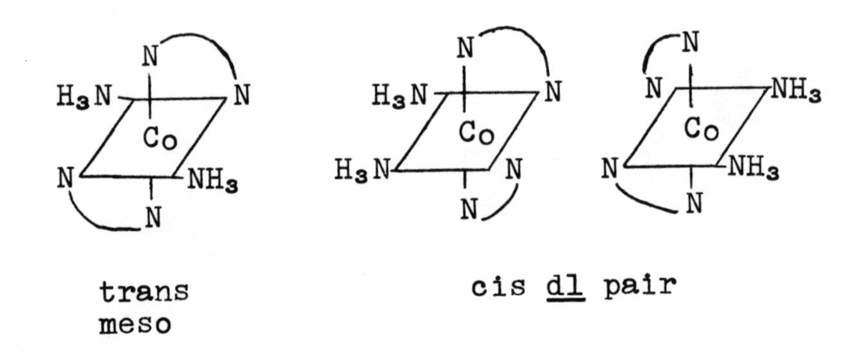

   i. 2 isomers (cis, trans: both inactive)

   j. 2 isomers (a *dl* pair)

   k. 2 isomers (cis, trans: both inactive)

   l. No isomerism. Only one compound.

   m. 2 isomers (a *dl* pair)

   n. No isomerism. Only one compound.

   o. 2 isomers (a *dl* pair)

   p. 3 isomers:

2. Cont.

trans
meso

cis dl pair

q. 8 isomers:
1. all cis
2. trans-cis-cis-cis-cis
3. trans-trans-cis-cis-cis
4. trans-cis-trans-cis-cis
5. trans-trans-trans-cis-cis
6. trans-trans-cis-trans-cis
7. trans-trans-trans-trans-cis
8. all trans

r. 16 isomers (4 meso; 6 *dl* pairs):

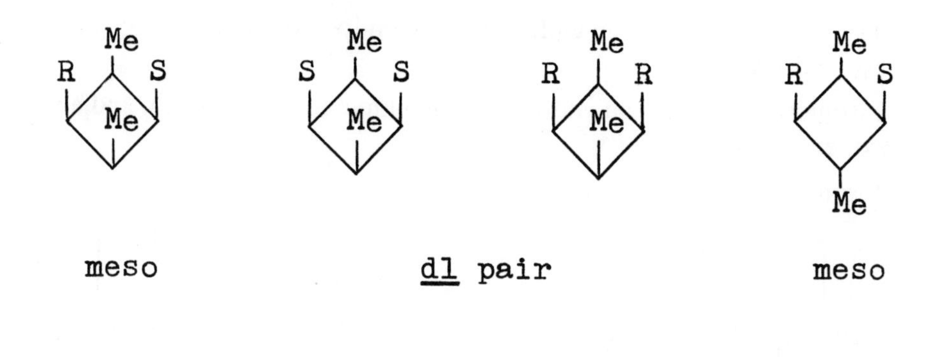

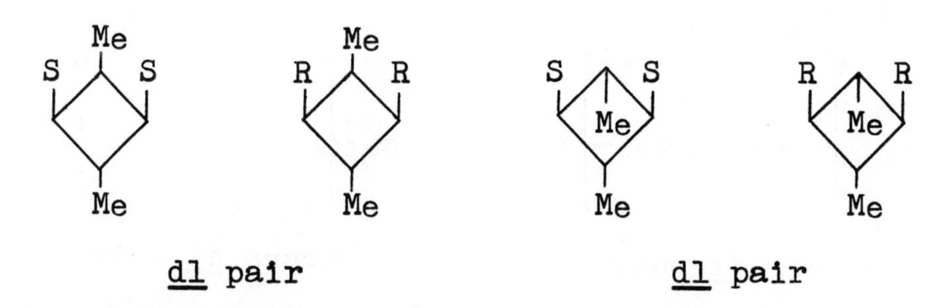

2. Cont.

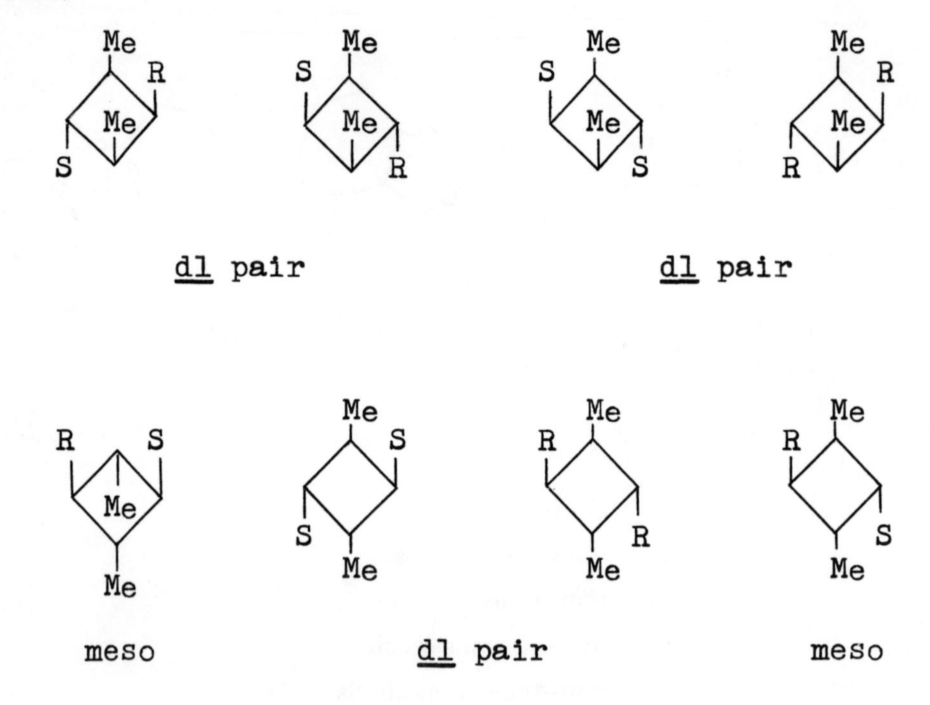

Note: The last structure shown has a center of symmetry.

s. 8 isomers (4 *dl* pairs: a cis-cis; a cis-trans; a trans-trans; and a trans-cis *dl* pair)

t. 3 isomers (a cis meso and a trans *dl* pair)

u. 8 isomers (4 dl pairs: an endo-endo; an endo-exo; an exo-exo; and an exo-endo *dl* pair)

v. 4 isomers (all meso: an endo-endo, an endo-exo; an exo-exo, and an exo-endo isomer)

w. 4 isomers:

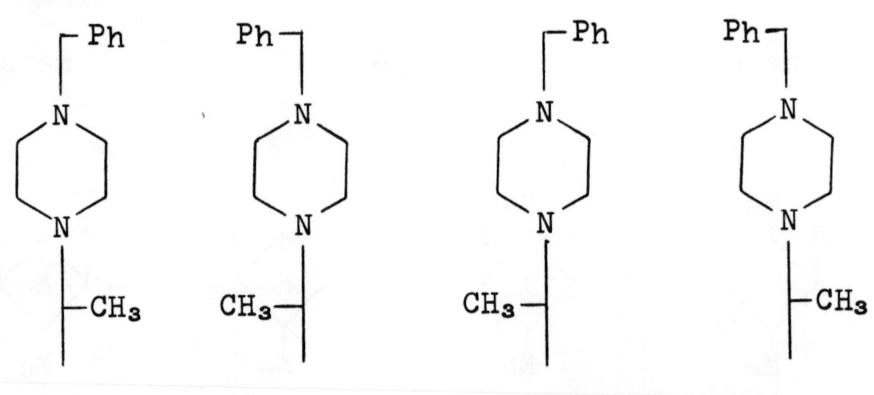

x. 10 isomers (2 meso and 4 *dl* pairs):

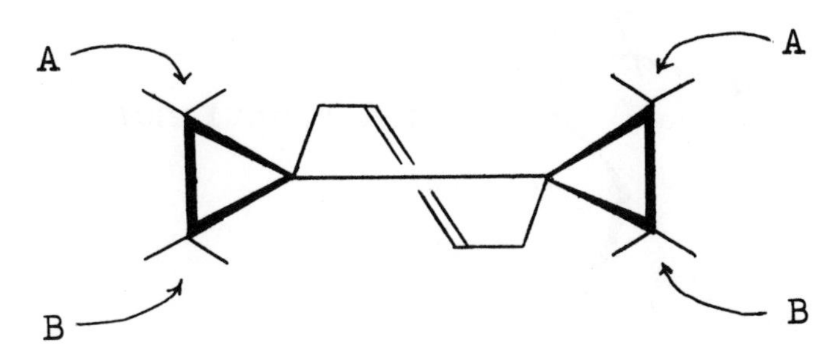

The CN groups can be in position A or position B, and in either case can be exo or endo. So we have:

1. A-exo, A-exo: meso
2. A-endo, A-endo: meso
3,4. A-exo, A-endo: *dl* pair
5,6. A-exo, B-exo: *dl* pair
7,8. A-exo, B-endo: *dl* pair
9,10. A-endo, B-endo: *dl* pair

y. 6 isomers (3 dl pairs):

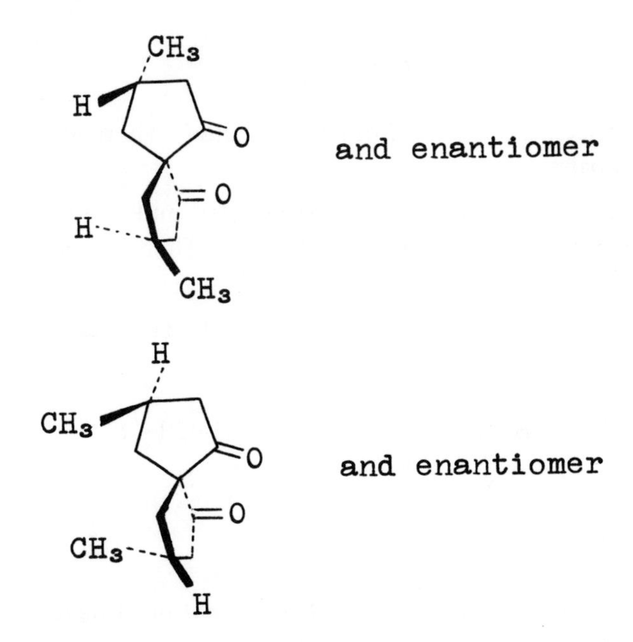

2. Cont.

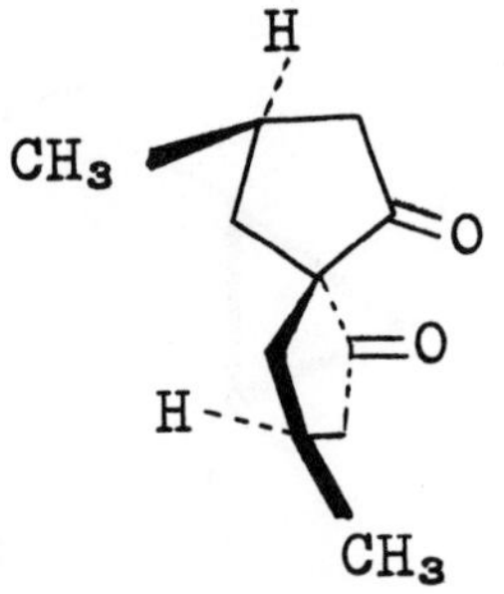

and enantiomer

It would seem that there should be a fourth *dl* pair: with the methyl group below the plane of each ring; but this is the same as the last *dl* pair shown.

z. There are 16 isomers of the structure where the methyl group on the three-membered ring is on the same side of the bridge as the methyl group near the bottom (as shown in the problem). Each of the three methyl groups could be exo or endo, so that there are 8 structures, each of which is nonsuperimposable on its mirror image, making 16 isomers in all. In addition, there are 16 more isomers where the methyl group on the 3-membered ring is on the opposite side of the bottom methyl group, i.e.:

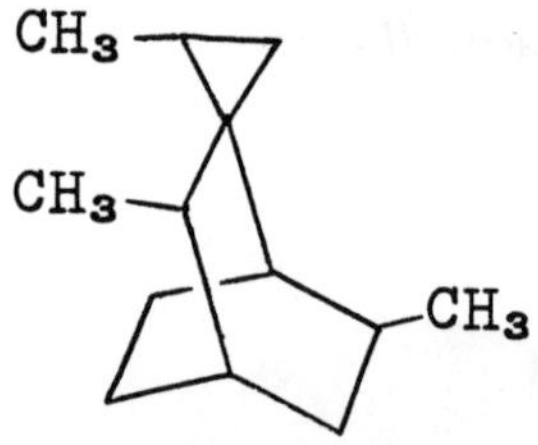

3. a. 4 asymmetric carbons – 16 isomers predicted. There are actually 10 (2 meso, 4 dl pairs).

b. 5 asymmetric carbons (actually 4 asymmetric and one pseudoasymmetric) – a maximum of 32 isomers predicted. There are actually 16 (4 meso, 6 *dl* pairs).

c. 3 asymmetric carbons – 8 isomers predicted. There are actually 4 (2 *dl* pairs):

```
R-CH-R      S-CH-S      R-CH-R      S-CH-S
  |           |           |           |
  R           S           S           R

      dl pair                 dl pair
```

The central carbon cannot be asymmetric since it must have at least 2 S groups or 2 R groups.

d. 4 asymmetric carbons – 16 isomers predicted. There are actually 5 (2 *dl* pairs and a meso):

3. Cont.

R_4-C S_4-C R_3-C-S S_3-C-R R_2-C-S_2

dl pair dl-pair meso

The meso isomer has neither a plane nor a center of symmetry. It has a 4-fold alternating axis of symmetry.

e. 3 asymmetric carbons – 8 isomers predicted. There are actually 8 (4 *dl* pairs).

f. 3 asymmetric carbons – 8 isomers predicted. There are actually 8 (4 *dl* pairs).

g. 3 asymmetric carbons (actually 2 asymmetric and 1 pseudoasymmetric)– a maximum of 8 isomers predicted. There are actually 4 (2 mesos and a *dl* pair).

h. 4 asymmetric carbons (actually 2 asymmetric and 2 pseudoasymmetric – C-2 and C-5) – a maximum of 16 isomers predicted. There are actually 8 (4 meso and 2 *dl* pairs).

i. 2 asymmetric carbons – 4 isomers predicted. There are actually 2 (a *dl* pair). The other 2 are precluded by the bridge.

j. 4 asymmetric carbons – 16 isomers predicted. There are actually 16.

k. 8 asymmetric carbons – 256 isomers predicted. There are actually 256.

l. 6 asymmetric carbons – 64 isomers predicted. There are actually 8 (4 *dl* pairs). The OH groups can be exo-exo; exo-endo; endo-endo; endo-exo. Each is nonsuperimposable on its mirror image. All other isomers are precluded by the bridges.

m. 4 asymmetric carbons (C-1, C-4, C-5, and C-8) – 16 isomers predicted. There is actually only 1. The molecule has a plane of symmetry. All other isomers are precluded by the bridges.

n. All 8 carbons are asymmetric – 256 isomers predicted. There are actually only 2 (a *dl* pair).

o. In this case too all 8 of the cubane carbons are asymmetric. Yet there are not 256 isomers, but only 1. The molecule has a center of symmetry.

p. 3 asymmetric carbons – 8 isomers predicted. There are actually 2 (the OH exo the bridge, and the OH endo to the bridge). Both are superimposable on their mirror images.

4. R: b, c, e, g, h, i, k, o.

S: a, d, f, j.

l. Carbon i – R; carbon ii – R

m. Carbon i – R; carbon ii – S

4. Cont.

n. Carbon i – R; carbon ii – R

p. Carbon i – S; carbon ii – R

q. Carbon i – S; carbon ii – S; carbon iii – R

5. *Z:* a, c, e, f.

E: b, g.

d. 2,3-bond: *E;* 4,5-bond: *E*

h. 3,4-bond: *Z;* 5,6-bond: *E*

6. a. stereoselective

b. partially stereospecific

c. nonstereoselective

d. partially stereoselective

e. stereospecific

f. nonstereoselective

g. partially stereospecific

h. stereoselective

7.

a. 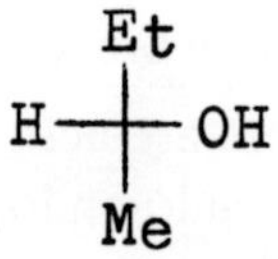

b. 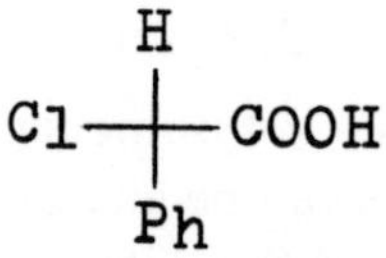

c. 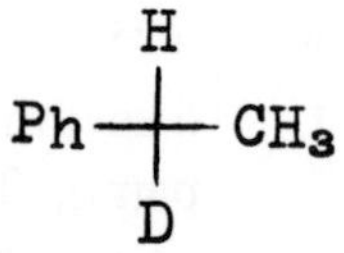

d. 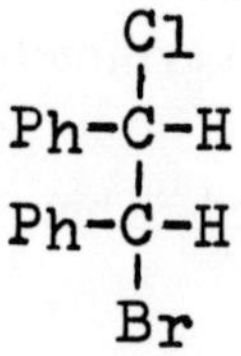

e. 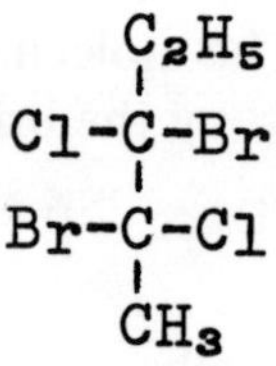

8.

a. 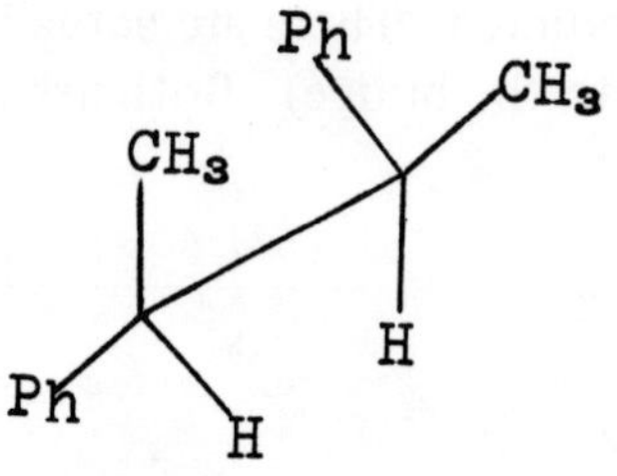

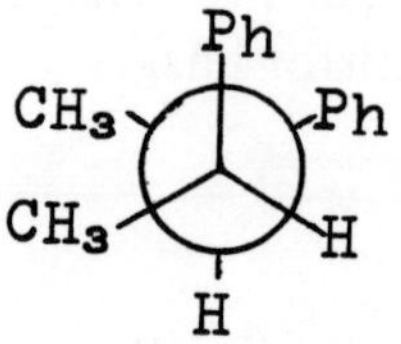

8. Cont.

b.

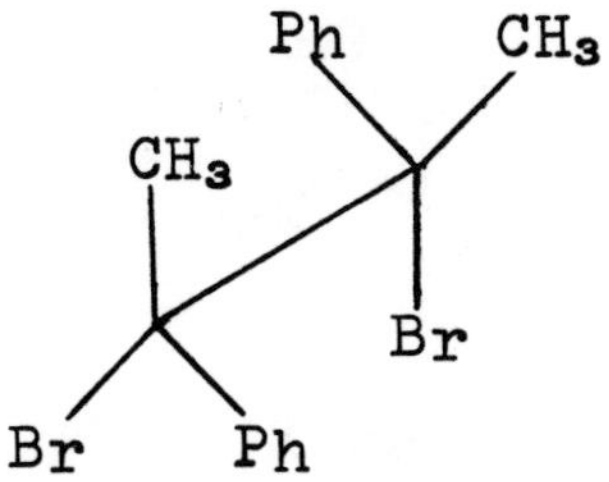

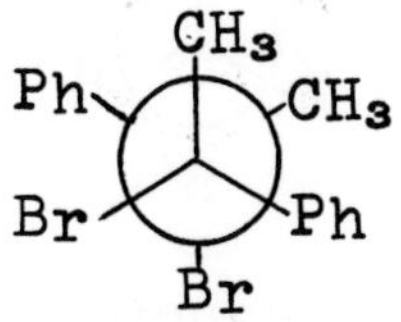

c.

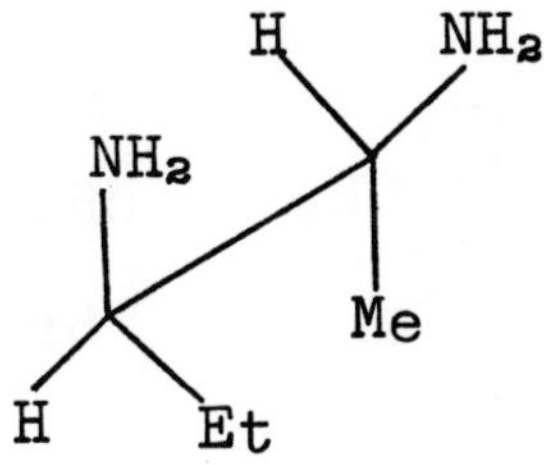

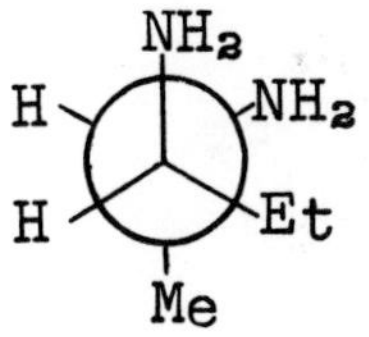

d.

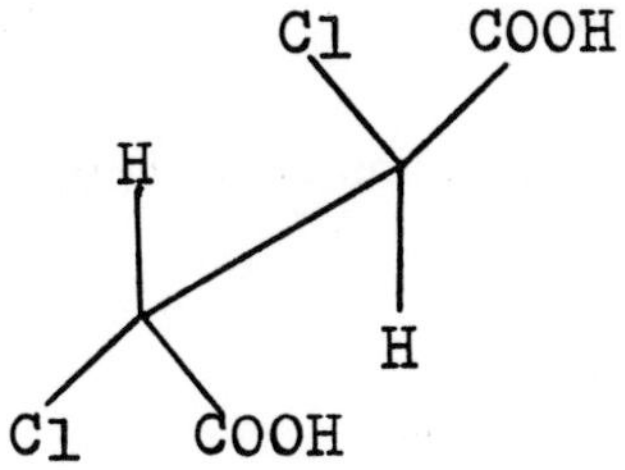

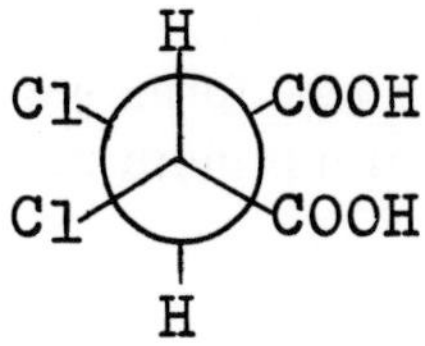

9.

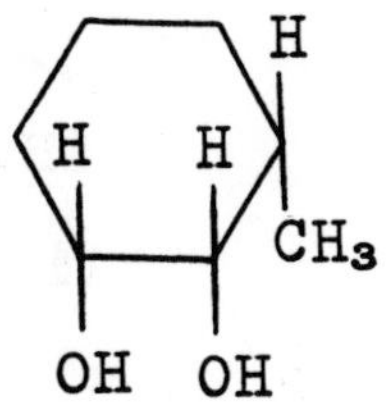

10.

	Part A	Part B
a.	v	v
b.	iii	i
c.	iv	iii
d.	iii	*

*iv for the 2 position and ii for the 4 position.

11. a.

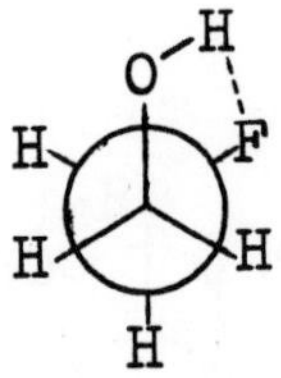

gauche 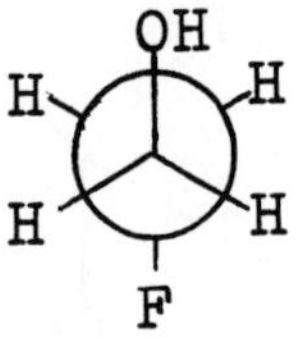 anti

The gauche is more stable because of hydrogen bonding.

b.

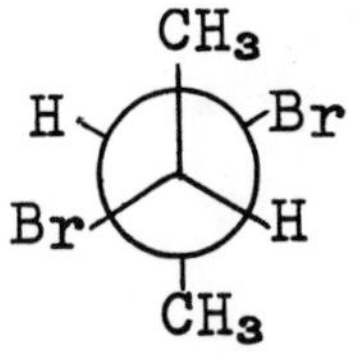

anti 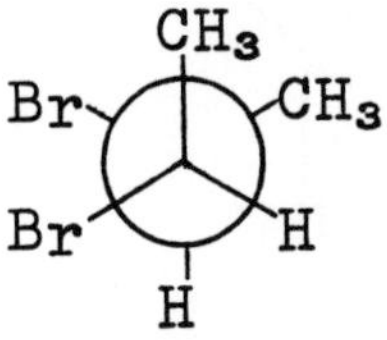 gauche 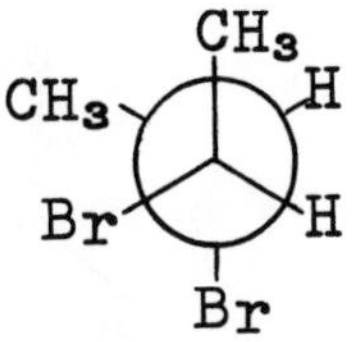 gauche

The two gauche forms are equivalent and hence have the same energy. The anti form is more stable.

c.

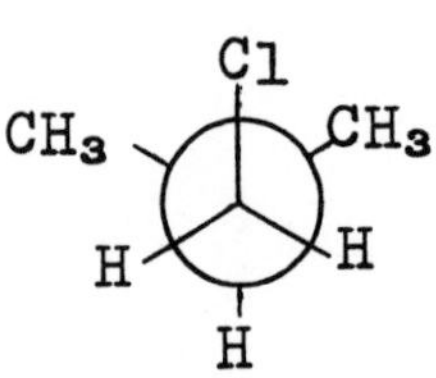

i 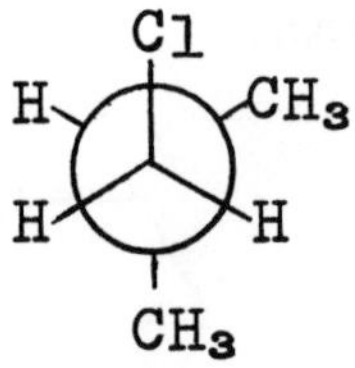 ii 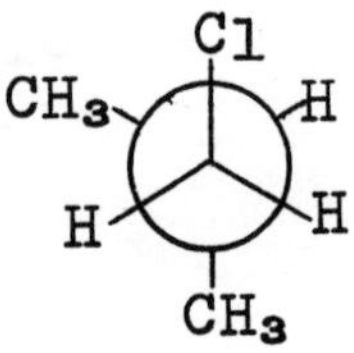 iii

i is less stable than ii or iii (which are equivalent) because of greater crowding.

d.

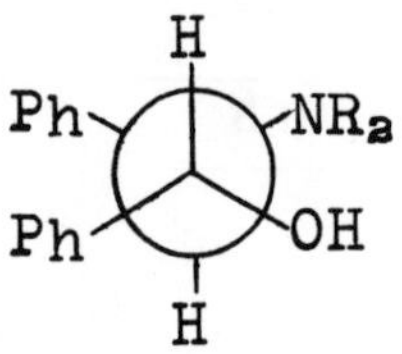

i 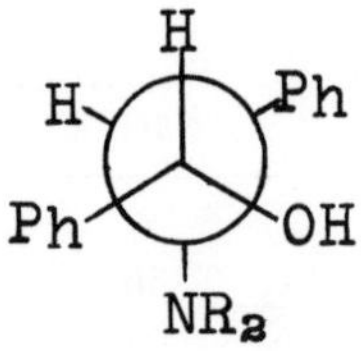 ii 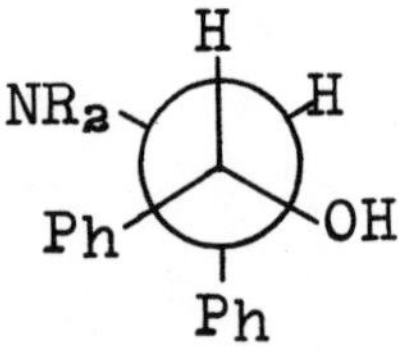 iii

iii has no hydrogen bonding between the NR_2 and the OH, so it is the least stable. i is more stable than ii because ii has 4 adjacent bulky groups.

(Meilahn and Munk, *J. Org. Chem.* **34**, 1440 (1969).)

e.

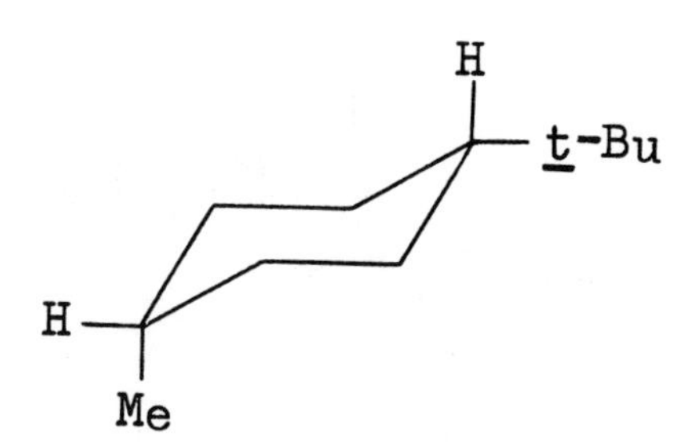

The *t*-butyl group has a much greater steric requirement than methyl, and so adopts the equatorial position.

f.

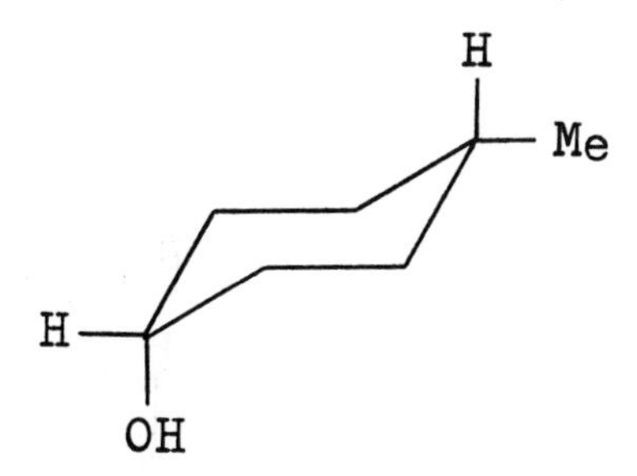

The methyl group is equatorial because it has a greater steric requirement than OH.

g.

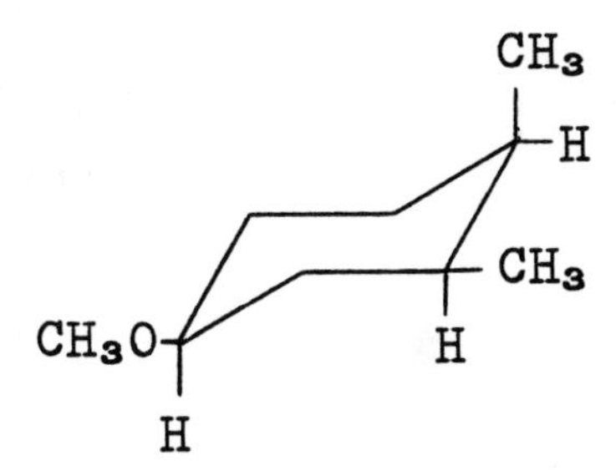

Since one methyl group must be axial and one equatorial, the molecule will adopt the conformation which allows the methoxy group to be equatorial.

h.

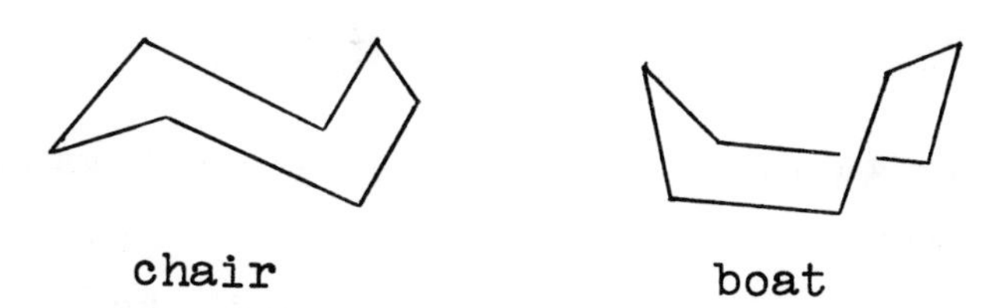

The chair is more stable.

11. Cont.

i.

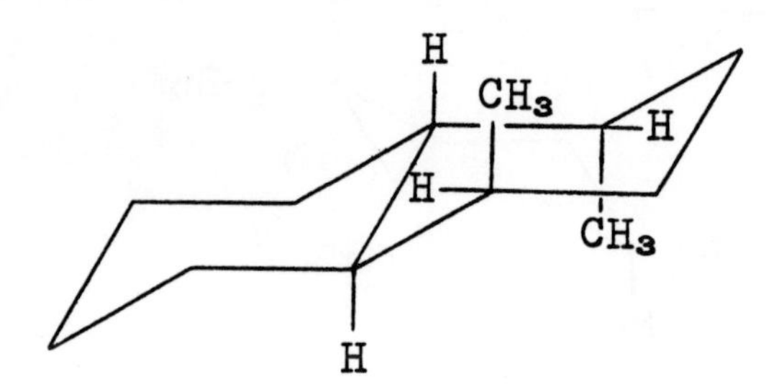

The methyl groups are both axial; in this rigid ring system they cannot be equatorial without disturbing the chair form of the ring.

12.

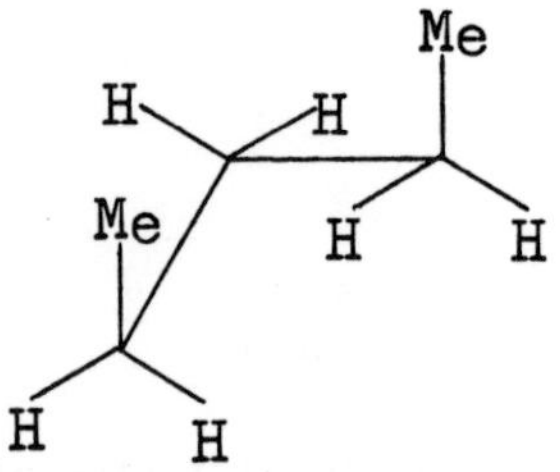

trans-trans
most stable

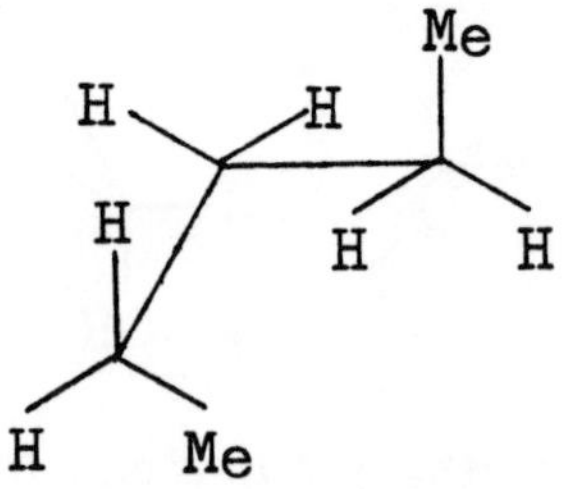

trans-gauche
and its
mirror image

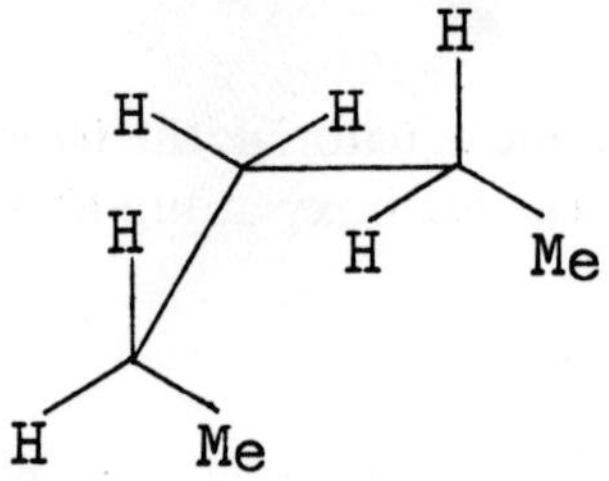

gauche-gauche
syn

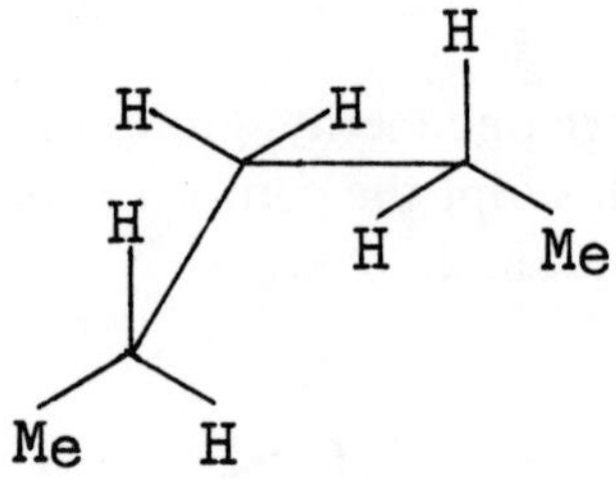

gauche-gauche
anti
and its
mirror image

13. There are ten isomers, in descending order of stability:

trans-anti-trans	*dl* pair	
cis-syn-trans	*dl* pair	about equal stability
cis-anti-trans	*dl* pair	
cis-anti-cis	*dl* pair	
cis-syn-cis	meso	
trans-syn-trans	meso	

For discussions, see Johnson, *J. Amer. Chem. Soc.* **75**, 1498 (1953), and Allinger, Gorden, Tyminski, and Wuesthoff, *J. Org. Chem.* **36**, 739 (1971).

14. Both transanullar and Pitzer (eclipsing) strain are found. The former results from interaction between H-1 and H-4, and the latter between H-2 and H-3 and H-5 and H-6. Pitzer strain is more important in this case.

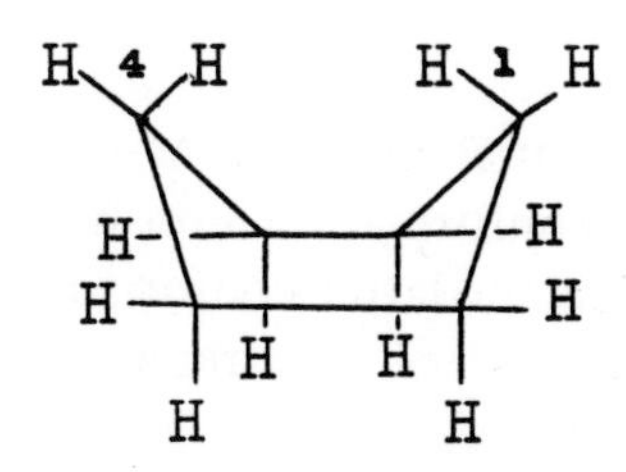

20. In i, the natural compound has the methyl groups in ring E both equatorial, and a change to a trans ring junction would make them both axial, so it does not happen. In ii the methyl groups must be a-e in any case, and a change to the more stable ring junction cannot affect this.
(Corey and Ursprung, *J. Amer. Chem. Soc.* **78**, 183 (1956).)

22. For i, in the diaxial conformation, both substituents are in octants which make negative contributions to the Cotton effect; while in the diequatorial conformation, both make positive contributions. Therefore the fact of a weak positive effect establishes the diequatorial conformation, which of course is the more stable.

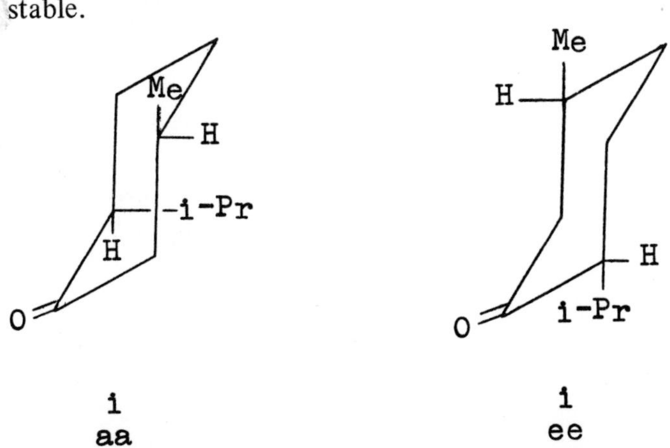

22. Cont.

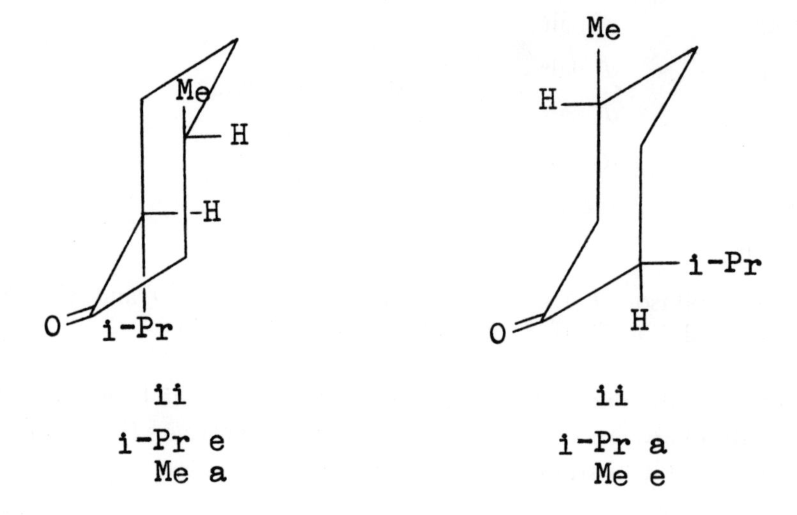

For ii, when the i-Pr is equatorial and Me is axial, a negative Cotton effect is predicted, but in the other conformation a positive Cotton effect is expected. Therefore the compound has the latter conformation, which is perhaps surprising, since i-Pr has a greater steric requirement. However, it is possible that twist forms are participating also.

23. The methyl group is above the plane of the ring: the ring junction is trans. If the methyl group were below the plane the Cotton effect would be negative (models will help see this).

Djerassi, Riniker, and Riniker, *J. Amer. Chem. Soc.* **79**, 1506 (1957).

ANSWERS TO PROBLEM SET 3

ACIDITY, MECHANISMS, AND REACTIVITY

5. Use an optically active migrating group, as in Ph-CH(Me)-CON_3. Racemization would mean that the group completely breaks away. Retention of configuration would mean it does not (the latter is the actual result).

6. Use a crossover experiment. For example, the reaction could be run in the presence of phenol. If intermolecular, some *p*-cresol would be found, otherwise not.

7. A double labeling experiment is necessary. Label one ring (for example by putting in a methyl group) and label one of the C≡C carbons (with ^{14}C).

8. Compare the rate of a ^{15}N labeled compound with that of an unlabeled compound. If path b then there should be an isotope effect, since the N–C bond is broken in the rate-determining step.

(See Ayrey, Bourns, and Vyas, *Can. J. Chem.* **41**, 1759 (1963).)

9. a. Replace one H by D. Then use optically active PhCHDCl. Look for inversion (S_N2) or racemization (S_N1).

   b. The S_N2 mechanism is bimolecular: the rate is dependent on the concentrations of halide and of OH^-. The S_N1 is unimolecular, with the rate dependent only on the concentration of halide. Therefore, a measurement of the dependence or independence of the rate on OH^- concentration will help decide.

   c. In the S_N1 mechanism a positive charge is developed in the transition state. This means that electron-withdrawing groups on the ring will slow the reaction. Therefore, measure reaction rates with meta and para substituents on the ring and calculate ρ. A negative ρ would indicate S_N1; a positive ρ would be evidence for S_N2.

10.
$$\frac{-d[NO]}{dt} = k_1[NO]^2 - k_2[N_2O_2]$$

$$\frac{d[N_2O_2]}{dt} = k_1[NO]^2 - k_2[N_2O_2] - k_3[N_2O_2][O_2]$$

Setting $\frac{d[N_2O_2]}{dt}$ equal to zero:

10. Cont.

$$k_1[\mathrm{NO}]^2 = k_2[\mathrm{N_2O_2}] + k_3[\mathrm{N_2O_2}][\mathrm{O_2}]$$

$$[\mathrm{N_2O_2}] = \frac{k_1[\mathrm{NO}]^2}{k_2 + k_3[\mathrm{O_2}]}$$

Substituting into the first equation:

$$\frac{-d[\mathrm{NO}]}{dt} = k_1[\mathrm{NO}]^2 - \frac{k_1k_2[\mathrm{NO}]^2}{k_2 + k_3[\mathrm{O_2}]}$$

$$= \frac{k_1k_3[\mathrm{NO}]^2[\mathrm{O_2}]}{k_2 + k_3[\mathrm{O_2}]}$$

If $[\mathrm{N_2O_2}]k_2 \gg [\mathrm{N_2O_2}][\mathrm{O_2}]k_3$

then $k_2 \gg [\mathrm{O_2}]k_3$; and

$$\frac{-d[\mathrm{NO}]}{dt} = \frac{k_1k_3[\mathrm{NO}]^2[\mathrm{O_2}]}{k_2}$$

If the contrary assumption, then $[\mathrm{O_2}]k_3 \gg k_2$; and

$$\frac{-d[\mathrm{NO}]}{dt} = k_1[\mathrm{NO}]^2$$

11. a.

$$\text{rate} = k_1[\text{ket}][\mathrm{B}] - k_2[\text{ion}][\mathrm{BH^+}]$$

applying the steady state assumption:

$$\frac{d[\text{ion}]}{dt} = k_1[\text{ket}][\mathrm{B}] - k_2[\text{ion}][\mathrm{BH^+}] - k_3[\text{ion}][\mathrm{Br_2}] = 0$$

$$[\text{ion}] = \frac{k_1[\text{ket}][\mathrm{B}]}{k_2[\mathrm{BH^+}] + k_3[\mathrm{Br_2}]}$$

substituting into the first equation:

11. Cont.

$$\text{rate} = k_1[\text{ket}][\text{B}] - \frac{k_2 k_1 [\text{ket}][\text{B}][\text{BH}^+]}{k_2[\text{BH}^+] + k_3[\text{Br}_2]}$$

simplifying,

$$\text{rate} = \frac{k_1 k_3 [\text{B}][\text{ket}][\text{Br}_2]}{k_2[\text{BH}^+] + k_3[\text{Br}_2]}$$

b. If the first step is very much slower than the second, then:

$$k_3[\text{ion}][\text{Br}_2] >> k_1[\text{ket}][\text{B}]$$

and, since

$$k_1[\text{ket}][\text{B}] = k_2[\text{ion}][\text{BH}^+]$$

then

$$k_3[\text{ion}][\text{Br}_2] >> k_2[\text{ion}][\text{BH}^+]$$

and

$$k_3[\text{Br}_2] >> \text{k}_2[\text{BH}^+]$$

and the rate equation simplifies to:

$$\text{rate} = k_1[\text{B}][\text{ket}]$$

In this expression, [B] represents the sum of all bases present. Since the first step is slow, all bases will affect the rate.

c. If step 2 is much slower than step 1, then a similar analysis shows

$$k_2[\text{BH}^+] >> k_3[\text{Br}_2]$$

11. Cont.

and the rate equation simplifies to

$$\text{rate} = \frac{k_1[\text{B}]k_3[\text{ket}][\text{Br}_2]}{k_2[\text{BH}^+]}$$

Note that this equation includes $[\text{BH}^+]$ as well as [B]. Now the equilibrium constant for step 1 is

$$\text{K} = \frac{k_1}{k_2} = \frac{[\text{BH}^+][\text{ion}]}{[\text{B}][\text{ket}]}$$

so that

$$[\text{ion}] = \frac{k_1[\text{B}][\text{ket}]}{k_2[\text{BH}^+]}$$

ignoring the slow loss of ion by step 2.

K_a (acidity constant) for the ketone is

$$\text{K}_\text{a} = \frac{[\text{ion}][\text{H}^+]}{[\text{ket}]}$$

Combining the last two equations

$$\text{K}_\text{a} = \frac{k_1[\text{B}][\text{H}^+]}{k_2[\text{BH}^+]}$$

and

$$\text{rate} = \frac{\text{K}_\text{a}k_3[\text{ket}][\text{Br}_2]}{[\text{H}^+]}$$

11. Cont.

since

$$K_w = [H^+][OH^-]$$

$$\text{rate} = \frac{K_a k_3 [OH^-][\text{ket}][Br_2]}{K_w}$$

Thus, OH^- is in the rate expression, and other bases only affect the reaction to the extent that they are converted to OH^- in the solvent water.

12. a. 3 1 4 2
    b. 2 1 3 4
    c. 1 2 3 4
    d. 2 4 1 3
    e. 16 8 13 1 2 7 15 10 17 18 4 12 5 9 19 3 14 6 11

13. a. 4 3 1 2 5
    b. 5 2 4 3 1
    c. 2 3 1 4

14. a. 2 is stronger, because of resonance in the conjugate base, not found in the conjugate base of cyclohexanol.
    b. 1 is about 30 times stronger: steric inhibition of resonance (by the C-8 hydrogen) of the COOH group in 1.
    c. 1 is stronger, for a similar reason.
    d. 1 is much stronger, because the conjugate COO^- group can hydrogen-bond with the OH group.

15. a. 1 is stronger because the larger groups cause some steric inhibition of resonance between the nitrogen and the ring (it is this resonance which causes aniline to be a weaker base than ammonia).
    b. 2 is a weaker base because of steric hindrance to solvation.
    c. Pyrrole (2) is a much weaker base because it needs its unshared pair of electrons to form part of the aromatic sextet and thus does not have them available for bonding with H^+. If it did bond with H^+ then loss of this pair would destroy the aromaticity. Pyridine (1) on the other hand has an aromatic sextet without the unshared pair.

15. Cont.

d. Guanidine (2) is a much stronger base, because the canonical forms of the conjugate base are equivalent, resulting in a much lower energy than for the conjugate base of urea (1).

16. a and c cannot be Lewis acids. The others, in descending order of Lewis acid strength are: d e b f.

19.

$$\log k = \sigma\rho + \log k_0$$

$$k_0 = 1.60 \times 10^{-4}$$

$$\log k_0 = -3.796$$

$$\log k = (0.232)(3.78) - 3.796$$

$$k = 1.21 \times 10^{-3}$$

DeWolfe, *J. Amer. Chem. Soc.* 82, 1585 (1960).

20. $\rho = -5.78$. See the graphs on p. 4984 of Brown and Okamoto, *J. Amer. Chem. Soc.* 80, 4979 (1958).

ANSWERS TO PROBLEM SET 4
ALIPHATIC NUCLEOPHILIC SUBSTITUTION

6.

$$\underset{\mathbf{i}}{RCOOH} \xrightarrow{CH_2N_2} RCOOCH_3 \xrightarrow{\overset{\ominus}{C}H_2SOCH_3} R\text{-}\underset{\underset{O}{\|}}{C}\text{-}CH_2SOCH_3$$

$$\xrightarrow{Al\text{-}Hg} R\text{-}\underset{\underset{O}{\|}}{C}\text{-}CH_3 \xrightarrow[\text{of ketal}]{\text{hydrolysis}} \mathbf{ii}$$

Ikezaki, Wakamatsu, and Ban, *Chem. Commun.* **1969**, 88.

11.

$$\mathbf{i} + HC\equiv CCH_2OH \xrightarrow{BF_3} \overset{O}{CH_2\text{—}CHCH_2OCH_2C\equiv CH}$$

$$\xrightarrow[HOCH_2CH_2Cl]{BF_3} \begin{matrix} CH_2OCH_2CH_2Cl \\ | \\ CHOH \\ | \\ CH_2OCH_2C\equiv CH \end{matrix} \xrightarrow{\text{H-N(morpholine)O}} \mathbf{ii}$$

Gautier, Miocque, Fauran, and Douzon, *Bull. Soc. Chim. Fr.* **1967**, 3190.

16.

$$\mathbf{i} + OH^- \longrightarrow \text{(cyclohexene oxide)} + Ph_3P{=}CHCOOEt \longrightarrow \mathbf{ii}$$

Osterberg, *Org. Syn.* **I**, 185; Denney, Vill, and Boskin, *J. Amer. Chem. Soc.* **84**, 3944 (1962).

21.

$$\mathbf{i} \xrightarrow{OH^-} \text{(1-methylcyclohexene oxide, } CH_3\text{)} \xrightarrow{LiAlH_4} \mathbf{ii}$$

Mousseron, Jacquier, Mousseron-Canet, and Zagdoun, *C. R. Acad. Sci.* **235**, 177 (1952).

29.

$$\text{i} \xrightarrow{PCl_5} p\text{-}CH_3C_6H_4SO_2Cl \xrightarrow[NaHCO_3]{Na_2SO_3} p\text{-}CH_3C_6H_4SO_2^- \xrightarrow{Me_2SO_4} \text{ii}$$

Field and Clark, *Org. Syn.* **IV**, 674.

30.

$$\text{i} \xrightarrow{2HBr} Br\text{-}(CH_2)_{10}\text{-}Br \xrightarrow{2NaCH(COOEt)_2}$$

$$(EtOOC)_2CH\text{-}(CH_2)_{10}\text{-}CH(COOEt)_2 \xrightarrow[2.\ \Delta]{1.\ OH^-} HOOC\text{-}(CH_2)_{12}\text{-}COOH$$

$$\xrightarrow[2.\ ThCl_2 \quad 3.\ \Delta]{1.\ NaOH} \text{ii}$$

Chuit, *Helv. Chim. Acta* **9**, 264 (1926); Ruzicka, Stoll, and Schinz, *Helv. Chim. Acta* **9**, 249 (1926).

33.

$$\text{i} + (EtCHClCO)_2O \longrightarrow$$

[acylated tetracyclic indole ammonium ion: $\oplus$N-C(=O)-CH(Cl)-Et, attacked by $\ominus$O-C(=O)-CH(Cl)-Et] — not isolated

$$\longrightarrow$$ [ring-expanded indole: N-C(=O)-CH(Cl)-Et in the ring, O-C(=O)-CH(Cl)-Et ester substituent, N-H] $$\xrightarrow[H_2O]{OH^-} \text{ii}$$

Dadson, Harley-Mason, and Foster, *Chem. Commun.* **1968**, 1233.

38.

i $\xrightarrow[Ag_2O\text{-}DMF]{MeI}$ (Me; CH_2COOMe; $PhCH_2O$; OMe; O) $\xrightarrow[2.\ H^+]{1.\ OH^-}$

(Me; CH_2COOH; $PhCH_2O$; OMe; O) $\xrightarrow[Et_3N]{COCl_2}$ (Me; CH_2COCl; $PhCH_2O$; OMe; O)

$\xrightarrow[\text{MgX}]{EtOOC\text{-}CH\text{-}CONH_2}$ (Me; OH; $C\text{-}CONH_2$; COOEt; $PhCH_2O$; OMe; O)

$\xrightarrow[Me_2SO]{CH_3SOCH_2^-}$ (Me; OH; $CONH_2$; $PhCH_2O$; OMe; O; OH) $\xrightarrow{HCl}$ ii

Gurevich, Kolosov, Korobko, and Popravko, *J. Gen. Chem. USSR* **38**, 56 (1968).

40.

$$\text{i} \xrightarrow{CH_3I} ArCH_2\overset{\oplus}{N}Me_3 \xrightarrow[H_2O]{Na\text{-}Hg} \text{ii}$$

Nesmeyanov, Perevalova, Shilovtseva, and Beĭnoravichute, *Dokl. Akad. Nauk SSSR* **121**, 117 (1958); *CA* **53**, 323g (1959).

43.

$$\text{i} \xrightarrow{\text{MeOH}} \text{HOOCCH}_2\text{CH}_2\text{COOMe} \xrightarrow{\text{SOCl}_2} \text{ClCOCH}_2\text{CH}_2\text{COOMe}$$

$$\xrightarrow{\text{CH}_2\text{N}_2} \text{N}_2\text{CH-C(=O)-CH}_2\text{CH}_2\text{COOMe} \xrightarrow{\text{HCl}} \text{ClCH}_2\text{-C(=O)-CH}_2\text{CH}_2\text{COOMe}$$

$$\xrightarrow{\text{potassium phthalimide-}^{15}\text{N (}^{15}\text{N}^{\ominus}\text{ K}^{+}\text{)}} \text{Phth}^{15}\text{N-CH}_2\text{-C(=O)-CH}_2\text{CH}_2\text{COOMe} \xrightarrow{\text{HCl}} \text{ii}$$

Neuberger and Scott, *J. Chem. Soc.* **1954**, 1820.

46.

$$\text{i} \xrightarrow{\text{Et}_2\text{NH}} \text{HS-CH}_2\text{-C(Me)(Cl)-C(=O)-NEt}_2 \longrightarrow \text{ii}$$

Orlov, Kuleshova, and Knunyants, *Bull. Acad. Sci. USSR, Div. Chem. Sci.* **1967**, 1365.

50.

$$\text{i} \xrightarrow{\text{NH}_2^-} \text{Ph-C(Cl)(Ph)-C(=O)-N}^{\ominus}\text{Ph} \longrightarrow \text{Ph}_2\text{C-C(=O)-NPh (}\alpha\text{-lactam)} \xrightarrow{\text{NH}_2^-}\ \text{a: iii; b: ii; c: iv}$$

α-lactam intermediate can open 3 ways

Paths a and b are each two-step tetrahedral mechanisms. Path c is an S_N2 step. In each pathway, protonation follows as the final step.

Sarel, D'Angeli, Klug, and Taube, *Israel J. Chem.* 2, 167 (1964).

57.

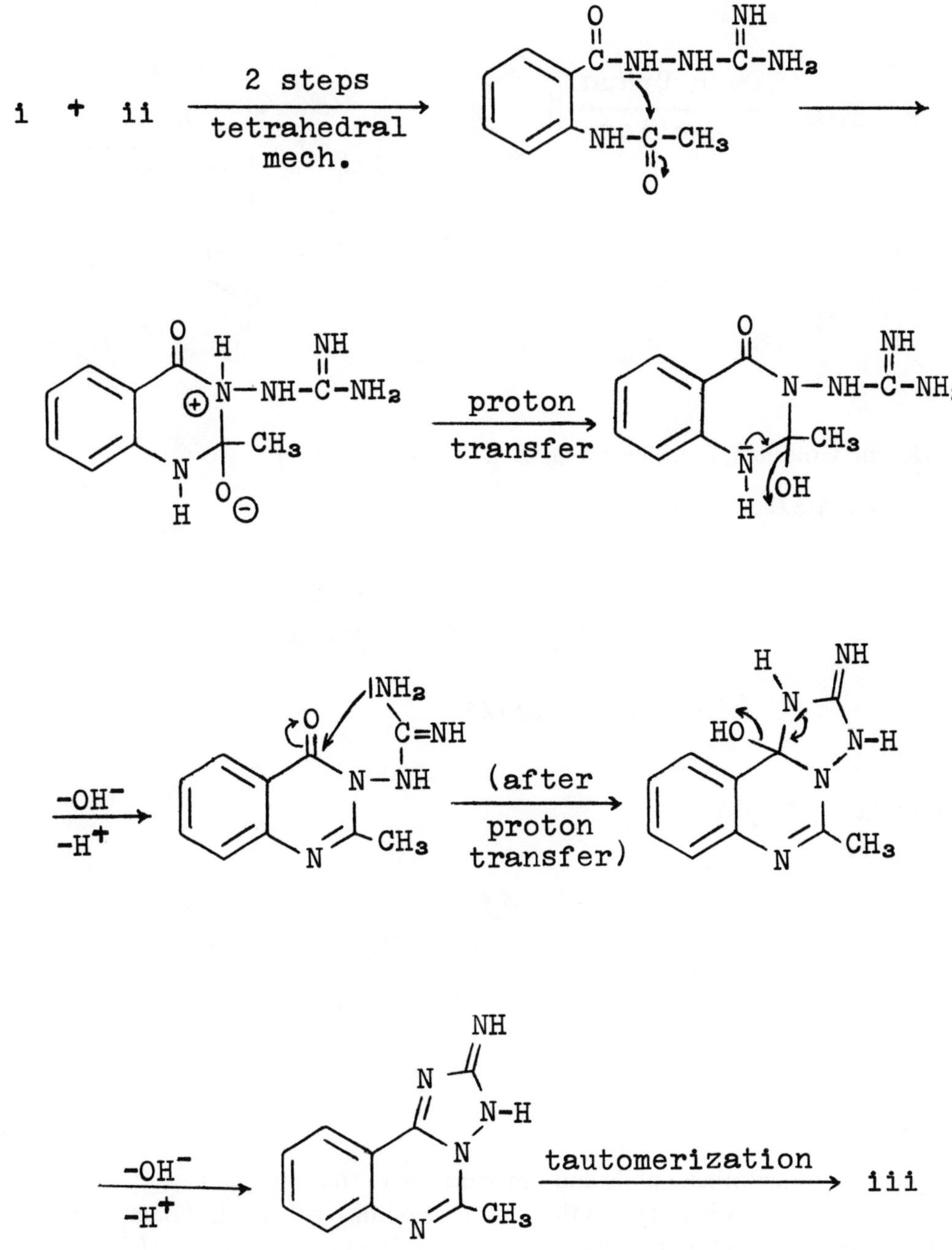

i + ii —2 steps / tetrahedral mech.→ [benzene ring with C(=O)-NH-NH-C(=NH)-NH₂ and NH-C(=O)-CH₃] →

→ proton transfer →

—$-OH^-$ / $-H^+$→ —(after proton transfer)→

—$-OH^-$ / $-H^+$→ —tautomerization→ iii

Ried and Valentin, *Chem. Ber.* **101**, 2106 (1968).

65.

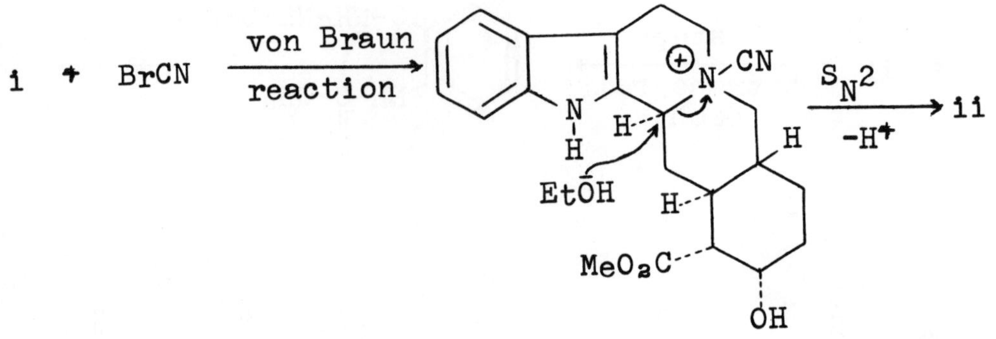

i + BrCN $\xrightarrow[\text{reaction}]{\text{von Braun}}$ [structure] $\xrightarrow[-H^+]{S_N2}$ ii

Albright and Goldman, *J. Amer. Chem. Soc.* **91**, 4317 (1969).

68. In each case the mechanism is

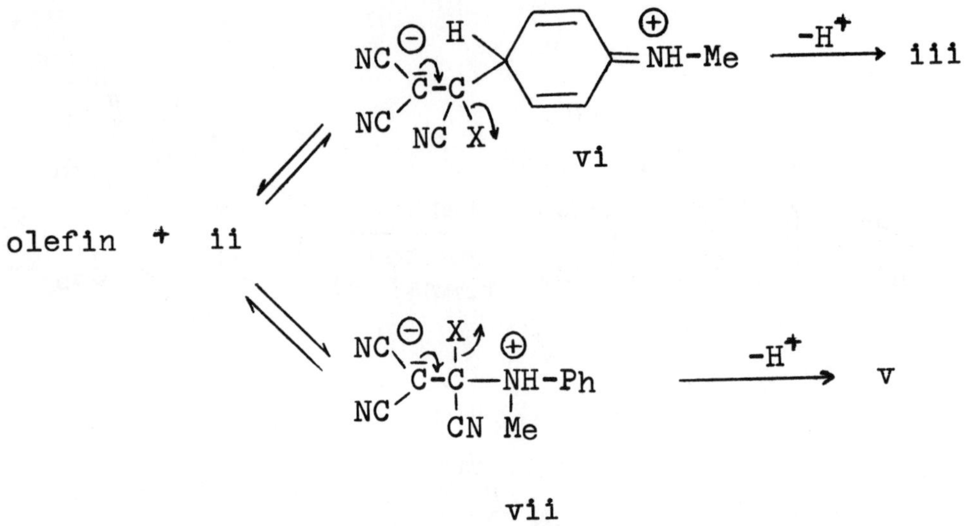

olefin + ii ⇌ vi $\xrightarrow{-H^+}$ iii

olefin + ii ⇌ vii $\xrightarrow{-H^+}$ v

ii is an ambident nucleophile and can attack with the nitrogen or with the ring. Attack by the ring is faster, so that vi is formed faster than vii, though vii is thermodynamically more stable. With a good leaving group such as Cl^-, vi goes to product much faster than it goes back to i and ii. However, CN^- is a much poorer leaving group and in this case there is time for equilibrium to be established between vi and vii. Consequently, iii is a product of kinetic control, and v of thermodynamic control.

Dickinson, Wiley, and McKusick, *J. Amer. Chem. Soc.* **82**, 6132 (1960); Gompper, *Angew. Chem. Intern. Ed.* **3**, 560 (1964); p. 563.

73.

$$R-CH(CH_2Cl)-O-C(=O)-N(\text{aziridine}) \longrightarrow \text{[oxazolidinone aziridinium ion, R, } N^{\oplus}\text{]} + Cl^- \longrightarrow ii$$

Culbertson and Dietz, *Can. J. Chem.* **46**, 3399 (1968).

75.

$$NH_2-CN + OMe^- \longrightarrow \bar{N}H-CN + CH_3-C(=S)-SEt \xrightarrow[\text{mechanism}]{\text{tetrahedral}}$$

$$CH_3-C(=S)-NH-CN \xrightarrow{OMe^-} \underset{i}{CH_3-C(S^-\ Na^+)=N-CN} \xrightarrow{CH_3I} \underset{ii}{CH_3-C(SCH_3)=N-CN}$$

Hartke and Seib, *Tetrahedron Lett.* **1968**, 5523.

77.

$$i \xrightarrow{H^+} Ph_2C(\overset{\oplus}{O}H_2)-C\equiv C-Br \xrightarrow{|\bar{B}r|^-} 82\%\ \underset{ii}{Ph_2C=C=CBr_2}$$

Kollmar and Fischer, *Tetrahedron Lett.* **1968**, 4291.

79.

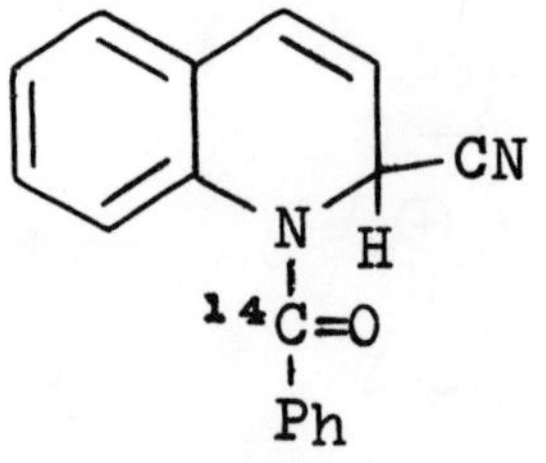

i

Reissert compound

ii

$Ph\text{-}^{14}CHO$

Murray and Williams, "Organic Synthesis with Isotopes," p. 626, Interscience Publishers, Inc., New York, 1958.

83. Two possibilities fit all the data:

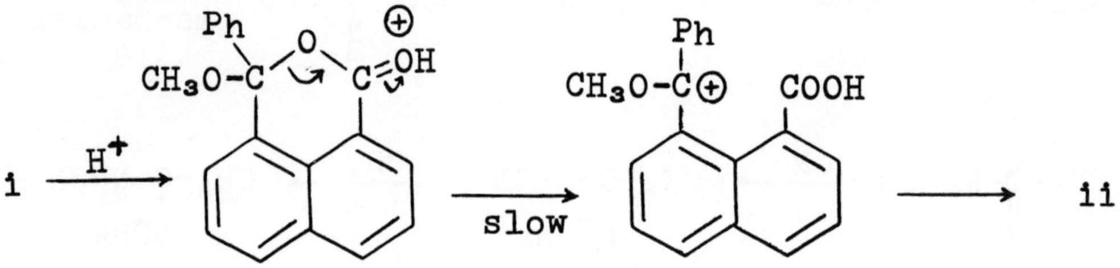

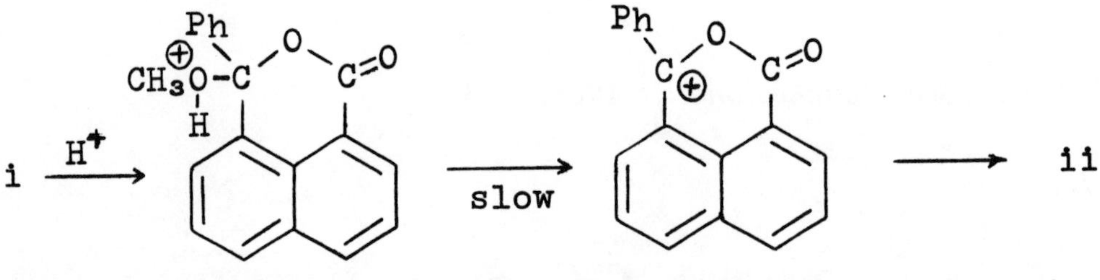

i $\xrightarrow{H^+}$ → slow → ii

i $\xrightarrow{H^+}$ → slow → ii

Fact 1 is unimportant. Facts 2 and 4 indicate that H_2O does not take part in the rate determining step. Fact 3 is consistent with protonation of the substrate before the rate determining step. Fact 5 shows carbonium ion character in the transition state on the carbon to which the phenyl group is attached.

Weeks and Zuorick, *J. Amer. Chem. Soc.* **91**, 477 (1969).

87. Neighboring group participation by methoxy leads to the same intermediate in both cases:

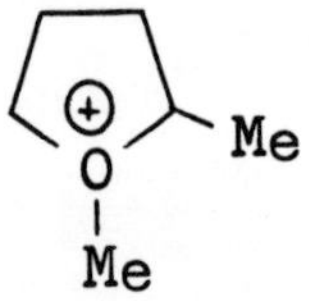

Allred and Winstein, *J. Amer. Chem. Soc.* **89**, 3991 (1967).

90. The rate determining step must involve cleavage of the C-X bond, which means that it cannot be the ordinary tetrahedral mechanism:

$$\mathrm{Ph{-}C(=O){-}X} + \mathrm{YH} \xrightarrow[k_1]{\text{slow}} \mathrm{Ph{-}C(\overset{\oplus}{Y}H)(O^{\ominus}){-}X} \xrightarrow[k_2]{\text{base}} \mathrm{Ph{-}C(Y)(O^{\ominus}){-}X} \xrightarrow[k_3]{} \mathrm{Ph{-}C(=O){-}Y}$$

If this were the mechanism, then the rate would have been greatest for X = F, since the inductive effect of the F would make the positive charge on the CO carbon larger than the other groups and it would attract the nucleophile most readily. Therefore the mechanism is that given above, except that the last step (k_3) is rate determining (which is favored by the authors of the paper), or there is an S_N1 mechanism.

Semenyuk, Oleinik, and Litvinenko, *Org. Reactiv. (USSR)* **4**, 308 (1967).

92. The amines are better solvated in the more polar solvent (especially by hydrogen bonding) and are thus less free to attack the substrate.

Okamoto, Fukui, Nitta, and Shingu, *Bull. Chem. Soc. Jap.* **40**, 2354 (1967).

97. Steric inhibition of resonance. The unfilled orbital of the cation must be perpendicular to the plane of the ring in order to overlap, but two ortho methyl groups hinder this.

Charlton and Hughes, *J. Chem. Soc.* **1956**, 850.

98. a.

$$\mathbf{i} \longrightarrow \mathrm{R'O{-}\overset{\oplus}{C}(R){-}OR'} + \mathrm{X^-} \xrightarrow[S_N2]{} \mathrm{R'O{-}C(=O){-}R} + \mathrm{R'X}$$

relatively stable cation

b. X should be F, the poorest leaving group among the halogens; R′ should be a group which undergoes S_N reactions poorly: aryl, vinyl, neopentyl, etc.

Scheeren, *Tetrahedron Lett.* **1968**, 5613.

ANSWERS TO PROBLEM SET 5

AROMATIC ELECTROPHILIC SUBSTITUTION

4.

$$\text{i} \xrightarrow[\text{arsenic acid, } H_2SO_4]{\text{glycerol}} \text{ii}$$

Skraup synthesis

Buu-Hoï, Dufour, and Jacquignon, *J. Chem. Soc., C* **1968**, 2070.

9.

PhOH $\xrightarrow[\text{base}]{CO_2}$ (o-hydroxybenzoic acid: OH, COOH) +

Kolbe-Schmitt

$^+N_2$–C$_6$H$_4$–S–C$_6$H$_4$–N_2^+ ⟶ conjugate acid of i

Lindsey and Jeskey, *Chem. Rev.* **57**, 583 (1957); Woodward, in Lubs, "The Chemistry of Synthetic Dyes and Pigments," p. 155, Reinhold Publishing Co., New York, 1955.

10.

i $\xrightarrow[\text{base}]{CO_2}$ (OH, OH, COOH) $\xrightarrow{Br_2}$ (Br, OH, OH, COOH)

Kolbe-Schmitt

$\xrightarrow{\text{decarboxylation}}$ ii

Sandin and McKee, *Org. Syn.* **II**, 100; Nierenstein and Clibbens, *Org. Syn.* **II**, 557.

13.

i $\xrightarrow[TiCl_4]{Cl_2CHOMe}$ (2,5-dimethylbenzaldehyde: Me, CHO, Me) $\xrightarrow[H_2SO_4]{HNO_3}$ (Me, CHO, NO_2, Me)

$\xrightarrow[Pd\text{-}C]{H_2}$ (Me, CH_2OH, NH_2, Me) $\xrightarrow[2.\ H_2\text{-}Pd\text{-}C]{1.\ HCl}$ ii

i $\xrightarrow[H_2SO_4]{HNO_3}$ (Me, NO_2, Me) $\xrightarrow[ClSO_3H]{(ClCH_2)_2O}$ ($ClCH_2$, Me, NO_2, Me) $\xrightarrow[Pd\text{-}C]{H_2}$ iii

Sato, Fujima, and Yamada, *Bull. Chem. Soc. Jap.* **41**, 442 (1968).

16.

i $\xrightarrow[AlCl_3]{\underline{n}\text{-}C_5H_{11}COCl}$ (Fe, $C(=O)\text{-}C_5H_{11}$) $\xrightarrow[HCl]{Zn\text{-}Hg}$ ii

Clemmensen

DeYoung, *J. Org. Chem.* **26**, 1312 (1961).

27.

i + Br–Br $\xrightarrow{-H^+}$ [Br, CH₃, O; NaSO₃, Br cyclohexadienone] $\xrightarrow[-SO_3,\ -Na^+]{H^+}$ ii

$\downarrow H_2O$

[Br, CH₃, O; HO, SO₃Na cyclohexadienone] $\xrightarrow{-NaHSO_3}$ iii

Aubort, *Helv. Chim. Acta* **51**, 2098 (1968).

29.

H–O–O–C(=O)–Ph

MeO, OMe, OMe (i) $\xrightarrow[-H^+]{2\ steps}$ [OH; MeO, OMe, MeO; O–O–C(=O)–Ph, H]

$\longrightarrow$ [⊕OH; MeO, OMe; MeO, OH] $\xrightarrow[-MeOH]{-H^+}$ ii

Friess, Soloway, Morse, and Ingersoll, *J. Amer. Chem. Soc.* **74**, 1305 (1952).

32.

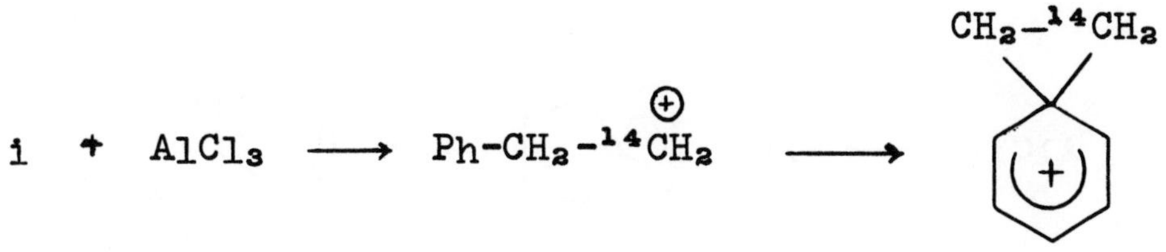

i + $AlCl_3$ ⟶ $Ph{-}CH_2{-}^{14}CH_2^{\oplus}$ ⟶

$CH_2{-}^{14}CH_2$

iii

iii has an approximately equal chance of being attacked by anisole at each of the 2 CH_2 groups.

Lee, Forman, and Rosenthal, *Can. J. Chem.* **35**, 220 (1957).

34.

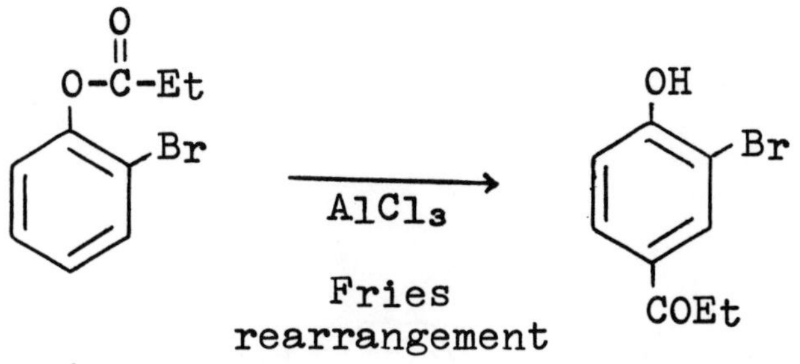

O-C(=O)-Et

Br

i

$AlCl_3$

Fries rearrangement

OH

Br

COEt

ii

Martin and Betoux, *Bull. Soc. Chim. Fr.* **1969**, 1079.

37.

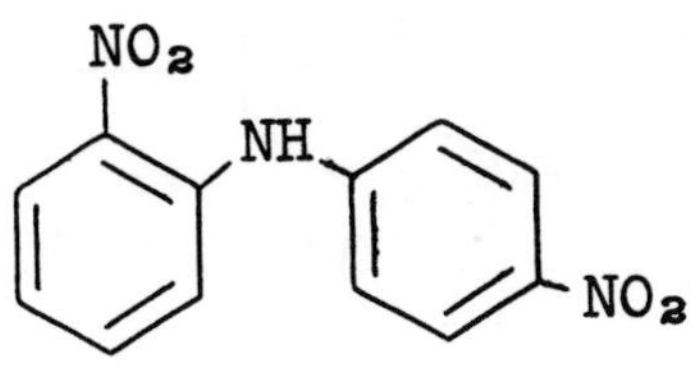

NO_2

NH

NO_2

i

The nitro group migrates to the ring without the deactivating group, and goes ortho, which is usual for this reaction.

Shine, "Aromatic Rearrangements", p. 236, American Elsevier Publishing Company, New York, 1967.

40.

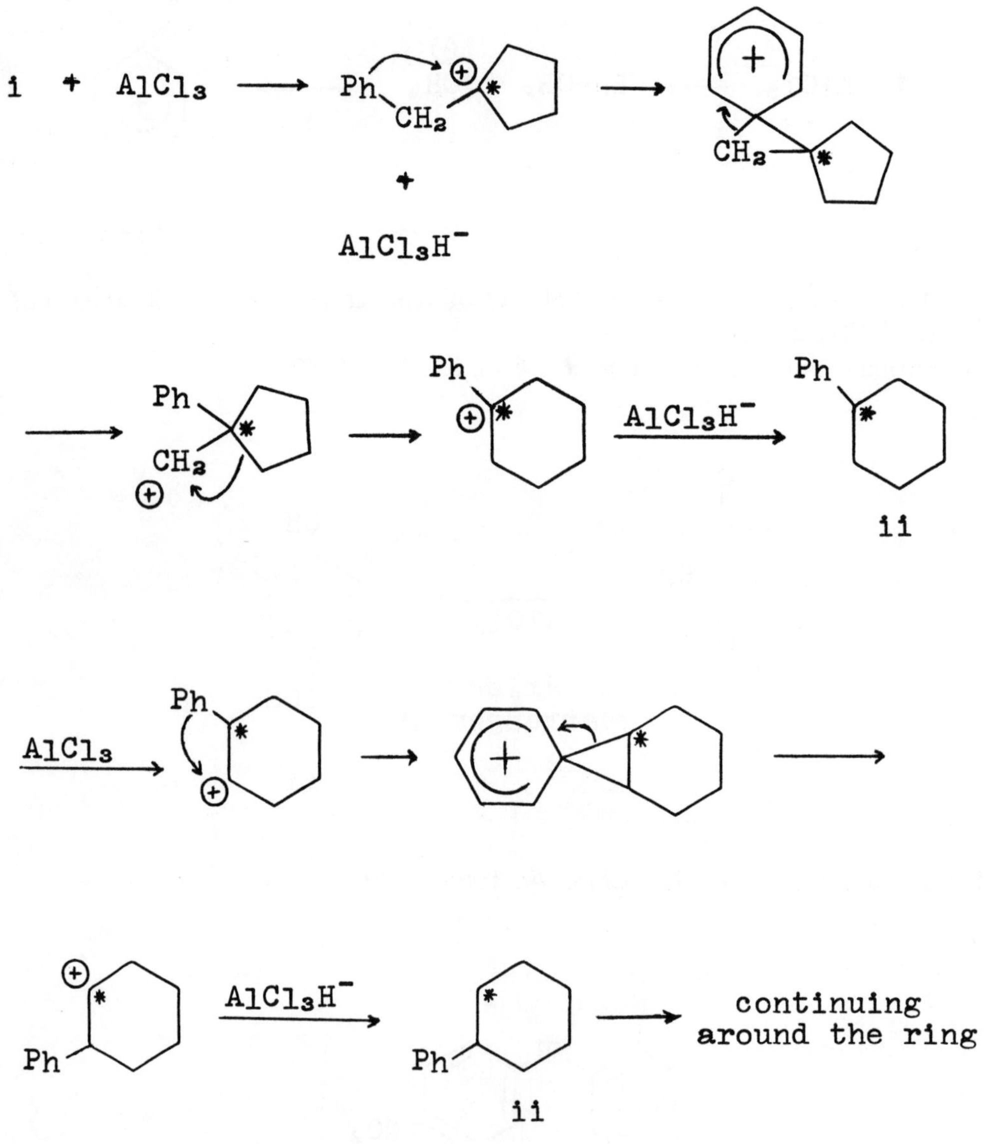

i + $AlCl_3$ ⟶ Ph–CH_2–(+)C* (cyclopentyl) + $AlCl_3H^-$ ⟶ [phenonium ion with CH_2 and cyclopentyl*]

⟶ Ph, CH_2^+ on C* of cyclopentane ⟶ Ph, (+)C* cyclohexyl $\xrightarrow{AlCl_3H^-}$ Ph–C* cyclohexane

ii

$\xrightarrow{AlCl_3}$ Ph–C* cyclohexyl (+) ⟶ [bridged phenonium ion with C* cyclohexane] ⟶

(+) C* cyclohexyl, Ph $\xrightarrow{AlCl_3H^-}$ C* cyclohexane, Ph ⟶ continuing around the ring

ii

Some of the steps may well be concerted.
Roberts and Fonken, in Olah, "Friedel-Crafts and Related Reactions," Vol. 1, pp. 842-844, Interscience Publishers, Inc., New York, 1963.

44. If the rearrangement is intermolecular, then the first step involves formation of NO^+ (either free or connected to a carrier such as HSO_4^-), which, when X = electron-withdrawing groups, is scavenged by the urea. The fact that some C-nitroso compound was still formed in the presence of excess urea indicates that for electron-donating groups, at least some rearrangement is intramolecular. If it were all intermolecular (as in the case of the electron-withdrawing groups) then the urea would have scavenged all of the NO^+ and no C-nitroso product would have been found.

Aslapovskaya, Belyaev, Kumarev, and Porai-Koshits, *Org. Reactiv. (USSR)* **5**, 189 (1968).

45. a.

Me

Me

600 x 5.5 = 3300

2420 x 5.5 = 13,310

2 x 3300 = 6600

2 x 13,310 = 26620

33220

$$\frac{33220}{6} = 5537 = 5.5 \times 10^3$$

Similarly for the others:

b. 2.2×10^3; c. 2.7×10^6; d. 1.6×10^9; e. 4.4×10^9.

Brown and Stock, *J. Amer. Chem. Soc.* **79**, 1421 (1957).

46.

ortho	$28.8 \times 0.569 \times 6/2$	49.2 o_f^{Me}
meta	$28.8 \times 0.028 \times 6/2$	2.4 m_f^{Me}
para	$28.8 \times 0.403 \times 6/1$	69.6 p_f^{Me}

The experimental values are from Olah, Kuhn, Flood, and Evans, *J. Amer. Chem. Soc.* **84**, 3687 (1962).

47. a.

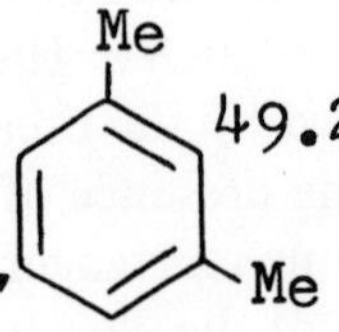

Me (1-position), Me (3-position)

$49.2 \times 49.2 = 2420$

$2.4 \times 2.4 = 5.8$

$49.2 \times 69.6 = 3420$

$$2 \times 3420 = 6840$$
$$1 \times 2420 = 2420$$
$$1 \times 6 = 6$$
$$9266$$

2-position $\frac{2420}{9266} = 26\%$

4-position $\frac{6840}{9266} = 74\%$

5-position $\frac{6}{9266} = 0\%$

b. A similar calculation for *o*-xylene gives: 3-position 41%; 4-position 59%. This example shows that partial rate factors do not always predict the approximate ratios.

The experimental values are from Clark and Fairweather, *Tetrahedron* **25**, 4083 (1969)

52. In descending order: furan, thiophene, pyrrole, benzene.

Clementi, Genel, and Marino, *Chem. Commun.* **1967**, 498; Linda and Marino, *Chem. Commun* **1967**, 499.

54. a. para; b. meta; c. para; d. no reaction (Friedel-Crafts reactions cannot be performed on rings which contain only meta-directing groups); e. ortho, para; f. 4-position; g. 4-position; h. 2-position; i. 2-position (steric effect); j. 6-position (ortho effect); k. 4-position; l. 4'-position; m. 2-position; n. 4-position; o. 1-position; p. 4-position; q. 9-position; r. 1. 1-position, 2. 2-position; s. 3-position; t. 2-position; u. first mole: 2-position, second mole: 6 or 7 position; v. 1-position; w. 1-position; x. 4-position; y. 2-position; z. 3 and 4-positions about equal; aa. 4-position; bb. 2-position; cc. 5-position; dd. 2-position; ee. 5-position; ff. 6-position; gg. 2-position.

References: a. Leiserson and Weissberger, *Org. Syn.* **III**, 183; b,c. Ingold, Shaw, and Wilson, *J. Chem. Soc.* **1928**, 1280; d. Patinkin and Friedman, in Olah, "Friedel-Crafts and Related Reactions," vol. 2, p. 109, Interscience Publishers, Inc., New

York, 1964; e. Yagupol'skii, Fialkov, and Panteleimonov, *J. Gen. Chem. USSR* **36**, 2120 (1966); f, q, v. Olah and Kuhn, in Olah, *op cit,* Vol. 3, pp. 1437, 1226, 1225; g, h, k, l, gg. Gore, in Olah, *op cit,* Vol. 3, pp. 170-171, 174-175, 59, 54-55, 84; i. Heintzelman and Corson, *J. Org. Chem.* **22**, 25 (1957); j. Kobe and Hudson, *Ind. Eng. Chem.* **42**, 356 (1950). m. Lewis, *J. Org. Chem.* **30**, 2798 (1965); n, o. Eaborn, Golborn, Spillett, and Taylor, *J. Chem. Soc., B* **1968**, 1112; p. Richer, Baskevitch, Erichomovitch, and Chubb, *Can. J. Chem.* **46**, 3363 (1968); q. (see f); r. Berg, Jakobsen, and Johansen, *Acta Chem. Scand.* **23**, 567 (1969); s. Streitwieser and Fahey, *J. Org. Chem.* **27**, 2352 (1962); t. Davies and Warren, *J. Chem. Soc., B* **1968**, 1337; u. Klanderman and Perkins, *J. Org. Chem.* **34**, 630 (1969); v. (see f); w. Bublitz and Rinehart, *Org. React.* **17**, 1 (1969), pp. 26-27; x, y. Fournari, Guilard, and Person, *Bull. Soc. Chim. Fr.* **1967**, 4115; z, aa. Östman, *Acta Chem. Scand.* **22**, 2754, 2765 (1968); bb. Clarke, Rawson, and Scrowston, *J. Chem. Soc., C* **1969**, 537; cc, dd. Abramovitch and Saha, *Advan. Heterocycl. Chem.* **6**, 229 (1966), pp. 241, 249; ee. Heindel, Ohnmacht, Molnar, and Kennewell, *J. Chem. Soc., C* **1969**, 1369; ff. Lomakin, *J. Gen. Chem. USSR* **38**, 654 (1968); and gg. (see g).

ANSWERS TO PROBLEM SET 6

ALIPHATIC ELECTROPHILIC SUBSTITUTION

1.

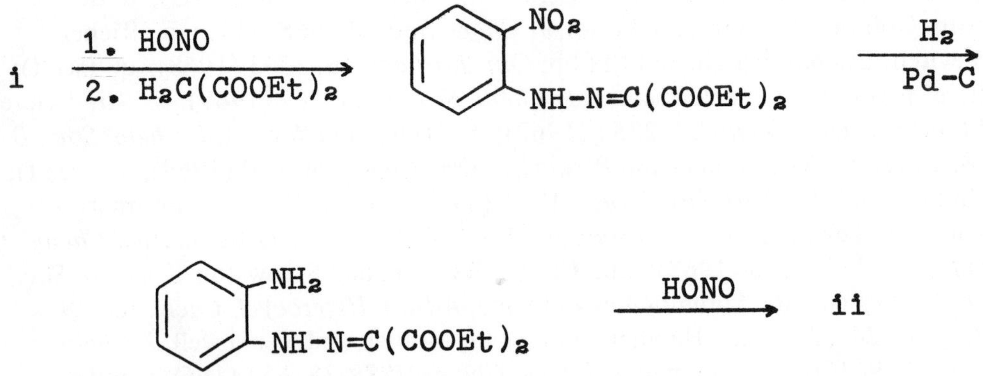

i $\xrightarrow[\text{2. } H_2C(COOEt)_2]{\text{1. HONO}}$ (o-nitrophenyl: NO_2, $NH{-}N{=}C(COOEt)_2$) $\xrightarrow[\text{Pd-C}]{H_2}$

(o-aminophenyl: NH_2, $NH{-}N{=}C(COOEt)_2$) $\xrightarrow{\text{HONO}}$ ii

Campbell and Rees, *J. Chem. Soc., C* **1969**, 742.

3.

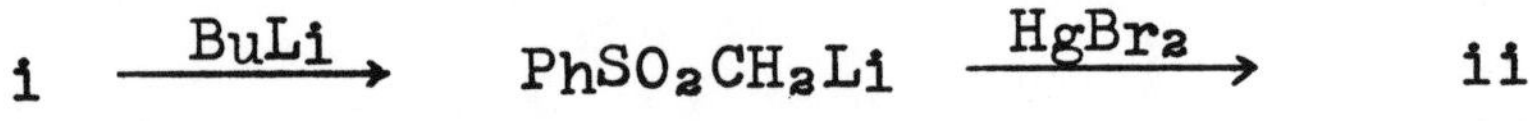

i $\xrightarrow{\text{BuLi}}$ $PhSO_2CH_2Li$ $\xrightarrow{HgBr_2}$ ii

Nesmeyanov, Kravtsov, Faingor, and Petrovskaya, *Bull. Acad. Sci. USSR, Div. Chem. Sci.* **1968**, 521.

4.

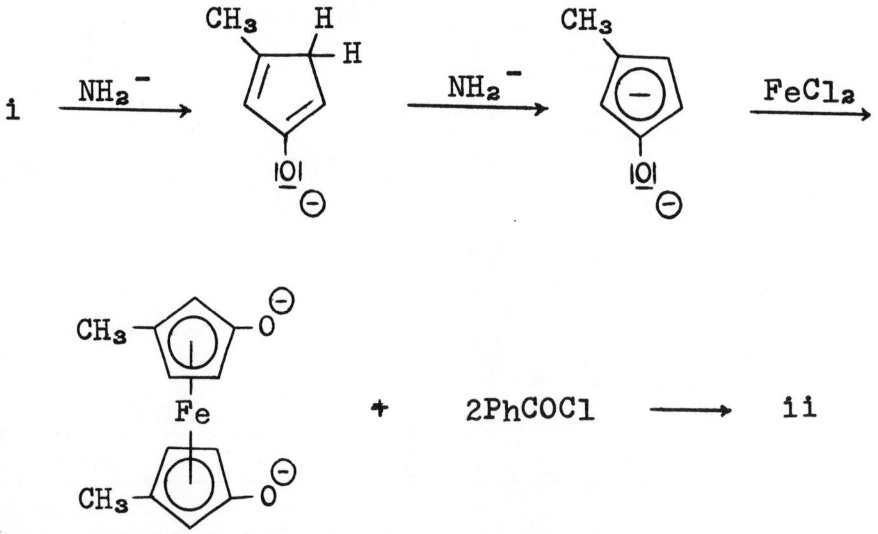

i $\xrightarrow{NH_2^-}$ (CH_3, H, H; $\overline{O}^{\ominus}$) $\xrightarrow{NH_2^-}$ (CH_3, ⊖; $\overline{O}^{\ominus}$) $\xrightarrow{FeCl_2}$

(CH_3, $O^{\ominus}$; Fe; CH_3, $O^{\ominus}$) + 2PhCOCl ⟶ ii

Benson and Lindsey, *J. Amer. Chem. Soc.* **79**, 5471 (1957).

11.

O

i — 1. $NaNH_2$ / 2. MeI → CH_3 CH_3 CH_3 CH_3 — $NaNH_2$ / C_6H_6 → ii

3 times

Haller-Bauer

Haller and Cornubert, *C. R. Acad. Sci.* **158**, 298 (1914).

15.

Li

i + BuLi ⟶ Fe + ClSiEt$_3$ ⟶ ii

Nametkin, Chernysheva, and Babaré, *J. Gen. Chem. USSR* **34**, 2270 (1964).

19.

Me

N-C-CH_2COOEt

O

i + $HOOCCH_2COOEt$ — PCl_5 → + NO_2

Me

N-C-CH_2-COOEt

O

H_2 / catalyst → NH_2 — HONO → Me N O COOEt N=N

Me

N O

tautom. → COOEt N-N H — hydrolysis / decarboxylation → ii

Rossi, Pirola, and Selva, *Tetrahedron* **24**, 6395 (1968).

21.

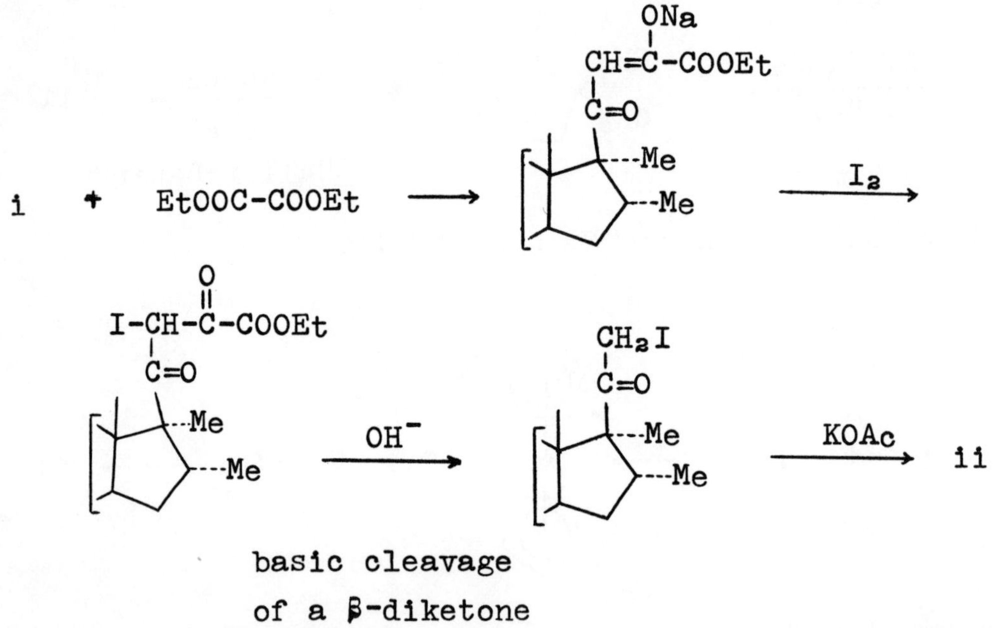

i + EtOOC-COOEt → [ONa, CH=C-COOEt, C=O; Me, Me] → I_2 → [O, I-CH-C-COOEt, C=O; Me, Me] → OH^- → [CH_2I, C=O; Me, Me] → KOAc → ii

basic cleavage
of a β-diketone

Muchnikova, Samsonova, Lur'i, and Maksimov, *J. Gen. Chem. USSR* 38, 450 (1968).

25.

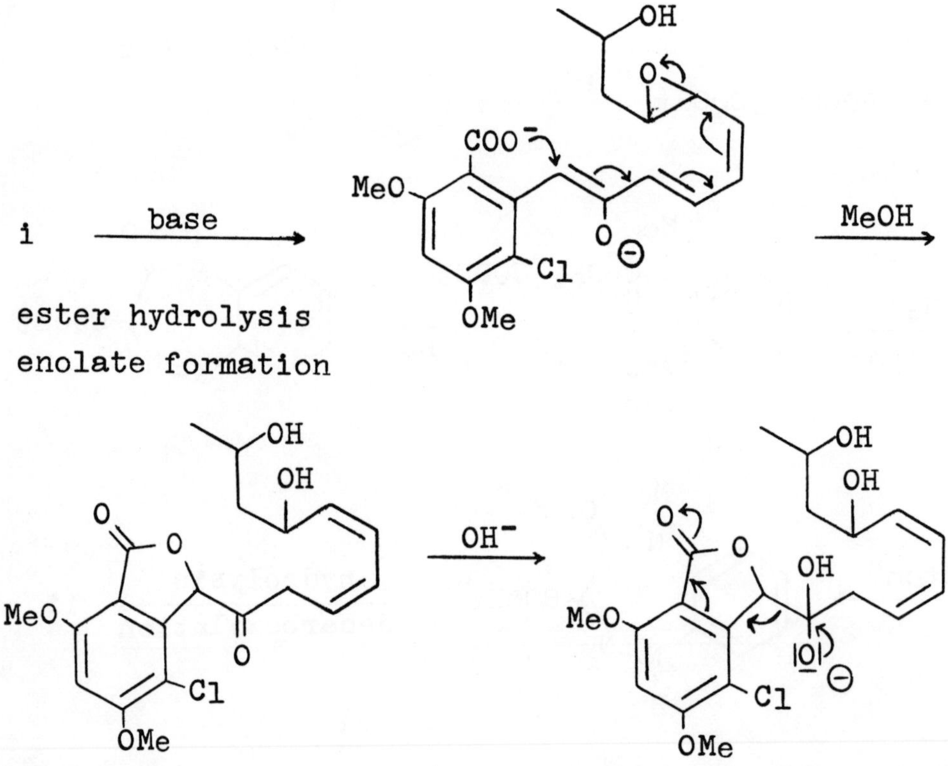

i → base → [OH, O, COO⁻, MeO, O⊖, Cl, OMe] → MeOH →

ester hydrolysis
enolate formation

[OH, OH, O, O, MeO, O, Cl, OMe] → OH^- → [OH, OH, O, O, OH, MeO, O⊖, Cl, OMe]

25. Cont.

→ [structure: anion with MeO, OMe, Cl substituents] $\xrightarrow{H^+}$ ii

Buckley, Ritchie, and Taylor, *Aust. J. Chem.* **22**, 577 (1969); Mirrington, Ritchie, Shoppee, Sternhell, and Taylor, *Aust. J. Chem.* **19**, 1265 (1966).

28.

Cl–CH₂ (furan) + CN⁻ → CH₂= (dihydrofuran, H, CN) $\xrightarrow{\text{tautom.}}$ ii

i

Runde, Scott, and Johnson, *J. Amer. Chem. Soc.* **52**, 1284 (1930).

33.

i + ii → [iminium, Br⁻; CH₂–C(=CH₂)–COOEt; EtOOC COOEt] $\xrightarrow{-HBr}$ [enamine; CH₂–C(=CH₂)–C(=O)–OEt; EtOOC COOEt]

→ [C-OEt enolate; COOEt; COOEt; iminium] $\xrightarrow{EtOH}$ [COOEt (OEt) C=O; COOEt; O=] → iii

Stetter and Thomas, *Angew. Chem. Intern. Ed. Engl.* **6**, 554 (1967) [*Angew. Chem.* **79**, 529 (1967)]

37.

i + ii $\xrightarrow{\text{Haller-Bauer}}$ $Cl_3C-C(NH_2Ph^{\oplus})(O^{\ominus})-CCl_3$ $\xrightarrow{-H^+}$ iii

Sukornick, *Org. Syn.* **40**, 103.

41.

ii + NO^+ ⟶ [Ph-CH=C(Me)-C(Me)=$N^{\oplus}$(OH)-N=O] $\xrightarrow{-H^+}$ [Ph-CH=C(Me)-C(Me)=$N^{\oplus}$(O$^{\ominus}$)-N=O]

⟶ [ring closure intermediate: Ph, H, Me, Me, $N^{\oplus}$-O$^{\ominus}$, $N^{\oplus}$-O$^{\ominus}$] ⟶ [Ph, Me, Me, N-O$^{\ominus}$, $N^{\oplus}$-OH] $\xrightarrow[\text{transfer}]{\text{proton}}$ iii

iii is formed from i by the same mechanism.

Freeman, Gannon, and Surbey, *J. Org. Chem.* **34**, 187 (1969); Freeman and Gannon, *J. Org. Chem.* **34**, 194 (1969).

43.

Br^- + [α-bromo ketone, H^+] ⟶ [enol, HO] ⇌

[enol, HO] + Br–Br ⟶ [cation, HO, Br, ⊕] $\xrightarrow{-H^+}$ ii

Jones and Wluka, *J. Chem. Soc.* **1959**, 907.

48. iv gives mostly normal decarboxylation:

iv $\xrightarrow{\text{base}}$ [EtOOC, $C(=O)-\bar{O}|^{\ominus}$] $\xrightarrow{\Delta}$ [EtOOC, $\ominus$] $\xrightarrow{H^+}$ iii and v

However, the ion from i abstracts a proton from the ring:

[$\ominus$ O, C=O, H, COOEt] $\longrightarrow$ [$|\bar{O}|^{\ominus}$, C=C–OH, COOEt] $\xrightarrow{\text{tautom.}}$ [$|\bar{O}|^{\ominus}$, CH–C=O, COOEt]

$\xrightarrow{H^+}$ ii

Musso, *Chem. Ber.* **101**, 3710 (1968).

50. It is known (see p. 444 of *Advanced Organic Chemistry*) that the reaction (*sec*-Bu)$_2$Hg + HgBr$_2$ → 2 *sec*-BuHgBr proceeds with 100% retention. Therefore, if we treat ii with HgBr$_2$ we will get i with complete retention. We next measure the rotation of the i thus obtained. If it is $-5.05°$ (within experimental error) then complete retention took place in the Mg reaction. If it is $+5.05°$, then inversion took place there. If the result is between these values, then some racemization took place. (The actual value was found to be $-4.60°$, so it was mostly retention, but with some racemization.)

Jensen and Rickborn, "Electrophilic Substitution of Organomercurials", p. 116, McGraw-Hill Book Co., New York, 1968.

ANSWERS TO PROBLEM SET 7
AROMATIC NUCLEOPHILIC SUBSTITUTION

3.

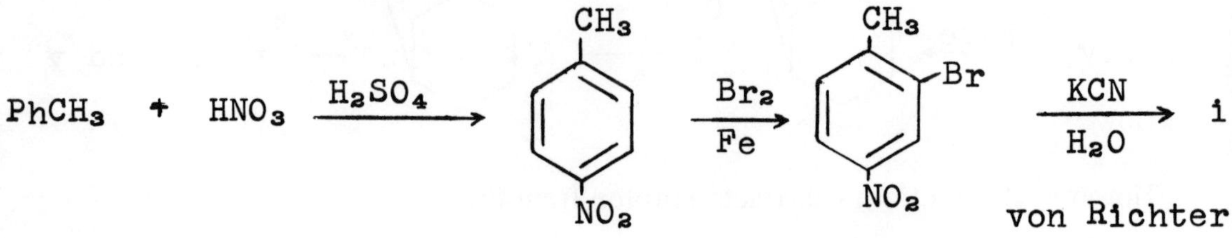

$PhCH_3$ + HNO_3 $\xrightarrow{H_2SO_4}$ [ring: CH_3 (top), NO_2 (bottom)] $\xrightarrow[Fe]{Br_2}$ [ring: CH_3 (top), Br, NO_2 (bottom)] $\xrightarrow[H_2O]{KCN}$ i

von Richter

7.

$PhNH_2$ + glycerol + $PhNO_2$ $\xrightarrow[\text{Skraup}]{H_2SO_4}$ [quinoline ring, N] $\xrightarrow[\text{Chichibabin}]{Ba(NH_2)_2}$ i

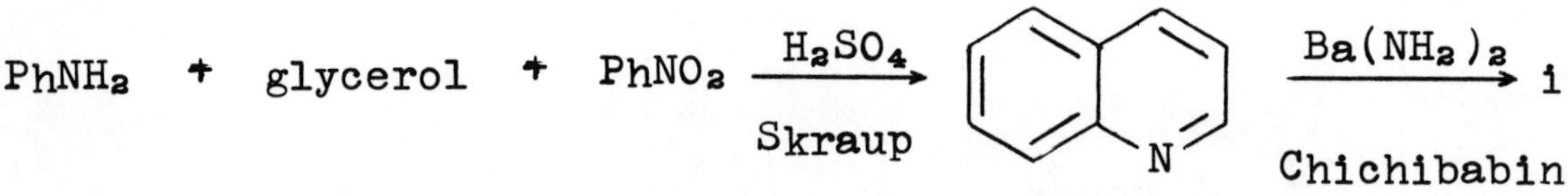

Clarke and Davis, *Org. Syn.* **I**, 478; Bergstrom, *J. Org. Chem.* 2, 411 (1937).

10.

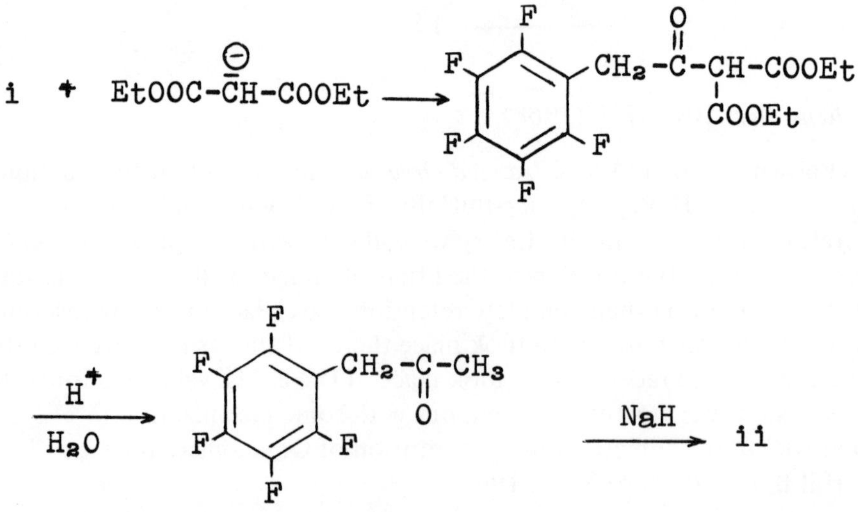

i + $EtOOC\text{-}\overset{\ominus}{C}H\text{-}COOEt$ $\longrightarrow$ [ring: F, F, F, F, F] $CH_2\text{-}\overset{O}{\overset{\|}{C}}\text{-}\underset{COOEt}{\underset{|}{C}H}\text{-}COOEt$

$\xrightarrow[H_2O]{H^+}$ [ring: F, F, F, F, F] $CH_2\text{-}\underset{O}{\underset{\|}{C}}\text{-}CH_3$ $\xrightarrow{NaH}$ ii

Brooke, *Tetrahedron Lett.* **1968**, 2029.

14.

$PhCH_3 \xrightarrow[H_2SO_4]{HNO_3}$ (p-nitrotoluene: CH_3, NO_2) $\xrightarrow[H^+]{Zn}$ (p-toluidine: CH_3, NH_2) $\xrightarrow{HOAc}$

(CH_3, NHAc) $\xrightarrow{Br_2}$ (CH_3, Br, NHAc) $\xrightarrow[2.\ OH^-]{1.\ H^+}$ (CH_3, Br, NH_2) $\xrightarrow[2.\ EtOH]{1.\ HONO}$ i

Johnson and Sandborn, *Org. Syn.* **I**, 111; Bigelow, Johnson, and Sandborn, *Org. Syn.* **I**, 133.

21.

i $\xrightarrow{base}$ (fluorenyl anion, ⊖; ⊕S–Me, Me) ⇌ (⊕S–$\bar{C}H_2$⊖, Me)

(CH₂–SMe, H) $\xrightarrow{tautom.}$ ii

Hilbert and Pinck, *J. Amer. Chem. Soc.* **60**, 494 (1938).

24. There are two successive Smiles rearrangements:

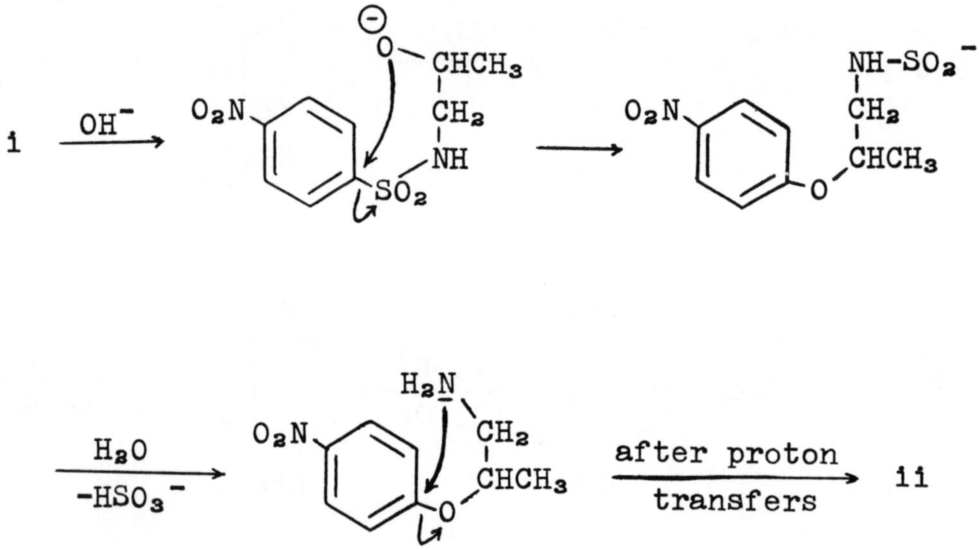

i $\xrightarrow{OH^-}$ O_2N–C_6H_4–SO_2–NH–CH_2–$CHCH_3$–O^- ⟶ O_2N–C_6H_4–O–$CHCH_3$–CH_2–NH–SO_2^-

$\xrightarrow[-HSO_3^-]{H_2O}$ O_2N–C_6H_4–O–$CHCH_3$–CH_2–H_2N $\xrightarrow{\text{after proton transfers}}$ ii

Kleb, *Angew. Chem. Intern. Ed. Engl.* 7, 291 (1968) [*Angew. Chem.* **80**, 284 (1968)]

26.

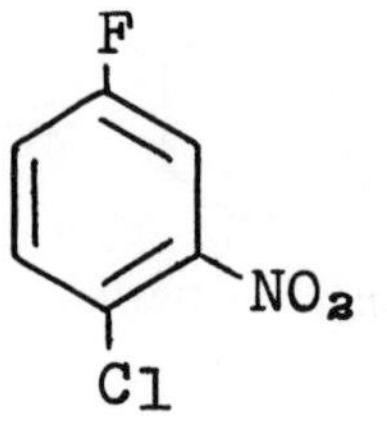

F

NO_2

Cl

i

Because of the activation by N_2^+, the F originally present has been replaced by Cl.

Suschitzky, *Angew. Chem. Intern. Ed. Engl.* **6**, 596 (1967) [*Angew. Chem.* **79**, 636 (1967)]

28. Label i with D in the position ortho to the bromine and the *tert*-butyl. If path a, the D will still be there in the product; if path b the product will have lost it. But in the latter case, it is necessary to establish that i did not lose D *before* being converted to ii (that is, recovered i must still contain all of its D); and that deuterated ii will not lose D under the reaction conditions. Path b was actually found.

van der Plas, Smit, and Koudijs, *Tetrahedron Lett.* **1968**, 9.

32. Carry out the reaction with a good leaving group, such as F^-, para to the N(NO)COPh group. If iii is formed at any time during the reaction, then the F will be displaced by the nucleophile $PhCOO^-$ because it will be activated by the N_2^+. When this experiment was actually performed, the F was displaced by the $PhCOO^-$.

Suschitzky, *Angew. Chem. Intern. Ed. Engl.* **6**, 596 (1967) [*Angew. Chem.* **79**, 636 (1967)]

33. iii > i > ii

Fendler, Fendler, Byrne, and Griffin, *J. Org. Chem.* **33**, 977 (1968).

34. Approximate rate constants at 50° in MeOH in the presence of a tenfold excess of *p*-nitrophenol: iv: 10^{-1}; vi: 10^{-5}; v: 10^{-7}; ii: 10^{-11}; iii: 10^{-16}; i: 10^{-20}.

Chan and Miller, *Aust. J. Chem.* **20**, 1595 (1967).

35. a. 3-Methoxyaniline
    b. 4-Fluoroaniline
    c. 3-Trifluoromethylaniline
    d. *m*-toluidine
    e. 3,4-dichloroaniline
    f. 3,5-dichloroaniline
    g. 4-chloro-2,5-dimethylaniline

References: a. Gilman and Kyle, *J. Amer. Chem. Soc.* **74**, 3027 (1952); b, c, d. Roberts, Vaughan, Carlsmith, and Semenow, *J. Amer. Chem. Soc.* **78**, 611 (1956); e, f, g. Wotiz and Huba, *J. Org. Chem.* **24**, 595 (1959).

ANSWERS TO PROBLEM SET 8

FREE-RADICAL SUBSTITUTION

3.

i + (2-thienyl)$(CH_2)_4$-OH $\longrightarrow$ (2-thienyl)$(CH_2)_4$-O-C=O
$(CH_2)_3$
COOH

$\xrightarrow{SOCl_2}$ (2-thienyl)$(CH_2)_4$-O-C=O
$(CH_2)_3$
COCl
$\xrightarrow{AlCl_3}$

O=C (thiophene, S) $(CH_2)_4$
$(CH_2)_3$ O
C
O

$\xrightarrow{\text{Raney Ni}}$ ii

Taits, Alashev, and Gol'dfarb, *Bull. Acad. Sci. USSR, Div. Chem. Sci.* **1968**, 550.

8.

i $\xrightarrow{2Cl_2}$ NH_2, Cl, COOH, Cl (benzene ring) $\xrightarrow[\text{2. } Cu^+]{\text{1. HONO}}$ ii

Atkinson, Murphy, and Lufkin, *Org. Syn.* **IV**, 872.

9.

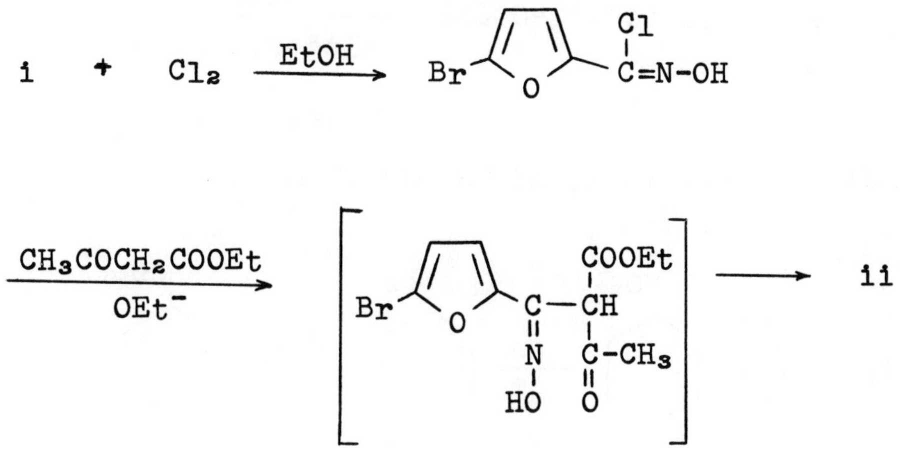

i + Cl_2 $\xrightarrow{EtOH}$ 5-bromo-2-furyl–C(Cl)=N–OH

$\xrightarrow[OEt^-]{CH_3COCH_2COOEt}$ [5-bromo-2-furyl–C(=N–OH)–CH(COOEt)–C(=O)–CH_3] ⟶ ii

Khisamutdinov, Pechenkin, and Aitova, *J. Gen. Chem. USSR* **36**, 1847 (1966).

15.

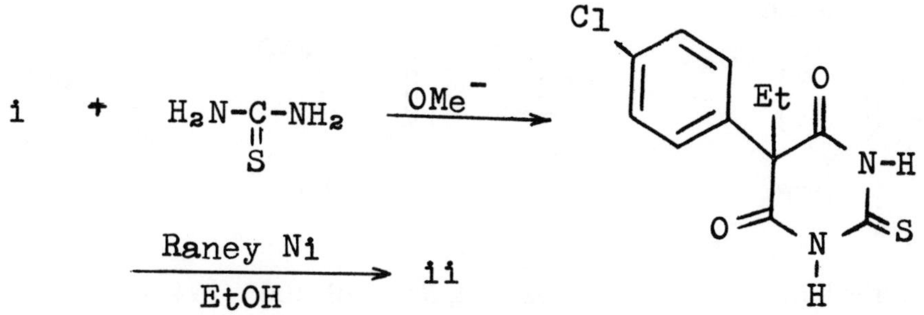

i + H_2N–C(=S)–NH_2 $\xrightarrow{OMe^-}$ 5-(4-chlorophenyl)-5-ethyl-2-thioxo-pyrimidine-4,6-dione (N-H, Et, O, S)

$\xrightarrow[EtOH]{Raney\ Ni}$ ii

Boon, Carrington, Greenhalgh, and Vasey, *J. Chem. Soc.* **1954**, 3263.

17.

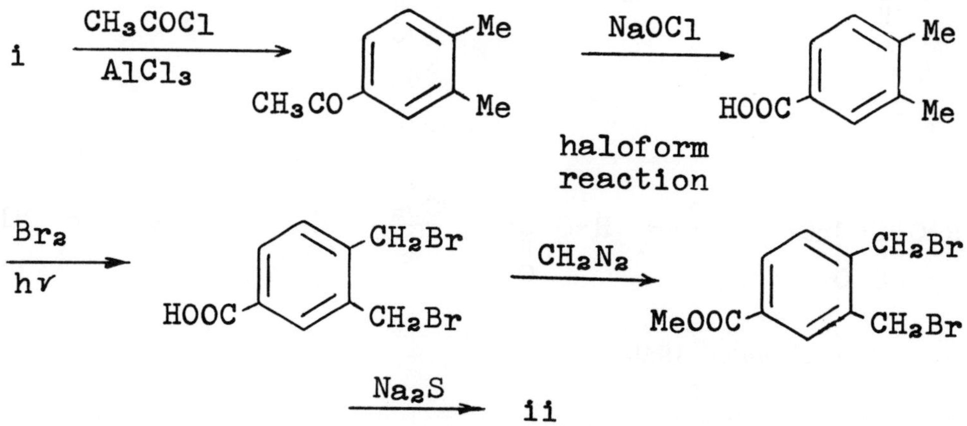

i $\xrightarrow[AlCl_3]{CH_3COCl}$ CH_3CO–C₆H₃(Me)(Me) $\xrightarrow{NaOCl}$ HOOC–C₆H₃(Me)(Me)

haloform reaction

$\xrightarrow[h\nu]{Br_2}$ HOOC–C₆H₃(CH_2Br)(CH_2Br) $\xrightarrow{CH_2N_2}$ MeOOC–C₆H₃(CH_2Br)(CH_2Br)

$\xrightarrow{Na_2S}$ ii

Wynberg, Feijen, and Zwanenburg, *Rec. Trav. Chim.* **87**, 1006 (1968).

20.

$$\text{i} + HOOC-(CH_2)_8-COOEt \xrightarrow[\text{electrolysis}]{\text{base}} \text{ii}$$

mixed Kolbe

Asano and Ohta, *J. Pharm. Soc. Jap.* **65**, No. 516A, 10 (1965).

21.

$$PhOCH_3 \xrightarrow[H_2SO_4]{HNO_3} \text{(p-}NO_2C_6H_4OCH_3) \xrightarrow[H^+]{Zn} \text{(p-}NH_2C_6H_4OCH_3) \xrightarrow{Ac_2O} \text{(p-}NHAcC_6H_4OCH_3)$$

NO2 / OCH3; NH2 / OCH3; NHAc / OCH3

$$\xrightarrow{HNO_3} \text{(NHAc, 2-}NO_2\text{, 4-}OCH_3) \xrightarrow{KOH} \text{(NH}_2\text{, 2-}NO_2\text{, 4-}OCH_3) \xrightarrow[2.\ CuBr]{1.\ HONO} \text{i}$$

NHAc / NO2 / OCH3; NH2 / NO2 / OCH3

Fanta and Tarbell, *Org. Syn.* **III**, 661; Samant, *Ber.* **75B**, 1008 (1942).

27.

$$R-O-\underset{\underset{O}{\|}}{C}-COOH + I_2 \longrightarrow R-O-\underset{\underset{O}{\|}}{C}-\underset{\underset{O}{\|}}{C}-O-I \longrightarrow$$

i

$$R-O-\underset{\underset{O}{\|}}{C}-\underset{\underset{O}{\|}}{C}-O\cdot \xrightarrow{-CO_2} R-O-\underset{\underset{O}{\|}}{C}\cdot \xrightarrow{-CO_2} R\cdot + I_2 \longrightarrow RI$$

ii

Goosen, *Chem. Commun.* **1969**, 145.

29.

i $\xrightarrow{\text{cat.}}$ (phenyl radical, ortho to) $CH_2CH_2CH_2COOH$ $\xrightarrow[\text{abstraction}]{\text{internal}}$

$PhCH_2CH_2\dot{C}HCOOH$ $\xrightarrow{\text{dimerization}}$ iii

$PhCH_2CH_2\dot{C}HCOOH$ $\xrightarrow{\text{abstraction of H}}$ ii

Beckwith and Gara, *J. Amer. Chem. Soc.* **91**, 5691 (1969).

32. $HOOH + HONO \rightarrow HOONO + H_2O$

$HOONO \rightarrow HO\cdot + NO_2$

$RH\ (i) + HO\cdot \rightarrow R\cdot + H_2O$

$R\cdot + HOONO \rightarrow ROH\ (ii) + NO_2$

$R\cdot + NO_2 \rightarrow RNO_2\ (iii)$

$R\cdot + NO_2 \rightarrow RONO$

$RONO + H_2O_2 \rightarrow RONO_2\ (iv)$

Inoue, Sonoda, and Tsutsumi, *Bull. Chem. Soc. Jap.* **36**, 1549 (1963).

36.

$i \longrightarrow \cdot CMe_3 + Ph\text{-}C(=O)\text{-}O\cdot \longrightarrow Ph\cdot + CO_2$

$Ph\cdot$ or $Ph\text{-}C(=O)\text{-}O\cdot + PhCHO \longrightarrow Ph\text{-}\dot{C}(=O) + C_6H_6$ or $PhCOOH$

$Ph\text{-}\dot{C}(=O) + PhCHO \longrightarrow Ph\text{-}\dot{C}H\text{-}O\text{-}C(=O)\text{-}Ph \xrightarrow{\text{dimerization}} ii$

Sosnovsky and Yang, *J. Org. Chem.* **25**, 899 (1960).

39. The difference is apparently steric. According to this explanation, SO_2 is too large to react with tertiary radicals. Tertiary radicals are still preferentially formed, but they abstract hydrogen from a primary (or secondary) carbon which can react with SO_2.

Asinger and Ebender, *Ber.* **75**, 344 (1942); Scott and Heller, *J. Org. Chem.* **20**, 1159 (1955); Asinger, Geiseler, and Hoppe, *Chem. Ber.* **91**, 2130 (1958).

40. b > d > c > a

van Helden and Kooyman, *Rec. Trav. Chim.* **73**, 269 (1954).

41. a. α; b. 2; c. 4; d. 3; e. 1; f. 2.

Walling, "Free Radicals in Solution", pp. 360-368, John Wiley and Sons, Inc., New York, 1957.

ANSWERS TO PROBLEM SET 9

ADDITION TO CARBON-CARBON MULTIPLE BONDS

4.

i $\xrightarrow[NH_3]{Li}$ [bicyclic diene with CH_3 and COOH] $\xrightarrow[catalyst]{H_2}$ [bicyclic alkene with CH_3 and COOH] $\xrightarrow{CH_2I_2}$ ii

Birch reduction

Simmons-Smith reaction

In the catalytic hydrogenation step, the middle double bond is unaffected, because it is tetrasubstituted.

Sims and Selman, *Tetrahedron Lett.* **1969**, 561.

7.

$$\text{i} \xrightarrow[Pt]{H_2} [CH_3O(CH_2)_4]_3CH \xrightarrow{HBr} [Br(CH_2)_4]_3CH$$

$$\xrightarrow{NaC\equiv CH} [HC\equiv C(CH_2)_4]_3CH \xrightarrow[\text{catalyst}]{\text{Ziegler}} \text{ii}$$

Hubert and Hubert, *Tetrahedron Lett.* **1966**, 5779.

9.

$$\text{i} + Ni(CO)_4 + HC\equiv CH + CO + MeOH \longrightarrow \text{ii}$$

Carboxylation with insertion of acetylene (p. 649, *Advanced Organic Chemistry*).

Mettalia and Specht, *J. Org. Chem.* **32**, 3941 (1967).

12.

i + [maleic anhydride] → [Diels–Alder adduct] $\xrightarrow{p\text{-}O_2NC_6H_4COOOH}$ [epoxide] $\xrightarrow[H^+]{MeOH}$ [lactone, OH, COOMe] $\xrightarrow[2.\ H^+]{1.\ OH^-}$ [lactone, OH, COOH] $\xrightarrow[17\ hr]{TsOH}$ ii

Gastambide and Langlois, *Helv. Chim. Acta* **51**, 2048 (1968); Langlois and Gastambide, *Bull. Soc. Chim. Fr.* **1965**, 2966.

17.

i + H_2N–cyclohexyl → iii (1-cyclohexyl-2-(4-biphenylyl)-3-COOMe aziridine)

iii + H(COOMe)C=C(H)COOMe (iv) → ii

The first step is a Michael addition followed by an internal nucleophilic substitution. The second is a 1,3-dipolar addition to iv of a diion formed by opening of the 3-membered ring of iii.

Woller and Cromwell, *J. Heterocycl. Chem.* **5**, 579 (1968).

22.

$$HC\equiv CH \xrightarrow{NH_2^-} \overset{\ominus}{C}\equiv\overset{\ominus}{C} \xrightarrow{2BuBr} BuC\equiv CBu \xrightarrow[EtOH,\ HOAc-H_2O]{Ni(CO)_4} i$$

Bried and Hennion, *J. Amer. Chem. Soc.* **59**, 1310 (1937); Jones, Shen, and Whiting, *J. Chem. Soc.* **1951**, 48.

25.

i ⇌ (O, CH, OH) + $Ts-\overset{\ominus}{\underline{N}}-\overline{N}=\overset{\oplus}{\overline{N}}$ ⟶

(O, N, NI, H–C, N–Ts, O, H) ⟶ ii + H–C(=O)–N(H)–Ts

Regitz and Rüter, *Chem. Ber.* **101**, 1263 (1968).

27.

i ⟶ iv

iv + v ⟶ iii

Diels-Alder

ii ⟶ v + SO_2

reverse Diels-Alder

Hatch and Peter, *Chem. Commun.* **1968**, 1499.

34. These results are from Mousseron-Canet, Dalle, and Mani, *Tetrahedron Lett.* **1968**, 6037. Very similar results are found in Foote and Brenner, *Tetrahedron Lett.* **1968**, 6041. ii is formed by a normal Diels-Alder reaction with singlet oxygen. The formation of iii and iv might be explained by ene syntheses with oxygen, e.g.:

R, O=O, H, CH_2 ⟶ R, O–OH, CH_2

However, evidence has been presented [Fenical, Kearns, and Radlick, *J. Amer. Chem. Soc.* **91**, 7771 (1969)] that such reactions are not ene syntheses, but rather have the following mechanism:

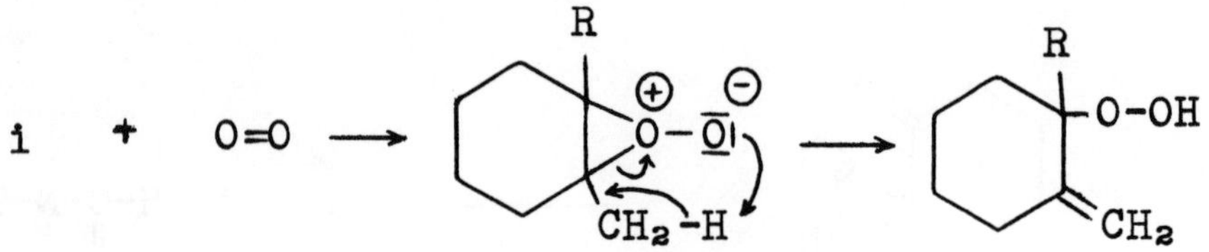

i + O=O ⟶ R, $\oplus$ O–O $\ominus$, CH_2–H ⟶ R, O–OH, CH_2

39. iii arises from normal free radical addion of PhSH.

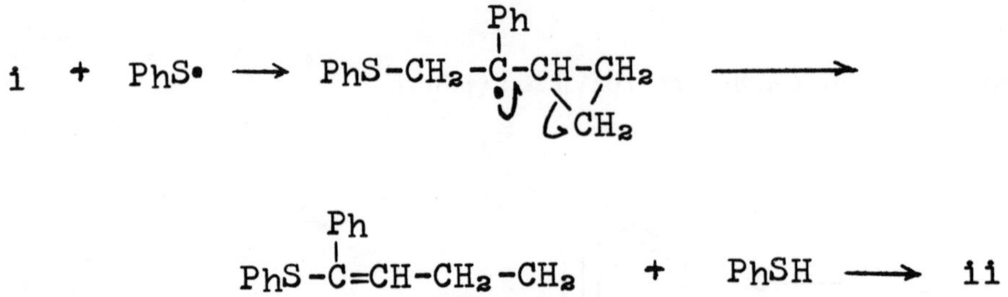

i + PhS• ⟶ PhS–CH_2–C(Ph)–CH–CH_2 (CH_2) ⟶

PhS–C(Ph)=CH–CH_2–CH_3 + PhSH ⟶ ii

Lishanskii, Vinogradova, Guliev, Zak, Zvyagina, Fomina, and Khachaturov, *Doklad. Chem.* (translation in English of *Doklad. Acad. Nauk SSSR*) **179**, 309 (1968).

43.

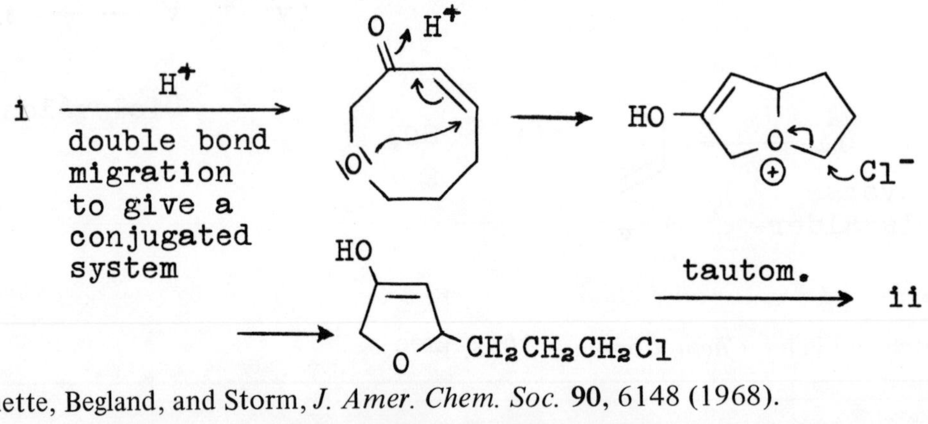

i $\xrightarrow{H^+}$ (double bond migration to give a conjugated system) O, H^+ ⟶ HO, O $\oplus$, Cl^-

⟶ HO, O, $CH_2CH_2CH_2Cl$ $\xrightarrow{\text{tautom.}}$ ii

Paquette, Begland, and Storm, *J. Amer. Chem. Soc.* **90**, 6148 (1968).

47.

$$\mathrm{i} + CH_2N_2 \longrightarrow \text{[pyrazoline intermediate: Ph, } CH_3\text{; arrows a, b]} \xrightarrow{a} Ph-C(=CH_2)-CH(CH_3)-\underline{N}=\overset{\oplus}{N}=\overset{\ominus}{\underline{N}}| \quad \mathbf{ii}$$

$$\xrightarrow{b} CH_2(-\underline{N}=\overset{\oplus}{N}=\overset{\ominus}{\underline{N}}|)-C(Ph)=CH-CH_3 \quad \mathbf{iii\ and\ iv}$$

Nair, *J. Org. Chem.* **33**, 2121, 4316 (1968).

51.

$$\mathrm{i} + \bar{C}Cl_2 \longrightarrow \underset{\mathbf{vi}}{\text{(Cl, Cl; Ph, Et cyclopropene)}} \xrightarrow[S_N]{CCl_3^-} \underset{\mathbf{vii}}{\text{(Cl, } CCl_2Cl\text{; Ph, Et cyclopropene)}}$$

$$\longrightarrow \text{(Cl, }\oplus\text{, Cl, Cl; Ph, Et cyclobutenyl cation)} \longrightarrow \mathbf{iii}$$

v is formed by hydrolysis of vi; iv is formed by migration of the other bond of vii. The possibility of:

$$\mathrm{ii} \longrightarrow Cl_2C{=}C{=}O + \mathrm{i} \longrightarrow \mathrm{iii} + \mathrm{iv}$$

was eliminated by the result given in the note, since $Cl_2C{=}C{=}O$ is certainly generated under those conditions.

Dehmlow, *Chem. Ber.* **100**, 3829 (1967).

52.

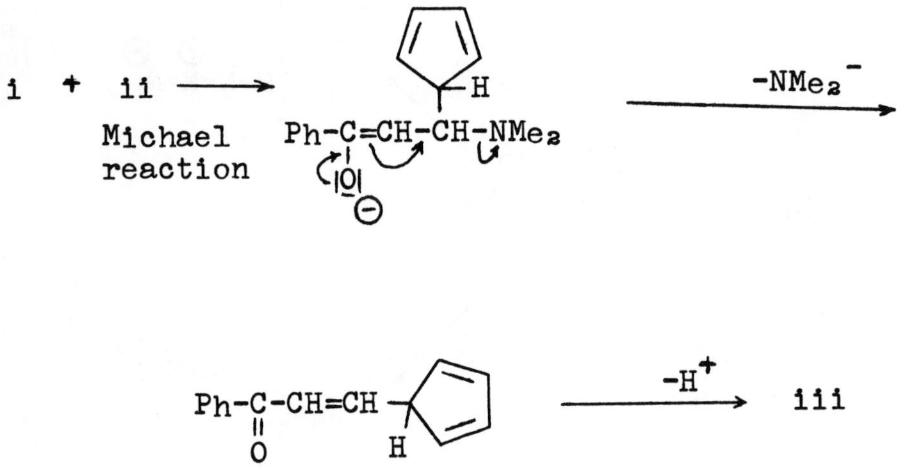

i + ii ⟶ (Michael reaction) Ph–C(–O⁻)=CH–CH(C₅H₅)–NMe₂ ⟶ (−NMe₂⁻) Ph–C(=O)–CH=CH–C₅H₅ ⟶ (−H⁺) iii

Nesmeyanov, Rybinskaya, and Korneva, *Bull. Acad. Sci. USSR, Div. Chem. Sci.* **1967**, 2523.

56.

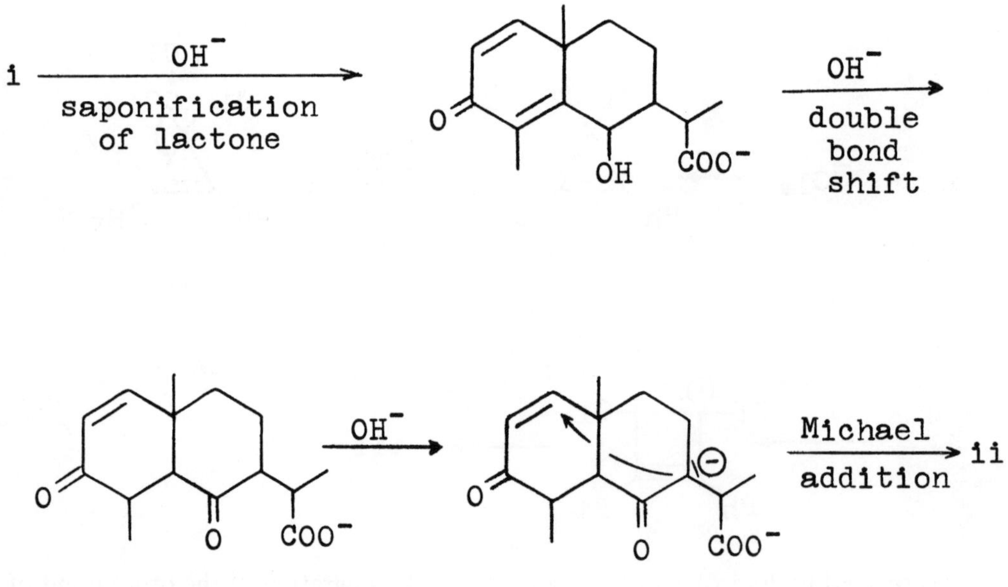

i ⟶ (OH^-; saponification of lactone) ⟶ (OH^-; double bond shift) ⟶ (OH^-) ⟶ (Michael addition) ii

Woodward, Brutschy, and Baer, *J. Amer. Chem. Soc.* **70**, 4216 (1948).

59.

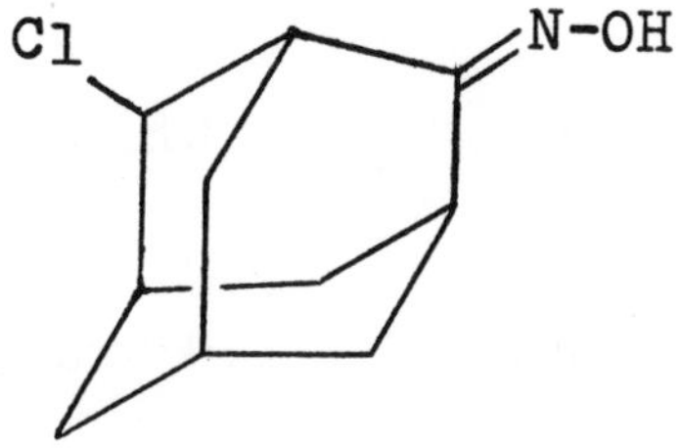

Cl N-OH

i

Analogous to the addition of NOCl to double bonds. The ring opens in this direction to preserve the stable adamantane structure.

Udding, Strating, and Wynberg, *Tetrahedron Lett.* **1968**, 1345.

61.

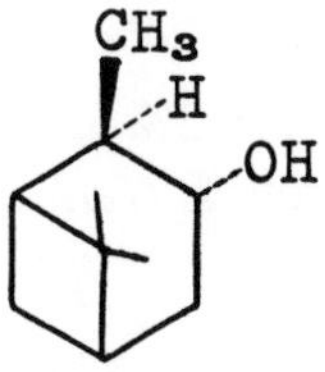

CH_3 H OH

i

Hydroboration gives anti-Markovnikov addition of H_2O from the less-hindered side of the molecule.

Brown and Zweifel, *J. Amer. Chem. Soc.* **83**, 2544 (1961).

62.

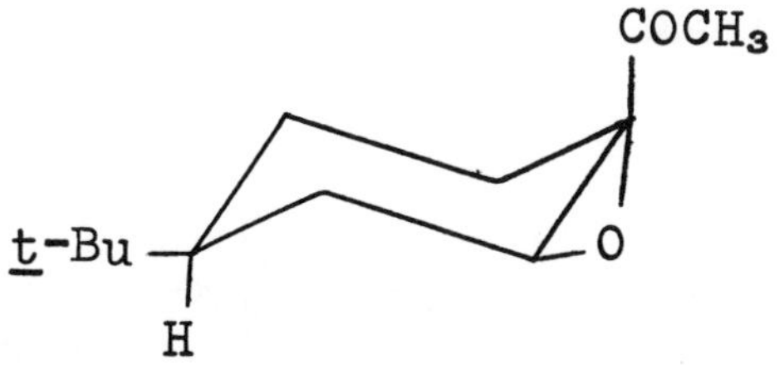

$COCH_3$ t-Bu O H

i

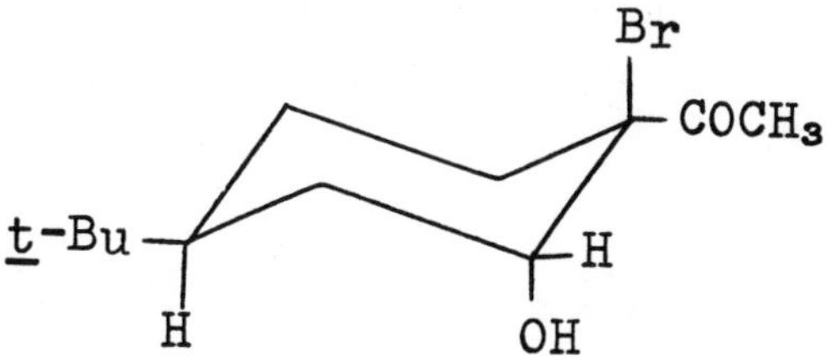

Br $COCH_3$ t-Bu H H OH

ii

Akhrem, Vladimirova, and Dobrynin, *Bull. Acad. Sci. USSR, Div. Chem. Sci.* **1967**, 1266.

69.

$EtSCH_2CH{=}CHCH_2SiMe_3$

i

1,4-addition product

$CH_3CH{=}CHCH(SEt)SiMe_3$

ii

1,4-addition product

$EtSCH_2CH_2CH{=}CHSiMe_3$

iii

1,2-addition product

a. Nucleophilic addition: 65% i, 25% ii, and 10% iii.

b. Free radical addition: 41% i; 2% ii; and 57% iii.

Sulimov, Sleta, Stadnichuk, and Petrov, *J. Gen. Chem. USSR* **38**, 207 (1968).

73. Yes. It is a 6 + 4 cycloaddition:

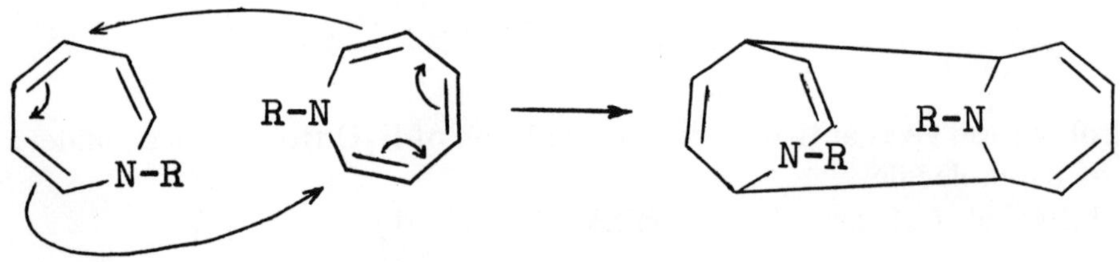

R-N N-R R-N N-R

i i ii

The reaction was carried out by Paul, Johnson, Barrett, and Paquette, *Chem. Commun.* **1969**, 6.

74.

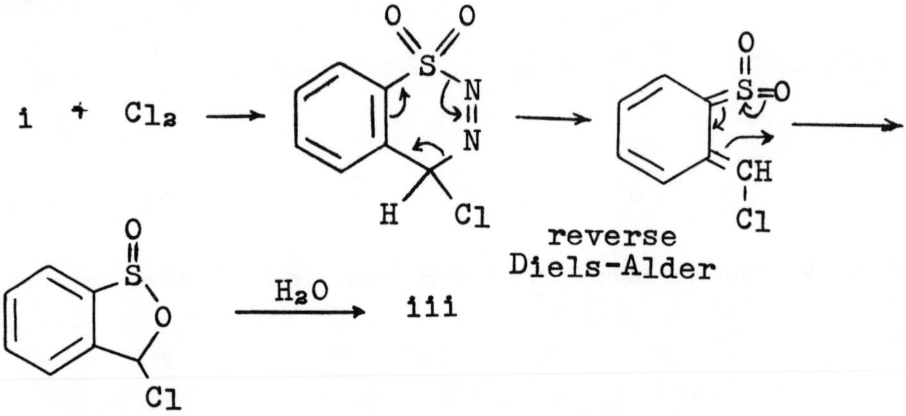

i + Cl_2 ⟶

O O S N N H Cl ⟶ O S=O CH Cl reverse Diels-Alder ⟶

O S O Cl ii $\xrightarrow{H_2O}$ iii

King, Hawson, Deaken, and Komery, *Chem. Commun.* **1969**, 33.

79.

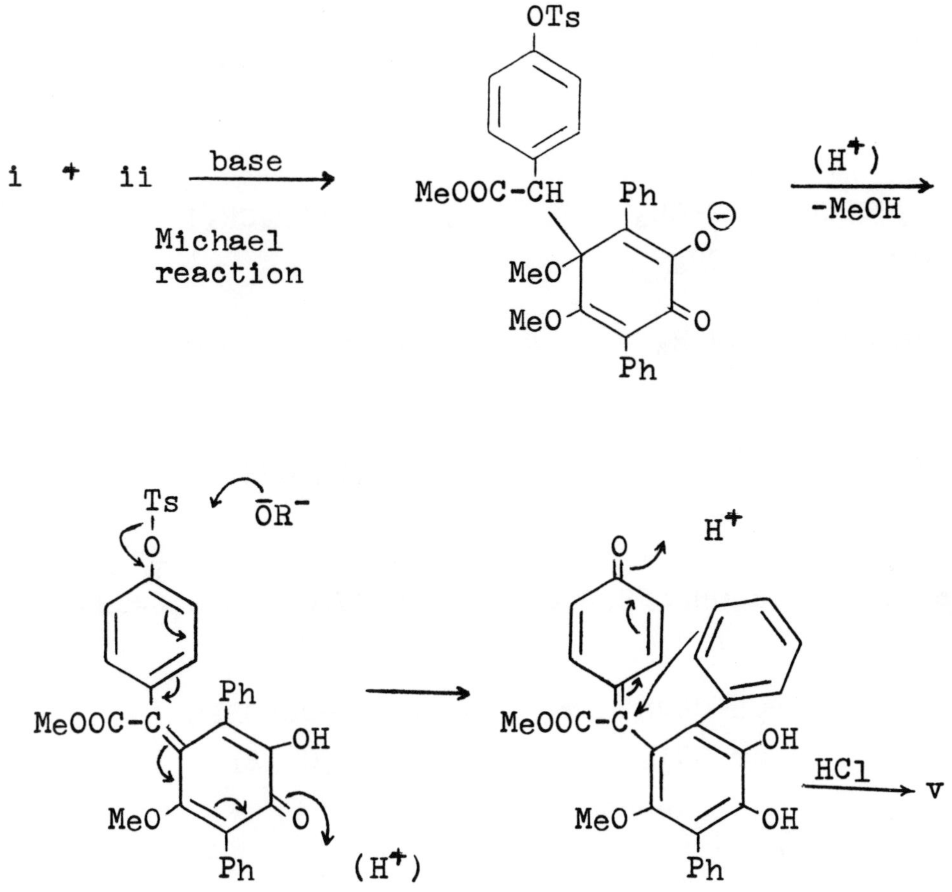

iii iv

Wanzlick and Jahnke, *Chem. Ber.* **101**, 3753 (1968).

82. The decomposition of ii is the reverse of a 4 + 2 Diels-Alder cycloaddition and hence allowed by the principle of orbital symmetry. The decomposition of i is the reverse of a 2 + 2 cycloaddition, and a concerted mechanism is disallowed. Rieber, Alberts, Lipsky, and Lemal, *J. Amer. Chem. Soc.* **91**, 5668 (1969).

83.

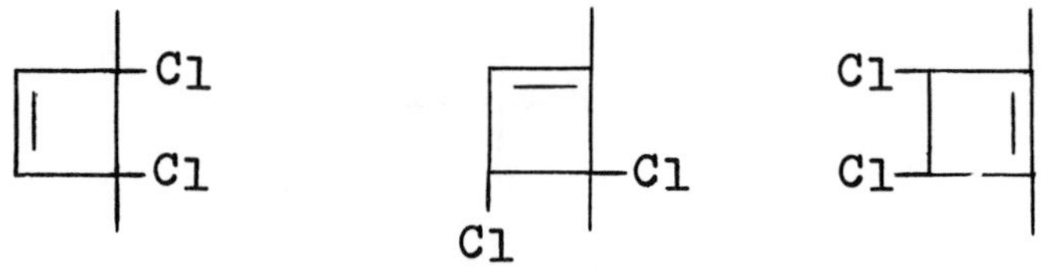

1,2 and 1,2-addition

1,2 and 1,4-addition

1,4 and 1,4-addition

Criegee, Eberius, and Brune, *Chem. Ber.* **101**, 94 (1968).

86.

a. $(CH_3)_2C(OH)CH_2CH_3$

b. $ClCH_2CH_2CHO$

c. $CF_3C{\equiv}CCH_2CH(Br){-}C(Br)(CH_3){-}CH_3$

d. $Me_3SiCH_2CH_2I$

e. *p*-$Me_3SiC_6H_4CH(Br)CH_3$

f. *p*-$Cl_3SiC_6H_4CH_2CH_2Br$

g. $CH_3{-}C(OEt)_2{-}CHO$

h. $CF_3CH{=}CHBr$

i. *p*-$CH_3C_6H_4CH_2CH_2SCH_2COOH$

j. *p*-$CH_3CONHC_6H_4C(Br){=}CH_2$

k. *p*-$O_2NC_6H_4{-}CH(Br){-}CH(CCl_3)Ph$ + *p*-$O_2NC_6H_4{-}CH(CCl_3){-}CH(Br)Ph$

about 60:40

l.

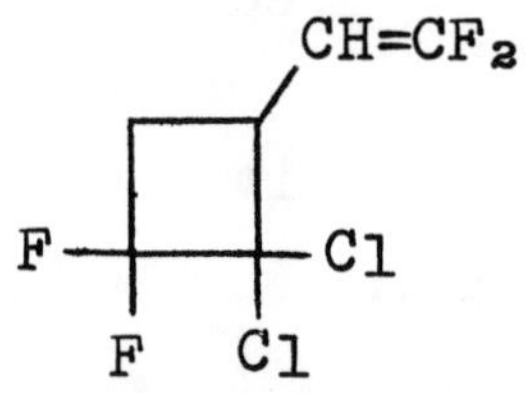
CH=CF2
F
Cl
F
Cl

m.

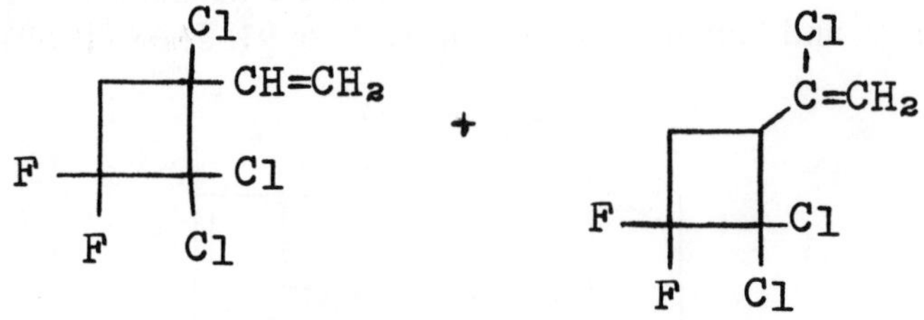
Cl
CH=CH2
F
Cl
F
Cl
+
Cl
C=CH2
F
Cl
F
Cl

ratio 9.8:1.6

n.

$Ar{-}C(=O){-}CH_2{-}CH(CN){-}Ar$

References: a. Levy, Taft, and Hammett, *J. Amer. Chem. Soc.* **75**, 1253 (1953); b. Moureu and Chaux, *Org. Syn.* **I**, 166; c. Konotopov, Porfir'eva, and Petrov, *J. Gen. Chem. USSR* **38**, 679 (1968); d. Sommer, Bailey, Goldberg, Buck, Bye, Evans, and Whitmore, *J. Amer. Chem. Soc.* **76**, 1613 (1954); e, f. Chernyshev, Krasnova, Khachaturov, and Moskalenko, *J. Gen. Chem. USSR*, **36**, 1626 (1966); g. Shostakovskii, Keiko, and Filippova, *Bull. Acad. Sci. USSR, Div. Chem. Sci.* **1967**, 368; h. Henne and Nager, *J. Amer. Chem. Soc.* **74**, 650 (1952); i. Walling, Seymour, and Wolfstirn, *J. Amer. Chem. Soc.* **70**, 2559 (1948); j. Grob and Cseh, *Helv. Chim. Acta* **47**, 194 (1964); k. Cadogan, Duell, and Inward, *J. Chem. Soc.* **1962**,4164; l. Lomas and Tarrant, *J. Org. Chem.* **34**, 323 (1969); m. Bartlett, Montgomery, and Seidel, *J. Amer. Chem. Soc.* **86**, 616 (1964); and n. Fuson and Bannister, *J. Amer. Chem. Soc.* **74**, 1631 (1952).

ANSWERS TO PROBLEM SET 10

ADDITION TO CARBON-HETERO MULTIPLE BONDS

1.

i + $2Ph_3P$ ⟶ [1,8-bis($CH_2\overset{\oplus}{P}Ph_3$)naphthalene] $\xrightarrow[OEt^-]{2PhCHO}$

[1,8-bis(PhCH=CH)naphthalene] $\xrightarrow{h\nu}$ ii

Houlton and Kemp, *Tetrahedron Lett.* **1968**, 1045.

5.

i + $CH_3CONHCH_2COOH$ $\xrightarrow[NaOAc]{Ac_2O}$ [azlactone: C_6D_5–CD= ; O ; N ; O ; CH_3] $\xrightarrow[acetone]{H_2O}$

Erlenmeyer azlactone synthesis

C_6D_5-CD=C(NHAc)-COOH $\xrightarrow[Pd-C]{D_2}$ C_6D_5-CD_2-CD(NHAc)-COOH $\xrightarrow[\text{enzyme resolution}]{\text{hydrol.}}$ ii

Blomquist and Cedergren, *Can. J. Chem.* **46**, 1053 (1968).

9.

i $\xrightarrow[\text{H}_2\text{O-MeOH}]{\text{NaOH}}$ (aldol condensation) $\xrightarrow[\text{H}^+]{}$ (for purification)

OCH3 O OH OCH2Ph OCH3 O

OCH3 O O O OCH2Ph OCH3 $\xrightarrow[\text{2. hydrolysis of acetal}]{\text{1. NH}_3\text{-H}_2\text{-Ni}}$ OCH3 NH2 OH OCH2Ph OCH3

$\xrightarrow[\text{NHMe}_2]{\text{AcCl}}$ ii

Wiesner, Kao, and Šantroch, *Can. J. Chem.* **47**, 2431 (1969).

13.

i + H_2SO_4 $\xrightarrow{\quad}$ (Ritter reaction) ⊕ $\xrightarrow{\text{NCCH}_2\text{CN}}$ ii

Sasaki, Eguchi, and Toru, *Bull. Chem. Soc. Jap.* **41**, 236 (1968).

16.

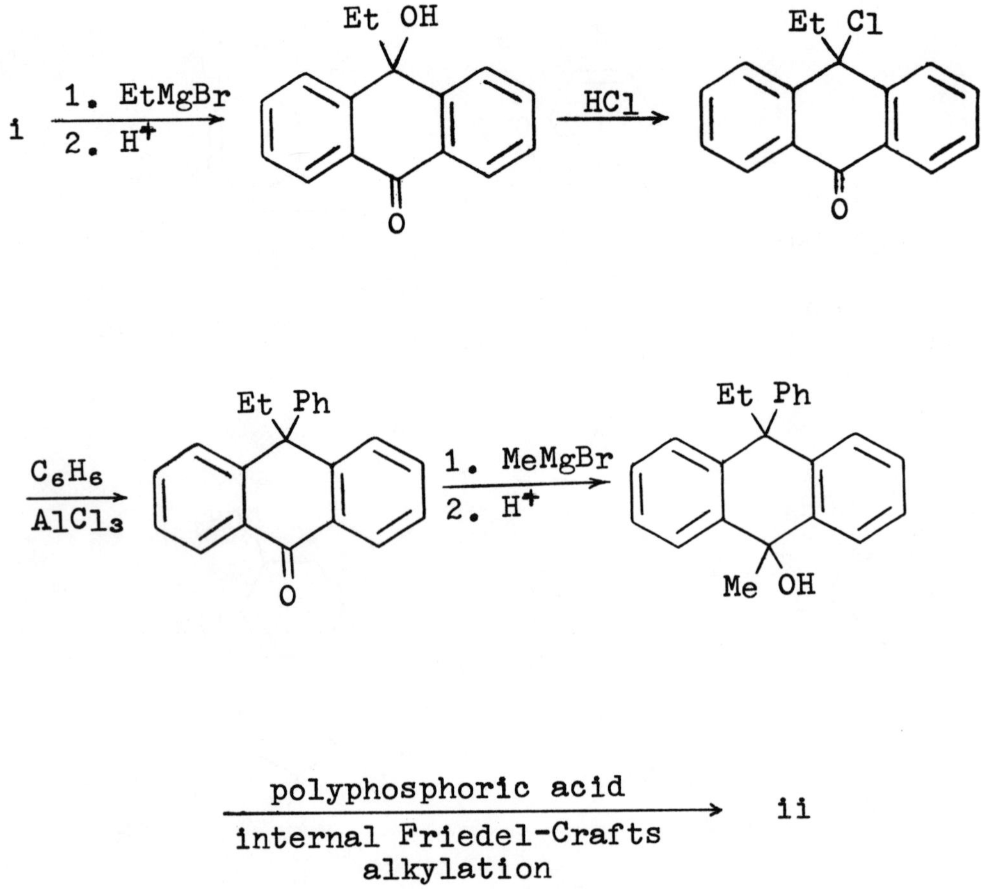

i $\xrightarrow[\text{2. } H^+]{\text{1. EtMgBr}}$ [10-ethyl-10-hydroxy-9-anthrone: Et, OH; C=O] $\xrightarrow{HCl}$ [10-chloro-10-ethyl-9-anthrone: Et, Cl; C=O]

$\xrightarrow[AlCl_3]{C_6H_6}$ [10-ethyl-10-phenyl-9-anthrone: Et, Ph; C=O] $\xrightarrow[\text{2. } H^+]{\text{1. MeMgBr}}$ [Et, Ph; Me, OH]

$\xrightarrow[\text{internal Friedel-Crafts alkylation}]{\text{polyphosphoric acid}}$ ii

Walborsky and Bohnert, *J. Org. Chem.* **33**, 3934 (1968).

20.

i $\xrightarrow[\text{benzoin condensation}]{CN^-}$ Ar-CH(OH)-C(=O)-Ar $\xrightarrow[\text{HCl hydrogenolysis}]{Sn}$ ii

Bösler, *Ber.* **14**, 323 (1881); Carter, Craig, Lack, and Moyle, *Org. Syn.* **40**, 16.

24.

i $\xrightarrow[\text{KCN}]{\text{HCHO}}$ (Strecker synthesis) 3-(hydroxymethyl)-1-(cyanomethyl)pyrrolidine (CH_2OH; N–CH_2–CN) $\xrightarrow[\text{2. NaCN}]{\text{1. TsCl}}$ 3-(cyanomethyl)-1-(cyanomethyl)pyrrolidine (CH_2CN; N–CH_2–CN) $\xrightarrow{t\text{-BuOK}}$ (Thorpe-Ziegler reaction)

bicyclic imino nitrile (CN, =NH) $\xrightarrow[\text{decarbox.}]{\text{hydrolysis}}$ bicyclic ketone (=O) $\xrightarrow[\text{EtOH}]{\text{Na}}$ iii; $\xrightarrow[\text{H}_2]{\text{PtO}_2}$ ii

Thill and Aaron, *J. Org. Chem.* **33**, 4376 (1620).

30.

i $\xrightarrow[\text{Pd-BaSO}_4]{\text{H}_2}$ (Rosenmund reduction) p-nitrobenzaldehyde (CHO, NO_2) $\xrightarrow[\text{NH}_2\text{OH·HCl, HCOONa}]{\text{HCOOH}}$ ii

Rosenmund and Zetzsche, *Ber.* **54**, 425 (1921); van Es, *J. Chem. Soc.* **1965**, 1564.

33.

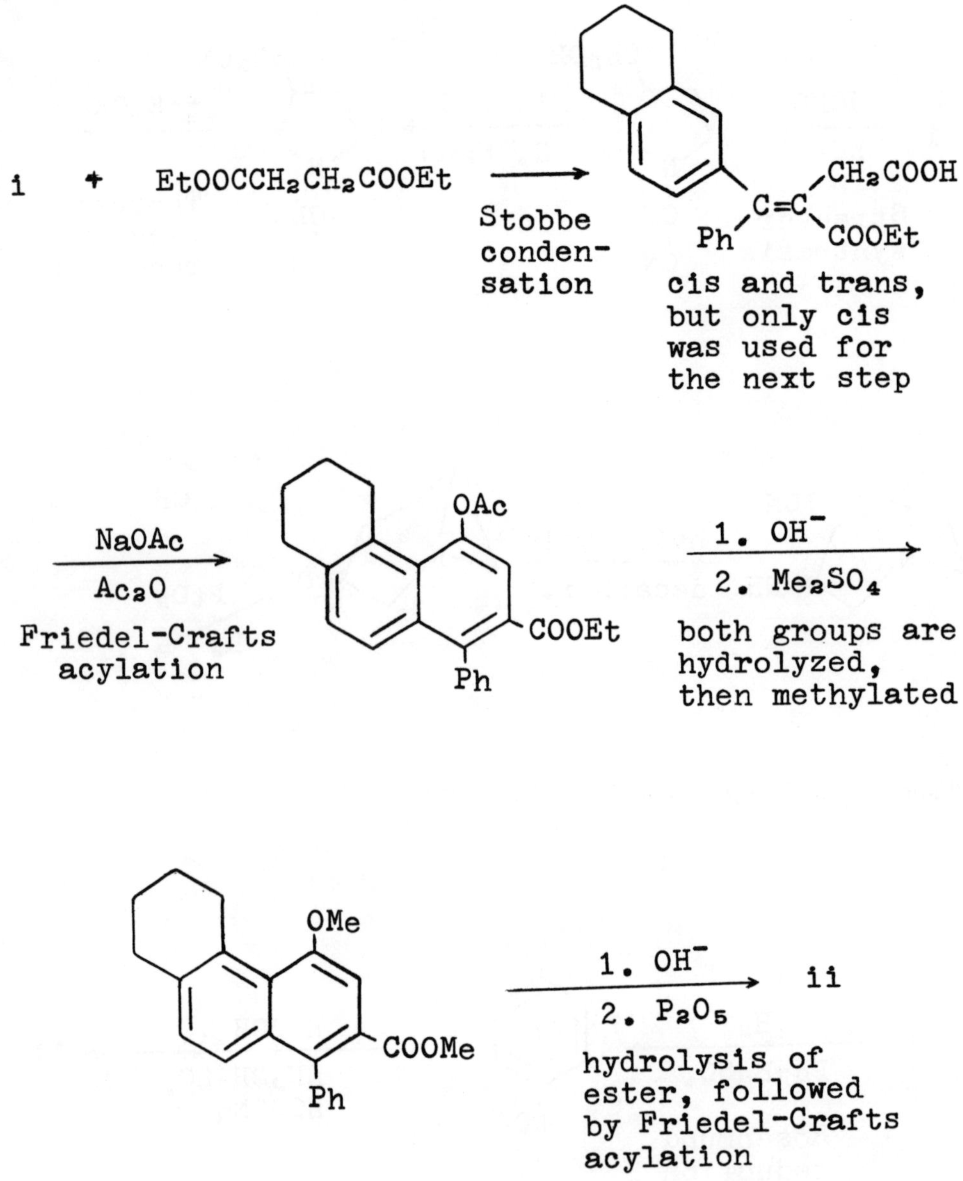

i + $EtOOCCH_2CH_2COOEt$ → (Stobbe condensation)

Ph, CH_2COOH, COOEt

cis and trans, but only cis was used for the next step

NaOAc / Ac_2O → Friedel-Crafts acylation

OAc, COOEt, Ph

1. OH^- / 2. Me_2SO_4 → both groups are hydrolyzed, then methylated

OMe, COOMe, Ph

1. OH^- / 2. P_2O_5 → ii

hydrolysis of ester, followed by Friedel-Crafts acylation

Awad, Baddar, Fouli, Omran, and Selim, *J. Chem. Soc., C* **1968**, 507.

38.

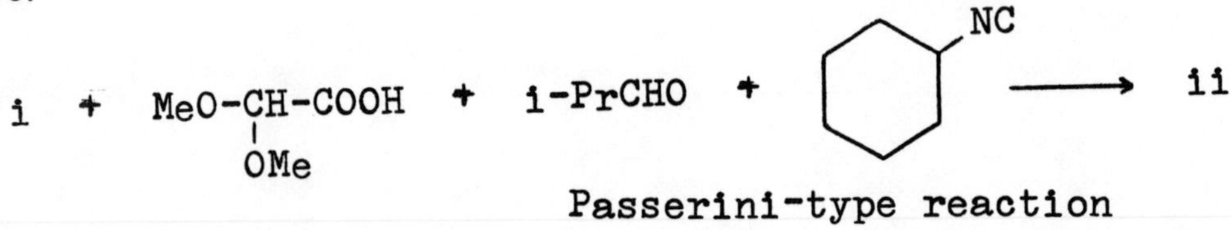

i + MeO-CH(OMe)-COOH + i-PrCHO + NC (cyclohexyl isocyanide) → ii

Passerini-type reaction

Gross, Gloede, Keitel, and Kunath, *J. Prakt., Chem.* [4] **37**, 192 (1968).

42.

$$\text{i} + CS_2 \xrightarrow{\text{base}} NC{-}\underset{\underset{{}^{\ominus}S\diagup\ \diagdown S^{\ominus}}{\overset{\|}{C}}}{C}{-}COOEt + BrCH_2CH_2Br \longrightarrow \text{ii}$$

Jensen and Henriksen, *Acta Chem. Scand.* **22**, 1107 (1968).

43.

$$\text{i} + HCl \longrightarrow \left[\begin{array}{c} CH_3CH{=}CHCH_2Cl \\ + \\ CH_3\underset{\underset{Cl}{|}}{C}HCH{=}CH_2 \end{array}\right] \xrightarrow{SCN^-} CH_3\underset{\underset{NCS}{|}}{C}HCH{=}CH_2 \xrightarrow[H_2O]{HCl} \text{ii}$$

Krueger and Schwarcz, *J. Amer. Chem. Soc.* **63**, 2512 (1941).

45.

$$\text{i} + HCONH_2 \xrightarrow[\text{Leuckardt reaction}]{} \text{9-(NHCHO)fluorene} \xrightarrow{OH^-} \text{ii}$$

NHCHO

Schiedt, *J. Prakt. Chem.* [2] **157**, 203 (1941).

50.

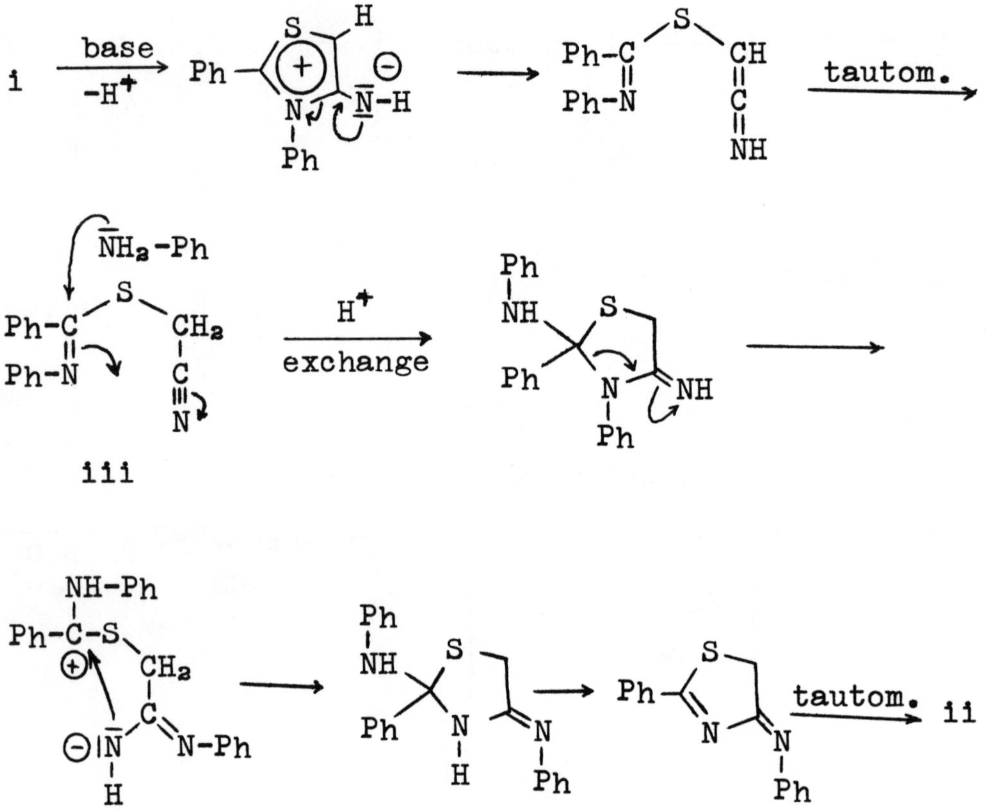

Sato and Ohta, *Bull. Chem. Soc. Jap.* **41**, 2801 (1968).

53.

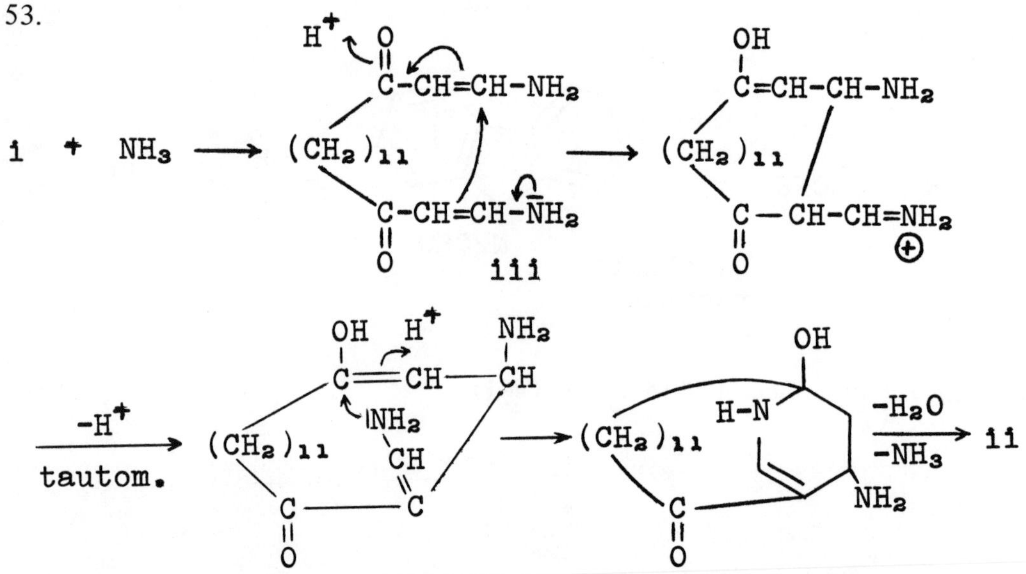

The above is the mechanism given in the paper: Gerlach and Huber, *Helv. Chim. Acta* **51**, 2027 (1968) (p. 2029). However, a simpler mechanism is:

iii

$\xrightarrow{-H_2O}$ $\longrightarrow$

$\xrightarrow[\text{HOAc}]{-NH_3}$ ii

55.

i + ii $\xrightarrow[\text{enamine formation}]{}$ $\xrightarrow[\text{tetrahedral mech.}]{\text{proton transfer}}$ iii

Wamhoff and Korte, *Chem. Ber.* **101**, 778 (1968).

59.

i + ii $\xrightarrow{t\text{-BuOK}}$ $ClCH_2CH_2C(=O)-C(Me)=$ ... COOMe

$\xrightarrow[\text{(vinylology)}]{\text{base}}$ Cl–CH_2–CH_2–C(=O)–C(Me)= ... COOMe $\longleftrightarrow$

59. Cont.

$\xrightarrow{}$ iii $\xrightarrow[\text{addn. of } COO^- \text{ to double bond}]{\text{hydrol. of ester}}$ iv

Mahalanabis, Mukhopadhyay, and Dutta, *Chem. Commun.* **1968**, 1641.

61.

i $\xrightarrow{H^+}$

proton transfer $\longrightarrow$ $\xrightarrow{-H_2O}$ $\xrightarrow{OAc^-}$ ii

This reaction is an acid-catalyzed aldel condensation.

Bélanger, Poupart, and Deslongchamps, *Tetrahedron Lett.* **1968**, 2127.

64. i and iv are products of a normal Prins reaction (*Advanced Organic Chemistry*, p. 711). ii is the product of a normal Lederer-Manasse reaction (hydroxymethylation of phenols: *Advanced Organic Chemistry*, p. 422). iii and v are the products of both reactions.

Griengl, Appenroth, Dax, and Schwarz, *Monatsh. Chem.* **100**, 316 (1969).

66.

i + ii —BF₃→ Et-C(-Me)(-O-BF₃⁻)-C≡N⁺-t-Bu + ⁻C≡N⁺-t-Bu →

(the C≡N⁺-t-Bu intermediate —H₂O→ iv)

t-Bu-N=C(-C≡N⁺-t-Bu)-C(Et)(Me)-O-BF₃⁻ —H₂O→ t-Bu-N=C(-C(OH)=NH-t-Bu)-C(Et)(Me)-OH —tautom.→

t-Bu-N=C(-C(=O)-NH-t-Bu)-C(Et)(Me)-OH —hydrolysis of C=N, -H₂O→ iii

Zeeh and Müller, *Justus Liebigs Ann. Chem.* **715**, 47 (1968).

71.

i + i-PrCHO → [⁻|O|-CH-i-Pr-N⁺ bicyclic intermediate with O, O, Et, Ph] → ii

Feinauer and Henckel, *Justus Liebigs Ann. Chem.* **716**, 135 (1968).

76.

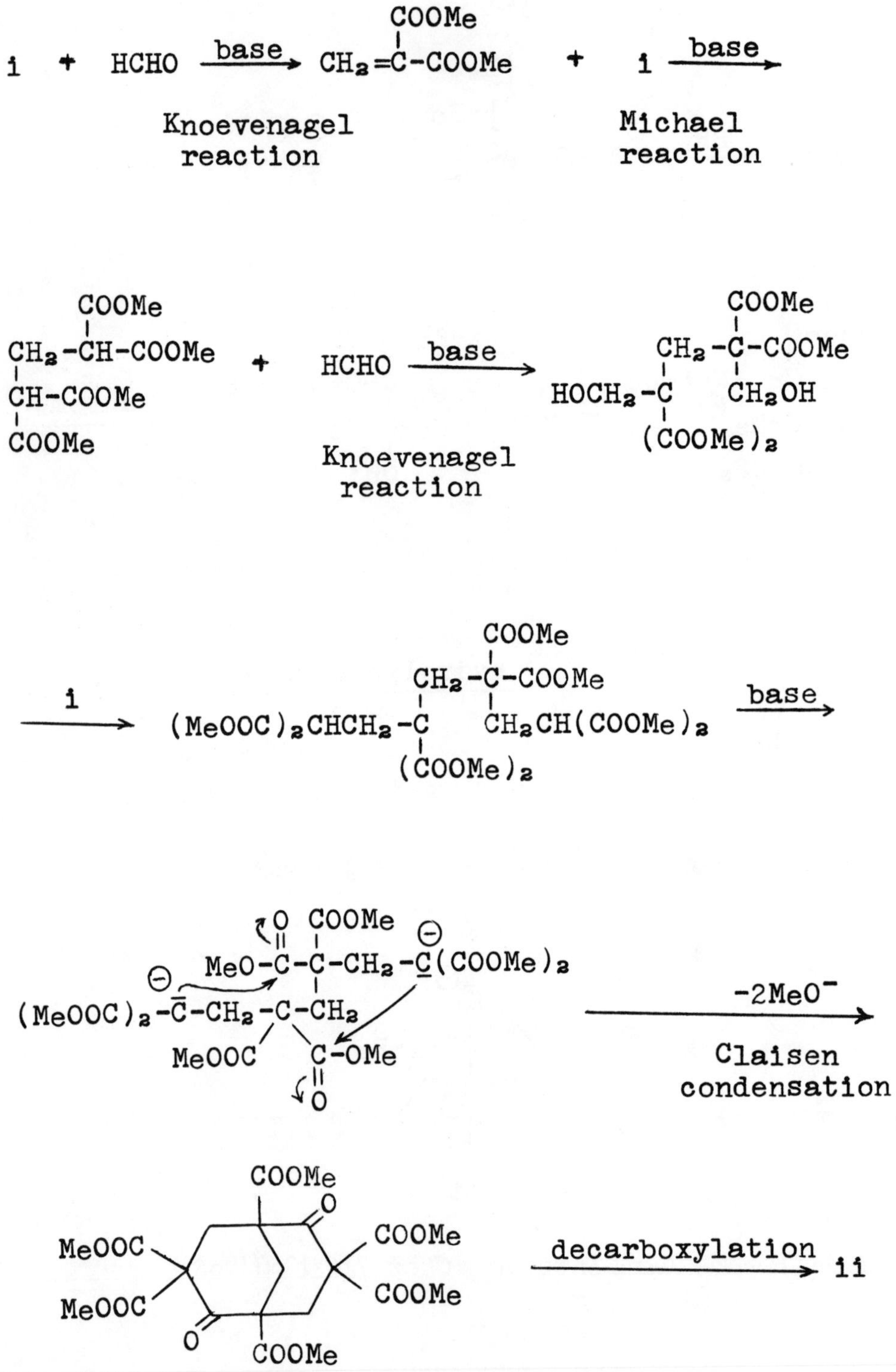

i + HCHO $\xrightarrow{\text{base}}$ $CH_2{=}C(COOMe)-COOMe$ + i $\xrightarrow{\text{base}}$

Knoevenagel reaction; Michael reaction

$CH_2(CH(COOMe)_2)-CH(COOMe)_2$ (i.e., $CH_2-CH-COOMe$ with COOMe above, and $CH-COOMe$ / COOMe below) + HCHO $\xrightarrow{\text{base}}$ $HOCH_2-C(COOMe)_2-CH_2-C(COOMe)_2-CH_2OH$

Knoevenagel reaction

$\xrightarrow{\text{i}}$ $(MeOOC)_2CHCH_2-C(COOMe)_2-CH_2-C(COOMe)_2-CH_2CH(COOMe)_2$ $\xrightarrow{\text{base}}$

$(MeOOC)_2\bar{C}^{\ominus}-CH_2-C(COOMe)(CH_2-)-C(=O)OMe$ … $MeO-C(=O)-C(COOMe)-CH_2-\underline{C}^{\ominus}(COOMe)_2$ $\xrightarrow[\text{Claisen condensation}]{-2MeO^-}$

Bicyclic diketone: MeOOC, MeOOC, COOMe, COOMe, COOMe, COOMe, O, O $\xrightarrow{\text{decarboxylation}}$ ii

Schaefer and Honig, *J. Org. Chem.* **33**, 2655 (1968).

85.

i → (1-benzylcyclohexyl cation, CH_2Ph, ⊕) $\xrightarrow[H_2O]{HCN}$ 1-benzyl-1-(NH-C(=O)-H)cyclohexane

ii

Normal Ritter reaction with HCN to give a formamide

Adlerová and Protiva, *Collect. Czech. Chem. Commun.* **33**, 2941 (1968).

89.

i + $POCl_3$ → $ClCH_2$-C(⊕)($OPCl_2$)-OH + ii → HO-C($OPCl_2$)(CH_2Cl)—N(⊕)(-Ph)=CH-Ph

$\xrightarrow{-H^+}$ HO-C($OPCl_2$)(⊖|CH-Cl)—N(-Ph)-CH(⊕)-Ph → HO-C($OPCl_2$)—N-Ph four-membered ring (Cl, Ph) → O=C—N-Ph four-membered ring (Cl, Ph)

iii

Ziegler, Wimmer, and Mittelbach, *Monatsh. Chem.* **99**, 2128 (1968).

90.

i + ii → (⊕N, C=O, |Ō-H) →

(N, O-H ⊕) $\xrightarrow{-H^+}$ (N, O, O)

iii

Ziegler, Kappe, and Kollenz, *Monatsh. Chem.* **99**, 2024 (1968).

96. In all 3 cases:

substrate

base

viii

case a
$R^2 = H$
(H^+)

ix

$-H_2O$

$-H^+$

iii

cases b and c
$R^2 = Me$
$-CNO^-$

vii
$(R^1 = H)$

case b only
H_2O
$-H^+$

(H^+)

hyd-rol.

$-H^+$

v

Intermediate ix can only lose water in case a. In cases b and c, therefore, ix reverts to viii. Case c ends at vii, but case b continues as shown.

Spence and Tennant, *Chem. Commun.* **1969**, 194.

100.

Carry out the reaction with compounds for which one mechanism is impossible:

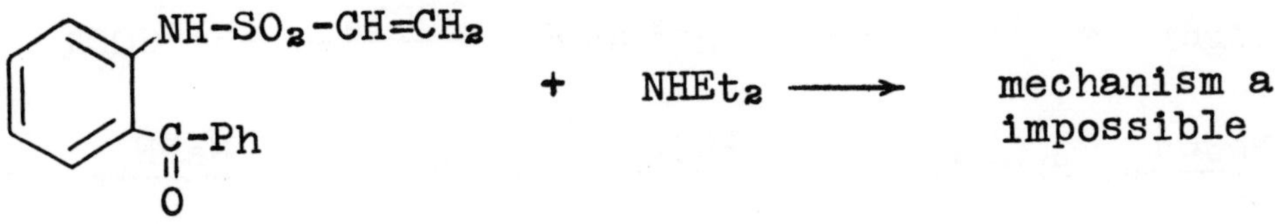

o-$(NH\text{-}SO_2\text{-}CH{=}CH_2)C_6H_4\text{-}C(=O)\text{-}Ph$ + $NHEt_2$ ⟶ mechanism a impossible

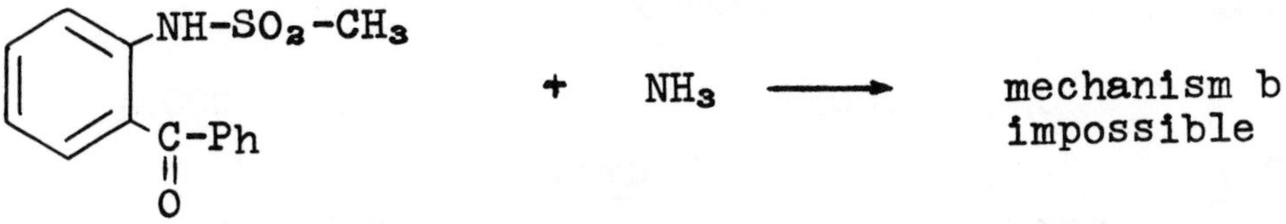

o-$(NH\text{-}SO_2\text{-}CH_3)C_6H_4\text{-}C(=O)\text{-}Ph$ + NH_3 ⟶ mechanism b impossible

These reactions were carried out, and the first was much faster than the second, indicating that mechanism b was probably operating in the original case.
Hromatka, Binder, and Knollmüller, *Monatsh. Chem.* **99**, 1062, 1124 (1968).

104.

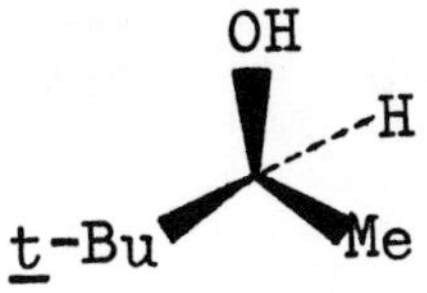

S-3,3-dimethyl-2-butanol

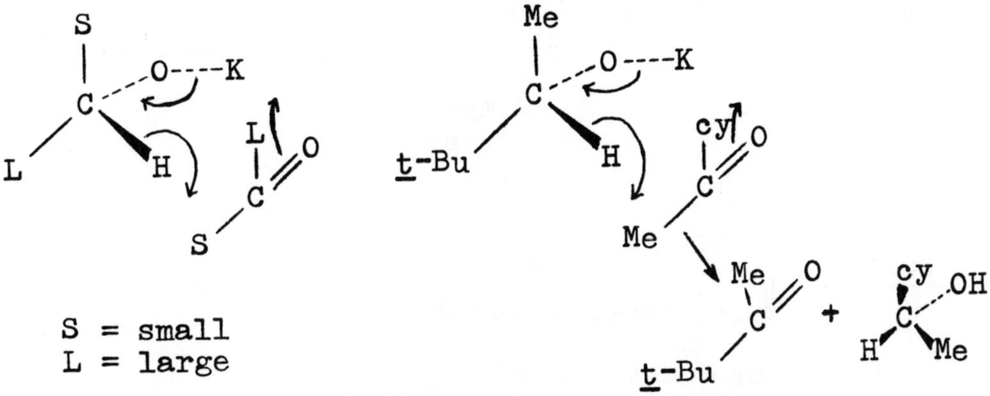

S = small
L = large

The diagram on the left shows how the reactants line up in the general case; that on the right shows the specific case under consideration. The product will predominantly have the S configuration.
Červinka, Bělovský, Fábryová, Dudek, and Grohman, *Collect. Czech. Chem. Commun.* **32**, 2618 (1967).

ANSWERS TO PROBLEM SET 11

ELIMINATIONS

3. $PhMgBr + {}^{13}CO_2 \longrightarrow Ph^{13}COOH \xrightarrow{SOCl_2} Ph^{13}COCl$

$\xrightarrow{Me_2Cd} Ph\text{-}^{13}C(=O)\text{-}CH_3 \xrightarrow{PCl_5} Ph\text{-}^{13}CCl_2\text{-}CH_3 \xrightarrow{NaNH_2}$ i

Casanova, Geisel, and Morris, *J. Amer. Chem. Soc.* **91**, 2156 (1969).

6. i $\xrightarrow{SO_2Cl_2}$ EtOOC–pyrrole(N–H)(CH₃)(COOEt)–CHCl₂ $\xrightarrow{\text{hydrol.}}$ EtOOC–pyrrole(N–H)(CH₃)(COOEt)–CHO

$\xrightarrow{NH_2OH}$ EtOOC–pyrrole(N–H)(CH₃)(COOEt)–CH=N–OH $\xrightarrow[Ac_2O]{NaOAc}$ ii

Hanck, *Chem. Ber.* **101**, 2280 (1968).

9. i + HCHO + $EtO\text{-}C(=O)\text{-}NH_2 \longrightarrow$ 3-(OCH₃)C₆H₄–$SO_2CH_2NHCOOEt$

Mannich reaction

$\xrightarrow{NOCl}$ 3-(OCH₃)C₆H₄–$SO_2CH_2N(NO)COOEt$ $\xrightarrow{Al_2O_3}$ ii

an N-nitroso-N-alkylurethan

Engberts, Zuidema, Zwanenburg, and Strating, *Rec. Trav. Chim.* **88**, 641 (1969).

14.

i + $CHBr_3$ $\xrightarrow{\text{base}}$ Br Br $\xrightarrow[\Delta]{\text{quinoline}}$

[Br Br or Br Br] $\xrightarrow[\text{MeOH}]{\text{Zn}}$ ii

Skatteböl, *Tetrahedron Lett.* **1961**, 167; Baird and Reese, *J. Chem. Soc., C* **1969**, 1808.

16.

i + HOBr $\longrightarrow$ CH_3 AcO Br OH $\xrightarrow{Pb(OAc)_4}$

H_2C O AcO Br $\xrightarrow{\text{Zn}}$ ii

Boord reaction

Schwarz, Zachová, and Syhora, *Collect. Czech. Chem. Commun.* **33**, 4337 (1968).

18.

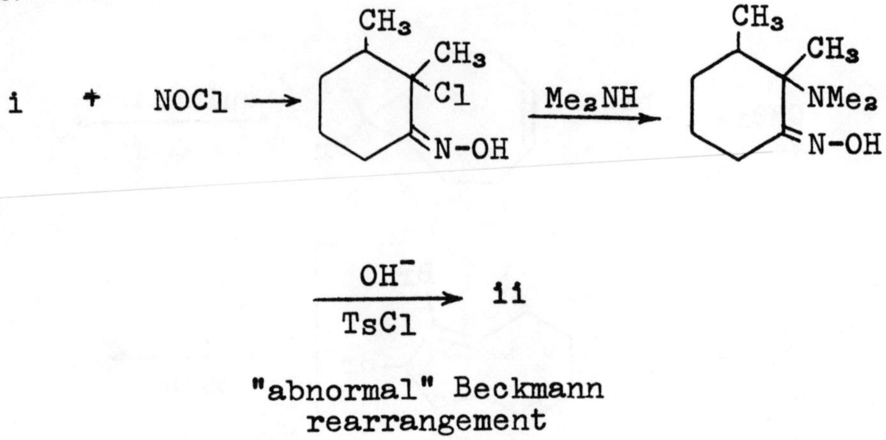

i + NOCl → [2-chloro-2,3-dimethylcyclohexanone oxime: CH_3, CH_3, Cl, N-OH] $\xrightarrow{Me_2NH}$ [2-dimethylamino-2,3-dimethylcyclohexanone oxime: CH_3, CH_3, NMe_2, N-OH]

$\xrightarrow[TsCl]{OH^-}$ ii

"abnormal" Beckmann rearrangement

Lunkwitz, Pritzkow, and Schmid, *J. Prakt. Chem.* [4] **37**, 319 (1968).

24.

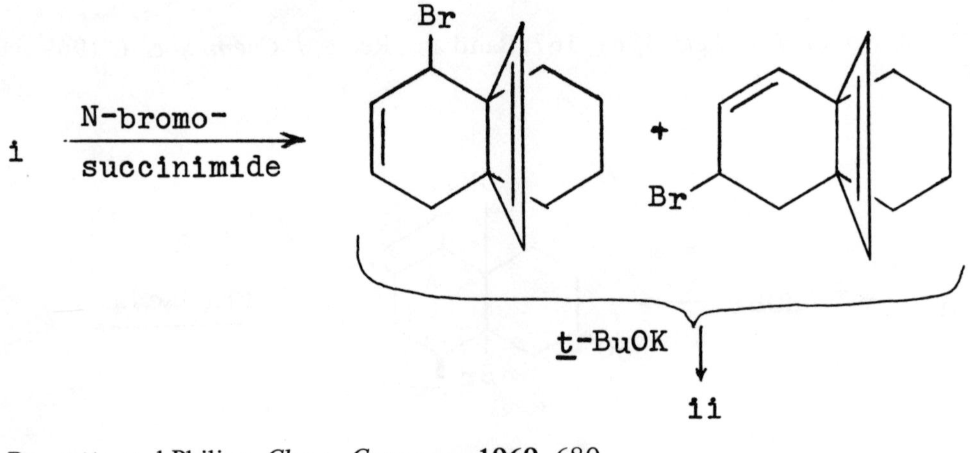

i $\xrightarrow[\text{succinimide}]{\text{N-bromo-}}$ [Br-substituted bicyclic diene] + [Br-substituted bicyclic diene, Br]

t-BuOK ↓

ii

Paquette and Philips, *Chem. Commun.* **1969**, 680.

26.

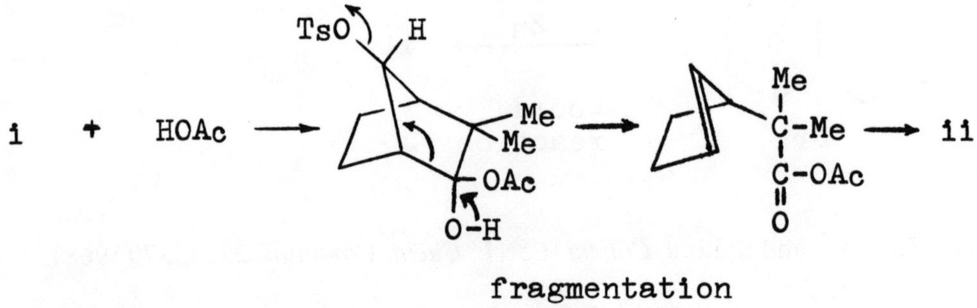

i + HOAc → [TsO, H, Me, Me, OAc, O-H] → [Me, C-Me, C-OAc, O] → ii

fragmentation

Gassman and Hornback, *Tetrahedron Lett.* **1969**, 1325.

30.

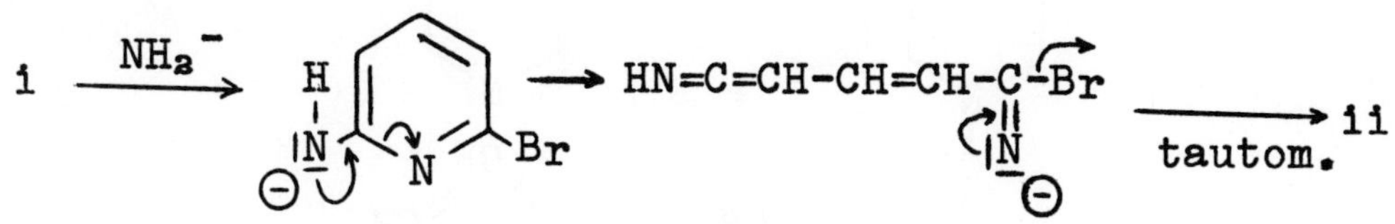

Streef and den Hertog, *Tetrahedron Lett.* **1968**, 5945.

34.

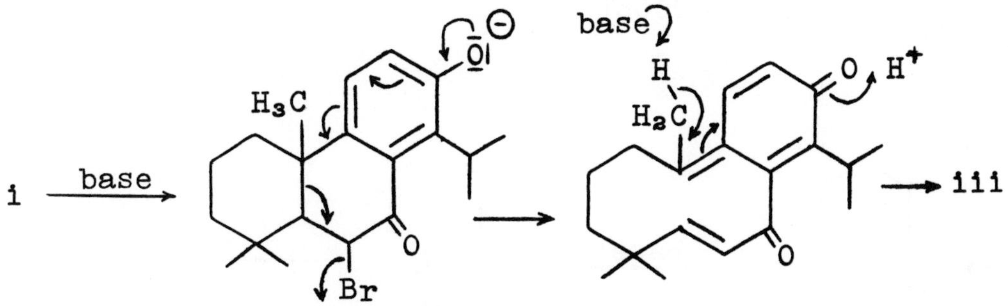

The simple E2 mechanism is hindered because there is no β-hydrogen trans to the bromine. The simple E1 mechanism is hindered because it would place a positive charge α to a C=O carbon. Because simple elimination by either mechanism is greatly slowed, the fragmentation process predominates.

Cambie and Gallagher, *Tetrahedron* **24**, 4631 (1968).

38.

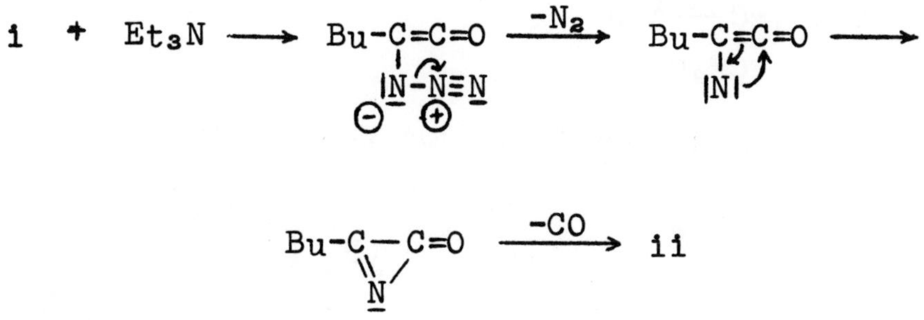

Hassner, Isbister, Greenwald, Klug, and Taylor, *Tetrahedron* **25**, 1637 (1969).

40.

reverse Michael reaction: E1cB

Smutny, *J. Amer. Chem. Soc.* **91**, 208 (1969).

43.

Newallis, Lombardo, and McCarthy, *J. Org. Chem.* **33**, 4169 (1968).

48.

Kuhn and Teller, *Justus Liebigs Ann. Chem.* **715**, 106 (1968).

52.

i —(1. Me_2SO_4; 2. OH^-)→ [Me, Me, $\oplus$N, H, OH^-, OMe, O, OMe] →

[OMe, O, CH_2–CH_2, NMe₂, OMe] —((H^+), ClO_4^-)→ [OMe, OH, $\overset{\oplus}{N}Me_2$ ClO_4^-, OMe]

ii

Fleischhacker, Passl, and Vieböck, *Monatsh. Chem.* **99**, 300 (1968).

56. On the basis of the E2 mechanism we assume trans elimination. (–)–Thujol and (+)–neoisothujol must have the OH cis to the CH_3, since otherwise the elimination could not place in that direction. Consequently, in the other two isomers the CH_3 and the OH must be trans, since this is the only possibility left. Since the configuration of the CH_3 group is not changed in the formation of (+)–β–isothujene or (+)–β–thujene, (–)–neothujol must be ii and (+)–isothujol must be iii. However, it is not possible from the data given to decide

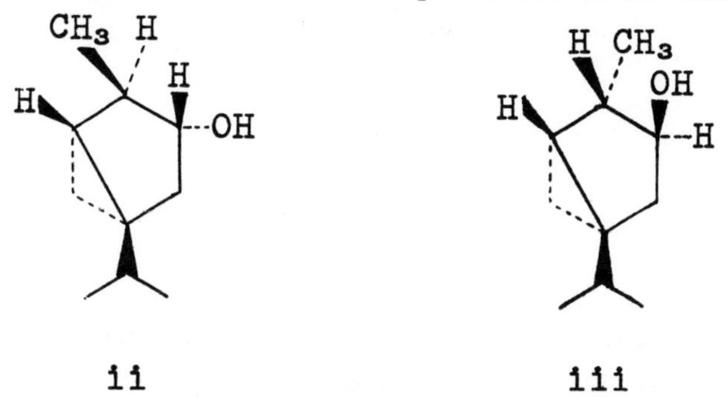

which of the cis structures corresponds to (–)–thujol and which to (+)–neoisothujol. The signs of rotation are no help, since there is no necessary correlation between direction of rotation of the plane of polarized light and configuration.

Banthorpe and Davies, *J. Chem. Soc., B* **1968**, 1339.

60. The formation of iii and v shows nothing. The fact that *cis*-i gave both ii and iv with all the deuterium present, and there was no isotope effect, shows that the hydrogen that departed (in the formation of both ii and iv) had to be trans to the nitrogen:

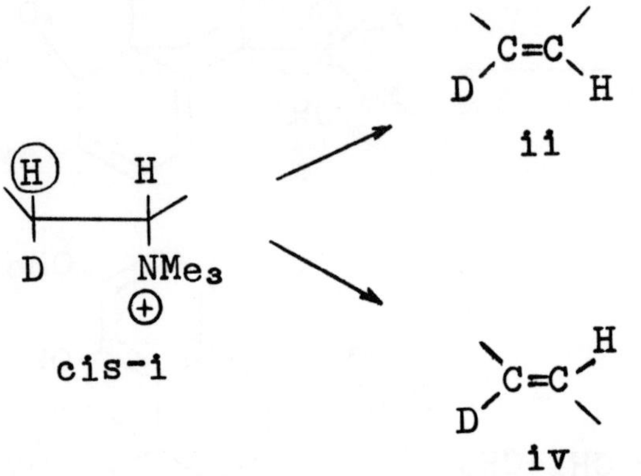

H H
D NMe_3 ⊕
cis-i

C=C D H
ii

H C=C D
iv

The results from *trans*-i are in accord with this statement, since both ii and iv had no deuterium, and there was an isotope effect. Since both cis and trans olefins were formed, then the cis olefin from either cis or trans-i was formed by the usual anti elimination, but the *trans olefin must have been formed by a syn elimination*, the molecule twisting around to this position.

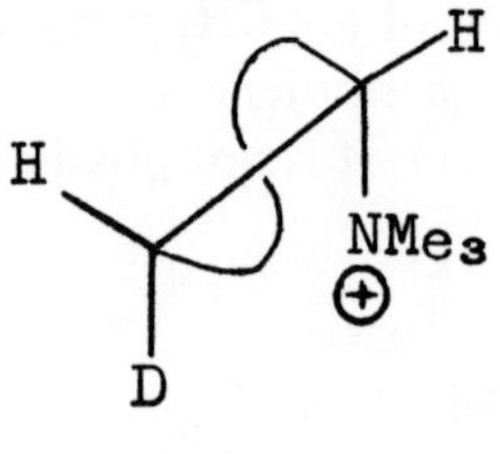

H H
NMe_3 ⊕
D

trans-i

An α', β elimination can be ruled out, since then the NMe_3 would have to be labeled, and it was not labeled in either case.

Závada, Svoboda, and Sicher, *Collect. Czech. Chem. Commun.* **33**, 4027 (1968); *Tetrahedron Lett.* **1966**, 1627.

ANSWERS TO PROBLEM SET 12

REARRANGEMENTS

4.

i $\xrightarrow{\Delta}$ $N_2CHCH_2CH_2CHN_2$ + [cyclohexanone] ⟶

[bicyclic ketone] $\xrightarrow{TsNHNH_2}$ NH—NHTs $\xrightarrow[\text{liq. } CH_3CONH_2]{Na}$ ii

Jacobson, *Acta Chem. Scand.* **21**, 2235 (1967).

8.

Me, Me, i $\xrightarrow{\Delta}$ Me, Me $\xrightarrow{t\text{-BuOK}}$ ii

Cope rearr.

double bond migrations

Heimbach and Schimpf, *Angew. Chem. Intern. Ed. Engl.* **7**, 727 (1968) [*Angew. Chem.* **80**, 704 (1968)]

10.

i + PCl_5 ⟶ [CH_2COOMe, N=C(Cl)–C_6H_4–Cl] +

[O^-, COOMe] ⟶ [CH_2COOMe, N=C(O–C_6H_4–COOMe)–C_6H_4–Cl] $\xrightarrow{\Delta}$ ii

Chapman rearr.

Schulenberg, *J. Amer. Chem. Soc.* **90**, 7008 (1968).

15.

i $\xrightarrow{NH_2OH}$ Pr-C(=N-OH)-Pr $\xrightarrow{Ac_2O}$ Pr-C(=N-OAc)-Pr $\xrightarrow[BF_4^-]{Et_3O^+}$ Pr-C=CH-Et, Et-N, O-C(=O)-CH$_3$

$\xrightarrow{\Delta}$ Claisen-type rearrangement → Pr-C(=N-Et)-CH(Et)-O-C(=O)-CH$_3$ $\xrightarrow{\text{hydrolysis of C=N}}$ ii

House and Richey, *J. Org. Chem.* **34**, 1430 (1969).

20.

i + N-Cl (succinimide) → HO (steroid) CH$_2$-N(Cl)-CH$_3$ $\xrightarrow[F_3CCOOH]{h\nu}$ ii

Hofmann-Löffler reaction

Hora and Šorm, *Collect. Czech. Chem. Commun.* **33**, 2059 (1968).

21.

i $\xrightarrow{}$ 1. HCl-HOAc hydrolysis 2. $AgNO_3$ 3. Br_2

Hunsdiecker reaction

Br, O, O, Br $\xrightarrow[H_2O]{KOH}$

Favorskii rearr.

HOOC, COOH $\xrightarrow{}$ 1. $AgNO_3$ 2. Br_2 → ii

Hunsdiecker reaction

Vogt, *Tetrahedron Lett.* **1968**, 1579.

23.

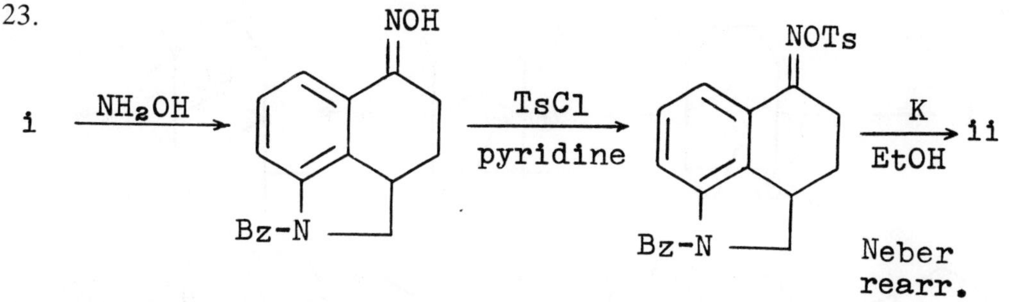

i $\xrightarrow{NH_2OH}$ [oxime, Bz-N] $\xrightarrow[\text{pyridine}]{TsCl}$ [NOTs, Bz-N] $\xrightarrow[EtOH]{K}$ ii

NOH

NOTs

Bz-N

Neber rearr.

Kornfeld, Fornefeld, Kline, Mann, Morrison, Jones, and Woodward, *J. Amer. Chem. Soc.* 78, 3087 (1956).

29

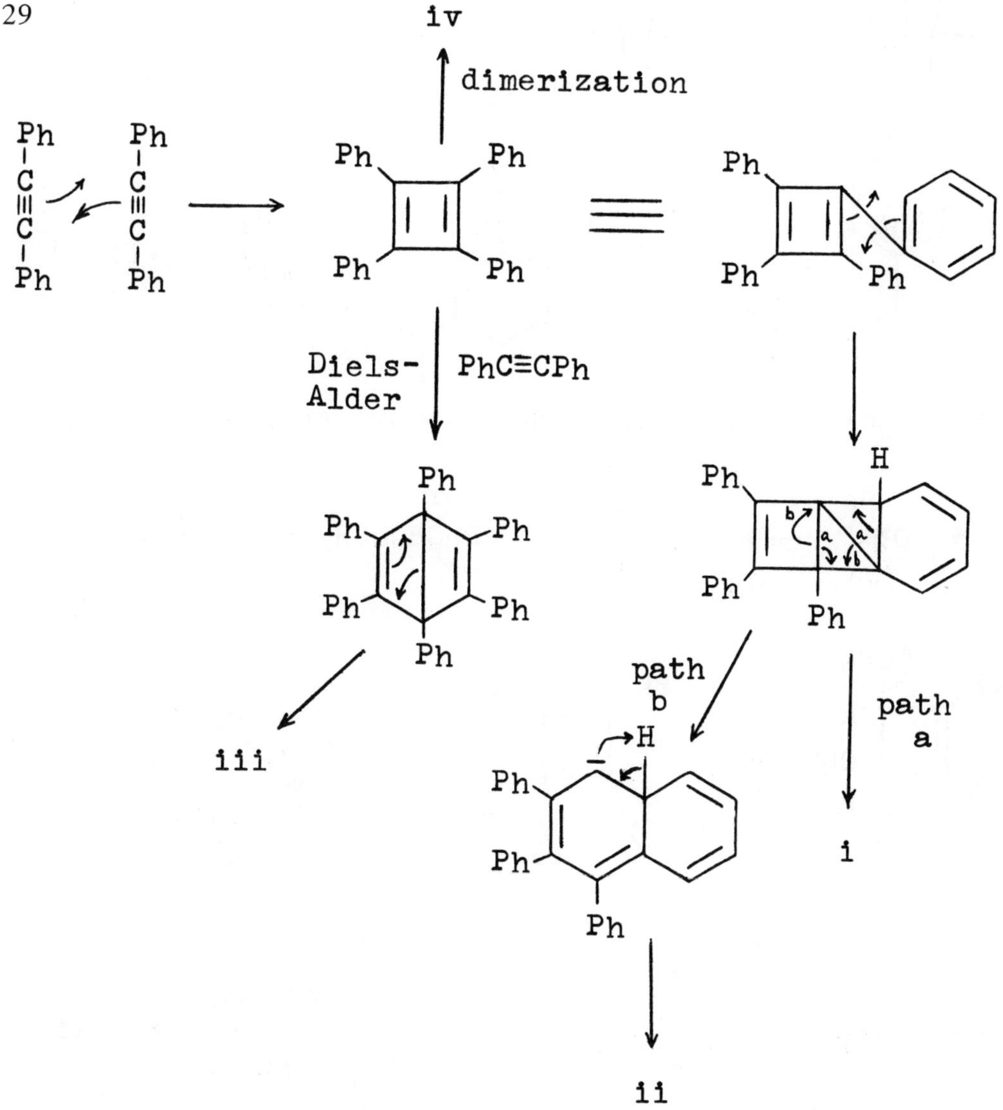

iv

dimerization

Ph Ph

Diels-Alder

PhC≡CPh

iii

H

path b

path a

i

ii

Büchi, Perry, and Robb, *J. Org. Chem.* 27, 4106 (1962).

32.

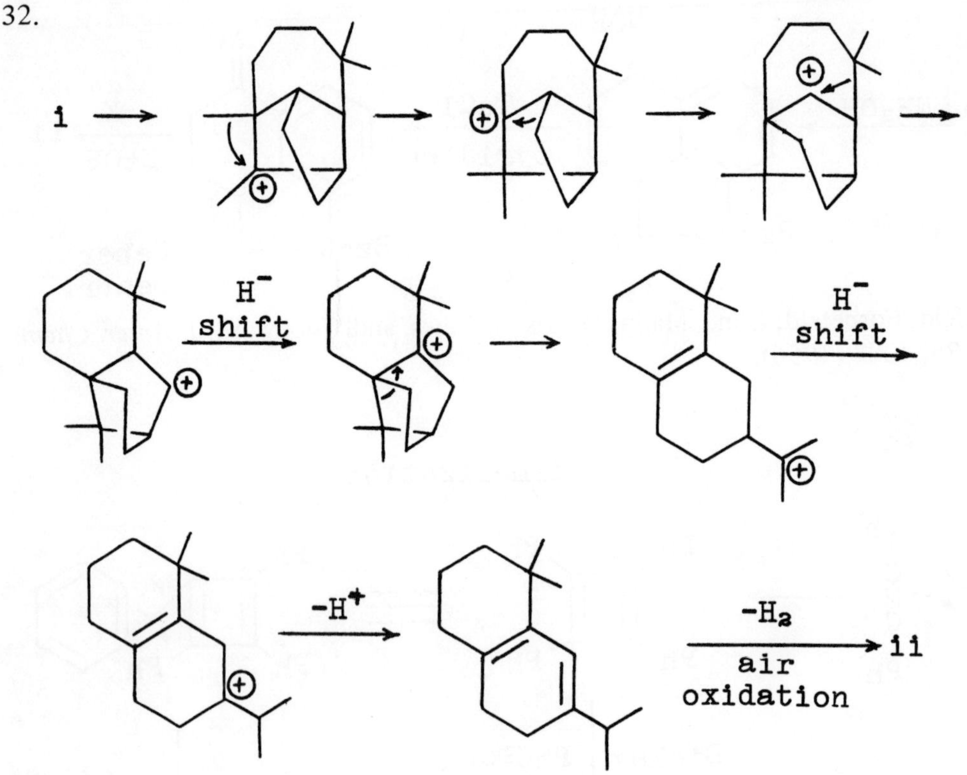

Kettenes, van Lierop, van der Wal, and Sipma, *Rec. Trav. Chim.* **88**, 313 (1969).

35.

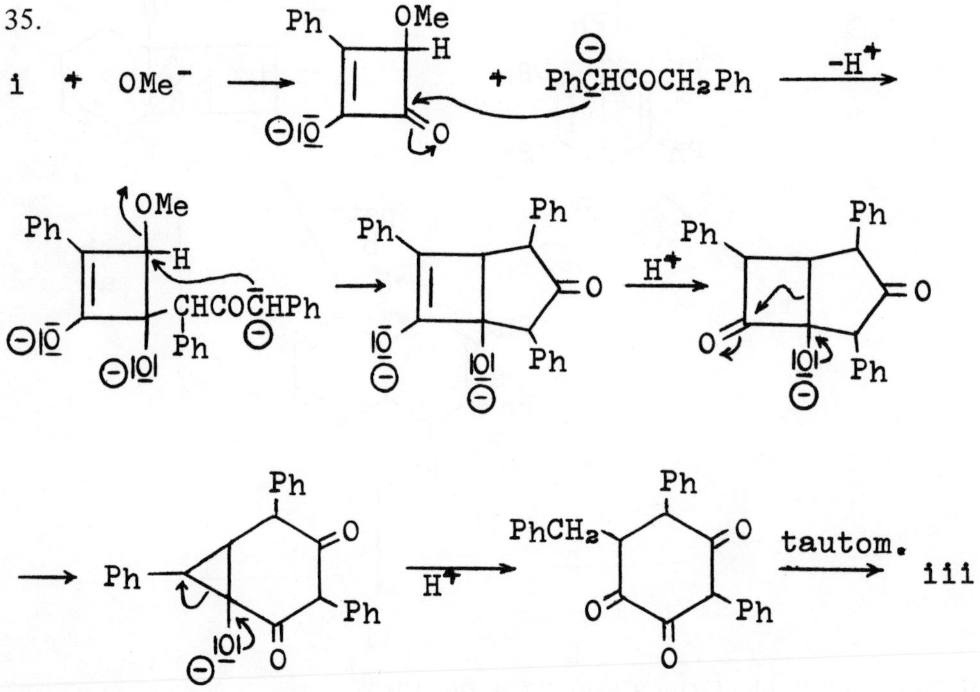

Ried and Kunkel, *Justus Liebigs Ann. Chem.* **717**, 54 (1968).

39.

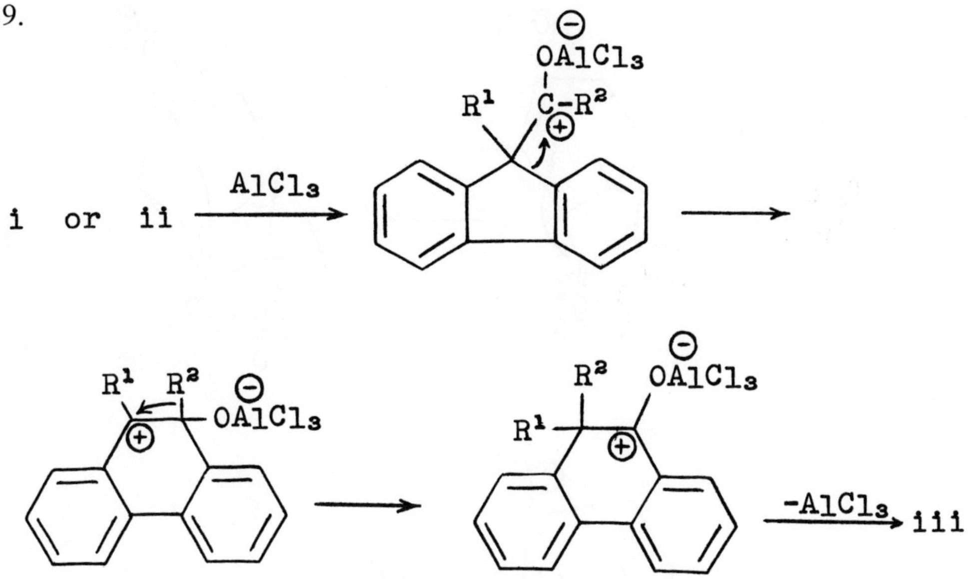

i or ii $\xrightarrow{AlCl_3}$ → → $\xrightarrow{-AlCl_3}$ iii

Rouzaud, Cauquil, Mathieu, and Boyer, *Bull. Soc. Chim. Fr.* **1968**, 4851.

43.

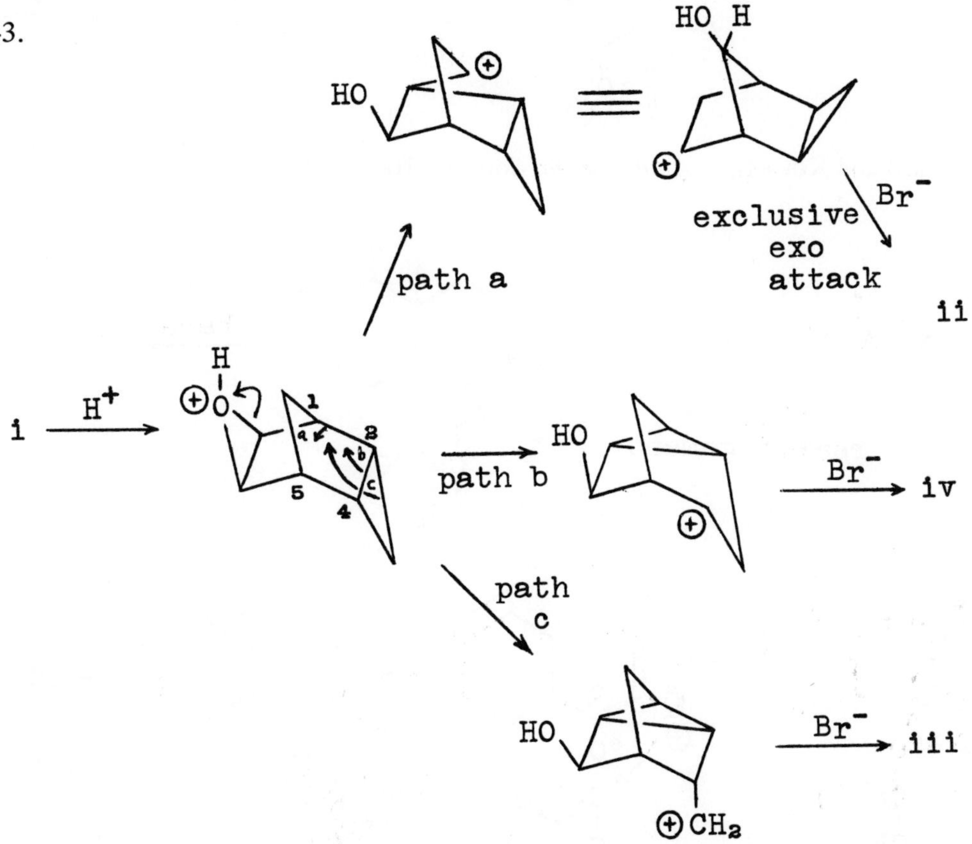

i $\xrightarrow{H^+}$ path a → exclusive exo attack, Br^- → ii; path b → $\xrightarrow{Br^-}$ iv; path c → $\xrightarrow{Br^-}$ iii

In path a, the 4,5 bond has an equal chance with the 1,2 bond of shifting. This will lead to the enantimer of ii, as drawn.

Henshaw, Rome, and Johnson, *Tetrahedron Lett.* **1968**, 6049.

47.

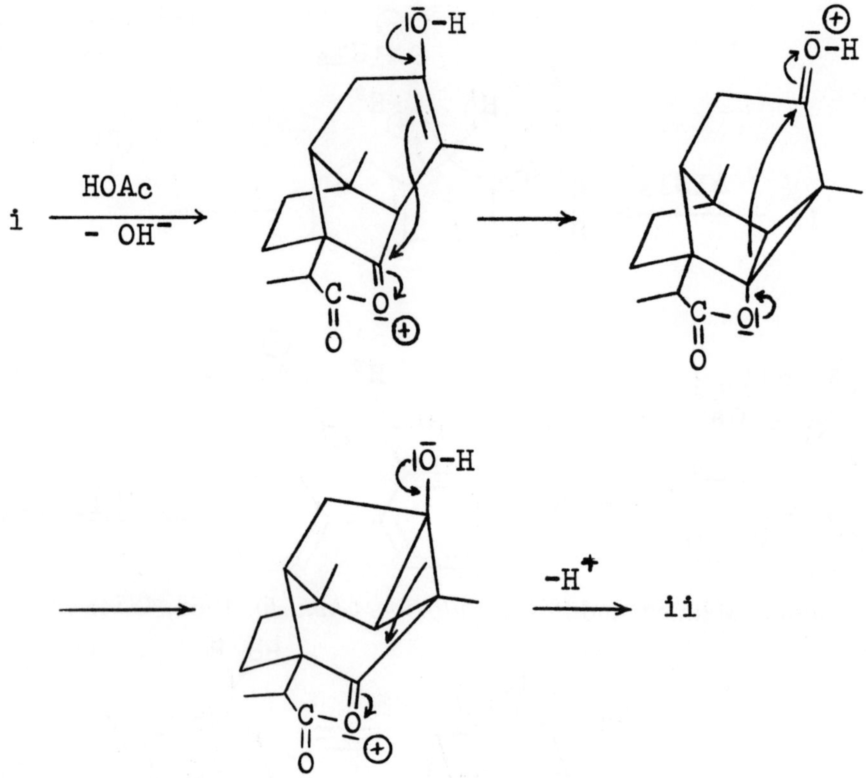

i $\xrightarrow[-\ OH^-]{HOAc}$

$\xrightarrow{-H^+}$ ii

Woodward and Kovach, *J. Amer. Chem. Soc.* 72, 1009 (1950).

52.

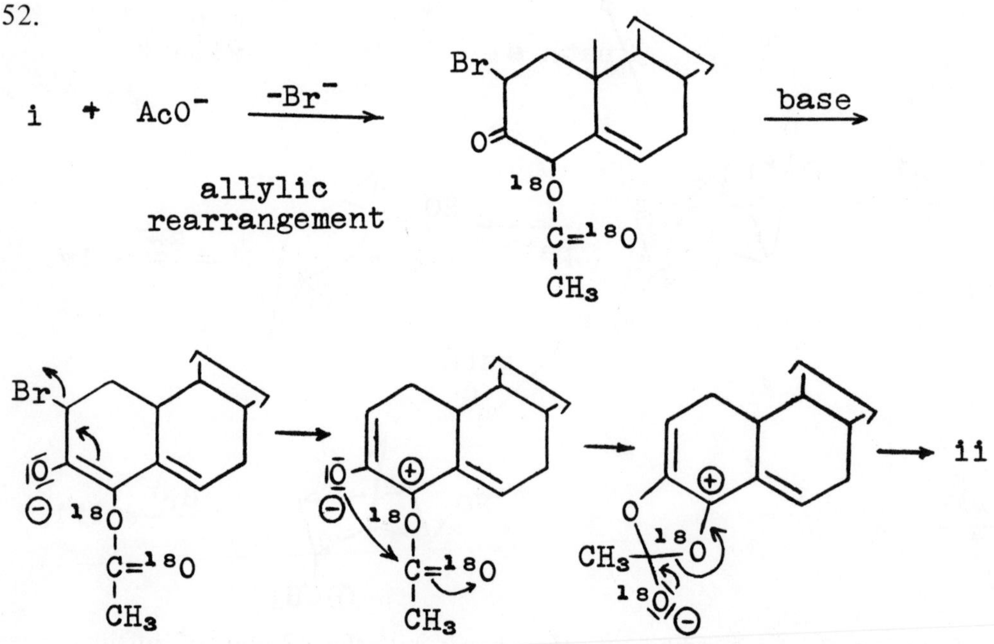

i + AcO^- $\xrightarrow{-Br^-}$

allylic rearrangement

$\xrightarrow{base}$

→ ii

Julian, Bauer, Bell, and Hewitson, *J. Amer. Chem. Soc.* **91**, 1690 (1969).

56.

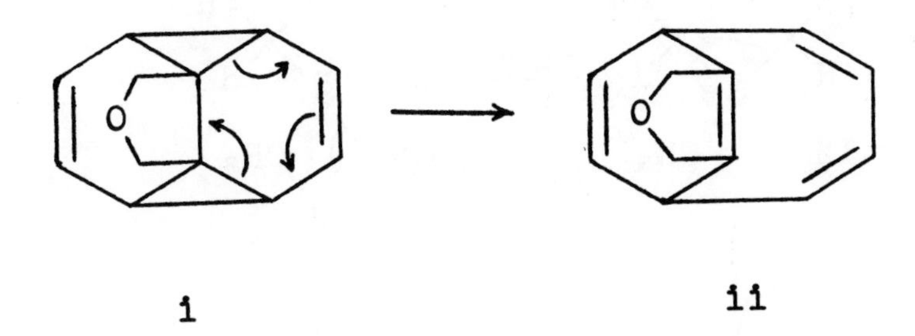

Babad, Ginsburg, and Rubin, *Tetrahedron Lett.* **1968**, 2361.

57.

i —tautom.→ Ph, H, C=C, O, |Ō–H, C=O, H⁺, CH_3 → Ph, H, O, O–H (+), CH_3, OH ⇌

Ph, H, H–O (+), O, CH_3, Ō–H → Ph, H, C=C, HO, O, C=OH (+), CH_3 —$-H^+$ / tautom.→ ii

Riehl and Fougerousse, *Bull. Soc. Chim. Fr.* **1968**, 4083.

62.

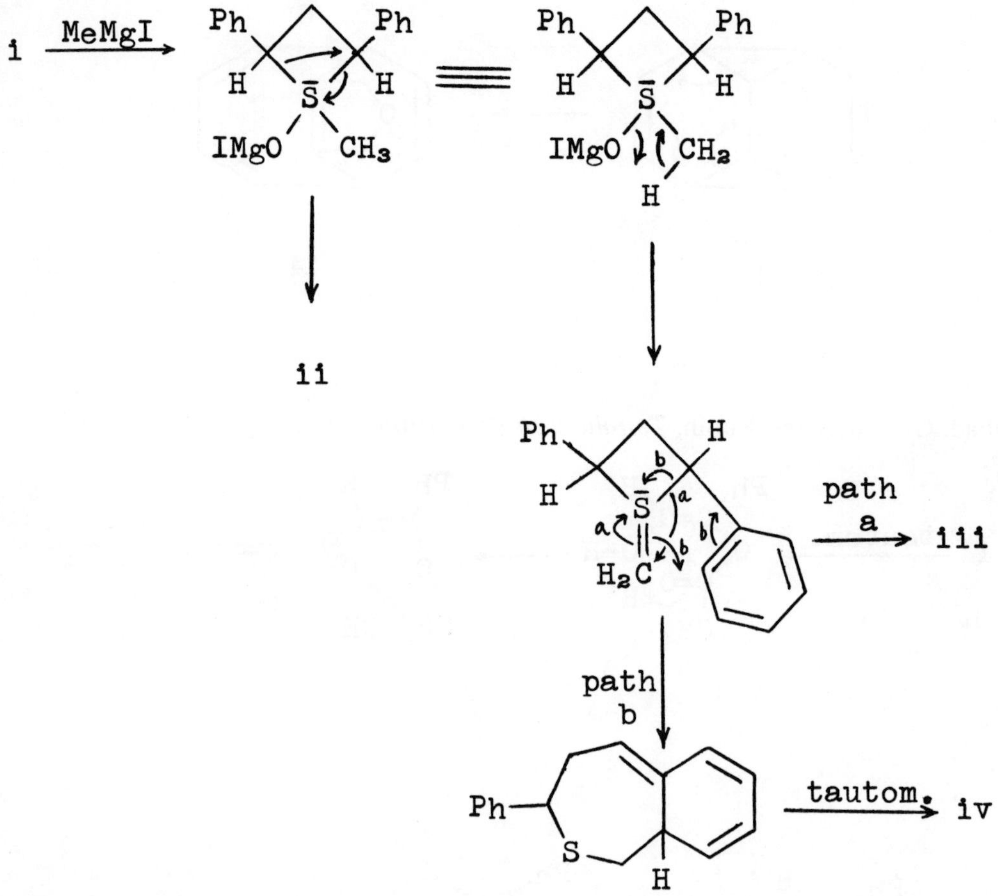
i
MeMgI
Ph
H
S
IMgO
CH3
CH2
ii
path a
iii
path b
H2C
tautom.
iv

Dodson and Hammen, *Chem. Commun.* **1968**, 1294.

64. Abnormal Claisen Rearrangement:

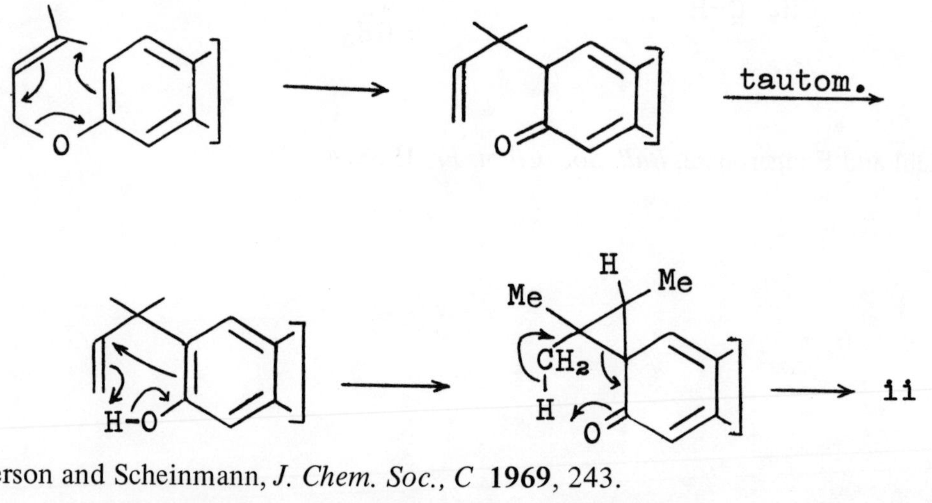
O
tautom.
H–O
Me
H
CH2
ii

Jefferson and Scheinmann, *J. Chem. Soc., C* **1969**, 243.

69.

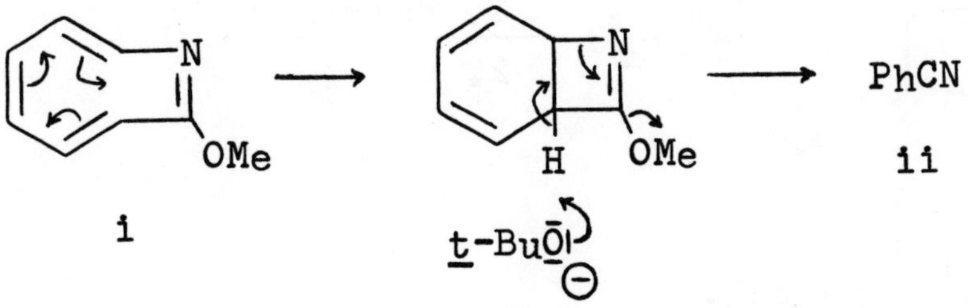

i → → PhCN

ii

t-BuO⁻

Paquette and Kakihana, *J. Amer. Chem. Soc.* **90**, 3897 (1968).

72.

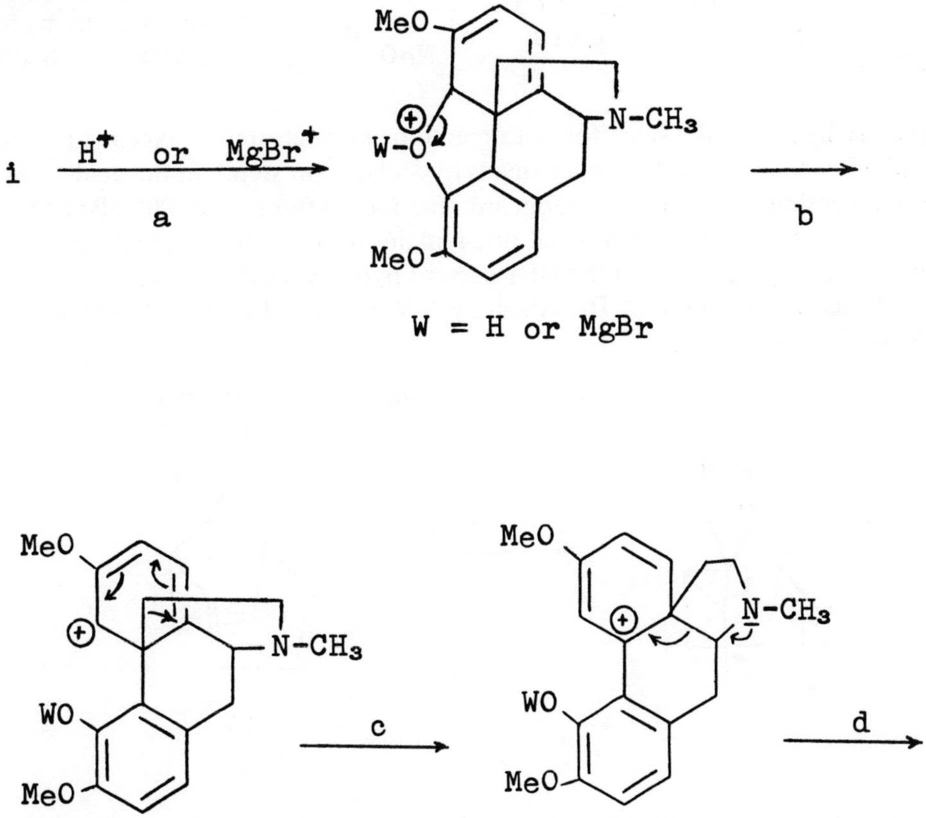

i —H^+ or $MgBr^+$ (a)→ ; W = H or MgBr —b→

—c→ —d→

72. Cont.

MeO
WO
MeO
$\overset{\oplus}{N}$-CH_3

1. PhMgBr
2. H^+
iii

H^+ hydrol. of imine

HO
HO
MeO
$NHCH_3$
CHO

upper ring turns on its axis

CH_3
NH
OH
CHO
HO
MeO

H^+
aromatic substitution ring closure
ii

The mechanism is the same for both reactions through step d, except that the enol ether function in the upper ring is hydrolyzed in step b in the acid-catalyzed reaction, but not in the Grignard reaction. After step d, PhMgBr adds to the imine so that the 9-membered ring remains intact in the product, but in the acid-catalyzed reaction the HC1 present hydrolyzes the imine.

Stork, in Manske and Holmes, "The Alkaloids," Vol. 2, pp. 196-198, Academic Press, New York, 1952.

74.

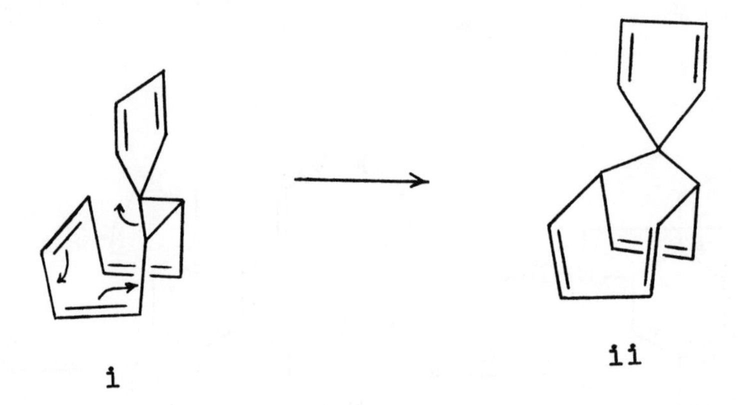

Note: i is a *stable cis*-divinylcyclopropane. Even upon heating it does not give the normal Cope rearrangement product, but instead the reaction shown above.

Schönleber, *Chem. Ber.* **102**, 1789 (1969).

79.

$$\begin{array}{l} CH_2-C\equiv C-CH_3 \\ | \\ CH_2-C\equiv C-H \end{array} \xrightarrow{\text{Cope rearr.}} \begin{array}{l} CH_2=C=C-CH_3 \\ \quad\quad\;\; | \\ CH_2=C=C-H \end{array} \longrightarrow \text{ii}$$

i

Huntsman and Wristers, *J. Amer. Chem. Soc.* **89**, 342 (1967); Coller, Heffernan, and Jones, *Aust. J. Chem.* **21**, 1807 (1968).

84.

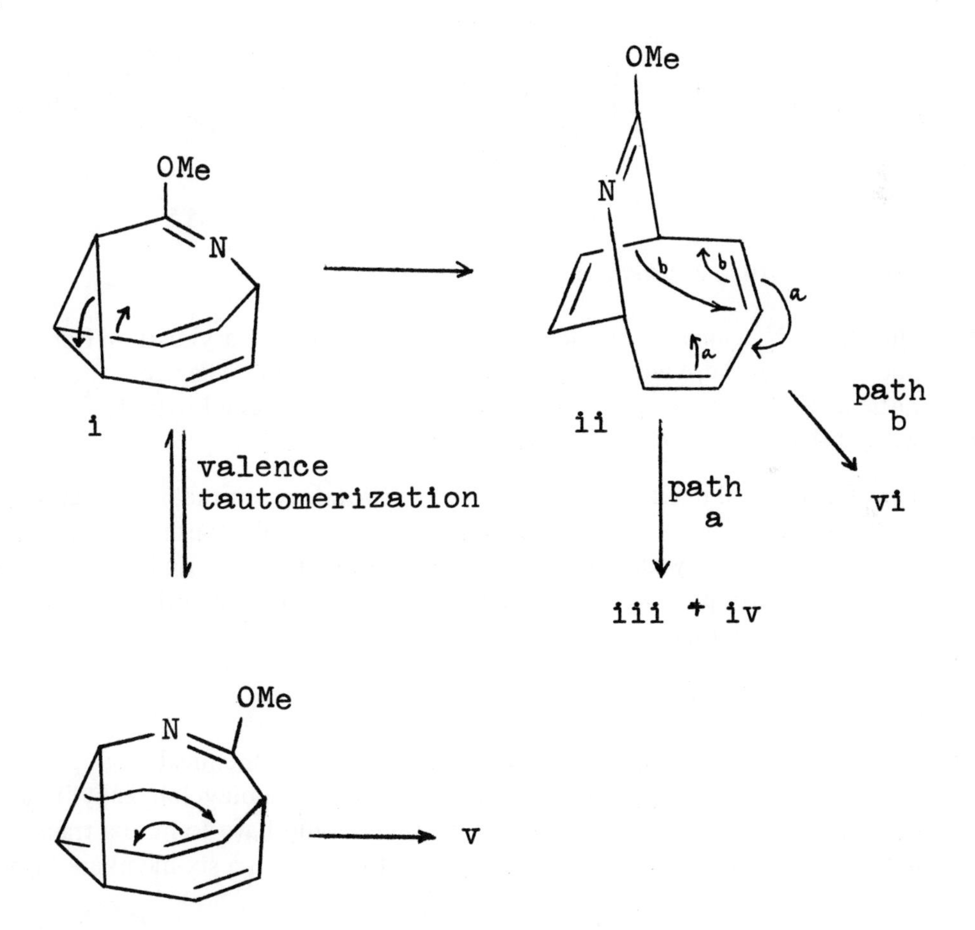

Path a is disrotatory, and iii is formed from upward-swinging hydrogens, and iv from downward-swinging hydrogens.

Paquette and Krow, *J. Amer. Chem. Soc.* **90**, 7149 (1968).

86.

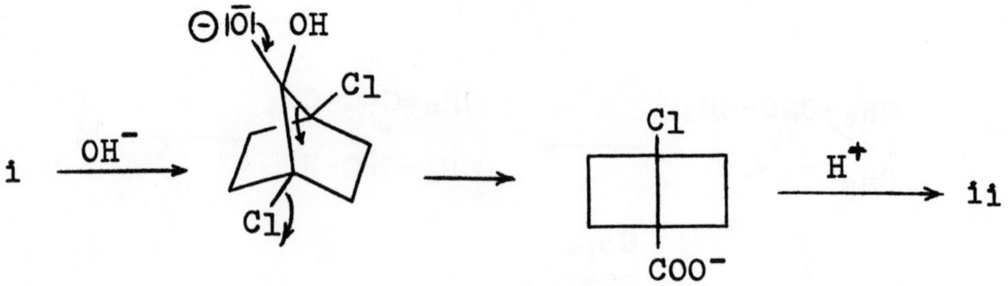

$$\text{i} \xrightarrow{OH^-} \ldots \longrightarrow \ldots \xrightarrow{H^+} \text{ii}$$

This is a normal Favorskii rearrangement, but the semibenzilic mechanism operates, since the normal mechanism is precluded.

Dauben, Chitwood, and Scherer, *J. Amer. Chem. Soc.* **90**, 1014 (1968).

89.

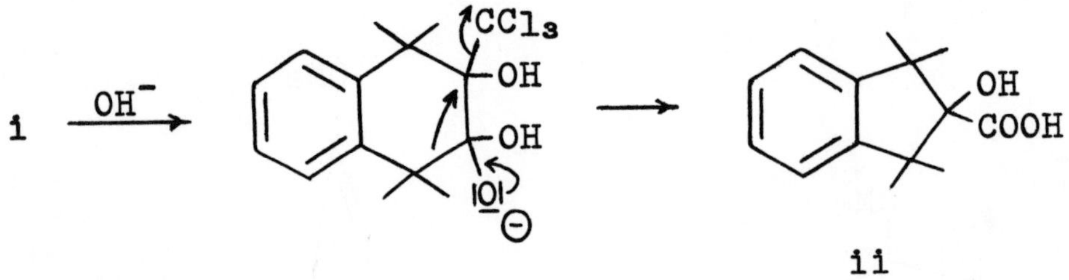

$$\text{i} \xrightarrow{OH^-} \ldots \longrightarrow \text{ii}$$

Gream and Paice, *Aust. J. Chem.* **22**, 1249 (1969).

92. The loss of N_2 cannot be concerted with cleavage of the aryl-carbon bond, since if it were there would have been an isotope effect. Therefore there is a carbene intermediate: either completely free or complexed with Ag^+.

Yukawa and Ibata, *Bull. Chem. Soc. Jap.* **42**, 802 (1969).

95. a. The thermal cyclization of ii must be disrotatory, so the methyl groups of iv are cis. The cyclization of i is conrotatory, so the methyl groups of iii are trans. b. i cyclizes more easily, since there is less steric hindrance: the methyl groups in a disrotatory process must approach each other, but not in a conrotatory process.

Huisgen, Dahmen, and Huber, *J. Amer. Chem. Soc.* **89**, 7130 (1967).

99. According to the principle of orbital symmetry, photochemical ring opening of a cyclobutene is a disrotatory process. If ii was to give bicyclohexenyl in a disrotatory process, one of the cyclohexene rings would have to have a trans double bond, which is of course, too strained to exist in a six-membered ring:

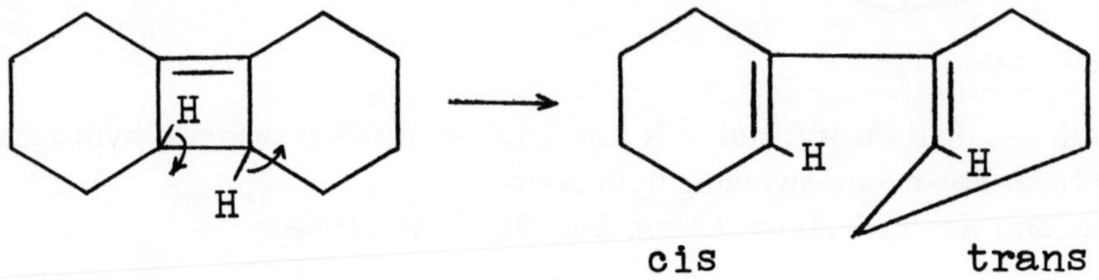

cis trans

Saltiel and Lim, *J. Amer. Chem. Soc.* **91**, 5404 (1969).

101.

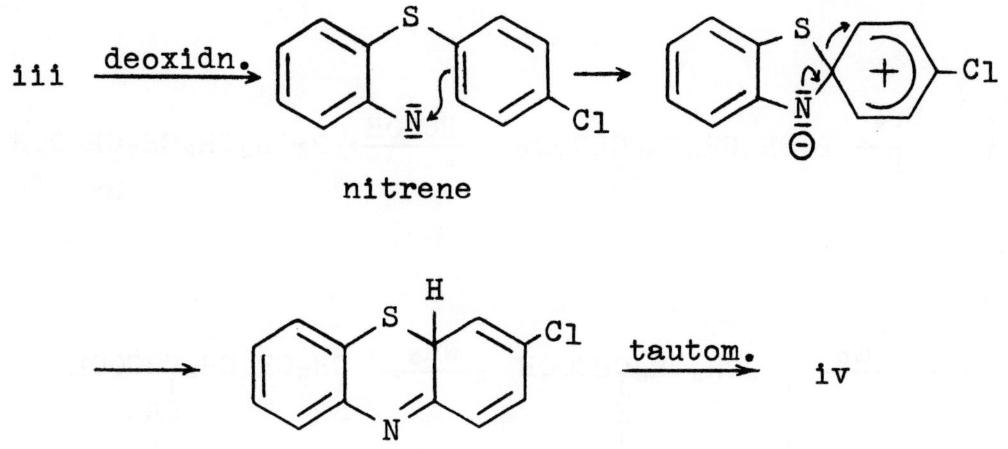

iii —deoxidn.→ [nitrene] → ... → ... —tautom.→ iv

nitrene

The reaction of i takes place by the same mechanism, but the product appears normal, since the ring originally containing the nitro group is unaffected by the rearrangement.

Cadogan, Kulik, and Todd, *Chem. Commun.* **1968**, 736.

ANSWERS TO PROBLEM SET 13

OXIDATIONS AND REDUCTIONS

1.

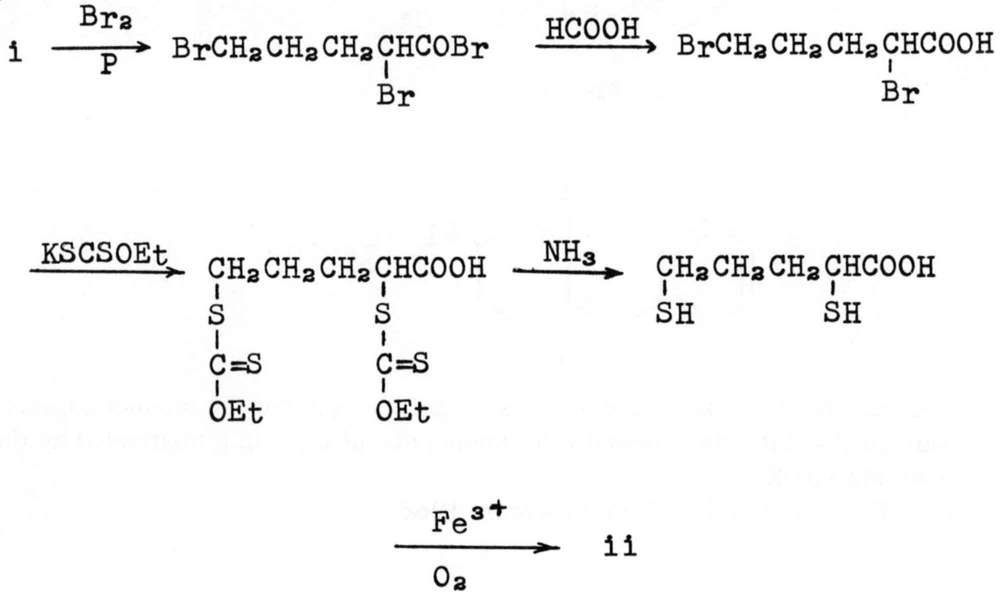

i $\xrightarrow[P]{Br_2}$ $BrCH_2CH_2CH_2CH(Br)COBr$ $\xrightarrow{HCOOH}$ $BrCH_2CH_2CH_2CH(Br)COOH$

$\xrightarrow{KSCSOEt}$ $CH_2(SC(=S)OEt)CH_2CH_2CH(SC(=S)OEt)COOH$ $\xrightarrow{NH_3}$ $CH_2(SH)CH_2CH_2CH(SH)COOH$

$\xrightarrow[O_2]{Fe^{3+}}$ ii

Claeson, *Ark. Kemi* **30**, 511 (1969).

4.

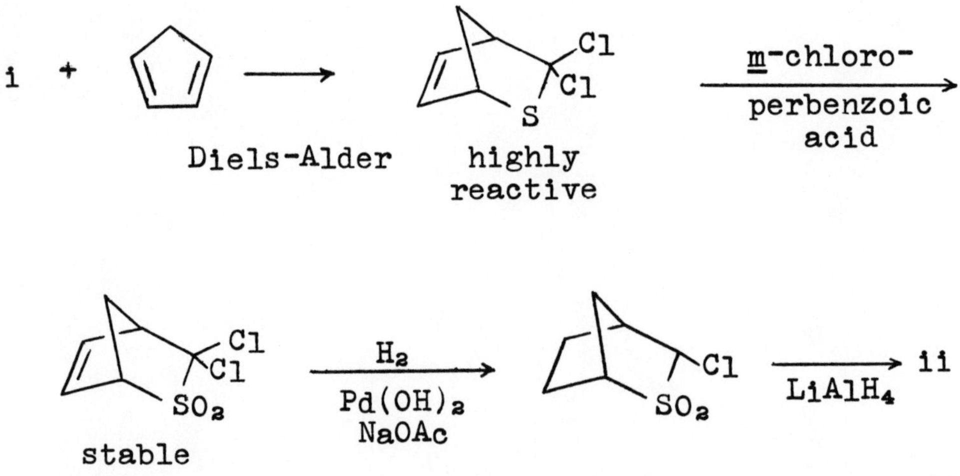

i + cyclopentadiene ⟶ (Diels-Alder) dichlorothiabicyclic adduct (Cl, Cl, S) — highly reactive $\xrightarrow[\text{perbenzoic acid}]{\underline{m}\text{-chloro-}}$

dichloro sulfone adduct (Cl, Cl, SO_2) — stable $\xrightarrow[Pd(OH)_2,\ NaOAc]{H_2}$ saturated chloro sulfone (Cl, SO_2) $\xrightarrow[LiAlH_4]{}$ ii

Diels-Alder

highly reactive

stable

Johnson, Keiser, and Sharp, *J. Org. Chem.* **34**, 860 (1969).

6.

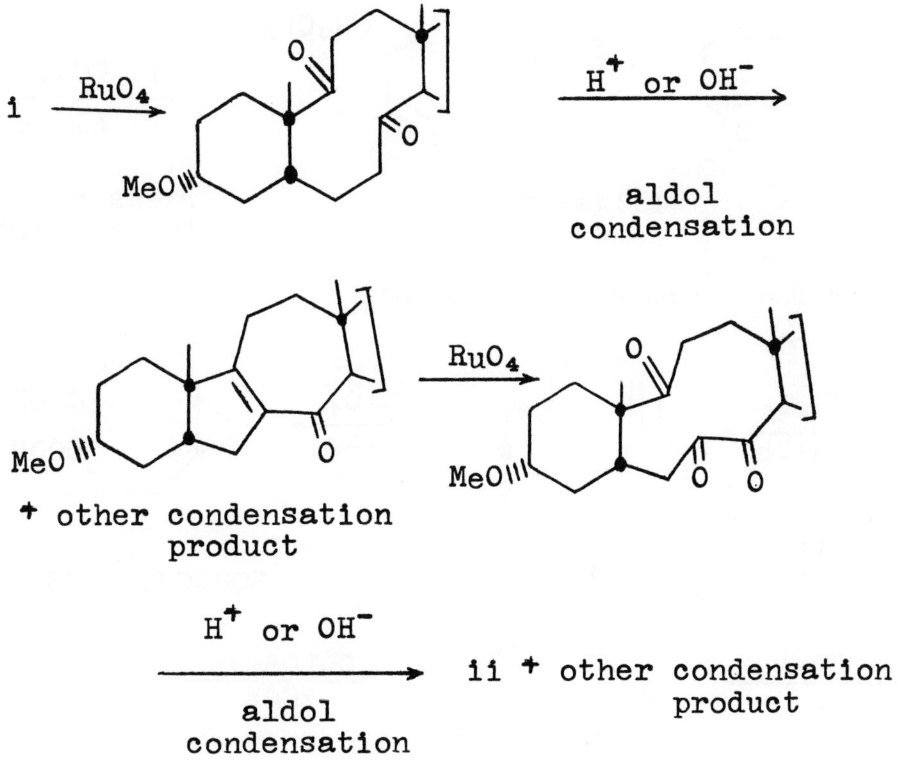

i $\xrightarrow{RuO_4}$ [structure, MeO] $\xrightarrow[\text{aldol condensation}]{H^+ \text{ or } OH^-}$

[structure, MeO] + other condensation product $\xrightarrow{RuO_4}$ [structure, MeO]

$\xrightarrow[\text{aldol condensation}]{H^+ \text{ or } OH^-}$ ii + other condensation product

Snatzke and Nising, *Justus Liebigs Ann. Chem.* **715**, 187 (1968).

12.

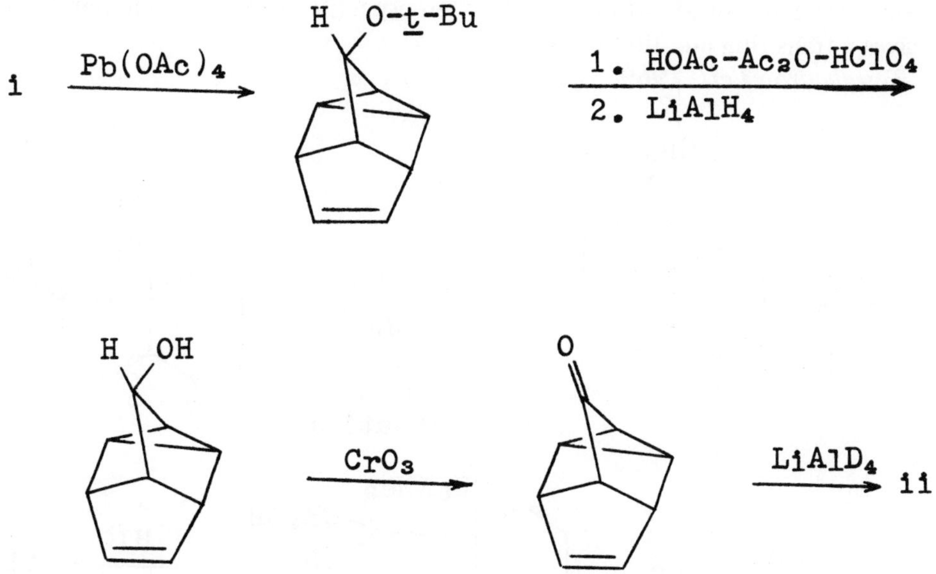

i $\xrightarrow{Pb(OAc)_4}$ [structure, H O-t-Bu] $\xrightarrow[2.\ LiAlH_4]{1.\ HOAc-Ac_2O-HClO_4}$

[structure, H OH] $\xrightarrow{CrO_3}$ [structure, O] $\xrightarrow{LiAlD_4}$ ii

Schleyer and Leone, *J. Amer. Chem. Soc.* **90**, 4164 (1968).

16.

i + H–N (morpholine) —S→ OMe, CH_2CH_2COOH —1. PCl_5 / 2. $AlCl_3$→ ii

Willgerodt reaction

Friedel-Crafts ring closure

Davis and Watkins, *Tetrahedron* **24**, 2165 (1968).

22.

i —CH_2I_2 / Zn-Cu→ iii —KIO_4 / $KMnO_4$→ COOH, COOH + $HOOCCH_2CH_2COOH$

Simmons-Smith reaction

—1. MeOH / 2. Na→ O, OH —$Cu(OAc)_2$ / HOAc→ ii

acyloin cond.

Since the Simmons-Smith reaction is stereospecific, a trans double bond in i gives a trans ring junction in iii.

Ashe, *Tetrahedron Lett.* **1969**, 523.

25.

i —MeI→ CH_3 CH_3 N⊕ I^- —*t*-BuOK / DMF→ NMe_2, CH_2, CH_2

Hofmann elimination

—HCOOOH→ CH_2NMe_2, CH_2OH, OH —HIO_4→ ii

Hora and Šorm, *Collect. Czech. Chem. Commun.* **33**, 2059 (1968).

28.

i $\xrightarrow{PbO_2}$ [structure] $\xrightarrow{PhN_3}$ ii

1,3-dipolar addition

Zefirov, Ivanova, Filatova, and Yur'ev, *J. Gen. Chem. USSR* **36**, 1893 (1966).

33.

i $\xrightarrow{SeO_2}$ [structure with CHO] $\xrightarrow{Pd\text{-}BaSO_4}$ [structure] $\xrightarrow[\text{2. oxidation}]{\text{1. hydroboration}}$

[ketone structure, O] $\xrightarrow{CH_3Li}$ [structure with CH_3, OH] $\xrightarrow[\text{2. pyrolysis}]{\text{1. conversion to ester}}$ ii

Burgstahler and Sticker, *Tetrahedron* **24**, 2435 (1968).

42.

i $\xrightarrow[Zn]{NaOH}$ [2-Br-C$_6$H$_4$–NH–NH–C$_6$H$_4$-2-Br] $\xrightarrow{HCl}$

benzidine rearrangement

H_2N–[3-Br-C$_6$H$_3$]–[C$_6$H$_3$-3-Br]–NH_2 $\xrightarrow[\text{2. CuBr}]{\text{1. HONO}}$ ii

Sandmeyer reaction

Snyder, Weaver, and Marshall, *J. Amer. Chem. Soc.* **71**, 289 (1949).

44.

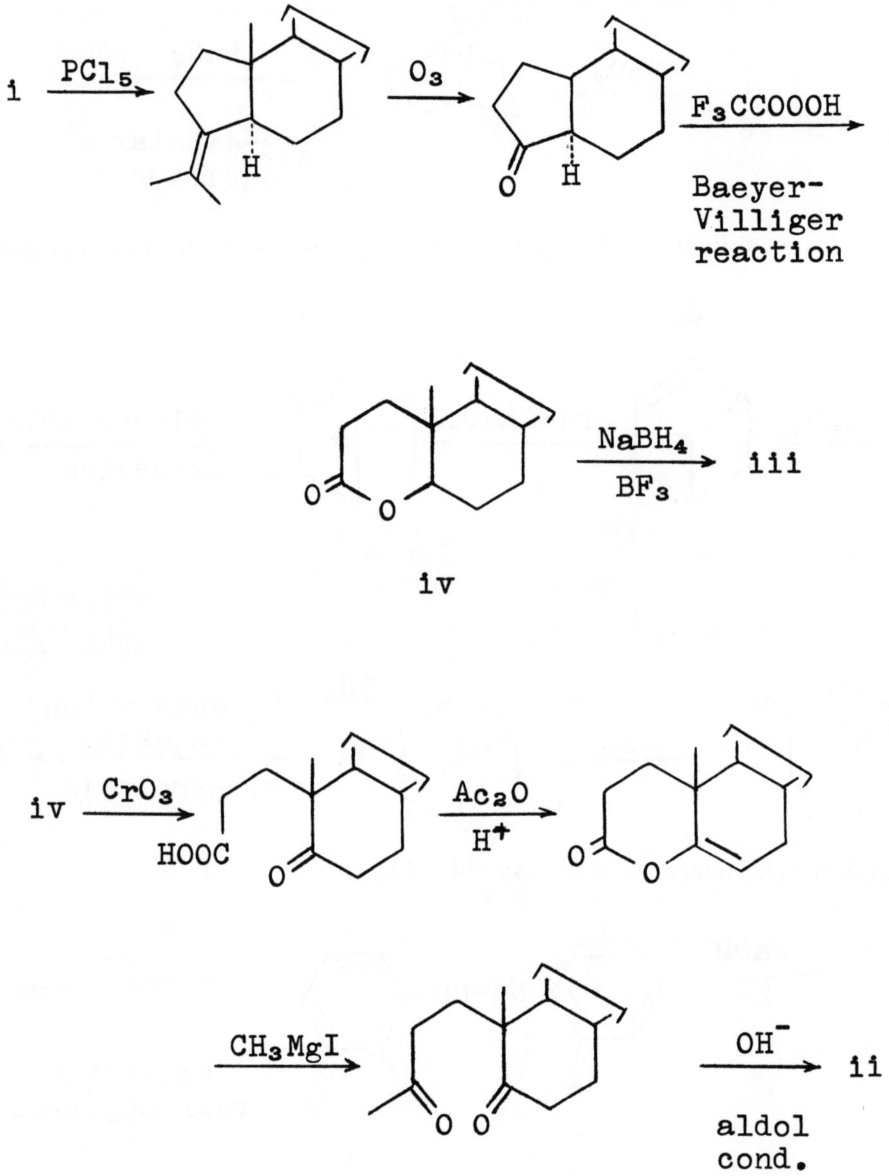
i
PCl5
H
O3
O
H
F3CCOOOH
Baeyer-
Villiger
reaction
O
O
NaBH4
BF3
iii
iv
iv
CrO3
HOOC
O
Ac2O
H+
O
O
CH3MgI
O O
OH-
ii
aldol
cond.

Pettit and Dias, *Can. J. Chem.* **47**, 1091 (1969).

49. The reagent which reduces i to ii is B_2H_6, which reduces amides but not esters.

Kornet, Thio, and Tan, *J. Org. Chem.* **33**, 3637 (1968).

52.

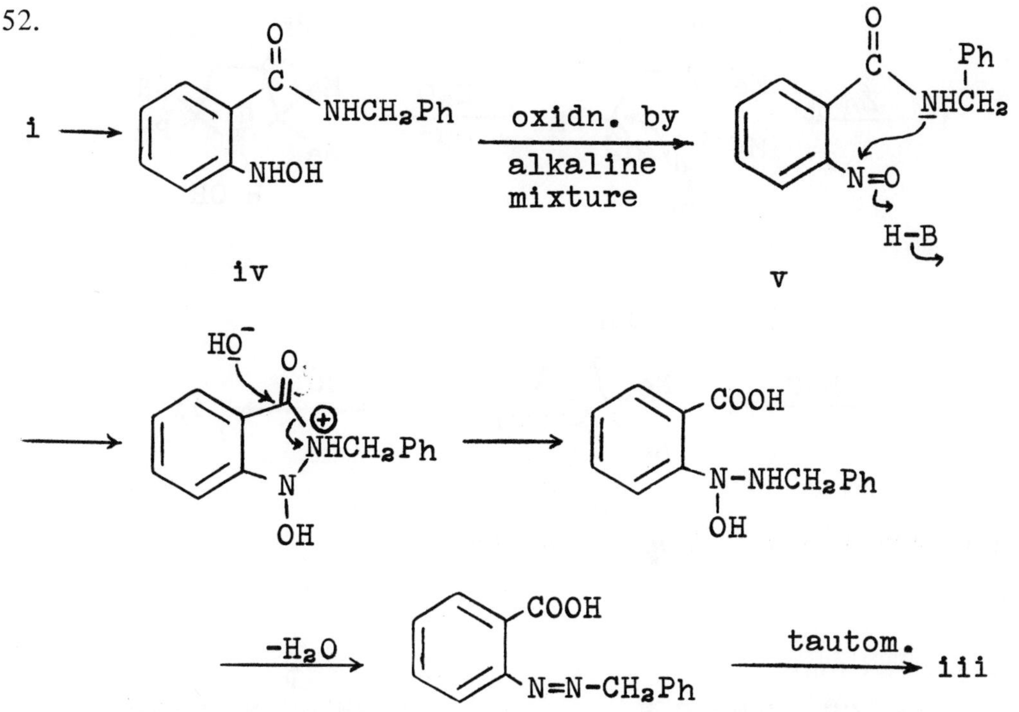

i → [benzamide with C(=O)NHCH$_2$Ph and NHOH substituents] —oxidn. by alkaline mixture→ [benzamide with C(=O)NHCH$_2$Ph and N=O substituents; H–B]

iv v

→ [HO$^-$ attack; $\oplus$NHCH$_2$Ph; N–OH] → [COOH; N–NHCH$_2$Ph; OH]

$\xrightarrow{-H_2O}$ [COOH; N=N–CH$_2$Ph] $\xrightarrow{\text{tautom.}}$ iii

ii is formed by a condensation of iv and v (*Advanced Organic Chemistry*, p. 907). For the mechanism of the conversion of i to iv, see the answer to Problem 96 in Problem Set 10 (case b).

Spence and Tennant, *Chem. Commun.* **1969**, 194.

56.

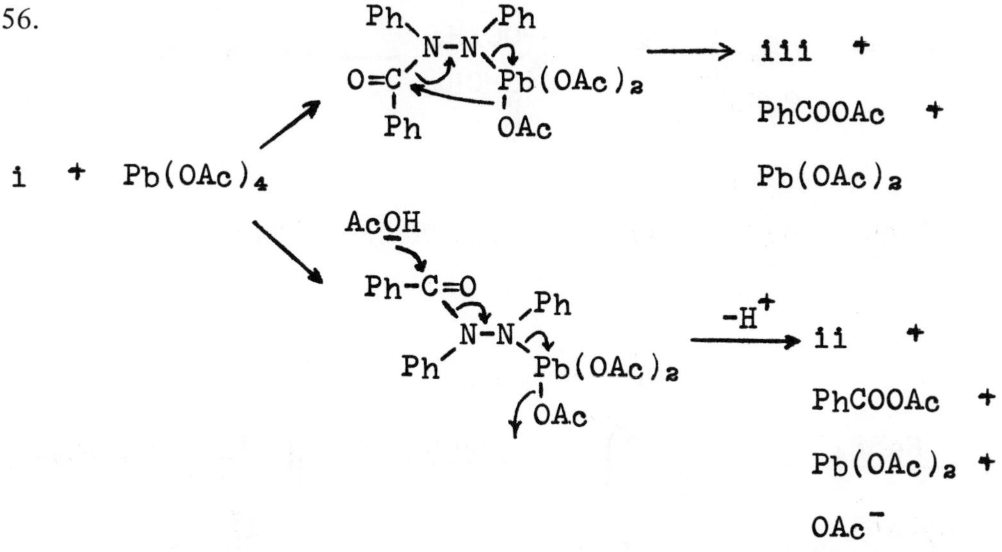

i + Pb(OAc)$_4$

↗ [Ph, Ph; N–N; O=C; Pb(OAc)$_2$; Ph; OAc] → iii + PhCOOAc + Pb(OAc)$_2$

↘ [AcOH; Ph–C=O; N–N; Ph; Ph; Pb(OAc)$_2$; OAc] $\xrightarrow{-H^+}$ ii + PhCOOAc + Pb(OAc)$_2$ + OAc$^-$

The upper pathway is significant only in the absence of an external nucleophile.

Gladstone, *J. Chem. Soc. C,* **1969**, 1571.

59.

i $\xrightarrow{Zn}$ [Me, Me, HN–OH, Ph, C=O–Ph intermediate] $\xrightarrow[-OH^-]{H_2O}$ [Me, Me, Ph, OH, Ph, $\overset{\oplus}{N}$, H OH]

$\xrightarrow{-H_2O}$ [Ph, Me, Me, $\overset{\oplus}{N}$=–Ph, OH] $\xrightarrow{-H^+}$ ii

Bapat and Black, *Aust. J. Chem.* **21**, 2483 (1968).

62.

i ⟶ $CH_3-C(=O)-\dot{N}-OC_4H_9$ $\xrightarrow[\text{dimerization}]{}$ $CH_3-C(=O)-N(-OC_4H_9)-N(-OC_4H_9)-C(=O)-CH_3$

⟶ $CH_3-C(=O)-OC_4H_9$ (ii) + $N(-OC_4H_9)=N-C(=O)-CH_3$ ⟶ ii + N_2

Cooley, Mosher, and Khan, *J. Amer. Chem. Soc.* **90**, 1867 (1968).

67.

i $\xrightarrow{NaBH_4}$ (normal reduction) [OH, benzene ring, $CH-CH_3$, OBH_3^-] $\xrightarrow{base}$ [$\bar{O}$, benzene ring, $CH-CH_3$, OBH_3^-] $\xrightarrow{a}$

67. Cont.

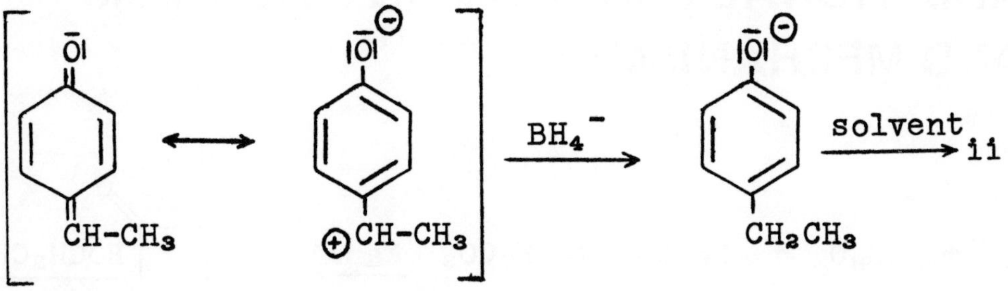

$\xrightarrow{BH_4^-}$ $\xrightarrow{\text{solvent}}$ ii

CH-CH$_3$ ⊕CH-CH$_3$ CH$_2$CH$_3$

The step marked a cannot take place with a meta hydroxy group.

Bell, *Aust. J. Chem.* **22**, 601 (1969).

68. This is a Dakin reaction (*Advanced Organic Chemistry*, p. 878) though ii is the tautomer of the expected phenol.

Fuks and Chiurdoglu, *Bull. Soc. Chim. Belges* **76**, 244 (1967).

72.

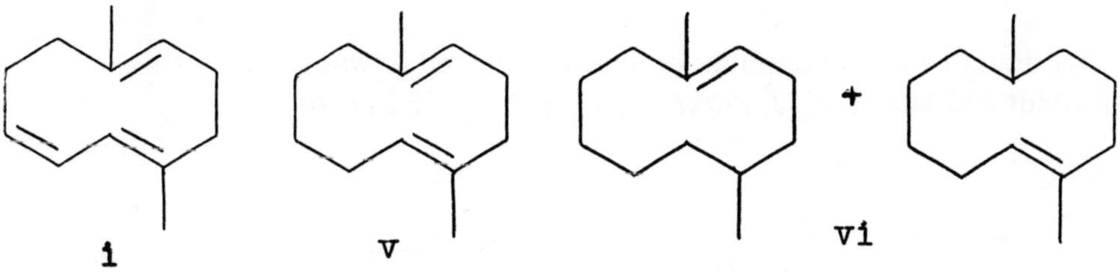

i v vi

Jones and Sutherland, *Aust. J. Chem.* **21**, 2255 (1968).

74. Label the C-O-C oxygen. If path a, then the label will be found in Ph$_3$PO; otherwise in the aldehyde or ketone. In the event, the label was found in the aldehyde or ketone, so path b is operating.

Carles and Fliszár, *Can. J. Chem.* **47**, 1113 (1969).

ANSWERS TO PROBLEM SET 14

ADDITIONAL PROBLEMS IN SYNTHESIS AND MECHANISMS

1.

i + CH_2O + Me_2NH + K_2CO_3 —Mannich→ iii (2-(dimethylaminomethyl)cyclobutanone, CH_2NMe_2) —$HOCH_2CH_2OH$→ ethylene ketal of iii (CH_2NMe_2) —1. CH_3I; 2. Ag_2O; 3. Δ→ ethylene ketal of 2-methylenecyclobutanone ($=CH_2$) —H^+→ ii

A direct Hofmann elimination on unprotected iii was unsuccessful. Mühlstädt and Meinhold, *J. Prakt. Chem.* [4] 37, 162 (1968).

5.

i + ClCO–C_6H_4–COCl —$AlCl_3$→

EtOOC–C_6H_4–CH_2–C_6H_4–C(=O)–C_6H_4–C(=O)–C_6H_4–CH_2–C_6H_4–COOEt

↓ Wolff-Kishner reduction

EtOOC–C_6H_4–CH_2–C_6H_4–CH_2–C_6H_4–CH_2–C_6H_4–CH_2–C_6H_4–COOEt

↓ 1. $LiAlH_4$ 2. PCl_5

$ClCH_2$–C_6H_4–CH_2–C_6H_4–CH_2–C_6H_4–CH_2–C_6H_4–CH_2–C_6H_4–CH_2Cl

↓ Na Wurtz reaction

ii

Inazu and Yoshino, *Bull. Chem. Soc. Jap.* **41** 647, 652 (1968).

11.

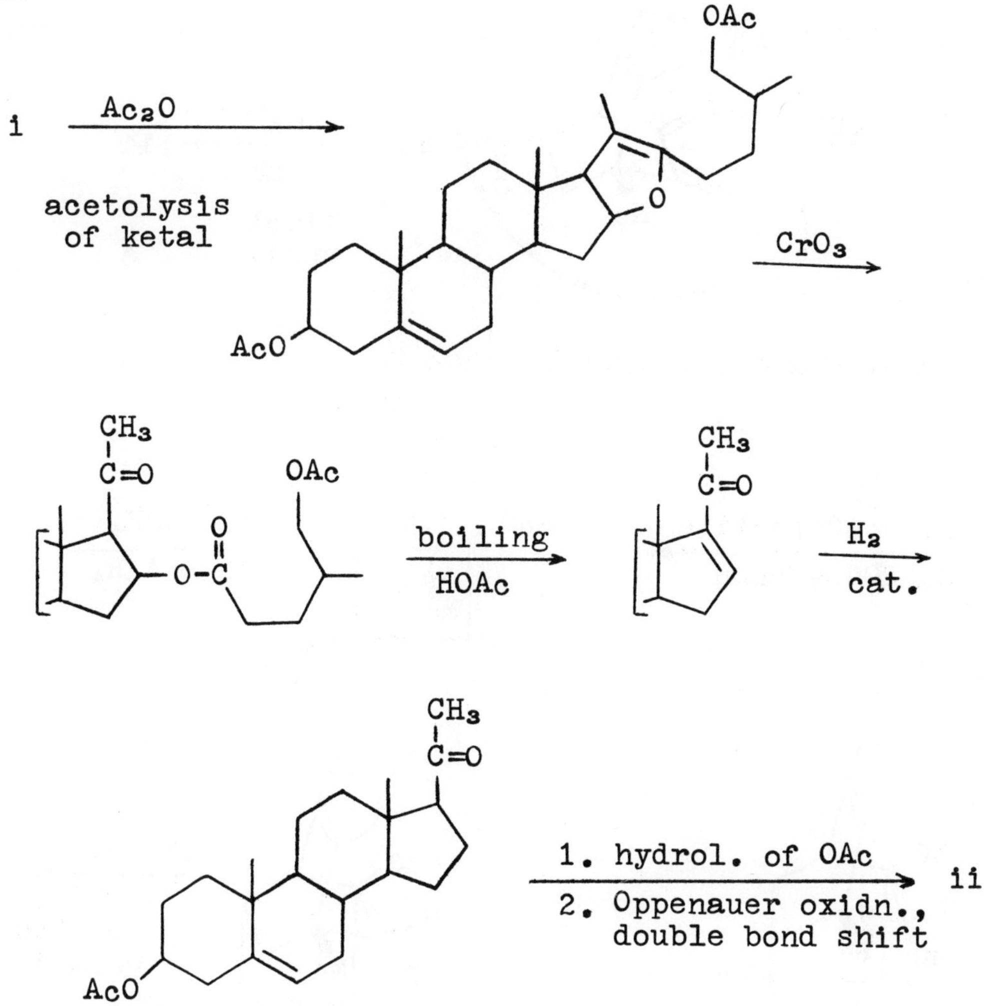

i $\xrightarrow{Ac_2O}$ (acetolysis of ketal) → [steroid structure with OAc, AcO] $\xrightarrow{CrO_3}$

[structure: CH_3, C=O, O, O-C, OAc] $\xrightarrow[HOAc]{boiling}$ [structure: CH_3, C=O] $\xrightarrow[cat.]{H_2}$

[structure: CH_3, C=O, AcO] → 1. hydrol. of OAc; 2. Oppenauer oxidn., double bond shift → ii

This is the Marker degradation process for the commercial production of progesterone.

Fieser and Fieser, "Steroids", pp. 546, 549-550, Reinhold Publishing Co., New York, 1959.

15.

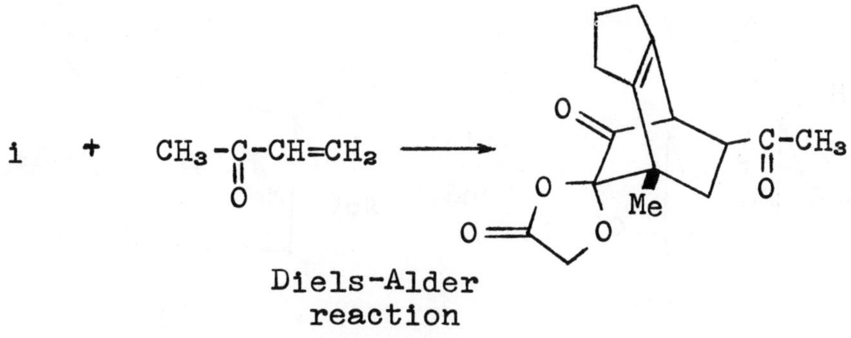

i + CH_3-C(=O)-CH=CH_2 → (Diels-Alder reaction) [structure: O, O, O, O, Me, C(=O)-CH_3]

15. Cont.

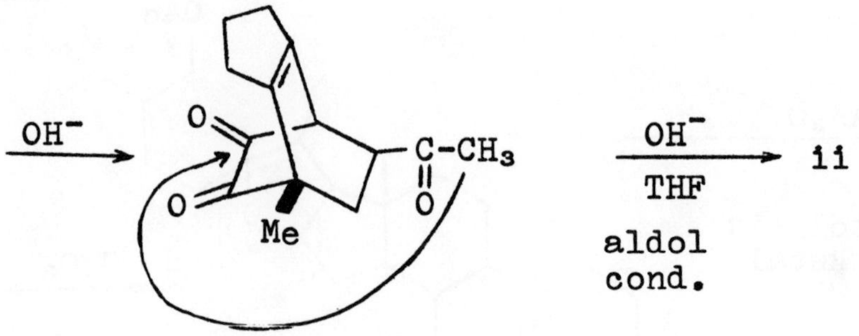

$\xrightarrow{OH^-}$ [structure with C(=O)-CH$_3$, Me] $\xrightarrow[THF]{OH^-}$ ii

aldol cond.

For the reference see the answer to problem 16.

16.

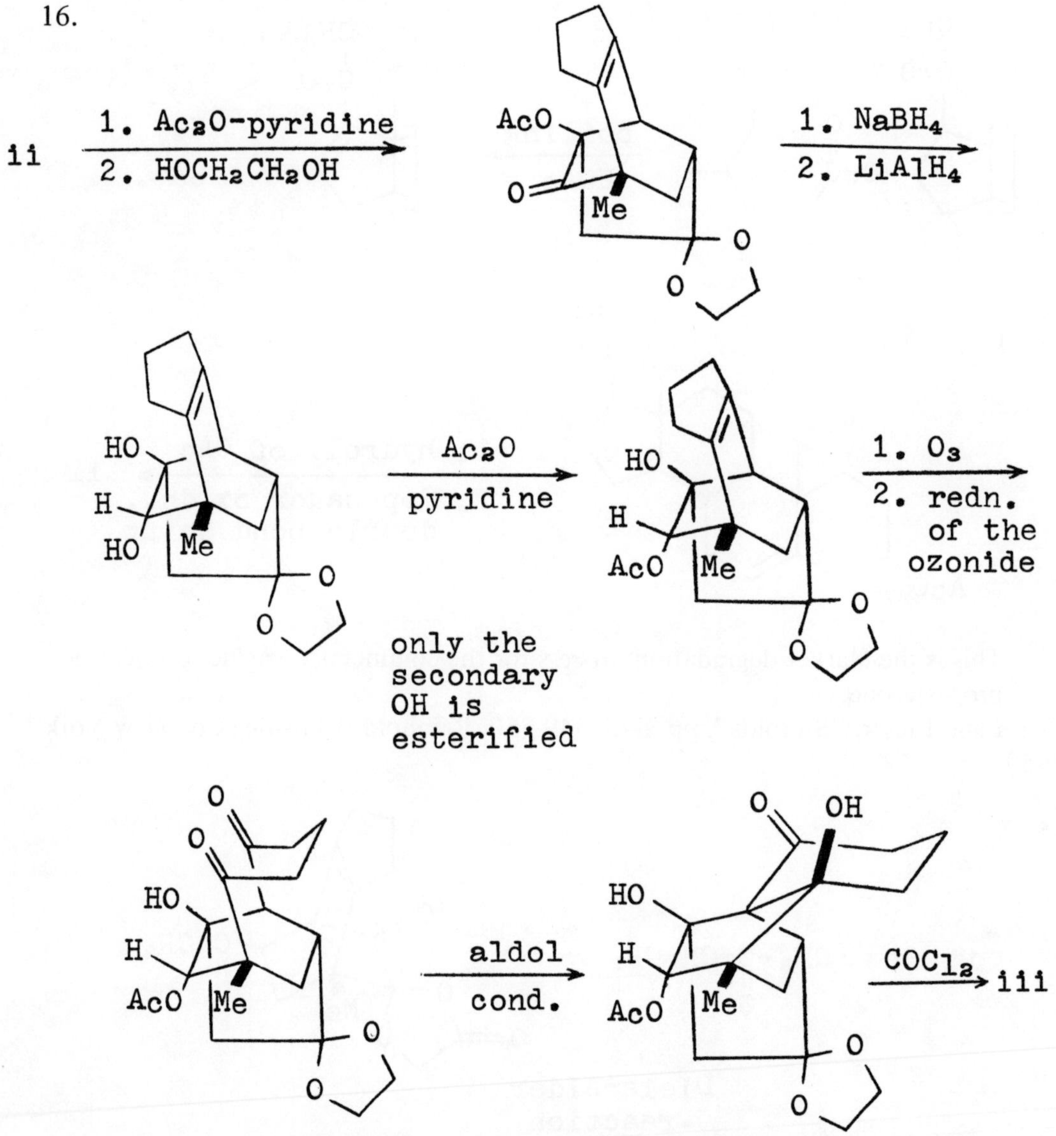

ii $\xrightarrow[\text{2. } HOCH_2CH_2OH]{\text{1. } Ac_2O\text{-pyridine}}$ [AcO, O, Me, ketal structure] $\xrightarrow[\text{2. } LiAlH_4]{\text{1. } NaBH_4}$

[HO, H, HO, Me structure] $\xrightarrow[\text{pyridine}]{Ac_2O}$ [HO, H, AcO, Me structure] $\xrightarrow{\text{1. } O_3 \text{ 2. redn. of the ozonide}}$

only the secondary OH is esterified

[O, O, HO, H, AcO, Me structure] $\xrightarrow[\text{cond.}]{\text{aldol}}$ [O, OH, HO, H, AcO, Me structure] $\xrightarrow{COCl_2}$ iii

Berney and Deslongchamps, *Can. J. Chem.* **47**, 515 (1969).

20.

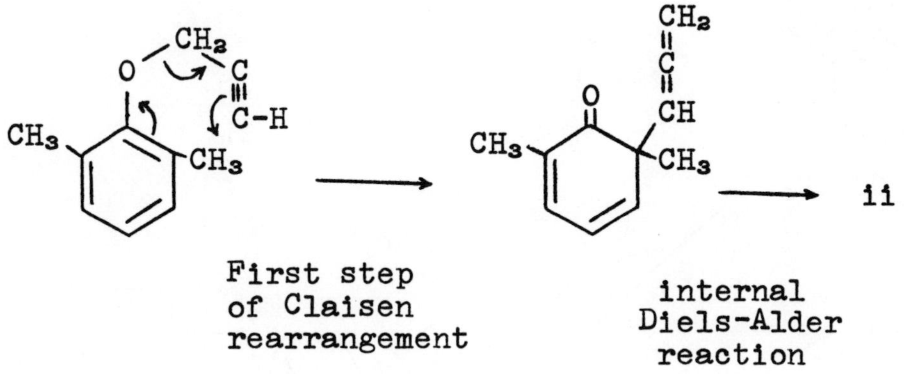

CH_3 CH_3 O CH_2 C C-H → CH_3 O CH_3 CH C CH_2 → ii

First step of Claisen rearrangement

internal Diels-Alder reaction

Zsindely and Schmid, *Helv. Chim. Acta* **51**, 1510 (1968).

29.

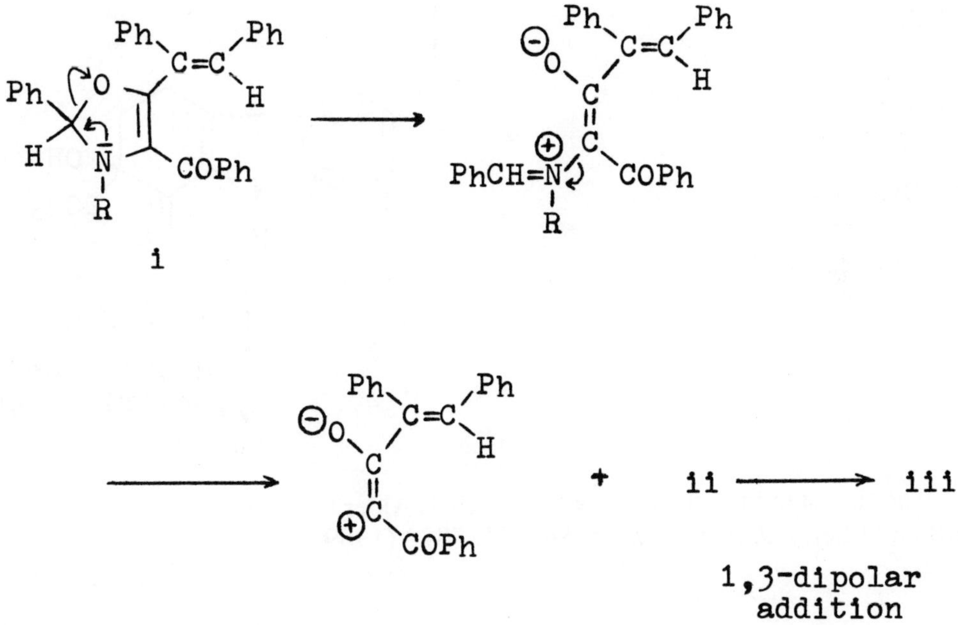

Ph O H N R C=C Ph Ph H COPh → PhCH=N⁺ R C C O⁻ C=C Ph Ph H COPh

i

→ O⁻ C C⁺ COPh C=C Ph Ph H + ii → iii

1,3-dipolar addition

Lown, Smalley, and Dallas, *Chem. Commun.* **1968**, 1543.

33.

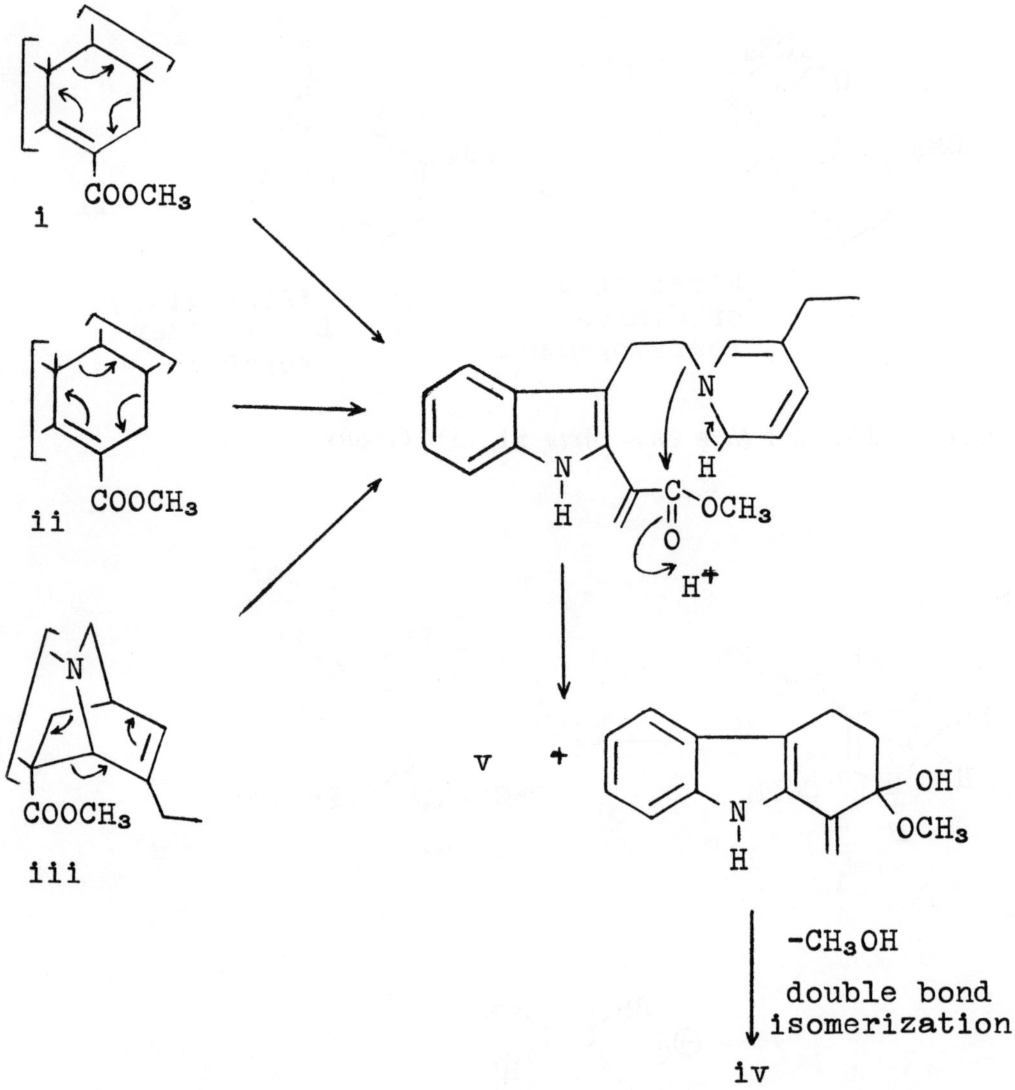

COOCH$_3$
i

COOCH$_3$
ii

N
COOCH$_3$
iii

N
N-H
C
OCH$_3$
O
H$^+$

v +

N-H
OH
OCH$_3$

-CH$_3$OH
double bond isomerization

iv

The first step in each case is a reverse Diels-Alder reaction. Scott and Cherry, *J. Amer. Chem. Soc.* **91**, 5872 (1969).

42.

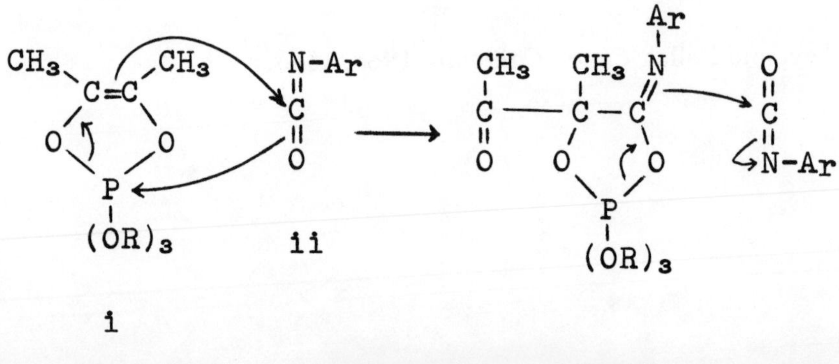

CH$_3$ C=C CH$_3$ O O P (OR)$_3$

N-Ar C O

ii

CH$_3$ CH$_3$ N-Ar C-C-C O O P (OR)$_3$

O=C=N-Ar

i

42. Cont.

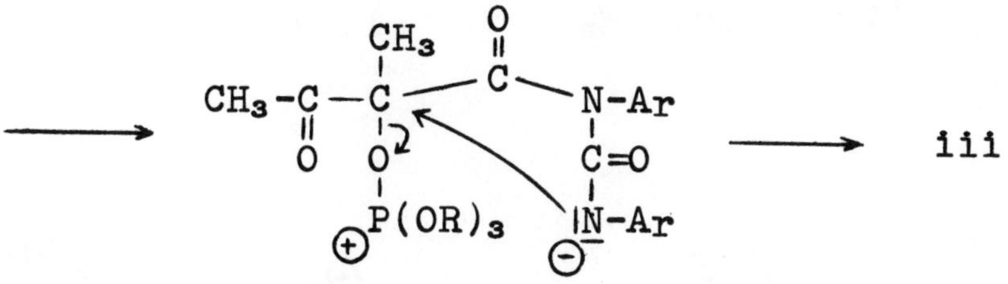

$\longrightarrow$ iii

Ramirez, Bhatia, Telefus, and Smith, *Tetrahedron* **25**, 771 (1969).

45

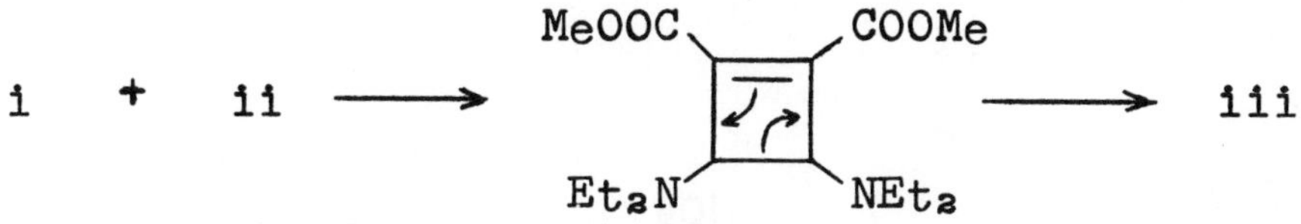

i + ii $\longrightarrow$ $\longrightarrow$ iii

Halleux and Viehe, *J. Chem. Soc., C*, **1968**, 1726.

48.

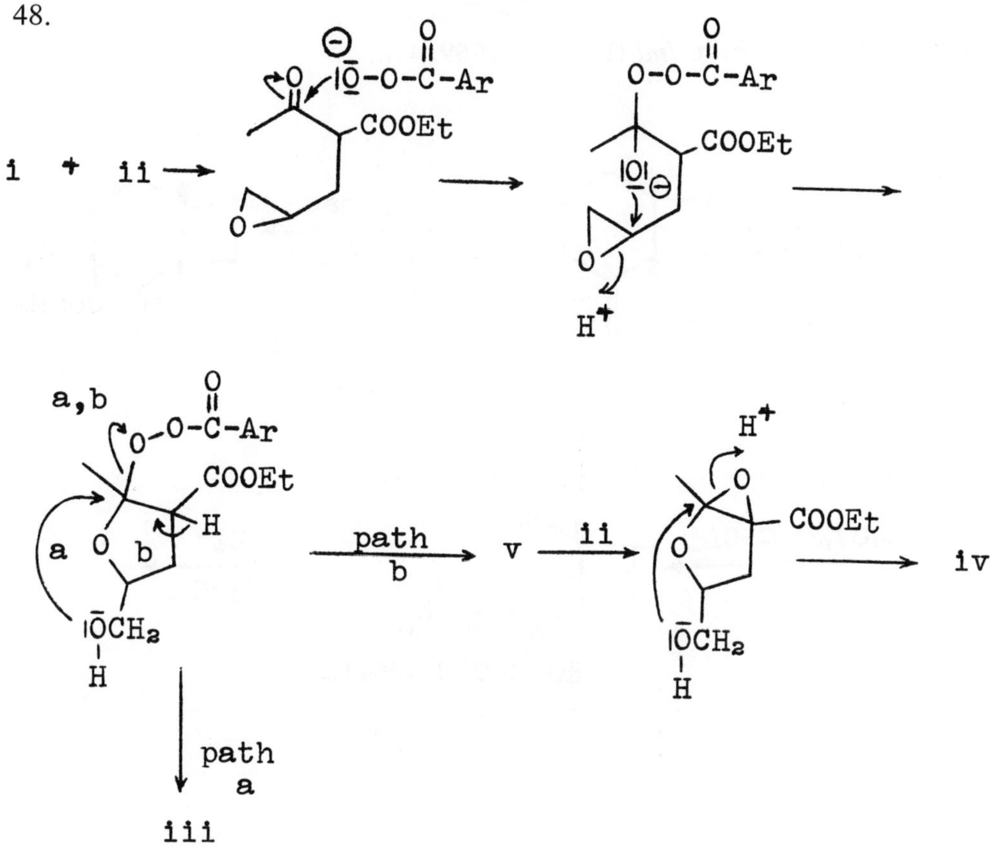

i + ii $\longrightarrow$ $\longrightarrow$ $\longrightarrow$

path b $\longrightarrow$ v $\xrightarrow{ii}$ $\longrightarrow$ iv

path a $\downarrow$ iii

Gaoni, *J. Chem. Soc., C* **1968**, 2925.

51.

$MeSO_2Cl \xrightarrow{base} SO_2{=}CH_2 + i \longrightarrow$

R_2N Me (H$^+$) O SO_2 NR_2 Me $\longrightarrow$ R_2N Me OH SO_2 NR_2 Me

$\longrightarrow$ R_2N Me OH SO_2 NR_2 Me $\xrightarrow[tautom.]{-H^+}$ ii

Stephen and Marcus, *Chem. Ind.* (London) **1969**, 416.

54.

A i $\xrightarrow[Et_2CO_3]{NaH}$ COOMe O $\xrightarrow[2.\ CH_3I]{1.\ base}$ CH_3 COOMe O

$\xrightarrow{(MeO)_2CHC{\equiv}CLi}$ CH_3 COOMe HO $C{\equiv}CCH(OMe)_2$ $\xrightarrow[Pd\text{-}C]{H_2}$

54. Cont.

HO $CH_2CH_2CH(OMe)_2$ — CH_3, COOMe $\xrightarrow[\text{aq. HOAc}]{CrO_3}$ ii

B ii $\xrightarrow{CH_3SOCH_2^-}$ Dieckmann condensation — OH, CH_3, =O, COOMe $\xrightarrow[\text{decarbox.}]{\text{hydrol.}}$ iii

C iii $\xrightarrow[\text{Raney Ni}]{H_2}$ OH, CH_3, OH $\xrightarrow{TsOH}$ H-O, CH_3, OTs, base

$\xrightarrow{CH_3SOCH_2^-}$ stereo-selective fragmentation — O, trans double bond formed $\xrightarrow[\text{Me}_2\text{SO}]{t\text{-BuO}^-}$ equilibration

O $\xrightarrow{Ph_3P{=}CH_2}$ iv

Wittig reaction

In sequence C, the catalytic hydrogenation produced a mixture of cis and trans isomers, but only the cis (shown above) was used in the next step.

Corey, Mitra, and Uda, *J. Amer. Chem. Soc.* **86**, 485 (1964).

56.

A i $\xrightarrow[\text{2. AlCl}_3]{\text{1. SOCl}_2}$ [ketone, Bz–N] $\xrightarrow{Br_2}$ [bromoketone, Bz–N, Br]

$\xrightarrow[\text{CH}_3\text{ (dioxolane) CH}_2\text{NHMe}]{}$ [ketone with CH_3 dioxolane CH_2N–Me side chain, Bz–N] $\xrightarrow[\text{hydrol.}]{\text{HCl}}$ ii

B ii $\xrightarrow[\text{MeOH}]{\text{NaOMe}}$ [enone, N–Me, H–N]

aldol cond.

$\xrightarrow[\text{2. NaBH}_4]{\text{1. Ac}_2\text{O}}$ [OH, N–Me, Ac–N] $\xrightarrow[\text{2. NaCN-HCN}]{\text{1. SOCl}_2}$ [CN, N–Me, Ac–N]

$\xrightarrow[\text{2. H}^+\text{ (hydrolysis)}]{\text{1. MeOH (formation of ester)}}$ iii

Kornfeld, Fornefeld, Kline, Mann, Morrison, Jones, and Woodward, *J. Amer. Chem. Soc.* 78, 3087 (1956).

60.

A i $\xrightarrow[2CH_2{=}CHCN]{\underline{t}\text{-BuOK}}$ (MeO, MeO)-aryl–C(CH_2CH_2CN)$_2$–$COCH_3$ $\xrightarrow[\text{2. EtOH-TsOH}]{\text{1. }OH^-}$

(MeO, MeO)-aryl–C(CH_2CH_2COOEt)$_2$–$COCH_3$ $\xrightarrow[\text{EtOH}]{\text{NaOEt}}$ ii

B iii $\xrightarrow{LiAlH_4}$ [HO, ketal (dioxolane), MeO, MeO, $CH_2CH_2CH_2OH$] $\xrightarrow{\text{oxalic acid, aq. MeOH}}$

ether formation

[MeO, OMe, O, O] $\xrightarrow{\text{HI-}Ac_2O}$ iv

cleavage of all 3 ether groups; reetherification

60. Cont.

C iv $\xrightarrow[\text{2. AgOAc}]{\text{1. MeI-}K_2CO_3}$ [structure: O, =O, MeO, $CH_2CH_2CH_2OAc$]

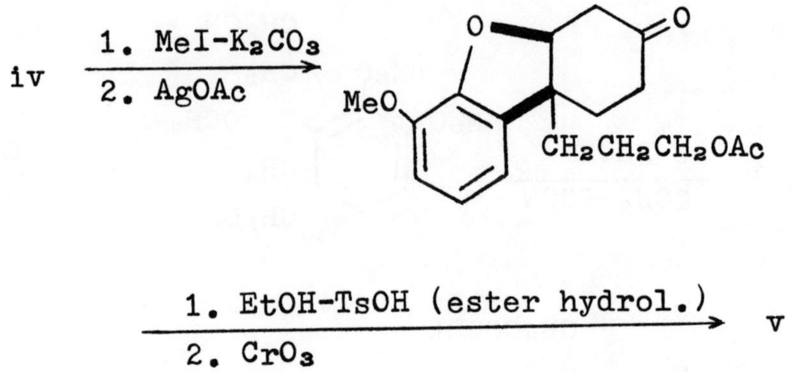

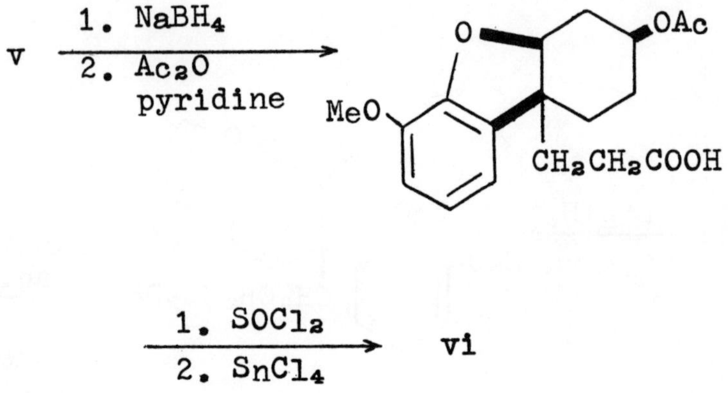

$\xrightarrow[\text{2. }CrO_3]{\text{1. EtOH-TsOH (ester hydrol.)}}$ v

D v $\xrightarrow[\text{2. }Ac_2O\text{ pyridine}]{\text{1. }NaBH_4}$ [structure: O, OAc, MeO, CH_2CH_2COOH]

$\xrightarrow[\text{2. }SnCl_4]{\text{1. }SOCl_2}$ vi

E vi $\xrightarrow[Cl_3CCOOH]{NaN_3}$ Schmidt reaction [structure: OAc, O, MeO, N-H, O] $\xrightarrow[\text{2. }LiAlH_4]{\text{1. MeI + NaH}}$ vii

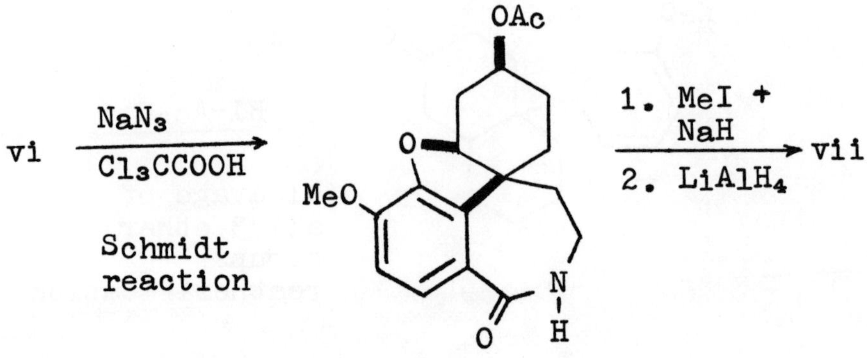

Misaka, Mizutani, Sekido, and Uyeo, *J. Chem. Soc., C* **1968**, 2954.

ANSWERS TO PROBLEM SET 15

NOMENCLATURE

1. a. 5,5-dimethylbicyclo[2.1.1]hexane
   b. 11-bromotricyclo[5.3.1.1^{2,6}]dodecane
   c. *exo*-8-methyltetracyclo[4.3.0.0^{2,4}.0^{3,7}]nonane
   d. pentacyclo[6.2.0.0^{2,7}.0^{3,10}.0^{6,9}]decane
   e. 2-oxa-7-thiatricyclo[4.4.0.0^{3,8}]decane
   f. 7,12,14-trioxapentacyclo[3.3.2.2^{6,10}.2^{8,9}.0^{2,4}]tetradecane
   g. 1,5,7,11-tetramethylpentacyclo[6.4.0.0^{2,7}.0^{4,11}.0^{5,10}]dodeca-3,6,9,12-tetraone
   h. tricyclo[5.4.0.0^{2,9}]undecane
   i. tetracyclo[5.2.2.0^{3,8}.0^{4,11}]undecane
   j. 3-methylhexacyclo[4.4.1.0^{2,5}.0^{3,9}.0^{4,8}.0^{7,10}]undecane
   k. 10,17-diazapentacyclo[7.4.3.1^{3,8}.1^{13,16}.0^{2,10}]octadecane
   l. 1,10-diazaheptacyclo[9.7.1.1^{2,10}.0^{3,8}.0^{6,19}.0^{12,17}.0^{15,20}]eicosane
   m. 3*H*-diazirine
   n. 3-methylselenetane
   o. 2-methyl-2*H*-1,2,3-oxadiazete
   p. 1,3,2-thiazaphospholidine
   q. 1,2,3-oxadiazol-3-ium bromide
   r. 4-bromo-1,2,5-oxathizaine
   s. 1,3,2-dioxaphosphepin
   t. 4,5,8,9-tetrahydro-1,2,3,5,8-trithiadiazonine
   u. 1,3-dioxa-7,8-dithiacycloundecane
   v. dibenzo[*b,g*]phenanthrene
   w. tribenz[*a,c,h*]anthracene
   x. dibenz[*b, f*]oxepin
   y. 4*H*-benz[*de*]isoquinoline
   z. 2*H*-pyrano[3,2-*g*]quinoline
   aa. 9*H*-indeno[2,1-*b*]pyridine
   bb. naphth[1,8-*cd*]-1,2-oxathiole
   cc. spiro[4.5]deca-1,6-diene
   dd. isopropylidenecyclohexane

1. Cont.

ee. 1,1-bis(4-methoxyphenyl)ethane

ff. 1,3-cyclopentanedisulfonyl chloride

gg. 2,5-bis(2-methylcyclopentyl)furan

hh. 1-(2-cyclopropylcyclobutyl)-3-(1,2-dimethylpropyl)cyclohexane

ii. 4,4′-bis(2-carboxyphenyl)-1,1′-binaphthyl

jj. 1,3,7,2-dioxathiastannecane

kk. 1-methyl-1,2,5,6,9,10-hexaaza-2,6,10-cyclododecatriene

ll. 2,2,6,6-tetraethylpyrazolo[1,2-*a*]pyrazole-1,3,5,7-tetraone

2.

a.

b.

c.

d.

e.

f.

g.

h.

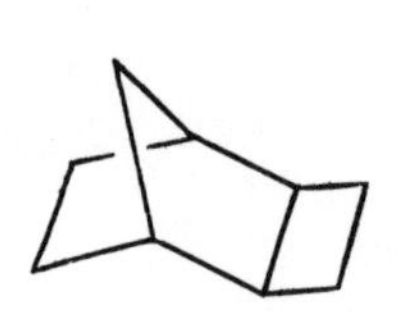

i.

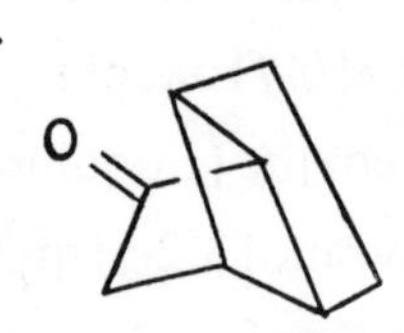

2. Cont.

j.

k.

l.

m.

n.

o.

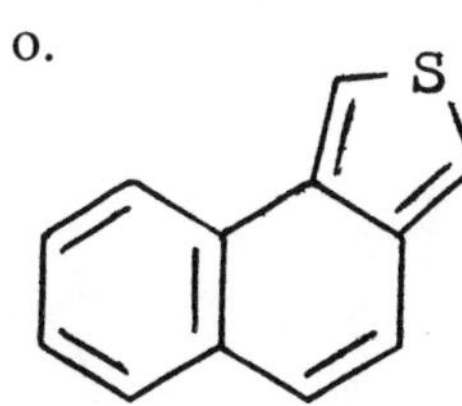

p.

q.

$$CH_3-\underset{\displaystyle OH}{\underset{|}{CH}}-\underset{\displaystyle OH}{\underset{|}{CH_2}}$$

ANSWERS TO PROBLEM SET 16

LITERATURE

1. a. Aberhalden and Schwab, *Hoppe-Seyler's Z. Physiol. Chem.* **148**, 20 (1925).
Beilstein, **25** E II 69 (not in H or E I).

b. Hofmann, Orochena, Sax, and Jeffrey, *J. Amer. Chem. Soc.* **81**, 992 (1959).
CA **53**, 14019e (1959).
Not in Beilstein through E III. Would be in Vol. **5**.

c. Markl and Lieb, *Agnew. Chem. Int. Ed. Engl.* **7**, 733 (1968); *Angew. Chem.* **80**, 702 (1968).
CA **69**, 106812f (1968).
Not in Beilstein through E II. Would be in Vol. **27**.

d. Hecht, *Ber.* **23**, 286 (1890).
Beilstein, **4** H 212, E II 665 (not in E I or E III).

e. Buchan and McCombie, *J. Chem. Soc.* **1931**, 137.
Beilstein, **6** E III 769 (not in H, E I, or E II).

f. Not in Beilstein through E III; would be in Vol. **6**.
Not in *CA* indexes through Volume **71** (1969).

g. Homeyer, U.S. Patent 2,399,118 (April 23, 1946).
CA **40**, P4085^{2,5} (1946).
Not in Beilstein through E II; would be in Vol. **27**.

h. Not in Beilstein through E III; would be in Vol. **5**.
Not in *CA* indexes through Vol. **71** (1969).

i. Holleman, *Rec. Trav. Chim.* **15**, 367 (1896).
Beilstein, **5** H 342, E I 168, E II 262, E III 762.

j. Mkryan, Mndzhoyan, and Gasparyan, *Izv. Akad. Nauk Arm. SSR, Khim. Nauki* **17**, 643 (1964).
CA **63**, 8183g (1965).
Not in Beilstein through E III; would be in Vol. **1**.

k. Djerassi and Scholz, *J. Org. Chem.* **13**, 697 (1948).
Beilstein, **7** E III, 3635 (not in H, E I, or E II).

l. Ruzicka, *Ber.* **50**, 1368 (1917).
Beilstein, **10** E I 292 (not in H or E II).

2. a. *Lietuvos TSR Mokslu Akademijos Darbai, Serija B: Chemija, Technika, Geografija*; Transactions of the Academy of Sciences of the Lithuanian SSR, Series B: Chemistry, Technology, Geography.

b. *Transactions of the Kentucky Academy of Science*

2. Cont.

   c. *Zeszyty Naukowe Uniwersytetu Jagiellon Skiego, Prace Chemiczne*; Scientific Bulletin of the Jagiellonian University, Chemistry Papers.

   d. *CA* does not abbreviate this title. Translation: Annual Report of the Tohoko College of Pharmacy.

   e. *Sbornik Praci Z Vyzkumu Chemickeho Vyuziti Uhli, Dehtu a Ropy*: Collected Works on Research in the Chemical Use of Coal Tar and Crude Oil.

3. a. *Liet. TSR* etc: NN, ICRL

   b. *Trans. Ky. Acad. Sci.*: DLC, NN

   c. *Zesz. Nauk.* etc: DLC, ICRL

   d. *Tohoku* etc: ICRL

   e. *Sb. Pr.* etc: ICRL

4. a. **15**H, 651, E II, 309.

   b. **7** E I, 316; E II, 536; E III, 3237 or 3238.

   c. **6** H, 902; E I, 442; E III, 4558.

   d. **3** H, 301; E I, 114; E II, 214 or 215.

5. a. Vol. **68** subject index p. 2823; Index of ring systems, p. 12.

   b. Not in Vol. **68**

   c. Vol. **68** subject index, p. 1230; Index of ring systems, p. 14.

   d. Not in Vol. **68**

6. a. The "Handbook of Chemistry and Physics" is the easiest place to find this, but the "Dictionary of Organic Compounds" also has it. Both give 218-219°.

   b. The "Handbook of Tables for Organic Compound Identification," 3d edition, lists a benzoate, m.p. 174°. The "Dictionary of Organic Compounds" also lists this derivative.

   c. The "Dictionary of Organic Compounds" lists a thiosemicarbazone, m.p. 245-7°.

   d. Isoquino[3,4-*b*] quinoxaline (Ring Index No. 4986).

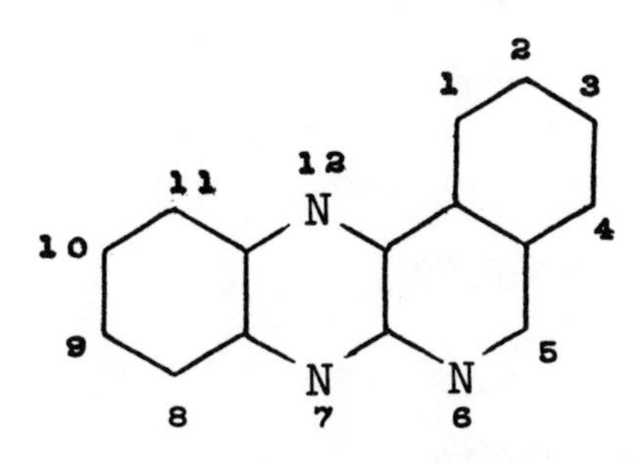

6. Cont.

e. Isonicotinic acid hydrazide (from the "Merck Index of Chemicals and Drugs," 8th edition).

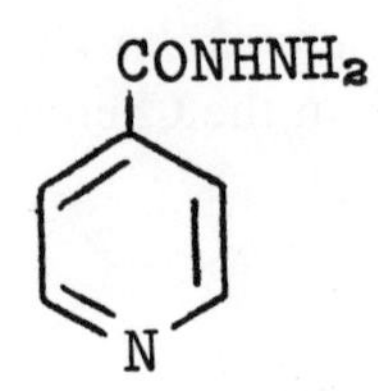

f. It is an anticholinergic, and an antispasmodic for the upper gastrointestinal tract (from the "Merck Index of Chemicals and Drugs," 8th edition).

g. The "Tables of Interatomic Distances and Configurations in Molecules and Ions," Supplement, gives 1.385 Å (p. 63s).

h. "Tables of Experimental Dipole Moments" gives 2.60 D (p. 379).

i. "Pressure-Volume-Temperature Relationships of Organic Compounds" predicts 65.6° (p. 69).

j. Kaufman, "Handbook of Organometallic Compounds" lists the value as 117-118° (p. 801).

k. Miyano, Abe, and Uno, *Chem. Pharm. Bull.* (Tokyo) **14**, 731 (1966). This reference obtained from Theilheimer, "Synthetic Methods of Organic Chemistry," Vol. 22, p. 175.

l. Fieser and Fieser, "Reagents for Organic Synthesis" gives the method (p. 977).

m. It is used to form solid salts with N-protected amino acids. Fieser and Fieser, "Reagents for Organic Synthesis," p. 231.

n. By bromination of *n*-valeraldehyde. Wagner and Zook, "Synthetic Organic Chemistry," p. 305.

o. i. 2-pentanol and benzene do not form an azeotrope.
ii. ethanol and cyclopentane form an azeotrope, b.p. 44.7°, containing 7.5% ethanol.
"Azeotropic Data," Vol. 1, pp. 152, 62.

7. a. *Org. Syn.* **III**, 656.
b. *Org. Syn.* **42**, 79.
c. *Org. Syn.* **I**, 211.
d. Not in *Org. Syn.* through Vol. **50**.
e. *Org. Syn.* **47**, 62.

AUTHOR INDEX

Aaron, H.S., 345
Aasen, A.J., 40
Abramovici, M., 234
Abramovitch, R.A., 315
Achiwa, K., 253
Ackerman, J.H., 26
Adachi, K., 222
Adam, G., 173
Adams, J.T., 55
Adams, R., 66, 67
Adlerová, E., 353
Adolphen, G.H., 188
Aeberli, P., 163
Ahmed, Q.A., 42
Aitova, É.F., 327
Akagi, K., 125
Akano, M., 152
Akhrem, A.A., 337
Alashev, F.D., 326
Alberts, J., 339
Albright, J.D., 304
Alden, C.K., 117, 131
Aldrich, P.E., 138
Allan, Z.J., 71
Alles, G.A., 104
Allinger, N.L., 69, 178, 291
Allison, C.G., 247
Allred, E.L., 306
Alt, K.O., 48
Anastassiou, A.G., 230
Anciaux, A.J., 90
Anderson, A.G., Jr., 37
Anderson, L.P., 116
Andrejević, L., 157
Anselme, J., 228
Anspon, H.D., 222
Appenroth, M., 350
Archer, W.L., 68
Arnold, H., 47
Arnold, R.T., 170
Arnott, R.C., 89
Arsenijevic, L., 157
Arsenijevic, V., 157
Arumugam, N., 218
Asano, M., 328
Asano, Y., 279
Ashby, E.C., 89
Ashe, A.J., 378
Asinger, F., 330
Aslapovskaya, T.I., 313
Atkin, R.W., 129
Atkinson, E.R., 326
Aubort, J., 310
Audier, H.E., 56
Awad, W.I., 346
Ayres, D.C., 248
Ayrey, G., 293
Ayyar, K.S., 137

Babad, E., 369
Babaré, L.V., 317
Bachi, M.D., 113
Baddar, F.G., 346
Badger, G.M., 8
Badger, R.A., 167
Baer, H., 336
Bagdi, J.F., 42
Bailey, A.S., 141
Bailey, D.L., 341
Baird, M.S., 202, 357
Bajwa, G.S., 196
Balasubramaniyan, P., 136
Balasubramaniyan, V., 136
Baldwin, J.E., 112, 228
Ballard, M.M., 65
Balquist, J.M., 158
Baltzly, R., 39
Bamkole, T.O., 171, 178
Ban, Y., 136, 299
Banerjee, A.K., 41
Banks, R.E., 164
Bannister, R.G., 341
Banthorpe, D.V., 361
Bapat, J.B., 382
Barclay, L.R.C., 76, 81
Barker, J.M., 105
Barnes, R.P., 41
Barnish, I.T., 44

Barr, T.H., 127
Barrett, J.H., 167, 338
Barth, G., 169
Bartlett, P.D., 341
Barton, D.H.R., 149, 183
Baskevitch, N., 315
Bassindale, A.R., 72
Battaglia, J., 115
Battiste, M.A., 9, 57, 69
Batzer, H., 146
Bauer, L., 368
Baum, K., 84
Baumgarten, H.E., 185
Baxter, C.A.R., 162
Beaujon, J.H.R., 218
Beckwith, A.L.J., 191, 329
Begland, R.W., 334
Beinoravichute, Z.A., 301
Bélanger, A., 350
Bell, C.L., 368
Bell, E.V., 67
Bell, K.H., 383
Bělovský, O., 355
Belsky, I., 66
Belyaev, E. Yu., 313
Bennett, G.M., 67
Bennett, P.J., 55
Benson, R.E., 316
Berchtold, G.A., 116
Berg, A., 315
Berger, M., 230
Berger, P.W., 142
Berger, T., 245
Bergstrom, F.W., 322
Berney, D., 37, 386
Berninger, C.J., 136
Betoux, J., 311
Bey, P., 202
Bhatia, I.S., 45
Bhatia, S.B., 389
Bhatti, P.J.S., 223
Bianco, E.J., 256
Bień, A.S., 227
Bigelow, L.A., 323
Binder, D., 355
Bindra, A.P., 276
Black, D.K., 114
Black, D.S.C., 382
Blagoev, B., 219
Blank, H.U., 122
Blatz, P.E., 136
Blickenstaff, R.T., 27
Blight, M.M., 102
Block, E., 163
Blomquist, A.T., 342
Bloss, K.H., 223
Blunt, J.W., 192
Bly, R.K., 156
Bly, R.S., 156, 162
Boerma, J.A., 45
Bogert, M.T., 217
Bohlmann, F., 85
Bohnert, T., 344
Boness, M., 40, 142
Bonner, T.G., 72
Boon, W.R., 327
Borch, R.F., 189
Borden, W.T., 234
Bordwell, F.G., 8, 43
Bosch, A., 137
Bosch, H.W., 103
Boskin, M.J., 299
Bösler, A., 344
Bossenbroek, B., 106
Bott, K., 112
Bourns, A.N., 293
Boussinesq, J., 229
Bowden, K., 188
Boyer, J., 367
Boyer, J.H., 227, 249
Bozak, R.E., 64
Bozzato, G., 142
Bozzini, S., 96
Bradsher, C.K., 130
Braithwaite, A., 164
Bramley, R.K., 194
Brändström, A., 105
Braünling, H., 93
Bravo, P., 45, 145
Breckwoldt, J., 27

Brenner, M., 334
Breslow, R., 9
Breuer, E., 245
Brewbaker, J.L., 129, 168
Brewer, J.P.N., 112
Brewster, J.H., 212
Bried, E.A., 333
Brienne, M., 27
Broadhead, G.D., 104
Brooke, G.M., 322
Brossi, A., 67
Brouwer, D.M., 190
Brown, C., 194
Brown, D.N., 108
Brown, G.W., 170, 276
Brown, H.C., 36, 186, 298, 313, 337
Brown, M., 235
Brown, P., 131
Brown, R.K., 137, 196
Bruck, P.R., 81, 141
Bruice, T.C., 61
Brune, H., 339
Brunet, J.J., 237
Brutschy, F.J., 336
Bryant, D.R., 41
Bublitz, D.E., 64, 315
Büchi, G., 81, 136, 222, 250, 365
Buchler, J.W., 126
Buchta, E., 38
Buck, C.E., 341
Buck, J.S., 39, 104
Buckley, D.G., 319
Buddrus, J., 69
Buggle, K., 119
Bull, J.R., 50
Bullock, M.W., 39
Bunnell, C.A., 122
Bunnett, J.F., 75, 322
Burgstahler, A.W., 138, 379
Burkett, H., 74
Burkhardt, J., 54
Burkoth, T.L., 6
Buschhoff, M., 88
Bushby, R.J., 31
Bushwick, R.D., 68
Butler, K., 256
Butler, L.C., 226
Buu-Hoï, N.P., 70, 308
Buurman, D.J., 203
Buysch, H., 173
Byck, J.S., 236
Bycroft, B.W., 124
Bye, T.S., 341
Byrd, D., 176
Byrne, W.E., 325

Cadogan, J.I.G., 341, 375
Cambie, R.C., 359
Campaigne, E., 68
Campbell, C.D., 316
Canonne, P., 62, 219
Cardnuff, J., 185
Carles, J., 383
Carlsmith, L.A., 325
Carlsson, S.Å.I., 105
Carrié, R., 241
Carrington, H.C., 327
Carter, P.H., 344
Casanova, J., Jr., 356
Caserio, M.C., 160
Caubere, P., 237
Cauquil, G., 367
Cava, M.P., 215, 258, 276
Cavé, A., 223
Čedergren, R.J., 342
Čeković, Ž., 106, 110
Červantes, A., 204
Červinka, O., 26, 355
Chambers, R.D., 96, 97, 247
Chan, T.L., 325
Chapman, O.L., 167
Chardonnens, L., 220
Charlton, J.C., 307
Chatterjee, S.S., 40
Chattha, M.S., 246
Chaux, R., 341
Cheburkov, Yu. A., 247
Chen, K.K.N., 115
Chernyshev, E.A., 341
Chernysheva, T.I., 317
Cherry, P.C., 388
Chitwood, J.L., 374

Chiurdoglu, G., 383
Chiusoli, G.P., 116
Christl, B., 126
Chubb, F.L., 315
Chuche, J., 229
Chuit, P., 300
Ciabattoni, J., 244
Claeson, G., 376
Claisse, J.A., 131
Clar, E., 67
Clark, D.T., 314
Clark, R.D., 81, 141, 300
Clarke, H.T., 322
Clarke, K., 315
Clark-Lewis, J.W., 152
Claus, C.J., 43
Cleland, G.H., 115
Clementi, S., 314
Clibbens, D.A., 308
Clough, J.A., 52
Cochoy, R.E., 202
Cocolas, G., 218
Cohen, D., 147
Cohen, M.H., 201
Cole, W., 39
Coleman, G.H., 67, 93, 222
Coleman, H.A., 196
Coller, B.A.W., 373
Collington, D.M., 169
Compagnon, P., 138
Conant, J.B., 225
Conia, J.M., 215, 223
Conover, L.H., 256
Conway, T.T., 238
Cook, C.D., 226
Cook, C.E., 142
Cookson, R.C., 131
Cooley, J.H., 382
Cope, A.C., 82, 172
Coppen, J.J.W., 102
Cordes, C., 103, 221
Corey, E.J., 103, 163, 165, 234, 253, 291, 391
Cornubert, R., 317
Corson, B.B., 66, 225, 315
Cose, R.W.C., 207
Courduvelis, C., 47
Cowsar, D.R., 213
Crabbé, P., 204
Craig, J.C., 344
Craig, T.W., 116
Cram, D.J., 30, 53, 87, 91, 119
Crandall, J.K., 55, 122
Crank, G., 188
Cremlyn, R.J.W., 62
Criegee, R., 339
Croisy, A., 70
Crombie, L., 237
Cromwell, N.H., 332
Crooks, S.C., 8
Cross, A.D., 220
Cross, B., 102
Crouse, D.M., 136
Crumrine, D.S., 113, 195
Cryberg, R.L., 80, 185, 208
Cseh, G., 341
Culbertson, B.M., 305
Cullen, W.R., 132
Culvenor, C.C.J., 41
Curby, R.J. Jr., 64

Dadson, B.A., 300
Daeniker, H.U., 258
Dahl, R., 246
Dahmen, A., 374
Dallas, G., 387
Dalle, J., 334
Damodaran, N.P., 106
D'Angeli, F., 302
d'Angelo, J., 117
Danion-Bougot, R., 241
Danishefsky, S., 121
Danks, L.J., 183
Da Rooge, M.A., 69
Dat-Xuong, N., 235
Daub, J., 184
Dauben, W.G., 374
Davidson, R.S., 81, 141
Davies, A., 74, 315
Davies, A.J., 191
Davies, A.M., 207
Davies, D.I., 51, 117, 131
Davies, H. ff. S., 361

Davies, J.S., 97
Davies, L.S., 96
Davies, V.H., 97
Davies, W., 41
Davis, A.W., 322
Davis, B.R., 109, 378
Davis, M.A., 118
Davis, R.A., 136
Dawson, C.R., 236
Dawson, D.J., 105, 196
Dax, K., 350
Deaken, D.M., 338
de Boer, T.J., 133, 231
de Groot, Ae., 45
Dehmlow, E.V., 151, 335
Dejardin, J.V., 169
Dejardin-Duchêne, M., 169
de la Mare, P.B.D., 36, 58
Delton, M.H., 30
den Hertog, H.J., 203, 359
den Hollander, W., 114
Denis, J.M., 223
Denney, D.B., 299
De Pooter, H., 85
Depovere, P., 216
DePuy, C.H., 107
Derieg, M.E., 135
Deslongchamps, P., 37, 350, 386
Dev, S., 39, 106, 235
Devanathan, V.C., 218
Devaprabhakara, D., 85
Devis, R., 216
DeWolfe, R.H., 298
DeYoung, E.L., 309
Deyrup, C.L., 57
Deyrup, J.A., 52, 155
Dias, J.R., 380
Dickinson, C.L., 304
Dietrichs, H. H., 218
Dietz, S., 305
Dilling, W.L., 46, 133
DiPasquo, V.J. 209
Dittmer, D.C., 158, 236
Djerassi, C., 292
Dobrynin, V.N., 337
Dobson, T.A., 118
Dodson, R.M., 370
Doering, W. von E., 277
Dolby, L.J., 6, 181, 200
DoMinh, T., 239
Döpp, D., 113, 195
Dorko, E.A., 35
Dornhege, E., 62
Douzon, C., 299
Doyle, T.W., 238
Dran, R.B., 66
Droescher, H., 233
Dubin, H.J., 153
Dubini, M., 116
DuBose, C.M., Jr., 162
Ducker, J.W., 146
Dudek, V., 355
Duell, E.G., 341
Dufour, M., 308
Dunn, G.L., 209
Dupin, J.F., 56
Durham, L.J., 224
Dutta, P.C., 43, 350
du Vigneaud, V., 81

Eaborn, C., 72, 315
Eastwood, F.W., 176
Ebender, F., 330
Eberius, W., 339
Ebnöther, A., 200
Eckhard, I.F., 112
Edamura, F.Y., 133
Edwards, J.A., 214
Edwards, J.M., 135
Effenberger, F., 95
Eglinton, G., 242
Eguchi, S., 158, 343
Eimura, K., 190
Eisenbraun, E.J., 188
Eistert, B., 191
Eiter, K., 40, 142
Elad, D., 124
Eliel, E.L., 26
Elix, J.A., 8, 276
Ellinger, C.A., 200
Emmons, W.D., 105
Emovon, E.U., 171, 178
Endo, S., 86
Engberts, J.B.F.N., 356

Engelhart, J.E., 82
Epstein, J.W., 113
Erichomovitch, L., 315
Eschenmoser, A., 114, 177
Esfandiari, S., 200
Evans, F.J., 341
Evans, J.C., 314

Fábryová, A., 355
Fahey, R.C., 315
Faingor, B.A., 316
Fairweather, D.J., 314
Fanta, P.E., 328
Fauran, C., 299
Feast, W.J., 116
Feijen, J., 327
Feinauer, R., 54, 351
Felix, D., 177
Felkin, H., 195
Fendler, E.J., 325
Fendler, J.H., 325
Fenical, W., 205, 210, 334
Fernandez, J.E., 247
Ferraris, M., 116
Fialkov, Yu. A., 315
Field, L., 300
Fields, D.L., 189
Fieser, L.F., 66, 225, 385
Fieser, M., 385
Filatova, R.S., 379
Filippova, A. Kh., 341
Finch, A.M.T., Jr., 233, 244
Findlay, J.A., 120
Fischer, H., 305
Fischer, U., 122, 276
Fishman, M., 243
Fitton, A.O., 170
Fleischhacker, W., 120, 361
Fliszár, S., 383
Flood, S.H., 313
Foglesong, W.D., 112
Foldi, A.P., 142
Fomina, O.S., 334
Fong, C.W., 89
Fonken, G.J., 312
Foote, C.S., 334
Forchiassin, M., 130
Forman, A.G., 311
Fornefeld, E.J., 365, 392
Forward, G.C., 162
Foster, G.H., 300
Fougerousse, A., 369
Fouli, F.A., 346
Fournari, P., 81, 315
Fox, A.S., 107
Fox, J.J., 122
Frame, R.R., 43
Frank, D., 169
Frank, R.L., 139
Franke, C., 280
Freeman, J.P., 105, 320
Fridinger, T.L., 147
Fried, J.H., 214, 220
Friedman, B.S., 314
Friedman, L., 30, 87
Friess, S.L., 310
Frints, P.J.A., 227
Fritz, H., 184
Fryer, R.I., 135
Fugiel, R.A., 167
Fujima, Y., 309
Fuks, R., 383
Fukui, K., 65
Fukui, S., 307
Fukumoto, K., 63
Funamizu, M., 40
Fuson, R.C., 341

Gagosian, R.B., 209
Gallagher, R.T., 359
Gannon, J.J., 320
Ganz, C.R., 102
Gaoni, Y., 389
Gapski, G.R., 50
Gara, W.B., 329
Garber, M., 39
Garbisch, E.W., Jr. 187
Gardner, J.E., 142
Garratt, P.J., 276
Gaskell, A.J., 83
Gassman, P.G., 51, 59, 80, 185, 208, 358

Gastambide, B., 332
Gates, M., 257
Gatsonis, C.D., 42
Gaudiano, G., 45, 145
Gautier, J., 138
Gautier, J.A., 299
Geisel, M., 356
Geiseler, G., 330
Genel, F., 314
Gensler, W.J., 42
Gerhart, F., 155
Gerlach, H., 348
Gertner, D., 66
Giardi, I., 98
Gibson, D.H., 107
Gibson, W.K., 89
Gilman, H., 325
Ginsburg, D., 369
Girard, J., 229
Giza, C.A., 26
Gladstone, W.A.F., 381
Gloede, J., 346
Godfrey, J.C., 121
Goldberg, G.M., 341
Goldberg, S.I., 182
Golborn, P., 315
Gol'dfarb, Ya. L., 326
Goldman, L., 304
Goldwhite, H., 46
Gompper, R., 304
Goosen, A., 328
Gopal, R., 97
Gorden, B.J., 291
Gore, P.H., 315
Gotthardt, H., 126
Goutarel, R., 223
Granger, R., 229
Grayson, C.R., 213
Gream, G.E., 374
Green, M.J., 180
Greenhalgh, N., 327
Greenwald, R.B., 359
Gregson, M., 218
Grethe, G., 186
Grey, T.F., 105
Gribble, G.W., 181
Griengl, H., 350
Griesbaum, K., 228
Griffin, C.E., 325
Griffin, G.W., 220, 230
Grigg, R., 194
Grimaud, J., 212
Grob, C.A., 341
Grohman, K., 355
Grohmann, K., 276
Gross, H., 346
Groth, W., 39
Grove, J.F., 102
Grübel, K., 164
Gruber, R., 237
Guermont, J., 235
Guerrieri, F., 116
Guilard, R., 81, 315
Guliev, A.M., 334
Gunning, H.E., 239
Gunter, M.J., 146
Günther, W.H.H., 81, 141
Gupta, R.K., 58
Gurevich, A.I., 301
Gustafson, D.H., 64
Gut, M., 41
Gutsche, C.D., 84
Gwynn, D.E., 47

Haack, K.B., 83
Haas, G., 280
Haeck, H.H., 140
Hafferl, W., 223
Hafner, K., 276, 277
Hagan, W.V., 155
Hall, D.M., 58
Haller, A., 317
Halleux, A., 234, 389
Hamann, K., 54
Hammen, P.D., 370
Hammett, L.P., 341
Hammond, W.B., 36
Hammons, J.H., 187
Hanck, A., 356
Hancock, R.A., 72
Hand, J.J., 39, 176
Hankinson, B., 147

Hardovin, J., 219
Hardwick, T.J., 107
Hargreaves, J.R., 88, 193
Harley-Mason, J., 300
Harris, T.M., 157
Harrison, I.T., 220
Harrison, K.G., 64
Hart, H., 129, 168
Hartke, K., 305
Hartshorn, M.P., 192
Hartung, A., 118
Hartung, W.H., 218
Harvey, G.R., 116
Harvey, J.T., 36
Haseltine, S., 153
Hassall, C.H., 97
Hassan, M., 36
Hassner, A., 359
Haszeldine, R.N., 164
Hata, Y., 218
Hatch, L.F., 333
Hatzmann, G., 278
Haug, T., 146
Haugwitz, R.D., 137
Hauser, C.R., 41, 44, 55, 94, 140, 167, 180
Hauser, K.L., 128
Hausigk, D., 67
Hautoniemi, L., 163
Hawson, A., 338
Hawthorne, D.G., 41
Haywood-Farmer, J., 57
Hazen, R.K., 66
Heaney, H., 112
Heathcock, C.H., 167
Hecht, S.S., 172
Heffernan, M.L., 373
Heilbronner, E., 114
Heimbach, P., 363
Heindel, N.D., 315
Heine, H.W., 125, 245
Heintzelman, W.J., 315
Hekkert, G.L., 103
Heller, M.S., 330
Henckel, E., 351
Henery-Logan, K.R., 147
Henne, A.L., 341
Hennion, G.F., 333
Henoch, F.E., 180
Henriksen, L., 347
Henshaw, B.C., 367
Henzel, R.P., 125, 245
Herbert, M., 235
Hermann, R.B., 69
Hershberg, E.B., 66
Herzberg-Minzly, Y., 113
Hetzel, F.W., 189
Heusler, K., 252
Hewitson, R.E., 368
Hewitt, G., 149
Hey, D.H., 108, 169
Hiatt, J.E., 165
Hickman, R.J., 191
Hickmott, P.W., 88, 193
Higgins, W., 180
Hilbert, G.E., 323
Hill, J.H.M., 69
Hiremath, S.V., 165
Hoang-Nam, N., 235
Hodgson, H.H., 102
Hoffman, T.D., 87, 91
Hoffmann, H.M.R., 121
Hoffmann, R.W., 117
Hofheinz, W., 250
Hogeveen, H., 189, 210
Holdrege, C.T., 121
Holker, J.S.E., 206
Hollister, K.R., 37
Holmquist, B., 61
Honig, L.M., 352
Hoover, J.R.E., 209
Hopkins, B.J., 88, 193
Hopkinson, A.C., 58
Hoppe, M., 333
Hora, J., 364, 378
Horiike, M., 128
Horn, U., 177
Hornback, J.M., 51, 358
Hornischer, B.B., 156
Houbiers, J.P.M., 103
Houlihan, W.J., 163
Houlton, P.R., 342

House, H.O., 364
Howarth, T.T., 157
Howe, E.E., 218
Hromatka, O, 355
Huang, H.H., 73
Hub, L., 26
Huba, F., 325
Huber, E., 348
Huber, H., 374
Hubert, A.J., 90, 331
Hubert, M., 331
Hudson, R.F., 175
Hudson, T.B., 315
Hughes, E.D., 307
Hughes, G.P., 119
Huisgen, R., 374
Huisman, H.O., 59
Hunig, S., 173, 217
Hunger, A., 258
Huntsman, W.D., 373
Hurlock, R.J., 170
Husbands, G.E.M., 276
Huston, R.C., 65
Huyffer, P.S., 113, 195
Hyeon, S.B., 209, 240

Ibata, T., 374
Iburg, W.J., 174
Ichihara, A., 53
Ichikawa, H., 209, 240
Ichikawa, T., 138, 225
Icke, R.N., 104
Ide, W.S., 104
Igolen, J., 93
Ikezaki, M., 136, 299
Illuminati, G., 98
Ilvespää, A.O., 117
Imamoto, T., 152
Immer, H., 42, 239
Imoto, E., 184
Inamasu, S., 128
Inazu, T., 384
Ingersoll, W.C., 310
Ingold, C.K., 314
Inoue, A., 40
Inoue, H., 168
Inoue, T., 329
Inouye, Y., 128
Inward, P.W., 341
Ireland, R.E., 105, 196
Iriarte, J., 204
Irie, T., 57, 279
Isbister, R.J., 359
Iskander, G.M., 189
Isler, O., 143
Isoe, S., 209, 240
Itō, N., 53
Itoh, T., 55
Ivanova, R.A., 379
Iwakura, Y., 52
Izumi, H., 40
Izzo, P.T., 276

Jackson, A., 83
Jackson, A.H., 72
Jackson, J.A., 97
Jacobson, I.T., 363
Jacques, J., 27
Jacquier, R., 299
Jacquignon, P., 70, 308
Jahnke, U., 339
Jakobsen, H.J., 315
Janković, J., 110
Jefferson, A., 370
Jemison, R.W., 152, 207
Jenne, M., 278
Jenny, E.F., 233
Jensen, F.R., 321
Jensen, K.A., 347
Jeremić, D., 106, 110
Jeskey, H., 308
Jirkovský, I., 243
Johansen, S.R., 315
Johnson, A.P., 124
Johnson, A.W., 238
Johnson, B.L., 367
Johnson, C.R., 376
Johnson, F., 138
Johnson, F.B., 183
Johnson, H.E., 84
Johnson, J.R., 319, 323
Johnson, S.M., 167, 338

Johnson, W.S., 291
Johnston, J.D., 256
Jones, A.J., 373
Jones, D.N., 180
Jones, E.R.H., 320, 333
Jones, F.N., 45
Jones, G., 96
Jones, R.G., 365, 392
Jones, R.S., Jr., 212
Jones, R.V.H., 383
Jones, W.M., 276
Jones, W.R., 206
Jongejan, H., 242
Josan, J.S., 176
Joshi, G.D., 233
Joule, J.A., 83
Joy, D.R., 121
Julia, M., 93
Julian, P.L., 39, 368
Jullien, J., 56
Junek, H., 156
Juppe, G., 223
Jurewicz, A.T., 30

Kaiser, E.M., 140
Kaiser, S., 219
Kakihana, T., 371
Kalsi, P.S., 45
Kalvoda, J., 213
Kametani, T., 63
Kane, V.V., 237
Kantor, S.W., 94
Kao, W., 343
Kappe, T., 70, 353
Kariya, Y., 86
Karnes, H.A., 199
Kato, H., 145, 155
Kato, T., 40, 138, 225
Katsumura, S., 209, 240
Katz, R., 221
Katzenellenbogen, J.A., 253
Kearns, D.R., 334
Keaveney, W.P., 230
Keiko, N.A., 341
Keiser, J.E., 376
Keitel, I., 346
Keith, D.D., 142
Kelkar, G.R., 165
Kelly, D.P., 228
Kelly, T.J., 142
Kemp, W., 342
Kende, A.S., 276
Kennewell, P.D., 315
Kesavan, V., 218
Kessar, S.V., 97, 223
Kettenes, D.K., 366
Keyton, D.J., 55
Khachaturov, A.S., 334, 341
Khan, M.A., 382
Khan, W.A., 54
Khisamutdinov, G. Kh., 327
King, J.F., 338
Kingsbury, C.A., 56
Kirk, D.N., 192
Kirmse, W., 88
Kirrman, A., 173
Kishida, Y., 119
Kitahara, Y., 40
Kitching, W., 89
Klamann, D., 246
Klanderman, B.H., 315
Klärner, F., 164
Kleb, K.G., 324
Kleeman, A., 89
Kline, G.B., 365, 392
Klingenberg, J.J., 224
Kloetzel, M.C., 140
Kloster-Jensen, E., 114
Klug, J.T., 302, 359
Klunder, A.J.H., 221
Knollmüller, M., 355
Knunyants, I.L., 302
Kobayashi, S., 156
Kobe, K.A., 315
Koga, G., 228
Kohan, G., 216
Kollenz, G., 353
Kölling, G., 67
Kollmar, H., 305
Kolosov, M.N., 301
Komery, J., 338
Konizer, G.B., 162

Konotopov, V.A., 341
Konstantinović, S., 110
Kooyman, E.C., 330
Koppel, G., 121
Kornet, M.J., 380
Korneva, L.M., 336
Kornfeld, E.C., 365, 392
Korobko, V.G., 301
Korst, J.J., 256
Korte, F., 152, 349
Kosower, E.M., 216
Koudijs, A., 324
Kovach, E.G., 368
Kovacic, P., 109
Kováts, E., 114
Koyama, T., 158
Kralt, T., 140
Krasnova, T.L., 341
Kravtsov, D.N., 316
Krbechek, L., 123
Krebs, A., 27, 176
Kreutzberger, A., 150
Kreutzkamp, N., 38, 157
Krishnamurti, M., 104
Krishna Rao, G.S., 137
Kroening, R.D., 46
Krow, G.R., 373
Krueger, J., 347
Krueger, W.E., 249
Kugita, H., 168
Kuhlmann, G.E., 236
Kuhn, R., 276, 360
Kuhn, S.J., 313, 315
Kuleshova, N.D., 302
Kulik, S., 375
Kulkarni, G.H., 165
Kulkarni, S.N., 233
Kumarev, V.P., 313
Kumari, D., 183
Kunath, D., 346
Kundu, N.G., 106
Kunesch, G., 239
Kunkel, W., 366
Künzle, F., 95
Küppers, H., 280
Kuran, W., 84
Kurz, M.E., 109
Kusama, O., 63
Kwart, H., 201
Kyle, R.H., 325

Lack, R.E., 140, 344
Ladwa, P.H., 233
Laidlaw, G.M., 26
Lallouz, E., 66
Lam, F., 182
Landon, W., 124
Landor, S.R., 114
Laney, D.H., 105
Langdale-Smith, R.A., 196
Langler, R.F., 120
Langley, W.D., 221
Langlois, M., 220
Langlois, N., 332
Laurent, A., 212
Lautenheiser, A.M., 127
Lawesson, S., 127
Lawson, A., 277
Leboeuf, M., 223
Lednicer, D., 167
Lee, C.C., 311
Lee, H.L., 186
Leeder, W.R., 132
Le Goffic, F., 93
Leiserson, J.L., 314
Lemal, D.M., 339
Leone, R.E., 377
Lepley, A.R., 54
LeQuesne, P.W., 127
Leung, C., 166
Levin, C., 226
Levine, R., 121
Levy, J.B., 341
Lewis, G.E., 8, 315
Lewis, K.G., 174
Liaaen Jensen, S., 40
Liede, V., 216
Liedhegener, A., 87
Liehr, J., 158
Liem, P.N., 235

Lier, E.F., 217
Lim, L.N., 374
Linda, P., 314
Lindsey, A.S., 308
Lindsey, R.V., 316
Lindy, L.B., 37
Lipsky, J.A., 339
Lishanskii, I.S., 334
Littlewood, P.S., 81, 141
Litvinenko, L.M., 307
Lockhart, L.B., 63
Loewenthal, H.J.E., 113
Logothetis, R.S., 136
Lohse, F., 146
Lomakin, A.N., 315
Lomas, D., 341
Lombardo, P., 140, 360
Long, D.W., 39
Long, F.A., 73
Longone, D.T., 82, 83
Los, M., 176
Lounasmaa, M., 93
Louwrier-de Wal, J., 133
Lown, J.W., 387
Lufkin, J.E., 326
Lunkwitz, K., 358
Lur'i, F.A., 318
Lythgoe, B., 81, 141

MacBride, J.A.H., 247
McCarthy, E.R., 140, 360
McCasland, G.E., 224
McCloskey, C.M., 222
MacDonald, P.L., 41
McDonald, R.N., 206
McGhie, J.F., 105, 183
McKee, R.A., 308
McKusick, B.C., 304
McLamore, W.M., 252
Mc Murry, J.E., 204
Madsen, P., 127
Magid, R.M. 213
Mahadevan, A.P., 102
Mahajan, R.K. 223
Mahalanabis, K.K., 43, 350
Mahendran, M., 238
Maheshwari, M.L., 45
Maksimov, V.I., 318
Malpass, J.R., 197
Malzieu, R., 187
Mani, J., 334
Mann, M.J., 365, 392
Manning, D.T., 196
Manson, J.M., 118
Mao, C., 180
Marchese, G., 54
Marcus, E., 390
Marcus, N.L., 127
Marino, G., 314
Maritz, F., 220
Marks, R.E., 191
Marples, B.A., 112
Marshall, C.D., 379
Marshall, J.A., 138, 174
Marshall, J.L., 59
Marshall, K.S., 200
Marsili, A., 193
Martin, R., 311
Marumoto, R., 156
Marvel, C.S., 81, 82, 225
Masui, M., 154
Mathieson, D.W., 120
Mathieu, A., 367
Mathur, K.B.L., 104
Matić, R., 106
Matsubara, S., 149
Matsui, M., 110, 222
Matsumoto, S., 238
Matsumoto, T., 53
Matt, J.W., 211
Mauger, E., 58
Mayer, H., 143
Mecke, N., 191
Meilahn, M.K., 288
Meinhold, H., 384
Meinwald, J., 202
Melloni, G., 244
Merk, W., 38
Merzoni, S., 116
Messinger, P., 38, 157
Mettalia, J.B., Jr., 331
Metzger, G., 170

Metzger, K., 146
Micetich, R.G., 84
Mihailović, M. Lj., 106, 110
Miles, D.H. 277
Millar, I.T., 147
Miller, A.H., 82, 238
Miller, E.J., 47
Miller, J., 325
Miller, J.A., 185
Milovanović, A., 110
Miocque, M., 138, 299
Mirrington, R.N., 319
Misaka, Y., 394
Mitchell, R.H., 276
Mitchell, R.W., 35
Mitote, T., 125
Mitra, R.B., 253, 391
Mittelbach. H., 353
Miwa, T., 149
Miyadera, T., 119
Mizutani, T., 394
Modena, G., 54
Modest, E.J., 115
Modler, R., 249
Moffett, R.B., 41
Molloy, B.B., 128
Molnar, J., 315
Monaco, D.J., 158
Mondelli, G., 116
Mones, J.D., 247
Montanari, F., 127
Montavon, M., 143
Montgomery, L.K., 211, 341
Moon, S., 102
Moore, H.W., 229
Moretti, I., 127
Morgenthau, J.L., Jr., 43
Mori, K., 110, 222
Mori, N., 279
Mori, T., 158
Morita, K., 156
Morris, R.N., 356
Morrison, D.E., 365, 392
Morse, B.K., 310
Mosher, M.W., 382
Mosher, W.A., 142
Moskalenko, V., 341
Moss, R.A., 56
Moureu, C., 341
Mousseron, M., 299
Mousseron-Canet, M., 299, 334
Moyle, M., 344
Muchnikova, A.M., 318
Mühlstädt, M., 384
Mukai, T., 192
Mukhopadhyay, S.K., 43, 350
Mullen, A., 67
Müller, E., 104, 351
Mulligan, P.J., 8
Munk, M.E., 288
Murphy, D.M., 326
Murphy, G.P., 157
Murphy, J.W., 214
Murray, A., III, 306
Musgrave, W.K.R., 97, 116, 247
Musso, H., 321

Nabeya, A., 52
Nadelson, J., 142
Naemura, K., 222
Nagai, T., 205
Nager, M., 341
Naidoo, B., 72
Nair, V., 335
Nakayama, M., 65
Nakazaki, M., 222
Nametkin, H.S., 317
Naso, F., 54
Nasutavicus, W.A., 138
Naumann, M.O., 224
Nauta, W.T., 92
Neale, R.S., 127
Nebel, C., 142
Neeter, R., 59
Nelson, P.H., 214
Nerdel, F., 169, 246
Nesmeyanov, A.N., 301, 316, 336
Neuberger, A., 302
Neugebauer, F.A., 278
Newallis, P.E., 140, 360
Newman, M.S., 29, 47, 189, 199
Nielsen, A.T., 153

Nierenstein, M., 308
Niess, R., 95
Niklaus, P., 43, 183, 200
Nilsson, A., 71
Nishiguchi, T., 52
Nising, W., 377
Nitta, I., 307
Noller, C.R., 66
Nonhebel, D.C., 277
Noyori, R., 158
Nozaki, H., 158
Nussim, M., 114

Oae, S., 231
Ochiai, M., 156, 231
Odom, H.C., 37
Ogawa, T., 110, 222
Ohnmacht, C.J., 315
Ohno, M., 128, 158, 253
Ohse, H., 46
Ohta, J., 328
Ohta, M., 145, 155, 348
Oka, H., 86
Okamoto, K., 307
Okamoto, Y., 36, 298
Okorodudu, A.O.M., 29
Olah, G.A., 313, 315
Oleinik, N.M., 307
Ollis, W.D., 207, 218, 258
Olsson, K., 71
Omran, S.M.A., 346
Orlov, A.M., 302
Orr, D.E., 183, 216
Orwig, B., 27
Osselaere, J.P., 169
Osterberg, A.E., 299
Östman, B., 315
Ototani, N., 40
Ouannès, C., 27
Oullette, R.J., 226
Ourisson, G., 202
Overman, L.E., 123
Owatari, H., 138

Padwa, A., 237
Pahls, K., 47
Paice, J.C., 374
Pandey, R.C., 39, 235
Pandit, U.K., 104
Panteleimonov, A.G., 315
Pappas, J.J., 230
Paquette, L.A., 167, 197, 219, 334, 338, 358, 371, 373
Parham, W.E., 116
Parr, J.E., 37
Partridge, J.J., 138
Passl, W., 361
Pasynkiewicz, S., 84
Patinkin, S.H., 314
Paudler, W.W., 50
Paukstelis, J.V., 250
Paul, I.C., 167, 338
Paulson, D.R., 122
Pauson, P.L., 104
Payne, G.B., 153
Pazdro, K.M., 227
Pechenkin, A.G., 327
Perevalova, E.G., 301
Perkins, W.C., 315
Perrin, C., 220
Perron, Y.G., 121
Perry, C.W., 365
Person, M., 81, 315
Pesaro, M., 142
Peter, D., 333
Petersen, J.M., 185
Peterson, L.I., 220
Petrov, A.A., 338, 341
Petrovskaya, L.I., 316
Pettit, G.R., 380
Philbin, E.M., 119
Philips, J.C., 219, 358
Phillips, W.G., 155
Pichat, L., 219, 235
Piers, K., 177
Pilgram, K., 46
Pinck, L.A., 323
Pincock, R.E., 57
Pinder, A.R., 37
Pines, H., 234
Pirkle, W.H., 94
Pirola, O., 317

Plepys, R.A., 46
Podesva, C., 120
Pojer, P.M., 42
Polonsky, J., 239
Ponsford, R., 237
Ponticello, I., 186
Poon, L., 243
Popravko, S.A., 301
Porai-Koshits, B.A., 313
Porfir'eva, Yu. I., 341
Porter, P.K., 225
Potsch, D., 47
Poupart, J., 350
Powers, J.C., 186
Prelog, V., 103, 221, 280
Preston, F.J., 127
Price, C.C., 102
Price, J.A., 44
Prinzbach, H., 122, 175, 276
Pritzkow, W., 358
Probasco, E.K., 187
Protiva, M., 353
Pyron, R.S., 276

Quast, H., 217
Quelet, R., 66, 117

Raap, R., 84
Rademaker, P.D., 59
Radlick, P., 205, 210, 334
Raman, H., 243
Ramirez, F., 389
Ramm, P.J., 206
Rance, M.J., 92
Ranganathan, S., 243
Rao, G.V., 61
Raphael, R.A., 242
Ratajczak, A., 53
Ratts, K.W., 155, 243
Rauhut, M.M., 322
Rawson, G., 315
Razdan, R.K., 237
Redemann, C.E., 104
Redman, B.T., 218
Reeder, E., 219
Rees, C.W., 108, 129, 169, 316
Reese, C.B., 357
Regitz, M., 87, 333
Regnault, A., 62, 219
Reich, H.J., 30, 119
Reichardt, P.B., 123
Reid, C.G., 109
Rewicki, D., 276
Richards, D.H., 115
Richer, J., 315
Richey, F.A., Jr., 364
Rickborn, B., 321
Riddle, G.N., 6
Rieber, N., 339
Ried, W., 89, 303, 366
Riehl, J., 369
Riehl, J.J., 160
Rigby, A., 170
Říhová, E., 175
Rinehart, J.K., 116
Rinehart, K.L., Jr., 64, 315
Riniker, B., 292
Riniker, P., 292
Rioult, P., 139
Ripoll, J., 202
Ripoll, J.L., 215
Risaliti, A., 96, 130
Ritchie, E., 42
Ritchie, E.R., 319
Robb, C.M., 81
Robb, E.W., 365
Roberts, J.C., 92
Roberts, J.D., 140, 160, 325
Roberts, R.M., 312
Robertson, A.V., 41
Robertson, D.E., 139
Robinson, B., 99
Robinson, F.P., 196
Roedig, A., 68
Rolle, F.R., 72
Rome, D.W., 367
Rosenkranz, G., 204
Rosenmund, K., 345
Rosenmund, P., 164
Rosenthal, A., 311
Rosenthal, I., 124
Rosowsky, A., 115

Ross, W.A., 105
Ross, W.J., 188
Rossi, J., 229
Rossi, S., 317
Rouzaud, J., 367
Rowland, A.T., 55
Rowley, P.J., 51
Rubach, G., 184
Rubin, M.B., 369
Rüegg, R., 143
Runde, M.M., 319
Rush, R.V., 230
Russell, A., 63
Russell, J.R., 224
Russell, P.L., 242
Rüter, J., 333
Ruzicka, L., 300
Rybinskaya, M.I., 336
Ryfors, L., 151

Saha, J.G., 315
Sakan, T., 149, 209, 240
Salmón, M., 217
Salomaa, P., 163
Saltiel, J., 374
Samant, B.V., 328
Sammes, P.G., 149
Samsonova, N.B., 318
Sandborn, L.T., 323
Sanders, E.B., 220
Sanders, L.A., 187
Sandin, R.B., 308
Sandvick, P.E., 123
Sanford, E.C., 76, 81
Šantroch, J., 343
Saraf, S.D., 44, 83
Sarel, S., 221, 245, 302
Sasaki, T., 158, 343
Sato, K., 309
Sato, S., 145, 155, 348
Satyanarayana, M., 246
Sauers, R.R., 103
Sawinski, J., 64
Sawyer, J.L., 39
Sayigh, A.A.R., 144, 163
Scanio, C.J.V., 174
Schade, W., 191
Schaefer, J.P., 352
Schamp, N., 85
Scheeren, J.W., 307
Scheinmann, F., 370
Schenker, K., 258
Scherer, K.V., Jr., 374
Schiedt, B., 347
Schildknecht, H., 278
Schimpf, R., 363
Schinz, H., 300
Schleyer, P. von R., 184, 217, 377
Schmid, G., 358
Schmid, H., 387
Schmidt, E.W., 104
Schmitt, J., 220
Schmitz, W.R., 139
Schmutz, J., 95
Schneider, K., 93
Schöllkopf, U., 155
Schönleber, D., 372
Schorno, K.S., 188
Schreiber, J., 177
Schubert, W.M., 74
Schudel, P., 142
Schulenberg, J.W., 49, 363
Schultz, E.M., 81
Schultz, H.P., 94
Schwarcz, M., 347
Schwartz, E.F., 128
Schwartz, M.L., 247
Schwarz, H., 350
Schwarz, V., 357
Schweininger, R.M., 135
Schweizer, E.E., 136, 158
Scilly, N.F., 115
Scott, A.I., 388
Scott, E.W., 319
Scott, J.J., 302
Scott, R.B., 330
Scribner, R.M., 201
Scrowston, R.M., 315
Seib, B., 305
Seidel, B., 341
Seikel, M.K., 65
Seip, D., 276

Sekido, M., 394
Selim, M.I.B., 346
Sellers, C.F., 100
Selman, L.H., 331
Selva, F., 317
Semenow, D.A., 325
Semenyuk, G.V., 307
Severn, D.J., 216
Seymour, D., 341
Shamma, M., 138
Sharp, J.C., 376
Shaw, F.R., 314
Shechter, H., 87, 106
Shelden, H.R., 229
Shen, T.Y., 333
Sherrod, S.A., 202
Shilovtseva, L.S., 301
Shimadzu, H., 156
Shine, H.J., 311
Shingu, H., 307
Shoppee, C.W., 319
Shostakovskii, M.F., 341
Shoupe, T.S., 55
Showell, J.S., 224
Shulman, J.I., 165
Shuttleworth, A.J., 141
Sicher, J., 362
Sigg, H.P., 198
Siman, J.F., 160
Sims, J.J., 331
Singh, J., 246
Singh, M., 97
Singh, T., 207
Sipma, G., 366
Sisti, A.J., 64, 106
Skattebøl, L., 113, 357
Skillern, L., 47
Sleiter, G., 98
Sleta, T.M., 338
Slouka, J., 82
Smalley, R.K., 387
Smit, P., 324
Smith, A.J., 89
Smith, C., 207
Smith, C.P., 389
Smith, J.R., 248
Smith, P., 72
Smith, P.A.S., 82
Smutny, E.J., 160, 360
Snatzke, G., 377
Snyder, G.A., 26
Snyder, H.R., 379
Soloway, A.H., 310
Sommer, L.H., 341
Sondheimer, F., 8, 114, 170, 252, 276
Sonoda, N., 329
Šorm, F., 364, 378
Sosnovsky, G., 329
Specht, E.H., 331
Speckamp, W.N., 59
Spence, T.W.M., 354, 381
Spicer, L.D., 39
Spillett, R.E., 315
Spring, D.J., 96
Staab, H.A., 93, 225
Stadler, D., 87
Stadnichuk, M.D., 338
Stammer, C.H., 58
Stansfield, F., 189
Stauner, T., 220
Stein, R.A., 83
Steinberg, H., 231
Stener, A., 96
Stephen, J.F., 390
Steppel, R.N., 206
Sterk, H., 156
Sternbach, L.H., 219
Sternhell, S., 319
Stetter, H., 128, 319
Stevens, M.F.G., 150
Sticker, R.E., 379
Stirling, C.J.M., 47
Stock, L.M., 313
Stoll, A.P., 43, 183
Stoll, M., 300
Stone, J.A., 130
Storey, R.A., 97
Stork, G., 372
Storm, P.C., 334
Stout, R., 64
Strating, J., 337, 356

Strausz, O.P., 239
Streef, J.W., 359
Streitwasser, A., 315
Strong J.G., 43
Stuart, F.A., 222
Stuber, F.A., 144
Stull, A., 217
Subba Rao, H.N., 106
Suda, K., 154
Suehiro, T., 190
Süess, R., 200
Suga, K., 114, 162
Sukornick, B., 320
Sulimov, I.G., 338
Sumiki, Y., 222
Suquet, M., 220
Surbey, D.L., 320
Suschitzky, H., 100, 324, 325
Suter, A.K., 121
Suter, C.M., 92
Sutherland, I.O., 207, 218
Sutherland, M.D., 383
Svoboda, M., 362
Swamer, F.W., 55
Swern, D., 224
Syhora, K., 357
Szkrybalo, W., 178

Tabor, T.E., 206
Taft, R.W., Jr., 341
Tagaki, W., 231
Taits, S.Z., 326
Takagi, H., 149
Takamizawa, A., 238
Takaya, T., 184
Takeda, A., 40, 86
Takeda, M., 168
Takeshima, T., 152
Takimoto, H., 123
Talbot, W.F., 93
Tambuté, A., 195
Tamm, R., 138
Tan, S.I., 380
Tanaka, M., 205
Tandon, J.P., 126
Tanida, H., 218
Tanida, T., 57
Tappe, H., 276
Tarbell, D.S., 44, 328
Tarrant, P., 126, 341
Tashiro, M., 55
Taub, D., 252
Taube, A., 302
Taylor, D.R., 164
Taylor, E.C., 359
Taylor, J., 139
Taylor, R., 315
Taylor, W.C., 42, 319
Tedder, J.M., 89
Teitel, S., 67
Telefus, C.D., 389
Teller, E., 360
Temme, G.H., III, 56
Tennant, G., 354, 381
Thil, L., 160
Thill, B.P., 345
Thio, P.A., 380
Thomas, A.F., 193
Thomas, H.G., 319
Ticozzi, C., 45
Tien, H., 226
Tien, J.M., 226
Ting, J., 226
Tjan, S.B., 231
Tobey, S.W., 138
Tochio, H., 52
Tochtermann, W., 280
Toda, F., 125
Todd, A.R., 108
Todd, K.H., 89
Todd, M.J., 375
Tökés, L., 220
Tokunaga, Y., 149
Tokura, N., 205
Topping, R.M., 60, 242
Torii, S., 40, 86
Torre, G., 127
Toru, T., 343
Triggle, D.J., 40
Trimitsis, G.B., 37, 41
Troxler, E., 103, 221
Troxler, F., 43, 183

Truce, W.E., 37
Truscheit, E., 40, 142
Tschudi, G., 257
Tsuge, O., 55
Tsuji, T., 57
Tsuruta, H., 192
Tsutsumi, S., 329
Tsuzuki, Y., 279
Tucker, B., 144, 163
Turner, A.B., 148
Turro, N.J., 36, 209
Tutt, D.E., 60
Tyminski, I.J., 291

Uda, H., 391
Udding, A.C., 337
Ulm, K., 246
Ulrich, H., 144, 163
Umani-Ronchi, A., 45, 145
Unde, N.R., 165
Ursprung, J.J., 291
Usković, M.R., 186
Uyeo, S., 394

Vagi, K., 120
Vaidyanathaswamy, R., 85
Valdez, C.M., Jr., 46
Valenta, Z., 216
Valentin, E., 130
Valentin, J., 303
van der Jagt, P.J., 114
van der Plas, H.C., 95, 234, 242, 324
van der Wal, B., 366
van Es, T., 345
van Helden, R., 330
van Lierop, J.B.H., 366
van Meeteren, H.W., 95, 234
VanMeter, J.P., 215
van Tamelen, E.E., 6, 138, 166
van Zanten, B., 114
Vasey, C.H., 327
Vatakencherry, P.A., 253
Vaughan, C.W., 325
Vaughan, J., 72
Vaughan, W.R., 233, 244
Venkatasubramanian, N., 61
Vestling, M.M., 155
Vialle, J., 139
Victor, R.R., 75
Vieböck, F., 120, 361
Viehe, H.G., 234, 389
Vill, J.J., 299
Vinogradova, N.D., 334
Vladimirova, I.D., 337
Vlattas, I., 220
Vogel, E., 164
Vogel, P., 175
Vögle, F., 225
Vogler, H.C., 210
Vogt, B.R., 364
von Schriltz, D.M., 140
Voong, S., 102
Vrba, Z., 71
Vyas, V.A., 293
Vystrčil, A., 175

Wache, H., 246
Wadia, M.S., 45
Wahren, M., 145
Waitkus, P.A., 220
Wakamatsu, T., 136, 299
Wakselman, C., 173
Walborsky, H.M., 128, 344
Walia, J.S., 246
Wall, M.E., 142
Walling, C., 38, 330, 341
Walls, F., 217
Wals, L., 92
Walton, D.R.M., 72
Wamhoff, H., 141, 152, 349
Wanzlick, H., 339
Ward, E.R., 102
Warren, K.D., 74, 315
Wasserman, H.H., 142, 202, 238
Watanabe, S., 114, 162
Watkins, W.B., 109, 378
Watts, W.E., 127
Wayne, R.S., 39
Weaver, C., 379
Weber, H.P., 198
Wechter, W.J., 159
Weeks, D.P., 306

Wehrmeister, H.L., 139
Weis, C.D., 48, 199
Weiss, U., 135
Weissberger, A., 314
Welch, G.J., 72
Welzel, P., 183
Wenkert, E., 137, 239
Wenzel, W., 68
Wepster, B.M., 6
Wertheim, E., 217
Wesslén, B., 151
Westen, H.H., 103, 221
Weston, A.W., 92
Weyerstahl, P., 169, 246
Whalen, E.J., 187
Whalley, W.B., 70
Whitehouse, R.D., 180
Whitham, G.H., 31, 102
Whiting, D.A., 162
Whiting, M.C., 333
Whitlock, H.W., Jr., 123
Whitmore, F.C., 341
Wibberly, D.G., 139
Wiberg, K.B., 107, 165
Wiberg, N., 126
Wiemann, J., 229
Wiesner, K., 216, 243, 343
Wilcox, C.F., Jr., 166
Wiley, D.W., 304
Willhalm, B., 175
Willi, A.V., 160
Williams, D.L., 306
Williams, G.J., 174
Wills, M.T., 37
Wilshire, J.F.K., 94
Wilson, I.S., 314
Wilson, J.W., 202
Wilson, N.D.V., 83
Wimmer, T., 353
Winn, M., 146, 172
Winstein, S., 306
Wiscott, E., 217
Wisegarver, B.B., 104
Withey, R.J., 175
Wluka, D.J., 320
Wohllebe, J., 187
Wolfe, S., 121
Wolfe, J.F., 37, 41, 44
Wolfstirn, K.B., 341
Wolinsky, J., 138
Woller, P.B., 332
Woo, P., 38
Wood, G., 38
Woodward, H.E., 308
Woodward, R.B., 207, 252, 256, 258, 336, 365, 368, 392
Work, S.D., 41
Wotiz, J.H., 325
Wright, G.J., 72
Wright, I.G., 166
Wright, J.B., 157
Wright, W.D., 83
Wristers, H.J., 373
Wróbel, J.T., 227
Wüest, H., 81, 136, 222
Wuesthoff, M.T., 291
Wulff, R.E., 247
Wynberg, H., 45, 103, 221, 327, 337

Yagupol'skii, L.M., 315
Yamada, A., 309
Yamaguchi, Y., 162
Yamamoto, K., 152
Yang, N.C., 329
Yankee, E.W., 53
Yanuka, Y., 221
Yao, A.N., 243
Yijima, C., 154
Yokoyama, M., 152
Yoshimoto, H., 184
Yoshino, T., 384
Yousif, G., 72
Yukawa, Y., 374
Yun, H.Y., 155
Yur'ev, Yu. K., 379

Zabkiewicz, J.A., 242
Zabriskie, J.L., 94
Zachová, J., 357
Zagdoun, R., 299
Zak, A.G., 334

Zambito, A.J., 218
Zaugg, H.E., 146, 172
Závada, J., 362
Zdero, C., 85
Zeeh, B., 147, 351
Zefirov, N.S., 379
Zeidman, B., 139
Zeiler, A.G., 50
Zetzsche, F., 345
Ziegler, E., 70, 353
Zilkha, A., 66
Zimmerman, H.E., 86, 113, 195
Zsindely, J., 387
Zuidema, G., 356
Zuorick, G.W., 306
Zuurdeeg, B., 95
Zvyagina, A.B., 334
Zwanenburg, B., 356
Zwanenburg, D.J., 327
Zweifel, G., 186, 337
Zymalkowski, F., 62